To Jackie, Kenneth, Amy, and Nicki —
you light up my life

R. G. A.

To A. Alan B. Pritsker —
mentor, colleague, and friend

C. R. S.

MODELING AND
ANALYSIS OF
MANUFACTURING SYSTEMS

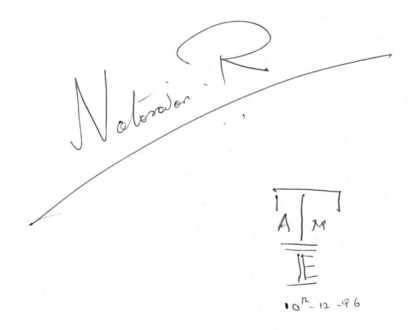

MODELING AND ANALYSIS OF MANUFACTURING SYSTEMS

Ronald G. Askin
The University of Arizona

Charles R. Standridge
Florida A&M University/Florida State University College of Engineering

JOHN WILEY & SONS, INC.
New York • Chichester • Brisbane • Toronto • Singapore

Acquisitions Editor:	Charity Robey
Marketing Manager:	Debra Riegert
Production Supervisor:	Charlotte Hyland
Designer:	Laura Ierardi
Cover Illustrator:	Marjory Dressler
Manufacturing Manager:	Andrea Price
Copy Editing Supervisor:	Deborah Herbert
Illustration Supervisor:	Jaime Perea

This book was set in ITC Garamond Light by Publication Services.

Library of Congress Cataloging in Publication Data:

Askin, Ronald G.
    Modeling and analysis of manufacturing systems / Ronald G. Askin,
Charles R. Standridge.
        p.   cm.
    Includes bibliographical references and index.

    1. Flexible manufacturing systems—Mathematical models.
2. Production management—Mathematical models.   I. Standridge,
Charles R.   II. Title.
TS155.6.A78   1993
658.5'001'5118—dc20   92-35014
                                                          CIP

Printed in Singapore

10 9 8 7 6 5 4 3 2 1

PREFACE

This book provides an introduction to the analysis of manufacturing systems using analytical and experimental models. It brings together useful models and modeling approaches that address a wide variety of manufacturing system design and operation issues. By gathering all of this information into one place, the text presents a concise, cohesive, and broad-based view of the application of modeling and analysis in the manufacturing realm.

Typically, students learn about modeling in operations research and simulation courses and about manufacturing systems, or their components, in manufacturing system courses. Some use of models is made in manufacturing courses, and manufacturing examples may be used in modeling courses. This organization of courses, however, is not likely to provide students with a clear, comprehensive picture of the utility of models in manufacturing system design and operation.

We believe that students need an understanding of how models address a wide range of manufacturing system design and operational issues and are therefore essential tools in many facets of the manufacturing system design process. Students need to be able to develop this understanding in a concentrated manner without having to take a voluminous number of courses. Thus, we believe that a single, focused course on the modeling of manufacturing systems is a valuable part of an industrial/manufacturing engineering or quantitative business curriculum. This book is intended for such a course.

The text discussion introduces students who are already familiar with operations research and simulation modeling to the application of these functions in manufacturing system design and operation. It summarizes basic information and shows how to use these techniques in the analysis of manufacturing systems. The text lays the foundation for more in-depth study of the modeling and analysis of individual manufacturing system design issues.

In addition, the text provides a comprehensive resource concerning current types of discrete part manufacturing systems, including recent developments like flexible manufacturing systems and cellular manufacturing. For each system type, a system description, discussion of important design and operations problems, and appropriate useful modeling techniques are included.

This book is best suited to the needs of upper division undergraduate or entry-level graduate students in industrial/manufacturing engineering or quantitative business programs. The undergraduate students will find that the text ties together the modeling and manufacturing aspects of their curricula as well as assists in formulating comprehensive senior design projects. Graduate students will find that the text is a foundation on which to build more specialized study. The book may also be used by practicing industrial/manufacturing engineers, systems analysts, operations researchers, and management scientists.

We assume that readers have had introductory courses in operations research, probability and statistics, and simulation. The operations research background should cover basic deterministic and stochastic techniques. It is helpful to have some understanding of production planning and inventory control, but it is not essential.

Our viewpoint for the modeling of manufacturing systems emphasizes the process of deciding what the inside of a manufacturing facility should look like and what it should do. We concentrate on general design and operating strategies for the control of the flow of material through the facility. Serial and discrete parts manufacturing systems in which batches of product flow from operation to operation and eventually depart as salable product are considered. We cover the major direct and support activities that are required to convert raw materials into finished products. The text includes descriptive material on the components and activities of manufacturing flow systems, definitions of specific decision problems, and modeling approaches.

Most chapters discuss a specific system type or activity. In general, each chapter begins with a description of the relevant system or activity, the equipment used, and the design or operational issues. We recognize that many of today's engineering and business students have only limited in-plant experience. Thus, chapter introductions are intended to supply necessary background material to put the subsequent models into context. We have also found that the use of videotapes, plant trips, and live classroom demonstrations is helpful. Typically, the remainder of the chapter is a set of decision problems and appropriate models. In several chapters, we have used a single manufacturing environment for the examples to illustrate both how models develop iteratively and how models vary based on the question of interest.

As we have already mentioned, this book covers a relatively broad set of systems and modeling approaches. Obviously it is not possible to cover the entire field of manufacturing systems problems, models, and solution techniques in one text. We have therefore endeavored to select topics that we believe are important problem areas or are conducive to illustrating useful modeling techniques.

Since modeling is as much art as science, we have felt justified in adding some of our own philosophy. The reader is encouraged to recognize the difference between the analytical facts, model abstractions, and the authors' philosophy. The latter should be accepted with increasing skepticism as the reader's own philosophy develops.

CONTENT AND USAGE

The book is divided into four parts. Part I consists of Chapter 1 and gives our philosophy of modeling and basic manufacturing principles. These principles constantly reappear in real life as well as in later chapters. We define what a model is and discuss the purposes of models. This chapter sets the context for the remainder of the book.

Technical information is provided in Parts II, III, and IV. Part II describes and models various types of material flow systems usually found in manufacturing. Part III discusses and gives models for support functions common to all types of material flow manufacturing systems. Thus, Parts II and III discuss models of the components of manufacturing systems. Part IV presents models that integrate all of the components and that assess how the system will operate with all of the components working together. In addition, Part IV presents case studies that show how multiple models work in a complementary fashion to aid in manufacturing system design.

Part II deals with material flow systems. We begin in Chapter 2 with configuration and balancing of assembly lines. A comparison of the advantages of assembly lines and small, job-enriched teams is given. COMSOAL and Ranked Positional Weight balancing heuristics are presented, followed by an efficient, implicit enumeration procedure for finding optimal line balances. Several practical issues are discussed, and we conclude with a look at mixed model lines and unpaced lines. Chapter 3 introduces the effect of random breakdowns and the curative power of buffers. A basic Markov chain model is used for two-stage, single-buffer lines followed by an approximate model for three-stage lines. Recently developed summary expressions for throughput in longer lines are presented for both random processing times and random station breakdowns. Chapter 4 examines the topic of scheduling. The effect of order release strategies and the importance of bottlenecks are first discussed. This is followed by several traditional algorithms for single- and two-machine scheduling and popular dispatching rules for general job shops. Chapter 5 is devoted to flexible manufacturing systems. After the basic system components are described, the hierarchical decision problems of systems design and systems operation (part selection and loading) are covered. We conclude with a brief look at the issues that are relevant for flexible assembly systems. Group technology is the subject of Chapter 6. The introduction describes the advantages and principles of cells. This is followed by an overview of coding schemes. The chapter presents a series of part–machine grouping models, which become increasingly more complex as more relevant information is included. Production flow analysis, clustering with similarity coefficients, and graph-theoretic approaches are discussed. A section is included for aiding in the assignment of parts to specific machines of a generic type to preprocess the data for grouping. Part II concludes with Chapter 7 on facility layout. Systematic layout planning is used to describe the overall problem. The quadratic assignment problem model of facility layout is presented, followed by the planar graph approach to forming block layouts. Recent work on automatic construction of new layouts with aisles is covered. A final section concerns location of one or more new facilities into an existing system.

Part III describes the support functions. Chapter 8 looks at task assignment and sequencing within a workcenter. The Traveling Salesman Problem and its applications are presented. Several models for integrating workcenter setup and task sequencing are also included. Chapter 9 covers material handling systems. Basic equipment and principles are presented followed by a section on equipment selection. Next comes a brief look at conveyor analysis and, finally, a complete model for the design and operation of AGV systems. Chapter 10 tackles the issue of storage and retrieval. Warehouses, and their appropriate usage, are described. We next describe the orientation of storage stacks. This is followed by techniques for assigning parts or pallets to locations in dedicated and open warehouses. The usefulness of storage classes is illustrated along with the importance of item complementarity. We conclude with a discussion of order picking.

Part IV discusses modeling techniques that help assess the assess the operation of entire manufacturing systems. We begin with analytical queueing models in Chapter 11, starting with single-station models and progressing to networks of related stations and their buffers. Both closed and open networks are covered. Difficulties in trying to relax model assumptions are demonstrated. Chapter 12 reviews simulation modeling and discusses its application to manufacturing systems. Simulation modeling approaches commonly used in manufacturing system analysis are discussed. Useful simulation analysis outputs are presented. Example models illustrate the unique benefits of simulation modeling. Chapter 13 presents case studies that show how multiple models contribute to the process of designing a manufacturing system, each at a different stage in the design process.

The broad range of topics discussed in the text allows flexibility in course design to meet individual program needs. A course introducing the application of modeling techniques to manufacturing system design and operation would be well served by Chapters 1 to 3, 5 to 7, and 9 to 13, omitting or reviewing material covered in other courses. More advanced sections marked with a "*" could be skipped for undergraduate courses. A survey course in operations management with emphasis on recent developments in manufacturing systems is obtained by using Chapters 1 to 2, 4 to 7, 12 to 13 and a sampling of topics from 9 to 11. The flow of material through a production facility provides a common thread throughout the book. Recognizing this, Chapters 6 to 7 and 9 to 10 provide a nucleus for the modern Master's level course in facility layout. Depending on the instructor and student interest, Chapters 2 to 3, 5, 8, 11, or 12 could be added to the course syllabus.

However readers choose to use this book, we hope that they will find it useful during that brief period prior to its obsolescence.

Ronald G. Askin
Charles R. Standridge

ACKNOWLEDGMENTS

We trust that the reader will judge this to be a clearly written text. We can unequivocally state that without the excellent example set by former instructors, such as Doug Montgomery, Lynwood Johnson, Alan Pritsker, Steve Roberts, John White, and Gary Whitehouse, the text would be less clear than it is. We thank Alessandro Agnetis, Yavuz Bozer, Martha Centeno, Richard Duncan, Les Frair, John Giffin, Jeffrey Goldberg, Sunderesh Heragu, Julie Higle, George Mitwasi, Avi Seidmann, and Rajan Suri for their helpful comments and suggestions. Ben Pourbabai taught a similar course at The University of Arizona prior to 1985. In selecting material and the level of presentation, we have benefited from his willingness to share his experience. Hughes Aircraft Company was gracious enough to provide support for this effort, both financially and through access to real problems.

R. G. A.
C. R. S.

CONTENTS

*"It ain't so much the things we know that get us into
trouble. It's the things we know that ain't so."*
—Artemus Ward

PART III   SUPPORTING COMPONENTS

PART IV   GENERIC MODELING APPROACHES

Appendix A.   REVIEW OF BASIC PROBABILITY

PART I
MANUFACTURING SYSTEMS AND MODELS

CHAPTER 1

MANUFACTURING MODELS

I'm very good at differential and integral calculus,
I know the scientific names of beings animalculous;
In short, in matters vegetable, animal, and mineral,
I am the very model of a modern Major-General.
– W. S. Gilbert, ***The Pirates of Penzance***

1.1
INTRODUCTION

The purpose of manufacturing, at least idealistically, is to enrich society through the production of functionally desirable, aesthetically pleasing, environmentally safe, economically affordable, highly reliable, top-quality products. Like inner peace, financial security, and responsible citizenship, these noble objectives are often conflicting. A more pragmatic definition of purpose would be to meet customer function, quality, and reliability wishes at minimum cost. The responsibility of the manufacturing management is to establish priorities and objectives and monitor performance. The manufacturing/industrial engineer determines how best to utilize the available inputs of labor, technology, capital, energy, materials, and information to achieve the objectives. This textbook provides insight into the use of the analytical and experimental models of the manufacturing system that aid engineering and manufacturing decision making.

Another purpose of manufacturing is to provide gainful employment to drive the economy. Throughout the twentieth century, however, the percentage of domestic U.S. employment in manufacturing has steadily declined. The decline was gradual, reaching 30 percent around 1960. The rate of decline then increased as the service sector gained influence. By 1980, only 21 percent of domestic U.S. employment was in manufacturing. The earlier reductions were largely accounted for by increasing productivity. In recent years, improved performance by our international trading partners has eroded the competitive position of U.S. manufacturing industry. One need only look at the Made in _____ label on one's automobile, stereo, camera, clothing, and other consumer goods to recognize this predicament. Instead of causing the loss of manufacturing jobs, robots and automated machines have become a necessary part of doing business. Without the productivity gains made possible by automation, even more jobs would be lost. The rapidly expanding economy of the 1960s meant that virtually any product produced could be sold at a profit. The incentive to improve dwindled and the declining rate of productivity growth was largely ignored. In today's

3

worldwide marketplace, that luxury is no longer affordable. Industry is learning that constant improvement is a prerequisite for continued existence.

Manufacturing can be classified as discrete parts or continuous processing. Discrete-parts manufacturing is characterized by individual parts that are clearly distinguishable such as circuit boards or engine blocks. Process industries operate on product that is continually flowing, the most obvious examples being oil refineries and other chemical industries. This book is written from the perspective of discrete-parts manufacturing. Nevertheless, many of the models could be used in process industries. Normally, however, process industries are capital intensive and concerned with capacity. Discrete manufacturing is mainly concerned with scheduling, materials control, and labor assignment. The system types do overlap. Mass production of discrete parts shares many of the characteristics of process industries, for example.

We can divide the manufacturing system into five interrelated functions. The functions are product design; process planning; production operations; material flow/facilities layout; and production planning/control. Information flow is the umbrella that drives the five functions, oversees their coordination, and measures compliance with corporate objectives. The information system interacts with accounting, purchasing, marketing, finance, human resources, and other administrative functions.

Product design is responsible for taking the inputs from marketing regarding customer desires and constructing the description of products that can be profitably manufactured to satisfy these desires. Historically, blueprints with annotated descriptions were used to describe products. Modern computer-aided design (CAD) systems have replaced blueprints for this task. The CAD model can be displayed on a graphics workstation along with any desired notes. Systems are capable of showing 3-D images of products or cross sections viewed from any vantage point and under any desired lighting. The model is stored in the computer as a set of edges that connect vertices located in space. Smooth surface patches that connect the edges can be defined through mathematical equations. The result is a 3-D object whose geometry and topology can be represented by points, data structures relating points, and parameters for the relevant mathematical expressions. Some CAD systems alternatively store the object as a set of primitive objects such as spheres, blocks (cubes), and cones, which are scaled and located in space. By combining and removing basic shapes, the product can be illustrated. The models can then be analyzed for mass and strength properties, also using computer-based mathematical tools such as finite element analysis. Figure 1.1 shows several views of a CAD model assembly.

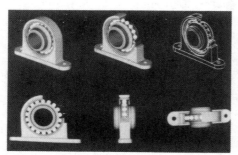 

Figure 1.1   Computer-aided product design models (photo courtesy of Auto-trol Technology).

Process planning entails the specification of the sequence of operations required for converting the raw material into parts and then assembling parts into products. Process planning requires an in-depth understanding of the availability and capability of production operations and the functional needs specified by the product designer. Machine selections must consider factors such as part size, demand rate, tooling costs, and power requirements. The planner must know, for instance, if a particular tool and machine can achieve a given tolerance specification. Can the hole be simply drilled, or must this be followed with a reaming operation? On a circuit board, can this component be inserted before wave solder, or must one mask and then add the component in post-solder operations? The finished process plan is a set of instructions specifying how the product should be manufactured, including the sequence of machine tools, the tools required, and the machine settings. Figure 1.2 shows a generic process plan. Each row contains the information the worker requires to manufacture the part. The location of detailed part drawings and the tape number for executing the plan on an NC machine may also be included.

Manufacturing operations are generally of either a fabrication or assembly nature. Fabrication refers to either the removal of material from the raw stock or a change in its form for the purpose of obtaining a more useful form. Plastic injection molding, aluminum extrusion, turning a diameter, drilling a hole, or bending a flange are examples. Assembly refers to the combination of separate parts or raw stock to produce a more valuable combined unit. Inserting a board to a PC chassis or adding the legs onto a table are examples. In practice it becomes convenient to relax our definitions slightly and to consider the typical manufacturing system as one involved in first fabricating parts and then assembling parts into products. Hence, the tasks of adding layers to a board that will then have holes drilled is all part of board fabrication. The later crimping of the leads on the axial components is considered part of assembly. Our interest is in taking a step back from the level of the individual processing step and looking at how material will flow through the system and how processes are linked to obtain the desired volume of production at the intended quality level. Issues such as tool wear and causes of solder defects are clearly important in manufacturing, but these issues are not the aim of this text.

Part Name   Shaft
Part No.   AS34967
Planner   John Doe
Date   9/11/90
Sheet   1 of 1

Department	Machine	Operation No.	Operation Description	Tool Name	Tool No.	Setup Time	Unit Time
120	Drill Press	100	Drill Cross Hole, 3/8"	Bit Fixture	D1415 P 967	.10 hrs	.002
120	Vert Mill	110	Mill Front Face	End Mill Fixture	GC111 S 3641	.15 hrs	.001
120	Lathe	120	Turn O.D. 1.540" ± .001	Cutter Fixture	HS 340 LC 967	.20 hrs	.014
⋮	⋮	⋮	⋮	⋮	⋮	⋮	⋮

Figure 1.2   Typical process plan format.

Although this text is not dedicated solely to material handling and facilities planning (see Tompkins and White [1984] or Sule [1988]), these topics are clearly relevant. Although such topics are generally taught from a generic system design viewpoint, our interest here is in the evaluation of specific system configurations. Material handling is concerned with the techniques used to transport parts, tooling, and scrap throughout the facility. Facility layout is concerned with the physical placing of production processes within the facility; the spatial relationships of the related processes; the delivery of required services such as compressed air, lighting, electricity, and HVAC (heating, ventilation, and air conditioning) to the work areas; and the removal of waste products such as paint fumes, chips, and coolant from those areas. Material flow system design must not be overlooked. Product designs can fail because of insufficient attention to layout and material handling.

Production planning, scheduling, and control constitute an important component of the total manufacturing system. Production planning is responsible for combining information on market demands, production capacity, and current inventory levels to determine planned production levels by product family for the medium to long term. This aggregate plan is then disaggregated through several steps to eventually obtain short-term schedules showing, for each work center, its goals for the next shift. The jobs assigned to the work center are then sequenced by the order in which they will be loaded onto the machines. The objective of this text is to determine how best to design the manufacturing system and predict its performance for a given production plan. We will not discuss the process of production planning, scheduling, and control. The interested reader should consult any of a number of texts on this topic, several of which are listed in the chapter references of this book.

Throughout most of this text we will assume that product design and process plans are known. The information system is assumed to exist as well. All of the models described in this book require reliable data inputs, which ideally can be extracted from the product design, process plans, and information system. As a practical matter, this is often not true. The gathering of model inputs is one of the most problematic aspects of any modeling project. Multiple sources of data may need to be used and inconsistencies between the data sources may need to be rectified. Accounting systems have been designed with a view toward filing financial statements for tax and stockholder purposes. Managerial decision making often takes second place, with top-level management receiving second-place honors with their desire for aggregated data. The lowly manufacturing engineers may often find themselves being told by the accountant "Sorry, we don't collect the data that way" when asking for a specific cost or productivity measure. The lack of data may cause a model to be modified or different modeling assumptions to be made. Standard guidelines for gathering model inputs do not exist. Therefore, model input gathering is beyond the scope of this book. Nevertheless, obtaining reliable inputs is crucial to modeling success; "garbage in, garbage out" (GIGO) rings as true today as ever. Ultimately, the modeler is responsible for the validity of the model. Therefore, it is incumbent on the modeler to understand the inputs the model wants and the inputs actually available and thus supplied. Definitions of parameters must be clear to avoid miscommunication between data supplier and user. Models should always be checked for robustness to inputs. Model parameters should be randomly varied and the sensitivity to parameters should be determined. Parameters for which small perturbations seriously affect outputs must be carefully scrutinized for accuracy. The authors have seen cases of models that were sensitive to parameters that were difficult to estimate accurately. As a consequence, model predictions of changes in output under proposed system modifications were off by more than 100 percent. The user would have been better off to simply assume that the proposed modifications would have no effect!

Figure 1.3 illustrates the role of the functions thus far discussed, which relate directly to the sequence of manufacturing activities. Although these functions have been presented as if they occur sequentially, current practice is moving toward **concurrent engineering** (also referred to as simultaneous or parallel engineering) wherein design teams attempt to integrate the design and development steps. In considering tooling, assembly, and routing during product design, it is hoped that we can reduce the time needed to bring a product to market and avoid designing products that cannot be efficiently manufactured.

Administrative functions such as accounting, order entry, finance, and sales are, of course, also essential. The entire manufacturing system is shown in Figure 1.4. Administrative functions are summarized but shown in enough detail to convey the fullness of the manufacturing system and its interrelationships. We have deliberately

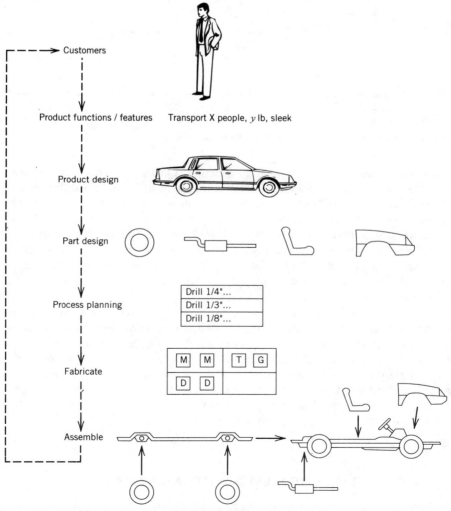

Figure 1.3   Product design through manufacturing scheme.

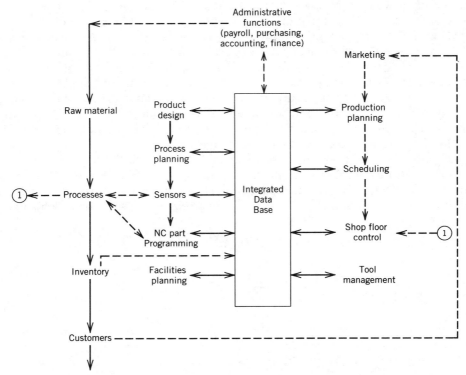

Figure 1.4   Manufacturing activities and information flow.

omitted quality assurance as a separate function. Quality is part of everyone's job and is integrated into each function. A key component in Figure 1.4 is the use of the integrated data base. The figure suggests a computer-integrated data base of the form being proposed for future computer-integrated manufacturing systems (CIMS). Data files may be distributed, but all users will have network access to the same data and information in a timely manner. A single organizational entity will have responsibility for accuracy and updating of each datum, but all functions to which this datum is relevant will have ready access. Such systems will be dependent on universal data format, transmission standards, and protocols.

For the purpose of building and using models, it is convenient to view the manufacturing system as interrelated sets of **tasks, materials, resources, products, plans,** and **events.** Plans include process plans (routings) and production plans. Materials are routed through resources such as workers, handling systems, and machines, thereby being transformed into products. Resources are assigned the tasks required to effect this transformation. Events are points in time when resources start and complete tasks. Status information on resources and materials/products indicates the state of the manufacturing system at any point in time.

1.2

TYPES OF MANUFACTURING SYSTEMS

Our objective in this section is to describe the different types of process configurations or facility layouts. First, however, we note that manufacturing systems are hierarchical in nature. It may be convenient for the reader to think of four levels.

At the top is the entire facility. The facility is composed of departments. It is the makeup of departments that we are concerned with in this text. A department may consist of the 10 milling machines in the plant or it may be the area where axles are made, housing different machine types. The choice of how departments are formed is referred to as layout type.

Departments contain work centers. A work center is one or more machines that are typically scheduled as a single entity. The work center may also contain machine controllers, sensors, robots, or other handling equipment. Reporting time and effort expended is normally done at the level of the work center, with the work center often having one employee. The bottom level is the individual piece of equipment, such as a machine tool, controller or robot.

Manufacturing systems may be classified by a number of characteristics. We have already discussed discrete versus process industries and fabrication versus assembly as potential classifiers. Raw material is another important characteristic—certainly plastic parts require totally different processes than sheet metal, and aluminum parts are processed differently from cast iron parts. For our purposes, the important issue is the approach used to physically group processes and configure the facility layout. Many of the models developed in this book can be applied to fabrication and assembly work centers and for any raw material type. However, each model is designed for one type of facility organization. The common configurations are **product, process, group technology,** and **fixed position.** The difference between the four approaches is most easily seen in the material flow system. Figure 1.5 shows material flow for product, process, and group layouts. Fixed position is used for large products such as ships, buildings, and airplanes because the size of the product makes it impractical to move the product between processing operations. All parts and processes, such as welding equipment, are brought to the product. In product, process, and group systems, the product moves to the process. Our interest is on the first three layout types, and consequently we will not discuss fixed position further.

Product layouts are designed for a specific product. Product layouts are sometimes referred to as flow lines because machines are oriented such that the product flows from the first machine to the second, from the second to the third, and so on down the line. Raw material enters the front of the line. Upon completing processing at the last machine, the raw material has been converted into a finished product. Product lines are unquestionably the most effective and efficient arrangement when justified by product mix and volume. Assembly lines and transfer lines are examples of product layouts and are dealt with in Chapters 2 and 3. Product layouts have the advantages of very low throughput time and low work-in-process inventories. Work-in-process (WIP) is the batches of parts and materials that have been released to the shop floor for manufacturing but have not yet been completed. In addition to the cost of the WIP inventory itself, we incur costs for storage, movement, obsolescence, damage, and record keeping. Product layouts are most effective in avoiding these costs. Forming a product layout implies dedicating the required production processes to the product. Many products do not have sufficient demand to justify a line. Flow lines are intended for mass production. Machines in flow lines are often designed specifically for the product and are not easily adapted to other products. This is not economically feasible unless the product has sufficient volume to absorb the cost of the rearrangement of the facility into a flow line and the full depreciation cost of the equipment while the line is in existence.

It has been estimated that more than 75 percent of manufacturing occurs in batches of less than 50 items. In such environments, machines must be able to perform a variety of production operations on a variety of parts. The traditional response has

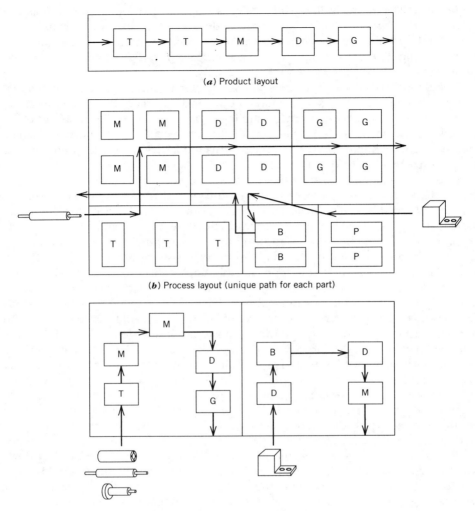

(*a*) Product layout

(*b*) Process layout (unique path for each part)

(*c*) Group technology (cellular) layout

Figure 1.5   Types of manufacturing layouts.

been to use a job shop (Chapter 4) or so-called process layout approach. Departments are composed of machines with similar capabilities that perform similar functions. Accordingly, this approach is also referred to as functional layout. For instance, lathes may form one department, milling machines a second, and punch presses a third. Successive batches assigned to a work center may require very different tooling and setup. Accordingly, highly skilled operators are typically required. In contrast to product layouts, process layouts are characterized by long throughput times and high WIP. The geographic spread in processes visited by each product complicates the setting of relevant priorities, resulting in work centers often doing the wrong job. It is difficult to find someone who will extol the virtues of process layouts. They are considered more of a necessary evil when product layouts cannot be justified. Local accumulation of process experience is one advantage. In custom job shops, with radically changing characteristics of orders, process layouts do facilitate process knowledge accumulation. Grouping of similar machines also allows for higher machine utilization, since extra capacity need not be spread out around the facility.

Table 1.1   General Characteristics of Layout Types

Characteristic	Product	Process	Group	Fixed Position
Throughput time	Low	High	Low	Medium
Work in process	Low	High	Low	Medium
Skill level	Choice	High	Medium-high	Mixed
Product flexibility	Low	High	Medium-high	High
Demand flexibility	Medium	High	Medium	Medium
Machine utilization	High	Medium-low	Medium-high	Medium
Worker utilization	High	High	High	Medium
Unit production cost	Low	High	Low	High

Group technology or cellular manufacturing can be used to convert otherwise process layout systems to pseudo product layout environments. Similar parts are grouped together in sufficient quantity to justify their own machines. A cell is then laid out to produce just this set of parts. Cells will be discussed more thoroughly in Chapter 6. It is probably fair to say that cellular manufacturing is potentially as important a technological innovation as numerical control and robotics. It may or may not be possible to arrange machines in a cell in a complete flow pattern, that is, where all parts follow the same sequence of machine visits. In either case, however, use of machines in a designated physical area for production of a specific set of parts facilitates scheduling and control and substantially reduces setup time (hence, batch sizes), material handling, WIP, and throughput time.

For each of the configurations—product, process, and cell—we have pointed out the strengths and weaknesses. These are summarized in Table 1.1. Each system may be viewed as the best alternative for its appropriate environment. The environment is most easily summarized by the volume–variety combination. The appropriate layout for combinations of product demand volume and variety of products or parts produced is shown in Figure 1.6. As a simple rule of thumb, given the choice, select

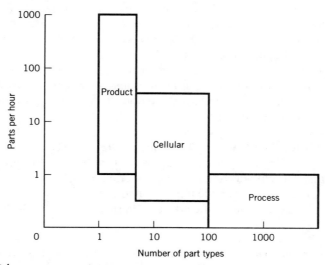

Figure 1.6   Volume versus variety.

cells over process layouts and flow line cells (essentially a product line is a flow line cell) over unstructured (in terms of material flow) cells. Last, we note that cells may be limited only by the ingenuity of the analyst, not any inherent limitation in the concept. At least, it seems fair to state that the potential of cells has not yet been fully explored.

1.3
PRINCIPLES OF MANUFACTURING SYSTEMS

Science and engineering disciplines depend on basic laws or principles. The study of manufacturing systems should be no exception. We will state several relevant principles below; the reader may note that in many instances these follow from basic scientific laws such as the First and Second Laws of Thermodynamics.[1] Unfortunately, specific principles of manufacturing systems are difficult to discern, and there has not been much effort in deriving widely accepted rules for system design and operation based on these principles.

The natural environment changes slowly. Physical phenomena such as gravity continue as constants through time and across space, facilitating recognition and description. Humans have been studying these static, omnipresent systems for thousands of years. Manufacturing systems, on the other hand, are relatively new, complex, and dynamic. Their performance varies with changes in human knowledge and needs instead of any inherent properties. Atoms, molecules, and elements come prepackaged. If you know the mass of an object, its gravitational force is known. However, two manufacturing systems with the same number of machines could have widely disparate production rates, throughput times, and quality. Fixed laws of nature guide the formation of materials under pressure and heat. The same is true for the transformation of materials with basic manufacturing processes. When designing systems, however, humans create artificial constructs for interpreting and integrating components. These include how machines are located and maintained, how parts are batched and dispatched, and how performance is measured. Even the terminology and framework with which we view the system is artificial and subject to change through time. Temperature can be measured absolutely and has a standardized, rigorous definition, but how should we consider machine utilization? Does it include time the operator spends searching for the proper tooling, making defective product, or operating the machine at a less than optimal speed? In practice, how do we know which definition is being used? Despite the lack of experience and standardized framework, it is important that we understand the principles of manufacturing systems, and we will begin to express our current understanding. Hopefully, the future will bring more wisdom and standard definition, thus allowing a fixed framework for system description and evaluation.

First Law (Little's Law). WIP = Production Rate × Throughput Time

Little's Law (Little [1961]) is perhaps the most widely recognized principle of manufacturing systems. The WIP levels and throughput time referred to are average values. The law applies at all levels: individual piece of equipment, work center, department, and system. We need only assume the system is in steady state. WIP is directly propor-

---

[1]The First Law of Thermodynamics states that energy is conserved in a system. The Second Law states that the entropy (disorder) of any system naturally increases through time.

tional to throughput time, the proportionality constant being production rate (since we are assuming steady state, the production rate is also the rate product enters the production system).

Although not that difficult to prove, we omit the proof of Little's Law. Instead, we offer the intuitive explanation of Figure 1.7. Imagine the system as a single process. Let the steady-state production rate be X. There are N jobs in the system. Imagine the system as consisting then of N spaces, each occupied by a job. Every $1/X$ time units a new job arrives to the system and each job in the system advances one place. How long will it take a job to get through the system? Spending $1/X$ time units at each of N spots, the time in the system will be $T = N(1/X)$ or equivalently, $N = XT$ in accord with Little's Law.

Little's Law has a very important consequence. Normally, increasing the WIP level by dispatching more material to the shop floor will increase both the production rate and the throughput time. As the production rate approaches capacity (at least one machine is fully utilized), production rate increases are diminished and further WIP increases result in longer throughput times. Likewise, if throughput time is deemed excessive, this cannot be cured by adding more work to the shop. Dispatching jobs earlier simply increases throughput time as the system becomes overloaded. If WIP levels are to be controlled, they should be controlled at the lowest level that satisfies demand. Managers whose "intuition" tells them otherwise are wrong.[2] Of course, we can, as we will see later, affect throughput times and production rates of specific part types by proper manipulation of input rates for the parts. Likewise, balancing work loads and synchronizing work flow as in an assembly line can increase the production rate for a given level of WIP, thus reducing throughput time. Here, the reader may wish to glance back to the previous section on layout types and note that throughput time and WIP had identical characteristics in each layout.

Second Law. Matter Is Conserved

Manufacturing systems exist to process materials from their raw to finished product state. Processing at a workstation often removes part of the material, as in metal cutting. Provision must be made to dispose of such chips and any consumable tooling. Good product is moved to the next workstation while rejects are either sent to scrap or rework. Our models should satisfy balance equations showing that the difference between material entering and leaving the workstation equals inventory accumulation. In the long term, a stable system cannot have any inventory accumulation; input must equal output.

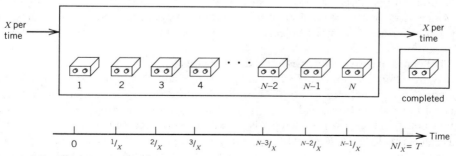

Figure 1.7   Illustration of Little's Law.

[2]We are reminded of the tongue-in-cheek definition of common sense: that which tells us the world must be flat. Valid intuition comes only from an engineering understanding of the relevant scientific principles.

This law holds at all levels, not only for a workstation but also for the entire plant and a microsphere surrounding the contact point of material and tool. An interesting extension is to note that energy is also conserved. This is important for determining environmental conditions at workstations but also for noting that humans only have so much energy to expend. We must choose wisely in selecting the tasks to undertake.

Third Law. The Larger the System Scope, the Less Reliable the System

Large systems are inherently difficult to design, coordinate, and maintain. A basic result from reliability theory is that if we have N (statistically) independent components in our system each with reliability $r_i, i = 1, \ldots, N$, then the probability the entire system is operating is $\Pi_{i=1}^{N} r_i$. We will call this probability the system availability, A. As $r_i \leq 1$, adding components can only reduce system availability. Of course, parallel components can be added and adjustments made to increase the individual r_i, but this only serves to diminish the rate of loss in availability as size increases; it cannot change the direction.

Suppose we double the scope of the system, that is, the number of components that must be operational for the system to function. The new system has $N' = 2N$ components. Further assume that each component has the same availability factor such that $r_i = r$ and $r_i' = r'$. The systems will have the same availability if $A_N = A_{N'}$ or $r^N = r'^{2N}$. This is achieved when $r' = r^{1/2}$. If $r = 0.90$, then we must have $r' \approx 0.95$. In this case, doubling the size of the system requires that we also cut the downtime of each component in half!

Before continuing our discussion of the size-based Third Law, it is convenient to introduce its time-based complement.

Fourth Law. Objects Decay

A machinist can tell you that machines cannot be kept constant. Bearings wear, replacement parts are not identical, and bugs appear. It was originally believed that flexible technologies currently being developed and implemented into the "factory of the future" would experience highly extended lifetimes. Indeed, such explanations are often used to justify new systems. Meredith [1987] concluded from a study of this idea that the new flexible systems wear out as quickly as the older machines. Flexibility allows adaptation to changing environments during the lifetime, but the lifetime varies little.

At the microscopic level, physicists know that particles decay. Coupling this with the recognition that all objects are subjected to external forces in their environment, from changing temperatures and dust to large occasional impacts such as a tool engaging a work piece or a new version of the computer operating system, we realize that both hardware and software objects will decay with time. We cannot remain static.

Fifth Law. Exponential Growth in Complexity

We are often confronted by the curse of dimensionality. If a system has M components, each of which can be in N states, then the system has N^M possible states. Each of these must be considered when designing and operating the system. Three components with two states each leads to eight possibilities, but six components with four states produces 4096 possible system states. Doubling M and N led to a 512-fold increase in the number of possibilities! With M components we have $M(M-1)/2$ possible linkages between components. Thus the number of interconnections between components in a network grows faster than linearly in system size as well.

Sixth Law. Technology Advances

Despite the Fourth Law, our history has been one of advancement. Comparatively speaking, the good old days weren't that good. Moreover, we seem to progress at ever-increasing rates. We leave it to the philosophers to decide if the process of natural decay is a blessing, allowing only the fittest to survive. Since nature is random, and distributions have two tails, the fittest that survive tend to be superior to the current norm. The new mean is better than the old mean. Species evolve to states that allow them to operate more successfully in their environment (or they become extinct).

The moral is that we must work toward constant improvement. Our capability to advance technology is the only blessing for engineers faced with the product law of reliability and natural decay.

Seventh Law. System Components Appear to Behave Randomly

Whether the world is inherently stochastic or simply too complex to be fully understood at the current level of human development, the effect is that events cannot be precisely predicted. We can write a Taylor's Tool Life expression relating tool life to cutting speed, but no one believes each tool will last exactly the same length of time. Cutting tools differ in hardness, work pieces differ, cutting fluid condition varies, and machines vary about the set speed. We will often decide to model systems as if they were deterministic, but what we really assume is that the stochastic variations will not significantly affect the solution. Accordingly, we will use parameters such as cost per hour for a machine to perform an operation. In actuality, we are in most cases using our best estimate of the mean value of that parameter.

We will frequently talk about identical, parallel servers. We all know that this is not valid or when choosing a line at the store we would not have to consider the relative efficiency of the cashiers, just the line length. Have you ever become frustrated with the speed of your line and switched? Machines are also not equivalent. PCs in a lab do not have the same maintenance requirements. In a machining environment with supposedly identical machines, operators soon learn which machines perform best on the difficult jobs. Expectations most likely exacerbate this occurrence through time.

Eighth Law. Limits of (Human) Rationality

Simon [1969] clearly points out the limits of human cognitive capability. We tend to have linear thinking concerning a single task at a time. Our short-term memory is limited to seven items, and our conceptual view is limited to experiences in our 3-D world. As mentioned earlier, engineering design is concerned with artificial constructs, which change with time, as opposed to the permanent characteristic of natural science. These limits, coupled with the realization that complexity grows at faster than a linear rate, lead to the conclusion that we must be willing to *satisfice* instead of demanding optimal solutions. We need to ask, "How good is good enough?", at least for now, and accept the answer.

Ninth Law. Combining, Simplifying, and Eliminating
Save Time, Money, and Energy

The advantage gained by combining and/or simplifying necessary tasks and eliminating unnecessary tasks cannot be overemphasized. Every activity consumes time, money, and energy. If a material handler can transport two loads with close pickup and destination points in a single trip, then doing so cuts product lead time and energy consumption as well as increasing the handler's productivity. The mere act of doing something implicitly assigns that activity importance' by stating it has priority over the multitude of constructive activities that went undone. Simplify, combine, and eliminate was a founding principle of scientific management in its early days, and it is just as valid today. Many of the recent effective trends in manufacturing adhere to this advice. Manufacturing cells are being constructed because they are simpler to build and operate than large systems and allow setups to be combined within product families or eliminated entirely. Kanban production control is simpler to operate than material requirements planning with large information systems and intricate product and shop status data reports. Design for manufacturability is aimed at simplifying the manufacture of products.

There is a popular belief in the business and engineering professions that systems integration and automation are the solutions to our competition problems. One particularly troublesome aspect of these laws is the implication that large software systems, the foundation of integrated manufacturing systems, will be unreliable. Indeed, it has long been recognized that software is often the major technical problem during system implementation (Meredith [1987]). The intangible nature of software makes quality assurance all the more difficult to establish. In automation, software becomes integrated into manufacturing operations. Changing machine speed and feed rates, for instance, or routing decisions requires adjusting software. Without complete, updated documentation the effects on these systems are impossible to predict. Interrelationships between software programs may be hidden deep in the code. In addition to the complexity of large software systems, it must be remembered that different individuals and organizational entities have different abilities and desires for information. System complexity, integration requirements, and development cost dictate the need for strict standardization in large systems. However, use of a standardized format may fail to fulfill the needs of certain users and naturally restricts the effectiveness of the software system.

The modern approach to manufacturing (information) system design, namely computer-integrated manufacturing systems, represents a battle against our laws. The object of CIMS is to maximize coordination and knowledge across system components. In this fashion the loss as we move between system components is minimized. The belief is that perfect communication can lead to a system with optimal performance. However, as we increase sophistication, the mere act of collecting and transmitting data as information consumes resources and requires time. This passage of time reduces timeliness of the data, and timeliness is a prerequisite for perfect communication. As the number of bits of information increases, the probability of error increases. External "energy" must be constantly applied to the system to collect and manage new data; otherwise, the system deteriorates with time. It is not our intent to discourage CIM efforts. On the contrary, improved information can lead to much better decisions, and computers are more reliable and efficient than humans. As a child you probably played the game of forming a circle, listening to a message from one neighbor, and then whispering it to the next. When the message returns, it has been distorted to the point that the original message is lost. The computer memory receiving these words as they are typed will remember their exact form long after the wording fades from the author's memory. Our point is that CIM approaches run against the laws of nature. The alternative strategy of trying to *simplify* systems and procedures is compatible with nature. If an integrated system is to be adopted, the key is to understand the crucial links and interactions between system components. One should plan the necessary communication between these components or otherwise let components maintain autonomy.

An important application of this discussion is the recognition that "If you can't make a simple machine work, you can't make a complex machine work." The initial response to a problem should not be to buy a better machine. Until the present technology is understood, we cannot adequately answer the question of its adequacy.

1.4
TYPES AND USES OF MANUFACTURING MODELS

Before describing the art and science of modeling, we present a word of caution to the analyst. An important concept to keep in mind when analyzing any system is the distinction between "efficiency" and "effectiveness." Efficiency refers to doing the task right, whereas effectiveness refers to doing the right task. Although both are important, the effective worker will always be prized. Efficient workers who are not effective may find themselves standing in the unemployment line shaking their heads and mumbling about the incompetent managers for whom they were forced to work. We can illustrate the difference with a single machine scheduling problem. The efficient engineer schedules jobs to guarantee maximum throughput and minimum waiting time at the machine. Unfortunately, the machine is one of many, and the parts it produces are used with many others in final assembly. Moreover, capacity may exceed demand. The effective worker determines what the true priorities are for the various part batches waiting at the machine and schedules parts to be completed as needed. The machine may even be left idle while batches sit if none are needed in the near future. In this example, the efficient planner left us with piles of unneeded inventory that occupy space, increase paperwork, and generally clutter the production process. At the same time successor workstations waited for the needed parts. The example indicates that in attempting to be efficient we must be very careful about how we identify our objectives. Suzaki [1987] identifies seven types of waste: (1) waste from overproduction, (2) waste of waiting time, (3) transportation waste, (4) processing waste, (5) inventory waste, (6) waste of motion, and (7) waste from product defects. In general, if an action does not directly add value to marketable product, the action is wasteful. Our objective should never be so myopic as to obscure the ultimate objective of meeting customer satisfaction in a profitable manner. As another example of effectiveness, consider the issue of setup reduction. We could be efficient and purchase a tool to tighten nuts automatically and more quickly when positioning dies on a machine. However, it might be cheaper to simply cut off any extra threads on the screw so that fewer revolutions are required to tighten the nut. Of course, if we save a minute here while the machine is required to spend an hour warming up the die, then how effective have we been? The suggestion to use prewarmed dies would be much more meaningful. We will not explore setup time reduction further. It is important to note, however, that this decision is made only because the objective of this book is manufacturing *system* design and analysis and we consider this a process or operation matter. The importance of setup reduction with its attendant reductions in throughput and inventory cannot be overemphasized.

The efficient modeler builds a mathematical description of the system and finds the optimal solution to that model. The effective modeler builds a mathematical model of the system, uses it to understand the important factors in the real system, finds a good solution to the model, and then modifies it with knowledge of relevant externalities that are not included in the model to find a very good solution to the real system! In summary, the effective worker is concerned with finding very good solutions to important problems, not in finding optimal solutions to trivial subproblems of the total system by ignoring the interactions between the various subsystems. The effective modeler models systems. Keep in mind that if we start at 80 percent of maximum (optimality), raising 50 percent of the system to 90 percent efficiency has a far greater impact than raising 10 percent of the system to 100 percent efficiency. Likewise, cutting 50 percent of cost in an area of 5 percent of total cost is only half as effective as saving 10 percent from a budget item that represents half the budget. The point is not that efficiency is bad; indeed, much of this book is concerned with efficiency. Traditionally, mass production systems have been very concerned with efficiency. Long product life cycles and capital-intensive mass production equipment have allowed and required the "right thing" to be defined early on in the system life. On the other hand, being effective has been the key ingredient for small-batch, custom-product operations. In this environment one must be constantly listening to the customer and providing what the market wants. The point is simply that effectiveness should be the prime consideration. Effectiveness requires the flexibility to adapt to the dynamic world. Flexibility will be a key issue in several later chapters.

This book purports to cover the development and use of models for design and control of manufacturing systems. Having already outlined what comprises a manufacturing system, we must define what we mean by "model." A model of a real system is a representation of that system in another medium, usually in a simplified form. Models can be either physical or mathematical abstractions of reality.

1.4.1 Physical Models

Physical models have been in widespread use for many years. We have all seen the models developed by architects to illustrate buildings being constructed. A model of this kind provides a visual aid for checking the desirability of potential designs and for ensuring proper construction. A picture *is* worth a thousand words, and much of the ambiguity and imprecision of verbal communication can be overcome with simple physical models.

Physical models can be two- or three-dimensional. Two-dimensional models include part blueprints and facility drawings. In facility layout, 2-D iconic models are often used. Cutouts (icons) of resources such as machines, workers, and service areas are moved around on a scaled outline of the facility until a satisfactory layout is found.

Three-dimensional models are becoming more popular. Many CAD systems now use 3-D solids modeling, for example. Laboratories in industrial and manufacturing engineering departments use small table-top machines or interlocking pieces to construct physical 3-D models of manufacturing systems that actually operate.

1.4.2 Mathematical Models

This text covers mathematical models. These models may reside on a computer or simply on a pad of paper. However, they all share the common factor that a set of mathematical equations or logical relationships is developed to describe the real system. Parameters of the models, such as standard production times, time between machine failures, and batch sizes, are estimated from accounting and other data.

Mathematical models differ from physical models in their use of decision variables. We must have some intended use for the model, which revolves around variables that we can control. These become the decision variables of the model. The key

CHAPTER 1   MANUFACTURING MODELS   **19**

to building useful models is to select the proper decision variables. This is closely related to problem definition and synthesis. As a general guide to determining the decision variables, the modeler should ask: What questions am I trying to answer? Decision variables might be the number of machines needed or the set of tasks assigned to a machine.

Mathematical models can be descriptive or prescriptive in nature. Simulation models tend to be descriptive. Given a set of values for the decision variables, we turn the model on and out comes an estimate of system performance. Mathematical programming models such as linear programming are prescriptive. Turn the model on and out comes the answer of how we should set the decision variables. An objective function such as cost exists. Subject to the specified constraints, the joint values for the decision variables that optimize the objective function are found. Descriptive models often provide an avenue for building very realistic models. Although a clever modeler can probably define a prescriptive model at the same level of detail, these models grow in size and become nonlinear very quickly when details are incorporated, making them virtually impossible to solve to (model) optimality. As we shall learn shortly, the ability to prescribe may be less important than the ability to describe performance over a wide range of input variable combinations.

Prescriptive models do not necessarily provide (model) optimal solutions. Problem size and the absence of efficient solution schemes often mean that a heuristic approach is needed. A heuristic approach attempts to use a rational method to find a good (nearly optimal) solution to the problem. We may or may not find the optimum, and even if we do find it, we may not be able to confirm that the solution found is optimal for the model. Bartholdi and Platzman [1988] summarized heuristics nicely with the explanation

> A heuristic may be viewed as an information processor that deliberately but judiciously ignores certain information. By ignoring information, a heuristic is freed from whatever effort might have been required to read the data and compute with it. Moreover, the solution produced by such a heuristic is independent of the ignored information, and thus unaffected by changes in that information. Of course the art of heuristic design lies in knowing exactly what information to ignore. Ideally, one seeks to ignore information that is expensive to gather and maintain, that is computationally expensive to exploit, and that contributes little additional accuracy to the solution.

In addition to often being robust to changes in the data, heuristics are generally much easier to develop and solve than optimization procedures. Figure 1.8 gives the basic trade-off found in many problems. Many real-world problems are simply too large and complex to solve in the time available. This is particularly true with operational control problems requiring frequent resolution. We can wait for a weekend for a computer to determine the optimal design of a new plant, but not for this afternoon's production schedule. Heuristic rules are easier to code and understand; thus, such models can be developed and implemented more quickly. The trick is to obtain the best model for the given resources and time allotment. Often the optimal solution will possess certain characteristics that can be exploited by the heuristics. One classic example is the dynamic lot sizing problem. We know that it is never optimal to carry inventory forward into a period unless the inventory can cover the entire period's demand. Otherwise, we needlessly incur holding costs in addition to the necessary process setup cost. Accordingly, heuristics consider producing only the demand for an integral number of periods whenever a setup is planned.

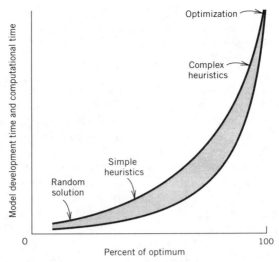

Figure 1.8 Heuristic versus optimization trade-off.

We should not infer that heuristic procedures are necessarily inferior to optimal algorithms. Heuristics are generally problem specific. The process of developing good rules for heuristic solution may provide some insight into the important factors in the problem. This insight can be very helpful in day-to-day management. If we had built a complex code based on standard optimization procedures and used it as a black box to solve the problem, this insight may not have been achieved. Moreover, it is good practice when using heuristics to also employ bounding techniques that let the modeler know how far the heuristic solution may be from optimal. Bounds can be based on either worst-case or average-case analysis. Worst case bounds are generally theoretical results that indicate the poorest result, relative to the optimal, at which the heuristic can ever terminate. These results hold for any problem instance (a problem instance is a specific set of data for which a solution can be found). Average case results are based on experimentation and indicate the expected deviation from optimality when the problem data fall within some specified range.

EXAMPLE 1.1

Suppose we have three jobs and three machines. Each machine is to be assigned a job. The objective is to find the least-cost assignment. Assignment costs are shown in Table 1.2.

Table 1.2 Assignment Costs for Example 1.1

Job	Machine		
	1	2	3
1	10	25	12
2	13	5	12
3	8	13	21

Suppose our heuristic says to take job 1 and assign it to the machine with minimum cost. Then remove this job and machine. Repeat the procedure until all jobs are assigned. Our solution would be

STEP 1. Assign job 1 to machine 1, since $10 < 12 < 25$.

STEP 2. Assign job 2 to machine 2, since $5 < 12$.

STEP 3. Assign job 3 to machine 3.

Total cost of this assignment is $10 + 5 + 21 = 36$. We can find a lower bound on cost by noting that each job must cost at least its lowest machine cost. Thus, job 1 must cost at least 10, job 2 must cost at least 5, and job 3 must cost at least 8. Therefore, any solution has a cost of at least $10 + 5 + 8 = 23$. Likewise, each machine must perform a job. Summing the minimum cost in each column we have another lower bound of $8 + 5 + 12 = 25$. Accordingly, any feasible solution must cost at least 25. Our solution costs 36. Our solution is therefore at most 11 over the optimum.

Note that the method used for finding the lower bound of 25 actually produced a feasible solution: each job is assigned and each machine has one job. This solution must therefore be optimal.   ■

Instead of categorizing mathematical models by their output (descriptive or prescriptive), we could categorize models by their computational form. With this taxonomy, models are **analytical** or **experimental.** Analytical models represent a more mathematical abstraction of the real system. A set of equations is produced that summarizes the aggregate performance of the system but does not describe the detailed events that occur. Examples are queueing theory, mathematical programming, and heuristics. Simulation models are experimental. Simulation models mimic the events that occur in the real system, allowing experimentation with operating parameters or control logic. Both computer simulation and the physical simulation models mentioned above fit into this group. Hybrid models may also be used. A simulation model of a complex system could be built where modules of the real system are replaced by simple analytical models such as single server exponential queues. The logic of the simulation model relates the individual modules together to replicate overall system occurrences.

1.4.3 Model Uses

Models are built for many purposes. Primary uses include the following:

1. Optimization—finding the best values for decision variables.
2. Performance prediction—checking potential plans and sensitivity.
3. Control—aiding the selection of desired control rules.
4. Insight—providing better understanding of systems.
5. Justification—aiding in selling decisions and supporting viewpoints.

Optimization has already been discussed in the context of prescriptive models. The model is constructed and then executed to determine the best settings for the decision variables. Problems such as selecting batch sizes and minimum cost shipping networks are good candidates for optimization. We again caution against the blind use of model output, however. Ultimate responsibility lies with the decision maker, not the model. It has been said that there are two types of troublesome decision makers when it comes to models (and computers); those who believe everything that comes out and those that believe nothing. The latter can be educated in time to trust the model; the former are helplessly exposed to the whims of fate.

Many real systems exhibit nonlinear behavior and require discrete (integer) variables. Models with these characteristics are often difficult to solve, and even though we may want an optimal solution, we may need to settle for a near optimal, heuristic solution. We include such models with optimization, since the purpose is still to find the best set of decision variable values possible. In extreme cases we may be satisfied at just finding a feasible solution, one that adheres to all specified constraints on the decision variables. Remember, being effective relates to finding a feasible solution to the right problem.

A word of caution regarding optimization. Formulating models for optimization typically becomes a matter of cost trade-offs. The standard economic order-quantity model, which compares inventory holding and setup costs, is a prime example. The creation of such models tends to legitimize the existence of these costs in the mind of the modeler. With this mindset we may overlook opportunities to reduce both costs, as has been accomplished through just-in-time techniques.

The second purpose is performance prediction. "What if" questions must constantly be answered. What if a machine breaks, what if a supplier can't supply, what if demand changes? Managers are not omniscient; plans must be constructed that are robust in the sense that recourse exists as events occur and conditions change. Continued success requires the generation of new ideas. Descriptive models such as simulations are naturally designed for this purpose. They are used to test out these ideas, separating the good from the bad. With descriptive models, we input values for the decision variables and out comes a measure of system performance. Performance prediction models may also be used during planning. Suppose marketing suggests a specific production plan. Our objective is to determine whether the schedule is feasible, and, if so, what the cost will be. By comparing the model output to budget and other constraints, we can evaluate the schedule.

Prescriptive models are also very important in performance prediction. Sensitivity analysis is an important activity for decision makers. By varying input parameters such as material cost, labor hours, and productivity factors, the effect of changes or uncertainty in parameter values can be assessed.

Control is the third purpose for modeling. Control policies may be derived from models. In scheduling a work center, should a shortest processing time (SPT) or earliest due date (EDD) rule be used? How do we decide when to release jobs to the shop floor and how to prioritize and route them once there? Models can be built that permit examination of system performance under different control policies. By evaluating system performance under a range of scenarios for each possible policy, a near-optimal policy can be selected.

The model-building process has considerable value. The lessons learned during model validation and respecification provide considerable insight into the real system. Suddenly it becomes clear which of the dissenting views regarding the true bottlenecks and important system relationships is correct. It may prove true that the knowledge gained by building the model increases insight into system performance sufficiently to obviate the need for thereafter using the model.

In the category of insight we include the use of models as diagnostic tools. Problems may be known to exist with the system, but the underlying causes are unknown.

Playing with the model can generate the circumstances under which the problems appear, exacerbate, and dissipate. Once the source of the problem is established, corrective action can be taken.

Finally, models can be used as effective sales tools. Simulations with animated graphics can go far toward convincing a skeptical manager or supervisor of the validity of the model. Indeed, there have been instances where analytical models were used to solve the problem and then simulation models built to sell the solution. It is not uncommon for simple graphical models to be built as a final step in modeling large, complex systems. A manager may not feel comfortable with a black box of several thousand equations. However, a simple schematic or animated simulation depicting the outcome from the complex model may be a very effective sales tool.

Common to all models is the importance of sensitivity analysis. For each of the primary uses, it is important to know the effect of changes in demand or machine reliability. In designing a layout we may want to examine several potential machine orderings to determine the resultant utilization on the material handling system. Engineering design must consider the dynamic and stochastic nature of the external (and internal) world. Accordingly, several potential designs should always be developed and evaluated over the range of possible scenarios.

1.4.4 Model Building

Model building is an art. Science comes into play more in model solution than in building. Model building iterates between the use of inductive and deductive reasoning, as shown in Figure 1.9. Once the problem is defined, we hypothesize the

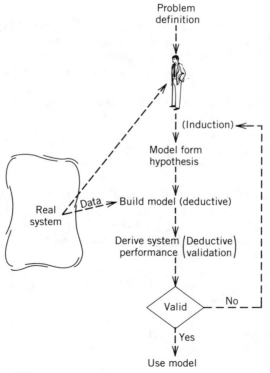

Figure 1.9   Model building.

important system aspects for addressing this problem. This step involves induction. We use our experience and intuition about how the system operates to distill the essential entities and their relationships. These entities and relationships are described mathematically or logically. This formalization of the system description is often deductive, employing physical laws such as conservation of inventory. Students too often believe that they can go out and build **the model** of the system. There is no such unique model of the system. The model must fit the question being asked. In complex environments several models may be built to address different aspects of the overall problem. Even for a single question, experienced modelers may well come up with different models. Hopefully, these models, if properly validated, would all produce similar solutions to the problem. It is generally best to have the simplest model that sufficiently describes the system. Simpler models are easier to build, maintain, and use; moreover, it is easier to estimate parameters for them. In addition, by not including extraneous factors that may have only by chance helped predict past performance, they are more likely to be good predictors of performance under new conditions.

Verification and validation are the next steps. Verification assures that the model specified on paper and that implemented on the computer are equivalent. Validation assures that the model, both in its specification and its computer implementation, corresponds to the system sufficiently well to produce results valid and useful for decision making. In addition, validation must assure that the data used in the project are appropriate, accurate, sufficient, and correctly transformed if necessary. Figure 1.10 summarizes these validation and verification concepts. The verification and validation processes often have overlapping or similar tasks. Some of these tasks are as follows:

1. Comparison of model and system structure. The system components as they exist or are designed are compared to their representation in the model.
2. Comparison of results and corresponding system data. The analysis results are compared to estimates of the same quantities resulting from system operation or design specifications.
3. Comparison of model and system behavior. The temporal behavior expressed in the model is compared to the temporal behavior seen in system operations or specified in system designs.
4. Comparison of model structure and results with the structure of and results obtained from another model of the same system.

A further discussion of verification and validation is given by Sargent [1988].

It is important to keep in mind the intended use of the model and to validate the model for that use before embarking on problem analysis. Extrapolation beyond the range of inputs for which the model was validated can lead to misleading results. Remember that models contain two sources of error: system approximation and solution approximation. Models only approximately describe the system, generally leaving out many minor details and quite often only describing the aggregate behavior of sets of major details. It is safest to decide in advance of model building what type of questions the model will be asked and to validate the model accordingly. Subsequent use of the model beyond that realm requires revalidation. Solution approximation refers to the use of heuristic procedures and numerical accuracy of implemented algorithms. Unproven heuristics and numerically unstable algorithms

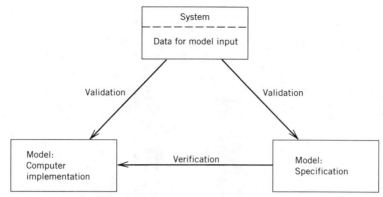

Figure 1.10   Model validation and verification.

can be arbitrarily bad. We leave this somber subject with the admonition *"do not go blindly into the night";* validate instead.

An important philosophical note concerns building unbiased models. Models should not be built to prove a particular viewpoint. The step of model validation should assure the accuracy of the model for its intended use. Those who build models with the expressed or hidden purpose of proving a prior belief run the risk of missing crucial model components and implementing poor solutions. While we recognize the political nature of all human organizations, the purpose of model building should be to discover truth, and not to distort it. In the war of manufacturing competition, such actions diminish the chances for success. Moreover, models soon lose their influence in this environment. Those who question this view should ponder the fate of the power of statistics to sway public opinion. Because of abuse of statistics and improper problem definition, public opinion is probably best summarized by the Disraeli quip: "There are three kinds of lies: lies, damned lies, and statistics."

As a final note, we should never forget that the model is a means, not an end. Engineering design involves the steps of problem definition, data synthesis and generation of alternative approaches, analysis of tentative solutions, evaluation of alternatives, recommendation, implementation, and monitoring/maintenance. Models tend to be used for analysis and evaluation. However, a model cannot lead to a solution that is outside its scope. For instance, consider the problem of inserting components into circuit boards. Perhaps the current system relies on manual insertion. The efficient engineer might try to speed up the manual process, but the effective engineer takes a step back and considers the overall system and possible alternative designs. In this case, alternatives may include a line of operators each responsible for different components (progressive hand assembly), robotic assembly, or high-speed insertion machines. Seifert [1988] presented Figure 1.11 to indicate the difference in costs for these approaches. Clearly, the engineer stuck in the mindset of micromotion would have missed some opportunity. The figure can be generated from simple models *if* the alternative system descriptions have occurred to the modeler. Modeling is not a replacement for thought and creativity; on the contrary, models and computer analysis should simplify and enhance the creative process by facilitating the comprehension of system behavior and relationships.

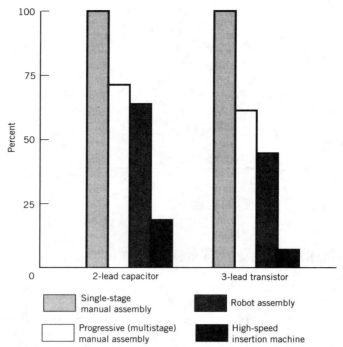

Figure 1.11　Relative costs for elective assembly approaches.

1.5
SUMMARY

Manufacturing systems can be characterized by their layout type. Layouts can be classified as being product oriented, process oriented, hybrid (product family inspired cells), or fixed position. Product layouts are generally the most efficient for repetitive manufacturing.

Several basic laws of nature impact manufacturing systems. Inventory level is the product of production rate and throughput time. Matter and energy are conserved. Systems become less reliable as more interacting components are added; however, adding parallel or redundant components can increase reliability. Objects decay with time, but we can nevertheless advance by taking advantage of technological gains. System designs should account for random behavior. Lastly, human capabilities are limited. It may often be best to simplify systems and eliminate unnecessary components.

Models form a rational basis for designing new systems and learning about existing systems. Playing with models of the system instead of the actual system allows fast acquisition of knowledge and avoids the risk of costly disruptions to the real system. Models can be used for system optimization, performance prediction, control, gathering insight into the system, or as a teaching tool. Models can be mathematical in nature, such as linear programming, or physical, such as a scale model of a plant showing aisles and machines. The key ingredient is to define the'problem and, hence, the purpose of the model. The modeler then constructs and validates a model with sufficient scope and complexity that it can aid in the detection and intellectual integration of important system components and interactions.

Modeling suffers two primary sources of information loss. First, the model is only a simplified abstraction of reality. Many of the details that affect actual behavior are not included in the model. We iteratively build and test models hoping to ensure that all significant (with respect to the model's purpose) factors are included. Second, models are only partially explored. Heuristics may end in nonoptimal solutions. Sensitivity analysis is necessarily limited in scope. Accordingly, we never extract all the information the model has to offer. We merely hope to learn enough to effectively design and operate the system under study.

REFERENCES

Bartholdi, John J., III, and Loren K. Platzman (1988), "Heuristics Based on Spacefilling Curves for Combinatorial Problems in Euclidean Space," *Management Science,* 34(3), 291–305.

Buzacott, John A. (1985), "Modeling Manufacturing Systems," *Robotics and CIM,* 2(1), 25–32.

Johnson, Lynwood A., and Douglas C. Montgomery (1974), *Operations Research in Production Planning, Scheduling and Inventory Control,* John Wiley & Sons, New York.

Little, J. D. C. (1961), "A Proof for the Queueing Formula $L = \lambda W$," *Operations Research,* 9(3).

Meredith, Jack R. (1987), "Automating the Factory: Theory versus Practice," *International Journal of Production Research,* 25(10), 1493–1510.

Sargent, Robert G. (1988), "A Tutorial on Validation and Verification of Simulation Models," in *Proceedings of the 1988 Winter Simulation Conference,* M. A. Abrams, P. L. Haigh, and J. C. Comfort, eds., Institute of Electrical and Electronics Engineers, San Francisco, 33–39.

Seifert, Laurence C. (1988), "Design and Analysis of Integrated Electronics Manufacturing Systems,"

in *Design and Analysis of Integrated Manufacturing Systems,* W. Dale Compton, ed., National Academic Press, Washington, D.C., 13–33.

Silver, Edward A., and Rein Peterson (1985), *Decision Systems for Inventory Management and Production Planning,* John Wiley & Sons, New York.

Simon, Herbert A. (1969), *The Sciences of the Artificial,* The MIT Press, Cambridge, MA.

Sule, D. R. (1988), *Manufacturing Facilities: Location, Planning and Design,* PWS-Kent Publishing Co., Boston, MA.

Suzaki, Kiyoshi (1987), *The New Manufacturing Challenge: Techniques for Continuous Improvement,* The Free Press, New York.

Tompkins, James A., and John A. White (1984), *Facilities Planning,* John Wiley & Sons, New York.

Vollmann, Thomas E., William L. Berry, and D. Clay Whybark (1988), *Manufacturing Planning and Control Systems,* Richard D. Irwin, Homewood, IL.

PROBLEMS

1.1. Why are manufacturing system models built and for what purposes are these models used?

1.2. What are the four basic types of layout?

1.3. Why do you think large systems are often organized and managed as a group of interrelated subsystems as opposed to one large, complete model? What are the implications of this organizational structure for model building?

1.4. A system is composed of eight independent subsystems. Each subsystem operates 95 percent of the time. What is the probability that all the subsystems are operating at any random point in time?

1.5. Visit a manufacturing plant in your vicinity (or recall a past visit). What are the parts and products

made at the plant? What type of layout is used? Sketch the layout of the facility indicating major manufacturing areas and support services. Superimpose the flows of material on this layout. Do the material flow pattern and layout seem well planned?

1.6. Consider a local fast food restaurant. List the tasks, materials, resources, products, plans, and events associated with this system.

1.7. The performance P of a system is highly dependent on the setting of a single parameter, say X. Table 1.3 shows the recent performance of the system along with the associated X settings.

    a. Plot the $P - X$ relationship. Does a simple $P = a + bX$ model appear reasonable?

Table 1.3   Historical Performance Values

P	X
10.0	0.0
14.9	1.0
19.6	2.0
24.1	3.0

b. Fit a model (either by regression or eye-hand approximation) to the system.

c. Predict performance if $X = 2.5, 10.0, 20.0, 30.0$.

d. The true system model is $P = 10 + 5X - 0.1X^2$. Compute the errors in your predictions for part c. What does this tell you about model validation?

1.8. You are the manager of a plant. You have a choice of using a heuristic solution to a model that realistically describes most of your operation or an optimal solution to a model that does not include several interactions within the plant that might be important. Discuss the implications of each decision and suggest a course of action.

1.9. Visit your local science/engineering library. Pick out a recent issue of *Industrial Engineering* or *Manufacturing Engineering* magazine. Browse through the issue reading at least one article on a manufacturing topic of personal interest and all the advertisements.

    a. Summarize the main point of the article in one or two sentences.

    b. List all the vendors and types of CNC machine tools advertised.

1.10. State the basic laws of manufacturing (and all) systems discussed in the chapter.

1.11. Explain why the basic law of reliability implies that complex systems will be hard to maintain.

1.12. Many supervisors believe they should keep large levels of work-in-process in front of each workstation so that machines are never "starved" for work. Recognizing that long-term production is dictated by demand, what is the actual effect of large in-process inventory levels?

1.13. During the 1970s and 1980s the United States lost much of its market share in many key industries, for instance, consumer electronics and automobiles. Do you think that U.S. manufacturing started doing a worse job during those 20 years? If not, what explanation can you give for the declining competitiveness?

1.14. For each of the following environments suggest an appropriate layout type.

    a. A machine shop that receives job orders for custom dies and fixtures from local manufacturers.

    b. An automobile body press shop that manufactures body parts for 1000 automobiles per day.

    c. A manufacturing plant that assembles electronic controls such as thermostats used in homes and household appliances.

    d. A manufacturing plant that produces several different types of parts such as gears, rods, and boxes in moderately high volume. Each part has several similar versions for sale.

    e. A plant that assembles, packs, and ships several thousand computer terminals of three different models daily.

1.15. A manufacturing plant produces 250 parts per day. Average work-in-process is 800 parts. Find the average throughput time for a part.

1.16. A plant keeps five days' worth of production in inventory. Suppose it is possible to increase production by 20 percent if the inventory level is doubled. What effect will this have on throughput time?

1.17. Write out a list of your weekly activities and indicate how much time you spend on each. If you wanted to spend an extra four hours per week studying, where would those hours come from?

1.18. Construct a simple model relating calorie intake, weight, and exercise. Assume a fixed number of calories per pound of weight and a fixed consumption of calories for a given activity level. Suggest strategies for losing weight. Suggest an experiment that would allow you to test this model.

PART II

MATERIAL FLOW SYSTEMS

The manufacturing system exists to produce goods that can be sold to customers. Facilities are constructed to accomplish that goal. Capital, energy, human, information, and raw material resources are acquired, transported, and consumed in transforming the material into value-added products. This part of the book is concerned with alternatives for locating physical resources within the plant and planning routes for the movement of material between the resources.

Chapter 2 addresses serial assembly systems, or assembly lines. Raw material enters the system and progressively moves through a series of adjacent workstations while being transformed into finished product. These tend to be the most efficient production systems but require reliable processes with only minimal variability in processing times at and between workstations and in the type of products produced. If technology is rapidly advancing, product life may be too short to justify construction of a line dedicated to one product. Conservation of material requires system design to include provision for supplying parts to all workstations along the assembly line and to remove wastes.

When reliability becomes a problem, buffers can be added between workstations. These systems are covered in Chapter 3. Although exact analysis is much more difficult, simple yet accurate approximations exist for predicting system production rates.

When the adverse effects of highly variable and dynamically changing requirements dominate, general job shop systems are used. Chapter 4 gives an introduction to the important issues and scheduling procedures for such environments. These systems tend to be the most flexible but least productive.

Technological advances in computers, electronics, and controls at the machine and system level have led to highly automated, flexible manufacturing systems, as described in Chapter 5. These systems provide a more productive alternative to job shops and help meet the demand for quick delivery of customized products when multiple, medium-volume part types are produced. Flexible manufacturing systems come with a new generation of planning problems. The complexity and varying time horizon aspects of the overall planning problem cause a hierarchical planning and control strategy to be used in practice. Chapter 5 illustrates how this hierarchy operates.

The basic reliability law of Chapter 1 tells us that small systems have advantages. The group technology/cellular manufacturing models of Chapter 6 attempt to bring the performance advantages of serial systems to the job shop environment. Expertise and equipment are organized by product line to maintain some flexibility and handle variability in requirements and technological changes while simplifying scheduling and reducing material handling. A variety of models are provided for cell formation to emphasize the notion that modelers have choices and better solutions may be obtained through models that require more data and solution effort.

Interdependencies between resources make the decision of where to locate resources a hard problem. Factors such as the distance, frequency, and batch size of material moves must be considered, but so must other factors such as safety and the effectiveness of communication as workers are separated. Nevertheless, scientific approaches exist for deciding where to place resources. Recent advances in models for the layout problem have led to useful tools. Chapter 7 describes the basic modeling approaches for this general facility layout problem.

CHAPTER 2

ASSEMBLY LINES: RELIABLE SERIAL SYSTEMS

KISS: Keep it simple, stupid!

2.1
INTRODUCTION

An assembly line is a set of sequential workstations, typically connected by a continuous material handling system. The line is designed to assemble component parts and perform any related operations necessary to produce a finished product. The complete assembly activity of the product is divided into productive **work elements** such as inserting an IC chip into a circuit card or attaching a side panel to a car body. A work element is the smallest unit of productive work, that is, an activity that adds value and must not be immediately followed by any other activity.[1] Each workstation is assigned a subset of these work elements. The product is passed down the line, visiting each station in sequence. Upon exiting the final station, the product is complete. The line is operated such that the stations are simultaneously busy. Upon completion of its assigned tasks on an item (unit of product), the station passes the item to the next station, obtains a new item from its predecessor station, and repeats its tasks.

Assembly lines rely heavily on the **Principle of Interchangeability** and the **Division of Labor**. Developed in the 1800s, the principle of interchangeability states that the individual components that make up a finished product should be interchangeable between product units. We should be able to take the firing pins from two rifles of the same model type, replace each in the other rifle, and still having two operational products. For the assembly line, we can have bins of interchangeable parts and select any one of these to insert in the current product unit. Division of labor embraces the concepts of work simplification, standardization, and specialization. Following these principles, complex activities are subdivided into elemental tasks, detailed work instructions are produced for rationally accomplishing each of these tasks independently, and the elemental tasks are assigned to different workers, who quickly become proficient in performing their repetitive operations. Together, interchangeability and division of labor facilitated mass production, allowed replacement parts to

---

[1]Picking up a screwdriver is not a work element because no permanent value is added. Inserting the screw(s) needed to connect the base of the part to the internal assembly would be a work element, however.

31

be used to lengthen a product's useful life, and made possible the pioneering work of Henry Ford and others in developing assembly lines. In the 1900s, assembly lines drastically reduced production cost and increased production volume. Many items previously available only to the wealthy were brought within the economic reach of the masses. The economic advantages of assembly lines are widely accepted. In fact, much of the manufacturing system research during the past half century has concentrated on increasing the set of environments in which continuous flow manufacturing concepts embodied in assembly lines could be applied.

Figure 2.1 shows one layout style for an assembly line. A conveyor delivers and removes work to each station (two-level conveyor systems may be used for bidirectional flow). An unload and storage spur is located at each workstation for unloading, storing, and returning product. A schematic model, equivalent for our purposes, is given in Figure 2.2.

Although lines are traditionally linear, U-shaped lines are growing in popularity. This allows for integration of material flow into the "spine" of the facility, facilitates worker and supervisory communication, and provides space efficiency. If carriers need to cycle back to the front of the line after their part is completed, move distances are reduced. Serpentine layouts can be used for lines with long lengths. The disadvantage of these shapes relates to potential difficulty in accomplishing the turns along the line for some products.

Foremost among the advantages of assembly lines is the ability to keep direct labor (or automated machines) busy doing productive work. In contrast, the typical job shop spends considerable time on setup activities such as acquiring and configuring raw material, tooling, and instructions. The dynamic nature of job shops often results

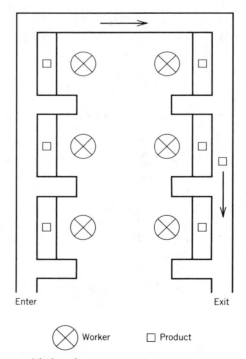

Figure 2.1   Possible assembly-line layout.

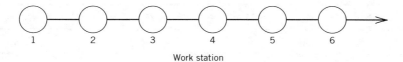

Work station

Figure 2.2   Simple schematic model—serial processing line.

in considerable waste of time owing to a lack of planning, standards, or instructions. Setup requirements are normally minimal in assembly lines since the tasks are repeated. Instructions are simply to repeat the previous activity. Thus, workers quickly reach effective points on the learning curve. The desired assembly line environment is one with very high volume production of a single item or similar family of items. Furthermore, the total work load should be equally divisible among the workstations. Work load refers to total item assembly time. Assembly line systems do not require large queues of work-in-process inventory. Consequently, there is less space required, plus lower inventory holding cost and shorter throughput time. Unlike costly batch production environments, where finished goods are stored to permit intermittent production and utilization is low, the high utilization, continuous nature of assembly lines allows matching of production and demand rates. Effectively, the batch size is one.

In practice, many items do not have sufficient demand to justify an assembly line, but a family of similar products might. Multiple product lines are then used. The line is periodically shut down and a changeover in line setup is made between products. The advantages of inventory reduction are lost in this case. We will shortly discuss how to assign tasks to workstations. This problem is solved separately for each of the multiple products. The line is then treated as a single workstation for planning production runs. Recent attention to just-in-time techniques has led to the use of **mixed lines** instead. In a mixed line, several products are allowed to be on the line (in different workstations) at the same time. The input mixture of products to the line must match product demand rates. The precise sequence of products entering the line should minimize workstation imbalances between model types. This issue is addressed in Section 2.4.

Several organizational issues must be addressed prior to actual planning of the assembly line. One issue concerns the number of lines to build. At one extreme is the use of a single line with each worker (workstation) performing only a very limited set of tasks on each unit of product. The opposite extreme uses as many single station lines as needed to meet demand. In a single station line, each worker assembles the entire product. In reality, we are free to choose a compromise design between these extremes. Human psychology and physiology as well as economics come into play when selecting the number of parallel lines. Table 2.1 lists several of the advantages and disadvantages of the use of multiple lines instead of a single assembly line.

In addition to the line-balancing advantages to be considered later, multiple lines allow workers to feel that they have accomplished something valuable, affording them the opportunity to take pride in their work. Product quality is often enhanced as well as worker retention increased in such environments. Problems are easily traced back to the source. This facilitates retraining or other corrective action. Moreover, if one station breaks down, only the workers on that line are idled. Minor changes in demand are easily accommodated by overtime or an added shift to one or more lines.

Table 2.1   Advantages/Disadvantages of Multiple Parallel Lines

Advantages	Disadvantages
Easier to balance work load between stations	Higher setup cost
Increased scheduling flexibility	Higher equipment costs
Job enrichment	Higher skill requirements
Higher line availability—worker independence	Slower learning
Increased accountability	More complex supervision

Multiple parallel lines have disadvantages as well. Additional equipment may be required. Workers changing tools or fixturing when work elements are added to their job description may raise labor cost per unit. Learning occurs more slowly, since workers are not repeating the same activity as frequently. In some instances technology may be the deciding factor. Machines performing at the speed required to meet demand may not exist or may be prohibitively expensive. If this is true multiple machines (lines) are needed.

As with most problems, multiple objectives exist. By far the most commonly used objective for analytical models is minimization of idle time. However, in practice, real-world issues of minimizing tooling investment, minimizing the maximum lift or strain by any worker, grouping of tasks requiring similar skills, minimizing movement of existing equipment, and meeting production targets cannot be overlooked.

Finished products are assembled from component parts. The coordination of component feeding into the total product assembly scheme and the use of buffers to increase productivity and flexibility must be considered. Consider the final assembly of an automobile. The engine, chassis, body, and interior are all large subassemblies that must be delivered to the proper location on the line for assembly. Each of these subassemblies was put together from a set of components in a previous line. Many small parts such as lamps, wiring and controls for the dash, and electrical assembly systems must be made available at the individual workstations. Figure 2.3 shows a general depiction of an assembly system. In many cases the parts and subassemblies have a natural definition but their coordinated delivery must be carefully planned. The placement of buffers has important implications for system effectiveness and inventory cost. If the total assembly system stopped every time a single part was unavailable or workstation failed, the line would rarely function. Buffers allow workstations to operate more independently, cushioning against machine failures, worker or part shortages, and production rate differences. In this chapter we look only at allocating work load to workstations in a completely reliable assembly system. Machines never fail and parts are always available. In later chapters we consider the effect of buffers and machine breakdowns on general assembly systems.

A choice between a paced and unpaced line must be made. In a paced line, each workstation is given exactly the same amount of time to operate on each unit of product. At the conclusion of this cycle time C, the handling system automatically indexes each unit to the next station. Paced lines can encourage human workers to maintain the proper pace and exactly balance production. However, unless slack time is built into each station, the randomness of performance occasionally will cause some items not to be completed. Extra time must be allowed in the fixed

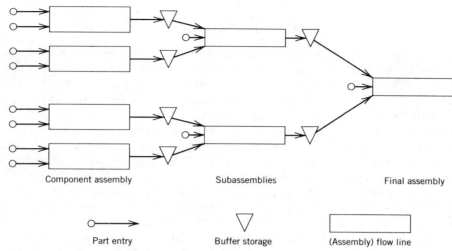

Component assembly   Subassemblies   Final assembly

Part entry   Buffer storage   (Assembly) flow line

Figure 2.3   General assembly system.

cycle time to cushion against task time variability. Small buffers may be needed in
nonautomatic assembly to avoid starving. Starving refers to a station having no input
product available for processing at the start of its cycle. Without buffers, if task times
vary, unpaced, or **asynchronous,** lines may be preferable. In unpaced lines, the
station removes a new unit from the handling system as soon as it has completed
the previous unit, performs the required tasks, and then forwards the unit on to the
next station. Parts need not be passed on incomplete. Likewise, when two adjacent
workers finish early, the second worker can begin the next part early and increase the
chance of finishing on time. As a practical issue, many lines have off-line rework areas
at the end of the line to allow finishing of tasks that were not completed during the
appropriate cycle. At the end of this chapter we will analyze the effect of the paced
versus unpaced decision in a more formal manner.

In many serial systems, each station along the line is in actuality a set of parallel,
identical workstations. Each workstation in a parallel set performs similar activities.
Normally, small input buffers are allowed at each workstation and in some instances
specific workstations may specialize in particular product models. If product units
are routed to individual workstations based on task requirements and the content of
input buffers, then the system is sometimes called a flexible flow line. Although the
total capacity of the parallel workstations at each stage should balance, the flexible
line approach allows some job enrichment, facilitates cycle time balancing, and allows
individual product model setups and production requirements to be managed. For
instance, consider a line to assemble personal computers. Stage 1 assembles the
boards, power supply, and cabling into the frame. Stage 2 does a quick quality check
to ensure that all boards are present and correct. Stage 3 is a burn-in test of the
assembled unit. The line assembles one hundred units per hour. Stage 1 takes an
average of eight minutes per unit, stage 2 takes two minutes, and stage 3 takes two
hours. To meet our goal we can use $8 \cdot 100/60 = 14$ parallel assembly stations,
$2 \cdot 100/60 = 4$ inspectors, and $120 \cdot 100/60 = 200$ burn-in test stations. Although
such systems do provide flexibility, they clearly also require sophisticated material
handling and routing control.

2.2

PROBLEM FORMULATION

The basic line-balancing problem assigns individual work elements to workstations. The objective is to minimize unit assembly cost. Assembly cost is composed of labor cost while performing tasks plus idle time cost. Since task times are fixed, we can simply concentrate on minimizing idle time. Constraints usually exist on production rate, sequencing of tasks, and grouping of tasks.

We assume that marketing requirements are for a production rate of P units per time, and we are currently considering m parallel lines. To meet this demand each line must turn out a completed unit every m/P time units. (We assume that all units are acceptable or P has been corrected for expected fallout.) The required time between completed units is our cycle time, and we satisfy demand by setting $C = m/P$. The time to perform task i will be denoted t_i. We must assure that no worker is assigned a set of tasks whose cumulative time exceeds C.

The order in which tasks are to be performed may be partially predetermined. For instance, in assembling an automobile, we cannot put the wheel cover on until the lug nuts are on, and we cannot tighten the lug nuts until the wheel is positioned. However, whether we put the wheel or the oil pan on first is irrelevant. We will let the set IP identify the assembly ordering constraints. $IP = \{(u,v): \text{task } u \text{ must precede } v\}$. It is only necessary to list the *immediate* predecessors in IP. That is, if r precedes u and u precedes v, we need only include the pairs (r,u) and (u,v) as (r,v) is automatically implied.

Zoning restrictions indicate which tasks must be assigned to the same workstation and which tasks must not be assigned to the same workstation. Tasks are defined to be self-containing, useful work elements. Nevertheless, because of safety, skill, or equipment requirements, we may wish to assign pairs of tasks to the same workstation. We let ZS be the set of task pairs that must be assigned to the same workstation. On the other hand, certain tasks may not be allowed in the same workstation. Consider the assembly of a microwave oven. The oven travels down the line with the front parallel to the direction of line travel. Certain tasks must be performed on the front of the oven and others on the rear. Workstations will be located on both sides of the line, but clearly some tasks are best performed from a specific side. We let ZD be the set of task pairs that cannot be performed in the same workstation.

To formulate the optimization problem we use binary indicators as decision variables. These variables indicate whether task i is assigned to station k,

$$x_{ik} = \begin{cases} 1, & \text{if task } i \text{ is assigned to station } k \\ 0, & \text{otherwise} \end{cases}$$

We will allow the possibility of a large number of workstations K. To minimize idle time we try to force tasks into the lowest numbered stations. Unused stations will be discarded. To accomplish this we define a sequence of cost coefficients c_{ik} such that $Nc_{ik} \le c_{i,k+1}; k = 1, \ldots, N - 1$, where N is the number of tasks. The formulation then becomes

$$\text{minimize} \sum_{i=1}^{N} \sum_{k=1}^{K} c_{ik} x_{ik} \tag{2.1}$$

subject to

$$\sum_{i=1}^{N} t_i X_{ik} \leq C \qquad k = 1, \ldots, K \tag{2.2}$$

$$\sum_{k=1}^{K} X_{ik} = 1 \qquad i = 1, \ldots, N \tag{2.3}$$

$$X_{vb} \leq \sum_{j=1}^{b} X_{uj} \qquad b = 1, \ldots, K \quad \text{and} \quad (u, v) \in IP \tag{2.4}$$

$$\sum_{k=1}^{K} X_{uk} X_{vk} = 1 \qquad (u, v) \in ZS \tag{2.5}$$

$$X_{ub} + X_{vb} \leq 1 \qquad k = 1, \ldots, K \quad \text{and} \quad (u, v) \in ZD \tag{2.6}$$

The objective function is structured such that it will always be advantageous to fill up lower numbered stations before opening a new station. Only those stations with at least one task assigned are constructed. We will let K^* be the number of stations (workers) required by the solution. A measure for comparing solutions is the proportion of time idle. This measure is called the **balance delay** $\equiv D$ and is computed as

$$D = \frac{K^*C - \sum_{i=1}^{N} t_i}{K^*C}$$

We can think of D as idle time over paid time, where idle time is paid time minus productive time. One problem with our objective is that it fails to recognize a secondary objective of allocating idle time equally to all stations. For reasons of production volume flexibility, variability in actual task times between units, and worker morale, we prefer solutions with work load as balanced as possible.

Constraints as in equation 2.2 ensure that the sum of task times for the set of tasks assigned to each workstation does not exceed the cycle time. The left-hand side (LHS) of the constraints look at each task. If the task is assigned to this station (k), then its task time is added to the sum for the station. The final sum is compared to the allowed cycle time C to ensure time feasibility.

Each task has a constraint of the form of equation 2.3. These constraints ensure that the task is assigned to exactly one workstation.

The next constraint type forces adherence to precedence restrictions. If task v is to be assigned to station b, then its immediate predecessor u must be assigned to some station between 1 and b. For instance, consider a case where task 2 must precede task 3 and a maximum of three workstations will be needed. For this one precedence restriction we need the three constraints:

$$x_{31} \leq x_{21}$$

$$x_{32} \leq x_{21} + x_{22}$$

$$x_{33} \leq x_{21} + x_{22} + x_{23}$$

Since task 3 must be assigned to one of the three stations, only one of the three equations will have a nonzero LHS. The other two constraints are automatically satisfied since the RHS is nonnegative. If task 3 is assigned to the first workstation ($x_{31} = 1$), then the first constraint forces task 2 to also be assigned to workstation 1 ($x_{21} = 1$). The worker does task 2 first and then does task 3. If task 3 is assigned to the second workstation, then the second constraint forces task 2 to be done in workstation 1 or 2. A similar argument holds for assigning task 3 to workstation 3 in the last equation.

Only zoning relations remain. Marriages are handled in expressions 2.5. Since each task can be assigned to only one home station, the relationships are consummated only if tasks u and v are assigned to a common station. Last, divorce constraints are satisfied by equations 2.6, which prohibit tasks u and v from associating in the same domicile.

The above formulation is complicated not only by the presence of binary integer variables but also by the nonlinearity and sheer size of the problem for many applications. Nonetheless, the formulation does give us a formal definition of the problem and a basis from which to work toward a solution. The first constraint leads us to a quick lower bound on the number of stations required. Total task time is $T = \sum_{i=1}^{N} t_i$. Maximum time per station is C. Thus, we must have at least $K^0 = \lceil T/C \rceil$ stations, where the ceiling function $\lceil \ \rceil$ implies the least greater integer. Should we find a feasible assignment with K^0 stations, we may stop looking. We also note from the formulation that if $(u, v) \in ZS$, then knowledge of X_{uk} automatically determines X_{vk}. Accordingly, tasks that must be performed in the same workstation can be aggregated into a single task. The immediate predecessors for the aggregated task consist of all those for the individual tasks. Any tasks that must fall in between the aggregated tasks can also be aggregated.

2.3
APPROACHES TO LINE BALANCING

We now describe three approaches for finding a solution. The first, COMSOAL, is a basic random solution generation method that can be used for many types of problems. The second is the widely documented Ranked Positional Weight heuristic, which quickly finds good solutions. Finally, we discuss an implicit enumeration scheme for finding optimal solutions. Many approaches to the line-balancing problem have been proposed. We believe these three are useful and illustrate the variety of options and data-handling techniques available. We assume that the required cycle time, sequencing restrictions, and task times are known. The first and third approaches construct more than one sequence. For these methods secondary objectives can be implemented for choosing between solutions requiring the same number of workstations.

2.3.1 COMSOAL Random Sequence Generation

The line-balancing problem consists of specifying an ordering for N tasks. $N!$ such orderings exist, far too many to explicitly consider each. The COMSOAL (Arcus [1966]) approach is to use a simple record-keeping system that allows a large number of possible sequences to be examined quickly. Sequences are generated by selecting at random from the set of available tasks and placing that task next in sequence. New workstations are opened when needed. We keep track of the current workstation receiving tasks, unassigned tasks, accumulated idle time in current solution ($IDLE$), and available time remaining in current workstation (c). Only tasks that satisfy all constraints are considered at each step. The sequence is discarded as soon as it exceeds the number of workstations in the current best solution (upper bound). If a complete sequence is generated that is better than the old upper bound, the sequence is saved and the bound updated. The modeler need not specify any beliefs about the form of the optimal solution but may, if so inclined, include heuristic beliefs, such as longer tasks should go first. These beliefs are included by giving such tasks a higher probability of being selected.

The efficiency of COMSOAL and, indeed, most computational approaches, depends on the data storage and processing structure. COMSOAL makes use of several lists to speed computation. The first overhead array to be developed contains the number of immediate predecessors for each task i, $NIP(i)$. The second overhead array $WIP(i)$ indicates for which other tasks i is an immediate predecessor. The N tasks make up array TK. During each sequence-generation procedure, we keep updated lists of unassigned tasks (A), tasks from A with all immediate predecessors assigned (B), and tasks from B with task times not exceeding remaining cycle time in the current workstation. This last list is referred to as F, the Fit List. The procedure for generating X trial solutions is as follows:

Procedure

1. Set $x = 0$, $UB = \infty$, C = cycle time, $c = C$.
2. Start new sequence: Set $x = x + 1$, $A = TK$, $NIPW(i) = NIP(i)$.
3. Precedence feasibility: For all $i \in A$, if $NIPW(i) = 0$, add i to B.
4. Time feasibility: For all $i \in B$, if $t_i \le c$, add i to F. If F empty, 5; otherwise 6.
5. Open new station: $IDLE = IDLE + c$. $c = C$. If $IDLE > UB$ go to 2; otherwise 3.
6. Select task: Set $m = card\{F\}$. Randomly generate $RN \in U(0, 1)$. Let $i^* = \lceil m \cdot RN \rceil$th task from F. Remove i^* from A, B, F. $c = c - t_{i^*}$. For all $i \in WIP(i^*)$, $NIPW(i) = NIPW(i) - 1$. If A empty, go to 7; otherwise go to 3.
7. Schedule completion: $IDLE = IDLE + c$. If $IDLE \le UB$, $UB = IDLE$ and Store Schedule. If $x = X$, stop; otherwise go to 2.

Although not the most mathematically eloquent approach, the COMSOAL technique has advantages. The technique is relatively easy to program. Second, feasible solutions are found quickly, and the greater the computational effort expended, the better the expected solution. Since in each sequence-generation iteration every sequence has a probability of at least $N!^{-1}$ of being generated, in the limit for number of sequences generated, the probability of finding the optimum goes to 1. Last, the basic idea can be quickly applied to many decision problems, the only requirements being that we can build solutions sequentially, preferably maintaining feasibility as we do so, and a function evaluation can be performed to rank candidate solutions.

We may consider the performance of COMSOAL. Let us assume that a unique optimal solution exists. For simplicity we also assume that no precedence constraints exist. (The argument is easily extendable, however. For instance, if $r < N$ is a more reasonable estimate of the number of precedent-feasible tasks from which to choose at each iteration, then we may simply replace $N!$ the number of possible sequences with $r! r^{(N-r)}$.) COMSOAL will generate X random sequences for investigation. There are $N!$ possible sequences, hence, the probability of generating the optimal sequence on any run is $p = 1/N!$. The distribution of sequence generations until the optimal is found is thus geometric with mean $= p^{-1}$ and variance $= (1 - p)p^{-2}$. If we generate X sequences we have a probability of $1 - (1 - p)^X$ of obtaining the optimal assignment of tasks to workstations. If we are willing to assume a distribution for the objective over the possible sequences, we can even extend these results to make probabilistic statements concerning how close we are to the optimum by using order statistics.

At this point you may be wondering why we would bother to generate random sequences, since we may needlessly reexamine the same sequence and waste computational resources generating random numbers. Why not just order the set of possible sequences and try the first X unique sequences? Would this not give us a better chance of finding the optimum? In theory yes, it would for our simplistic model. In fact, we will shortly look at such an approach. In practice, however, there is likely to be a large number of optimal or near optimal solutions. Many of these solutions will have similar-looking workstations or sequences. By generating random sequences we may "jump around" the set of possible sequences and are likely to find at least one of these good ones. If we start exploring by always looking at a minimally different sequence each time, we will generate very similar sequences; perhaps we may find all good sequences, but it is just as likely we will find all bad sequences.

EXAMPLE 2.1

Consider the assembly of the spring-activated toy car shown in Figure 2.4. To take advantage of a large local labor pool of persons wishing part-time day employment and part-time second jobs in the evening, two 4 hour-shifts will be used, 4 days a week for assembly. Each shift receives two 10-minute breaks. Marketing estimates have been used to obtain a planned production rate of 1500 units per week. With holidays, vacations, and other events, the plant averages only 4 working days per week. The set of tasks, times, and precedence constraints are given in Table 2.2. No zoning constraints exist.

Table 2.2   Toy Car Assembly Tasks

Task	Activity	Assembly Time (seconds)	Immediate Predecessors
a	Insert front axle/wheels	20	—
b	Insert fan rod	6	a
c	Insert fan rod cover	5	b
d	Insert rear axle/wheels	21	—
e	Insert hood to wheel frame	8	—
f	Glue windows to top	35	—
g	Insert gear assembly	15	c,d
h	Insert gear spacers	10	g
i	Secure front wheel frame	15	e,h
j	Insert engine	5	c
k	Attach top	46	f,i,j
l	Add decals	16	k

Solution

The precedence relations are easier to visualize in graphical form. Figure 2.5 summarizes the table. Each task is represented by a node. Associated task times

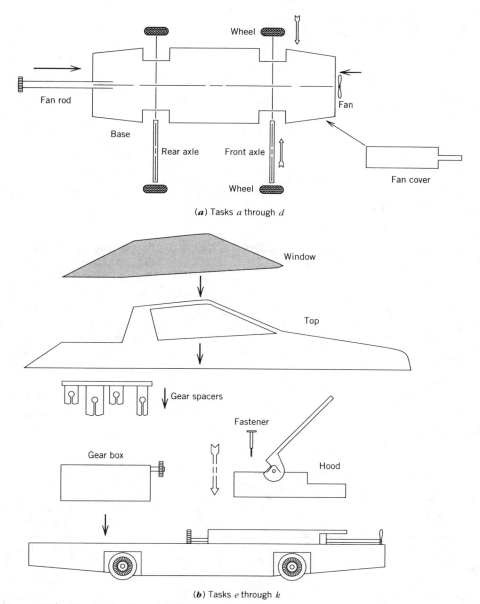

(*a*) Tasks *a* through *d*

(*b*) Tasks *e* through *k*

Figure 2.4   Toy car for Example 2.1.

are shown with each node. Directed arcs between nodes indicate immediate predecessor relations.

To illustrate the COMSOAL approach, we will generate a random sequence. First, we must determine the cycle time.

$$C = \frac{1 \text{ week}}{1500 \text{ units}} \times 4\frac{\text{days}}{\text{week}} \times 2\frac{\text{shifts}}{\text{day}} \times 220\frac{\text{minutes}}{\text{shift}} = 1.17\frac{\text{minutes}}{\text{unit}}$$

To meet demand we will use $C = 70$ seconds. Reference to Figure 2.5 indicates four potential first tasks ($a, d, e,$ or f). We generate a random number uniformly distributed between 0 and 1 ($U[0, 1]$). Our outcome is $R = .34$. Thus, since R is

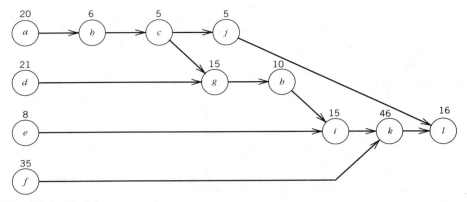

Figure 2.5 Model car precedence structure.

in the second quartile and d is the second task, d is placed first. We can continue in this fashion until a schedule is complete. Table 2.3 summarizes the results of our random-sequence generation. Each row represents the assignment of one task or the opening of a new station because list F is empty. After assigning all 12 tasks, we obtain a feasible solution. As indicated in Table 2.3, our first solution requires four stations. A quick check shows our lower bound to be $K^0 = \lceil \sum_{r=a}^{l} t_r / C \rceil = \lceil 202/70 \rceil = 3$. Thus, better solutions may exist.   ■

After reading the next two sections the reader may wish to consider how the basic COMSOAL procedure could be improved. Several end-of-chapter exercises explore this issue.

Table 2.3   Single COMSOAL Sequence Results

Step	List A	List B	List F	$U(0,1)$	Selected Task	Station (idle time)
1	a through l	a,d,e,f	a,d,e,f	.34	d	1(49)
2	a through l, -d	a,e,f	a,e,f	.83	f	1(14)
3	a,b,c,e,g,h,i,j,k,l	a,e	e	—	e	1(6)
4	a,b,c,g,h,i,j,k,l	a	—	Open station		
4	a,b,c,g,h,i,j,k,l	a	a	—	a	2(50)
5	b,c,g,h,i,j,k,l	b	b	—	b	2(44)
6	c,g,h,i,j,k,l	c	c	—	c	2(39)
7	g,h,i,j,k,l	g,j	g,j	.21	g	2(24)
8	h,i,j,k,l	j,h	h,j	.42	h	2(14)
9	i,j,k,l	i,j	j	—	j	2(9)
10	i,k,l	i	—	Open station		
10	i,k,l	i	i	—	i	3(55)
11	k,l	k	k	—	k	3(9)
12	l	l	—	Open station		
12	l	l	l	—	l	4(54)

2.3.2 Ranked Positional Weight Heuristic

The Ranked Positional Weight (RPW) technique is one of the best known heuristics. The procedure constructs a single sequence. A task is prioritized based on the cumulative assembly time associated with itself and its successors. Tasks are then assigned in this order to the lowest numbered feasible workstation. The logic rests on the fact that, other things being equal and unknown, the more tasks we have freed up and made eligible for assignment, the higher the likelihood of having at least one task available to fit in the remaining station idle time. This should result in less idle time per workstation and accordingly fewer workstations being required. The cumulative remaining assembly time also constrains the number of subsequent workstations required. RPW illustrates the important class of greedy, single-pass heuristics. The procedure is greedy in the sense that, instead of looking ahead, we pick the next task from a previously ordered list and assign it to the first feasible workstation. It is single pass in that the set of tasks is traversed only once.

The RPW procedure requires computation of the "positional weight" $PW(i)$ of each task. Let $S(i)$ be the set of successors of task i, that is, task $j \in S(i)$ implies that j cannot begin until i is complete. We then compute $PW_i = t_i + \sum_{r \in S(i)} t_r$. Tasks are ordered such that $i < r$ implies i not $\in S(r)$. Task r is then a member of $S(i)$ if and only if there exists a path of immediate successor relationships from i to r. Immediate successors $IS(i)$ are known from the inverse of the $IP(i)$ relationships. The procedure is then easily described as follows.

RPW Procedure

1. Task ordering: For all tasks $i = 1, \ldots, N$ compute $PW(i)$. Order (rank) tasks by nonincreasing $PW(i)$.
2. Task assignment: For ranked tasks $i = 1, \ldots, N$ assign task i to first feasible workstation.

Note that the positional weight procedure assures that a task's predecessors are always assigned before the task. Hence, for each task, precedence constraints are satisfied by assignment to any workstation at least as large (in numerical order) as that to which its immediate predecessors are assigned. Zoning and time restrictions are checked on placement. New workstations are opened as necessary.

EXAMPLE 2.2

Apply the RPW procedure to the car assembly problem of Example 2.1.

Solution

For our problem, positional weights are readily calculated using Figure 2.5. Start at the last task, l. PW_l is simply its task time. Proceeding backward, $PW_k = t_k + PW_l$. By working our way backward to task a, we obtain the values shown in Table 2.4. Note that for large problems this is not a simple task. Consider task c, for instance. We cannot simply add the positional weights of its immediate successors j and g to its own task time. This would double count t_k and t_l. Nevertheless, after the data are processed to obtain the positional weights, we can rank the tasks as shown in the table.

Table 2.4   Positional Weight Example Data

Task	PW	Ranked PW	Task	PW	Ranked PW
a	138	1	g	102	5
b	118	3	h	87	7
c	112	4	i	77	9
d	123	2	j	67	10
e	85	8	k	62	11
f	97	6	l	16	12

The next step involves iteratively assigning the tasks to the first feasible station. Assignment order is given by the rankings. It is advisable to make use of a list that indicates for each task the other tasks for which it is an immediate predecessor. For task c we would store its immediate successors j and g. Then on assignment of a task, we update an array $V(p)$ that stores the highest number station to which any of task ps predecessors are assigned. When assigning task p, we can begin at the station stored in $V(p)$ and search for the first time-feasible station for p.

We start with task a, assigning it to station 1. This leaves $c - t_a = 70 - 20 = 50$ seconds in station 1. Next task d is assigned. It has no predecessors and since sufficient time remains in station 1, d is assigned to station 1. Station 1 now has 29 seconds available. Continuing in this fashion we obtain the following sequence:

Station	Time Remaining	Tasks
1	70,50,29,23,18,3	a,d,b,c,g
2	70,35,25,17,2	f,h,e,i
3	70,65,19,3	j,k,l

The solution requires three workstations. As this is K^0, the solution is optimal. We were fortunate; the RPW procedure is heuristic and does not guarantee optimality. Consider for instance the line-balancing problem of Figure 2.6. Cycle time is 10. A two-station solution exists with tasks (a, c, d) in station 1 and (b, e) in station 2. The ranked positional weight procedure will produce the three-station solution (a, b), (c, d), and (e). We want task b to be located early because its positional weight exceeds that of c and d. However, combining b with a leaves one unit of idle time in the workstation that cannot be used. ∎

2.3.3* Optimal Solutions

Efficient schemes make possible analytical determination of optimal solutions for many real problems. In some instances, problems with 1000 or more tasks can be solved in a couple of seconds by computer. Imagine a decision tree containing all possible sequences of tasks that obey the precedence constraints. Once again, for

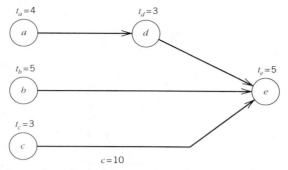

Figure 2.6   Precedence diagram showing nonoptimality of RPW.

convenience, tasks are ordered such that predecessors come first. A partial tree, showing the first two or three tasks in any sequence, is shown in Figure 2.7 for our model car example. The complete tree would have 13 columns, one for each task plus the initial node. The "0" node is called the root of the tree; the terminal nodes at the other end are called leaves. A unique path leads from the root to each leaf. Each leaf represents a complete sequence, and each feasible sequence is represented by exactly one leaf. Since every feasible sequence is in this tree, the optimal sequence of tasks must be in this tree. If somehow we could examine every path (leaf), we could be sure of finding the best solution. Unfortunately, unless there are many precedence constraints, this tree grows very large as the number of tasks increases. Nevertheless, we will explore every leaf, but we will do it cleverly so that we may still be home in time for dinner.

Tree Generation

A method is needed to allow both systematic exploration of the tree and efficient storage of our location and history during exploration. Backtracking is one such technique often used in searching trees. To the explorer, backtracking connotes retreating over a path already covered. We use the term in the same sense. We retreat because we have reached a dead end. We use the same path to avoid becoming lost or confused.

We will describe a depth-first backtracking order. This approach allows us to quickly complete a feasible solution. Although this quickly generated sequence is not expected to be optimal, who knows, since once in a while we all luck out. In either event, the sequence does present a feasible solution, and, hence, an upper bound.

The general applicability of our backtracking scheme is to an environment in which N sequential decisions are to be made. The available choices at stage n may depend on earlier stage decisions. In our case the stages correspond to selection of the next task in the assembly sequence. At each stage, choices are ordered (by task number for our problem). In generating the tree of possible sequences we are not concerned with workstations, only the order of performing the assembly tasks. Eligible choices at each decision stage are all tasks whose predecessors have already been added to the partial sequence. All sequences created will then be feasible with respect to the precedence restrictions. The tree initially grows by selecting the first alternative at each stage until we have reached a terminal leaf (a complete assembly sequence). This constitutes the depth-first aspect. Backtracking now takes over. We move back

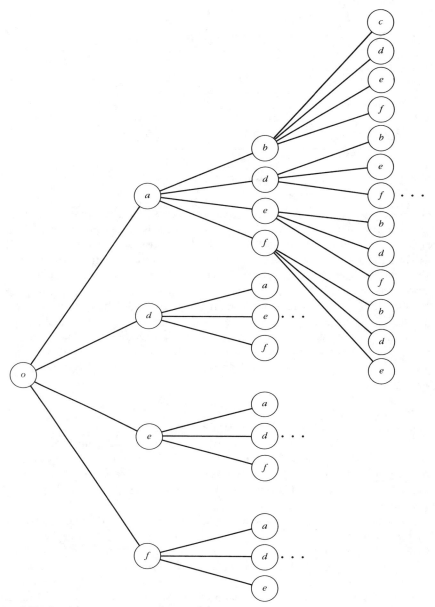

Figure 2.7   Partial sequence tree for model car.

up the tree until we reach the first node that has an unexplored branch. We then take this branch. We then continue moving forward, decision by decision, as far as possible (once again, depth-first). In our problem this means that we continue until another sequence is generated. At this point we backtrack once again. The process continues until all possible leaves are explored. Figure 2.8 shows the order in which nodes are created for a four-task problem with no precedence constraints. Letters represent the task added to the partial sequence at each step. Node numbers indicate the order in which nodes were created.

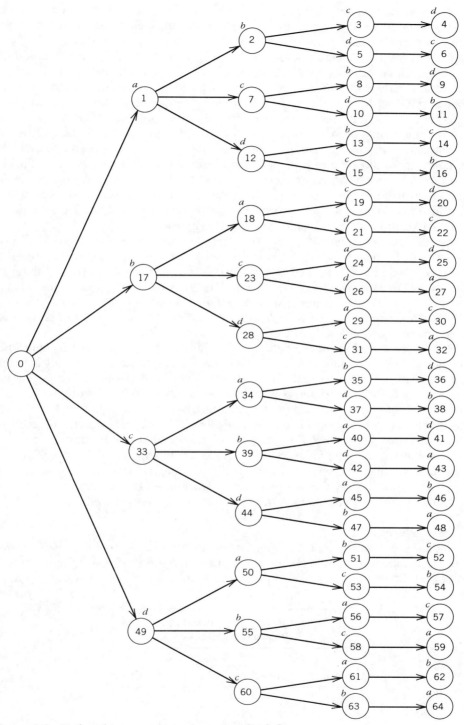

Figure 2.8   Backtracking tree generation using depth-first.

Given a sequence, the problem solution involves dividing the sequence into workstations without violating cycle time. As we progress through the sequence from beginning to end, we start a new station only when necessary. For instance, if at some point the current workstation has unused cycle time at least as large as the task time for the next task, that next task will be assigned to the current station. To do otherwise cannot reduce the number of workstations needed and therefore cannot yield a better solution. In fact, at any point in a partial sequence, if there exists an unassigned task that is time and precedence feasible, we would be foolish to close the current station without including that task. We use this fact to reduce the number of sequences that we must create. At any partial solution, if the next normal backtracking task step would require a new station be started while a "fittable" task lingers unassigned, then we skip over the creation of that new partial solution. A task is regarded as **fittable** if it satisfies three conditions. The task must fit in the remaining idle time of the station, it must be currently unassigned, and all its predecessors must be assigned. To illustrate, consider the example in Table 2.5.

Assume that cycle time is $C = 20$. The depth-first search has started the sequence $0 \rightarrow a \rightarrow b$. Normally c would be next. However, c will require a new station, since $t_a + t_b + t_c = 22 > C$. On the other hand, task d is fittable. As we have nothing to lose, we add d to the sequence next. Returning to c, we can add c then e and obtain the optimal two-station sequence $0 \rightarrow a \rightarrow b \rightarrow d \rightarrow c \rightarrow e$. We need not explore any sequence that starts $0 \rightarrow a \rightarrow b \rightarrow c$ because we know that we can do at least as well by inserting task d before c.

We now have an efficient tree-generation scheme for assembly-line balancing. This scheme has the added advantage that if we know what node we are currently at in the tree, we know which nodes have already been explored. This follows because nodes are created in a specific order in accord with task numbering. This feature significantly reduces the amount of memory required to search the tree. We can forget where we have been. Adding to previous notation, let p be the number of tasks assigned in the current partial sequence, TA_p the pth task assigned, c_k idle time in station k, i^* the task selected for assignment, A_i the workstation to which task i is assigned, and B the task to which we will backtrack. The procedure is given in the flowchart in Figure 2.9. We will apply the tree-generation algorithm of Figure 2.9 to our example problem, but first we consider further improvements in our procedure.

Tree Exploration

Finding the optimal sequence requires enumerating every possible path in this tree and computing the number of workstations required in each instance. This is still a prohibitive task for real-size problems. Fortunately, however, we may reduce

	Table 2.5 Five-Task Line Balancing Example	
Task	Task Time	Immediate Predecessors
a	5	—
b	5	a
c	12	b
d	10	b
e	8	c,d

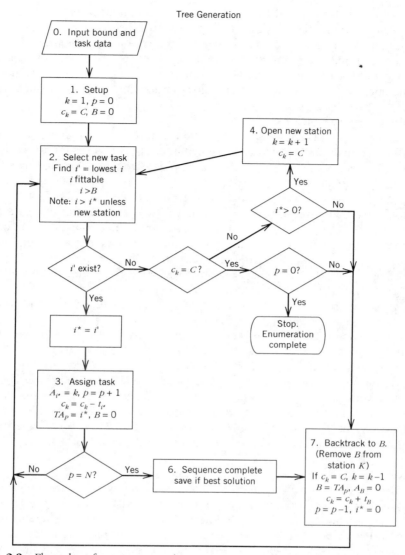

Figure 2.9 Flow chart for tree generation.

the work load considerably. In practice, we need not create each node of the tree. Suppose for a given problem we already have found a feasible sequence requiring eight workstations. If we can show that all solutions that place task 2 in workstation 2 will require at least eight workstations, then there is no need to consider any lower nodes that emanate from a node with this characteristic. Thus we prune the tree to reduce computations. Essentially, our approach will operate in this fashion, creating only those new nodes required to find or disprove the existence of a better solution. At any partial sequence node, if we can establish that all completions of this partial sequence are nonoptimal; then we will start backtracking immediately instead of completing a sequence.

To reduce computations we will develop several useful rules. Our first set of rules is used for simplifying the search process. The second set of rules each has the property that if a node satisfies one of the rules, then all paths emanating from that node can be discarded as either nonoptimal or at least being no better than a solution remaining in the tree. This process is referred to as fathoming the node. Once all nodes remaining are either fathomed or are leaves, we can stop searching. After listing these fathoming rules, we combine them with the node-enumeration (tree-generation) algorithm to obtain an algorithm for finding optimal solutions.

Problem Structure Rules

A basic optimality principle is that we should never close a workstation while "fittable" tasks remain. This principle was used in the previous section to reduce the size of the tree generated. It results from the problem structure. Our objective is such that given any sequence, we would necessarily use as few workstations as possible to perform the assembly. Two optimal policies are implied by this objective. One, never close a station while fittable tasks remain. Two, a sequence implies a set of workstations. That is, open a new workstation only when necessitated by time.

A second product structure rule deals with task time augmentation. Suppose the longest task is such that no other task can feasibly fit in the same station. This long task becomes a station, and we can assign it task time C to strengthen the above bounds. This holds true for all tasks i such that $t_i + \min_j t_j > C$.

Having a lower bound on the number of workstations needed is useful in determining if we have an optimal solution. The lower bound $K^0 = \lceil T/C \rceil$ was discussed earlier in the chapter. Other bounds are possibly stronger (larger). Suppose that we have a 10-task problem with 5 tasks having $t_i = .6C$ and 5 tasks with $t_i = .1C$. Total work time is only $3.5C$, implying $K^0 = 4$. However, no two of the 5 long tasks can ever be grouped together as this would violate cycle time. The lower bound is actually five workstations. Such bounds are potentially useful for problems with few tasks per station. This principle can be generalized and checked for other fractions. For instance, at most 2 tasks with task time exceeding $C/3$ can be assigned to any station. Thus, if 7 such tasks exist, we have a lower bound of four stations.

Fathoming Rules

We will now discuss fathoming rules. These rules allow us to rapidly prune the tree. When a node representing a partial sequence satisfies one of the fathoming rules, we can conclude that no completion of this partial schedule is needed. Either all completions are at least as expensive as an already completed sequence, or all completions are at least as expensive as another completion we will find later in tree exploration. This allows us implicitly to consider all leaves without explicitly enumerating many of them.

The rules are designed such that they need only be checked when a new station must be opened. Recall that we start with no tasks assigned, and then one-by-one add a new task to the end of the partial sequence until a complete sequence is obtained. A new station is "opened" whenever the current station cannot handle the next task in the sequence owing to cycle time violation. This condition corresponds to having zero fittable tasks in the current station. At this point, we stop and check to see if any of the fathoming rules allow us to exclude all possible completion of the partial sequence.

Rule 1: Task Dominance

Suppose a tentative station exists where one of its tasks, say i, could be feasibly replaced by a longer task, say j, and all the successors of i must also follow j.

If we substitute j for i, we reduce the remaining work load without losing any possible sequence completions. Whatever sequence we end up with by placing i in the current station, that sequence must have at least as many workstations as the sequence that places j in the current workstation and saves i for later. Since the tree must also contain this other sequence, which is at least as good, we can fathom the current partial sequence. The rule states, if $t_i < t_j, k \in S(i)$ implies $k \in S(j)$, and j is fittable if i is removed, then fathom the partial sequence that places i in the station but not j. Task j is said to dominate task i if the first two conditions hold. Task pairs with a dominance relationship can be found during the initiation phase of problem solution and saved. As a simple example suppose we have three tasks $a, b,$ and c with task times $(1, 2, 1)$, respectively, and the only precedence relations being that c requires both a and b. (The reader may want to sketch this problem.) In this case b dominates a, since $t_b > t_a$ and all of a's successors (c) are also successors of b. Cycle time must be at least 2 to accommodate task b. Suppose $c = 2$. If we place a first, then the station must be closed for lack of any fittable tasks. A three-station solution $(a), (b), (c)$ results. However, if b is placed first we have the two-station solution $(b), (a, c)$. The dominance of b over a along with the fittability of b into the first workstation upon a's removal would have allowed us to stop exploring the first branch as soon as we closed the station (a). Task dominance provides the foresight to know the solution (b) was coming and would be at least as good.

Rule 2: Station Dominance

Suppose at some point in the tree we attempt to form a station that is identical to a "first" station that was explored earlier. All possible subsequent sequence completions were considered from that earlier path, with at most the one station reordered. As no new sequences can be considered, no better solution can be found and the new node is fathomed.

Rule 3: Solution Dominance

We have already seen that K^0 gives a lower bound on the optimal number of stations. Once we obtain a complete sequence, which requires K^0 workstations, we can stop. This sequence must be optimal.

Rule 4: Bound Violation

Suppose the best complete sequence found thus far requires K workstations. Unless a $K - 1$ station solution can be found, we have an optimal sequence. We can determine upper bounds on the largest workstation to which each task can be assigned if we are to find a $K - 1$ (or better) station solution. Let A_i be the station to which task i is assigned. If we require at most $K - 1$ stations, the upper bound on A_i is

$$U_i = K - \left\lceil \left(\frac{t_i + \sum_{j \in S(i)} t_j}{C} \right) \right\rceil$$

Nodes containing at least one A_i outside these bounds can be pruned from the tree. At the close of each station we can check unassigned tasks. If any such task has an upper bound less than or equal to the order of the closed station, then this partial solution is fathomed.

Rule 5: Excessive Idle Time

Suppose total task time is 356 time units and we are looking for a solution of six or fewer workstations with cycle time of 60. The total allowable idle time for all workstations is $60 \cdot 6 - 356 = 4$. Thus, whenever cumulative idle time exceeds four time units or, in general $(K - 1)C - \sum_i t_i$, we may fathom that partial sequence.

These five rules can substantially reduce the effort required to search the tree. The procedure we describe is adapted from FABLE, Fast Algorithm for Balancing Lines Effectively, described by Johnson [1988].

FABLE Procedure

0. Generate an initial solution. Let K be the number of workstations for the current best solution.
1. Input data, check to increment t_i, compute upper workstation bounds for each task, check for possible dominance task pairs.
2. Enumerate the tree depth-first always adding the lowest numbered, untried, fittable task and
    a. when a station is completed, check fathoming methods and backtrack instead of opening a new station if bounds exceeded;
    b. update task upper bounds whenever a better solution is found; and
    c. stop if any complete solution equals the line bound (K^0).

EXAMPLE 2.3

Reconsider the model car assembly.

Solution

To begin, let us determine the aids for fathoming nodes. Product structure rules will be addressed first.

1. *Task time augmentation.* The longest task has $t_k = 46$ and the shortest has $t_c = 5$. Since $t_k + t_c \le C$, we cannot assume that any tasks require their own station.
2. *Solution lower bound.* We previously found the lower bound $K^0 = 3$. Let's see if a better bound exists. Only one task has a task time exceeding $C/2$. This will be of no help. Only two tasks (f, k) have tasks times exceeding $C/3$. This will be of no help either.

Next consider aids for fathoming nodes.

1. *Task dominance.* Several tasks can be replaced in tentative workstation assignments. Task pairs (r, s) satisfying $t_r < t_s$ and $IS(r)$ a subset of $S(s)$ include $(j, d), (j, e), (j, f), (j, g), (j, h), (j, i), (i, f), (e, a), (e, h), (e, g)$, and (e, d). We exclude pairs where r is a successor of s, since these cannot be helpful. We will keep an eye out for use of task dominance as we search through the tree.
2. *Task bound violation.* Let us assume our current best solution has four stations. Such solutions are easy to find, for instance, the natural ordering $(a, b, c, d, e), (f, g, h), (i, j, k), (l)$ is such a sequence.

Four stations is then an upper bound on the solution. We still need to search for a three-station solution. Rule 4: Bound Violations will be useful. The positional weights of Table 2.4 are used to compute upper bounds $U_i = 4 - \lceil PW_i / C \rceil$. Since $PW_a = 138, U_a = 4 - \lceil 138/70 \rceil = 2$. Continuing, we find an upper bound of three for tasks $j, k,$ and l and two for all other tasks.

Tree Generation. We now jointly apply the node generation and fathoming algorithms. Figure 2.10 shows the sequence of 20 nodes created. The label indicates the task added at each step. We begin by selecting (step 2 of flowchart) and assigning (step 3 of flowchart) task a. Cycle time is 70. Since $t_a = 20$, this leaves unused station time of $c_1 = 50$. Task b is assigned next, reducing c_1 to 44. This process continues up through the assignment of task e, leaving $c_1 = 10$. At this point j is selected as the next fittable task. Both f and g violate the remaining cycle time. Tasks h and i do not have all their predecessors assigned. Task j is fittable, however, and is assigned. No other tasks will fit in the remaining 5 seconds. We can try to fathom this node before continuing. No U_i bounds are violated, however. Likewise, no tasks are dominated. Task j cannot be replaced by its dominated tasks f or g because of time, or h or i because of violation of precedence relations. Task e cannot be replaced by its dominant task h because of the need to assign g before h. Since we are in the process of assigning tasks ($i^* > 0$ and $B = 0$), we open station 2 (flowchart step 4). Available cycle time is reset, that is $c_2 = 70$. Task f is the first fittable task and is accordingly selected and assigned. We can then continue assigning through task h. We could open station 3 and continue. However, a quick check shows an upper bound of 2 for unassigned task i. From Rule 4, we conclude that any completion of this partial sequence must have at least 4 workstations. Following step 2 of the FABLE procedure, it is time to backtrack.

Backtracking at h (flowchart step 7), we find that all later tasks are assigned (j) or require $h, (i, k, l)$. We then backtrack to g. Once again no later tasks can be assigned. We backtrack to f. Now g can be assigned and we move forward again. Notice that no nodes are created while backtracking. Moving forward we assign h and i before filling up the station. After node 12, no fittable task (i') can be found. Since we are moving forward ($i^* > 0$ and $B = 0$), we could consider opening a new station. However, we balk at opening a new station because f is unassigned and $U_f = 2$. Again, Rule 4 applies, and we must backtrack. (We could also fathom this node with Rule 1 since the included task i is dominated by the fittable task f.) We backtrack through $i, h, g, j,$ and e until we can start moving forward again by assigning g. This time our forward step produces a three-workstation solution. We had a lower bound of three workstations, and according to Rule 3 we may stop here, satisfied that we have found an optimum.

It is important for efficiency to remember that when backtracking, only tasks ordered later than the backtrack task are eligible for i'. This is why g was not selected when we backtracked to j. Only k and l could be considered, and these had unassigned predecessors. Thus, we backtracked one step farther to e before moving forward again. The reader who is unconvinced about the importance of clever fathoming rules should compare Figures 2.7 and 2.10. ■

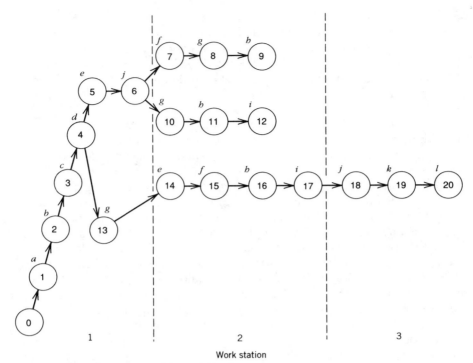

Figure 2.10 Nodes generated in optimization.

2.4
SOME PRACTICAL ISSUES

The problem formulation and models just discussed are simply abstractions of the real environment. We again emphasize the need to keep this in mind. It is not the intent of these procedures to replace the designer's creativity, but merely to assist in certain formal computational activities. For instance, when stations have only a small number of tasks each, the discrete nature of the problem tends to yield instances that are hard to solve and solutions, even "optimal" solutions, are poor in the sense that idle time is large. The problem is hard to solve, since the unused idle time required in many stations results in more than K^0 stations being required. Thus, it is difficult to verify optimality without examining the entire tree. The presence of significant idle time suggests that we consider using parallel lines with larger cycle times. However, the creative designer, who can look past the model, back to the original problem, may ask whether it is possible to group stations, that is, use more than one worker per station. Now, we still have one physical line, but we can sequentially form groups wherever an integer multiple of C can give a station with high utilization. Workers are cross-trained and can either cooperate on performing tasks or alternate certain tasks between successive product units. We may find a solution better than our model's "optimum." The foregoing procedures could be modified to use cycle times up to MC, where M is the maximum number of workers that can be grouped. At any point in the sequence-construction process that a station has assigned work load within some desired utilization threshold of a multiple of C, fix that station and start another. Suppose, for instance, that $C = 100$, the threshold is 5 percent, and we have the following sequence of activity times: $\{t_a = 65, t_b = 30, t_c = 75, t_d = 50, t_e = 70, t_f = 100\}$. Our model would normally require five workstations (a, b), (c), (d), (e), and (f). However, $K^0 = \lceil 390/100 \rceil = 4$. Grouping two workers together may allow shrinking the stations (c), (d), and (e) into 1 two-worker station. To know whether combining these two stations is technologically feasible, the modeler must understand the physical aspects of the problem environment, not just the numbers. In other words, the modeler must also walk and talk.

C has been fixed up to this point. In reality, we are only forecasting demand. Since demand determines C, the cycle time is not cast in stone. We can chisel away at the idle time by looking for a better C, one that reduces the balance delay. Ideally, C should be set such that $\sum_i t_i/C$ is an integer (or just less than an integer). This permits possible stacking of tasks into (nearly) full stations. Schedule changes such as overtime or second shifts are then combined to produce the desired output each period. This approach is relatively easy for labor-intensive lines. Capital-intensive lines are normally planned for round-the-clock production. Accordingly, we cannot increase schedule hours to offset an increased cycle time. Capacity planning for such lines must consider demand uncertainty and growth potential.

In the introduction we noted the importance of adding slack to task times to account for variability in performance times. Other than that brief comment we have not discussed randomness. Yet natural phenomena, and humans, in particular, are inherently random. Suppose the variance σ_i^2 of performance time for task i is known. Let $\sigma_{ij}(\rho_{ij})$ be the covariance (correlation) between task times i and j. For now, assume task times are independent, implying $\sigma_{ij} = 0$. S_k is the set of tasks assigned to station k. Let s_k be the random variable for the time required by station k for a cycle. From basic statistics we know that $E(s_k) = \sum_{i \in S_k} t_i$ and $V(s_k) = \sum_{i \in S_k} \sigma_i^2$. Distributional statements on s_k can also be made. If either t_i are normal or S_k has enough tasks to invoke the central limit theorem, then s_k is normally distributed. A common approach might be to require each station a probability of at least 99 percent of completing the assigned tasks each cycle. A standard normal table would then tell us that we must accommodate the mean performance time plus 2.33 standard deviations. For our assembly-line balancing problems this leads to a new procedure for determining whether a task is fittable. In addition to the task being unassigned and having its predecessors assigned, when added to station k the task must satisfy

$$E(s_k) + 2.33V(s_k)^{1/2} \le C$$

On the average, idle time for the operator will be $C - E(s_k)$, but the allowance of 2.33 standard deviations permits the paced line to flow without problem. Note that the probability that all workstations are completed on time may be only 0.99^K; thus, for long lines to operate smoothly even more slack time must be planned to absorb randomness. We can begin to see the advantages of unpaced lines, which can plan for average time, and methods standardization and training, which reduce task-time variances.

We have made an important assumption in the previous paragraph, namely, that task times are independent. This may well be true if variance is caused by events such as parts not mating and having to be discarded on occasion. However, if variability is caused by worker skill level, then it is reasonable to assume that performance times are positively correlated within a workstation. On the other hand, humans often tend to work faster when behind schedule, thus inducing negative correlation. If covariances (or correlations since $\rho_{ij} = \sigma_{ij}/\sigma_i^2\sigma_j^2$ where σ_{ij} is the covariance between task i and j performance times) can be estimated, we can use the result that $E(s_k) = \sum_{i \in S_k} t_i$ and $V(s_k) = \sum_{i \in S_k} \sigma_i^2 + 2\sum_{(i,j) \in S_k} \sigma_{ij}$.

Another approach to dealing with randomness is to accept the inevitability of some products not being completed on the line. A rework area is provided to add finishing touches. Note that if a task is not completed in the planned workstation, all successor tasks may also be delegated to the rework station. Thus, it can be quite expensive to overschedule the early workstations on the line since early operations may have many successors. Kottas and Lau [1981] describe a line-balancing procedure for this environment.

The foregoing procedures did not save alternate "optimum." In practice, we should. We can then select the alternative that satisfies other criteria such as minimizing the maximum time assigned to any station or maximizing the minimum time to improve morale. Likewise, tooling costs could be computed for each alternative for comparison.

The given data are probably not fixed. If manufacturing problems arise, a team of design engineers and process planners must work together. Redesign of the product might allow greater flexibility in precedence restrictions or modify assembly tasks or times. The product system development effort is best accomplished iteratively and cooperatively, not sequentially.

2.5
SEQUENCING MIXED MODELS

Prenting and Thomopoulos [1974] discuss the use of mixed model assembly systems. When mixed lines are utilized, several different product types can be assembled simultaneously on the line. A dispatch system carefully controls the order of entry of product types to the first station. Mixed lines are typically unpaced to allow stations to make up time lost on a work-intensive item on a subsequent low-work item. Products can generally be classified as one of two types. Type 1 products have a constant ratio of item task time to average item task time, that is, items are either longer or shorter than average in all stations. Type 2 items have independent station time requirements, that is, each product has its own long and short stations.

We assume that the desired product mix is known. Let q_j be the proportion of product type j, $j = 1, \ldots, P$ to be produced. The first step in our approach is to develop an assembly line balance for the weighted average product. Let t_{ij} be the time to perform task i on product type j and S_k the set of tasks assigned to workstation k. We can state an average feasibility condition as

$$\sum_{i \in S_k} \sum_{j=1}^{P} q_j t_{ij} \leq C \qquad k = 1, \ldots, K$$

This condition states that averaged across all items produced in the long term, no workstation is overloaded. The feasibility condition indicates that we need only solve one single product assembly-line balancing problem. In solving this problem we use task times of $t_i = \sum_{j=1}^{P} q_j t_{ij}$.

Sequencing products on the line now becomes the issue. For each item j we must produce Q_j items this period (shift). Let r be the greatest common denominator of all Q_j. We desire to construct a repeating cycle comprised of $N_j = Q_j/r$ units of product type j, $j = 1, \ldots, P$. The cycle will be repeated r times to satisfy period demand. $N = \sum_{j=1}^{P} N_j$ items are produced each cycle. We would like a cycle that both smooths the production rate of each item type and prevents excessive idle (delay) time at workstations because of mix-induced starving of workstations. A workstation is starved if on completion of all its tasks there is no item available for it to work on because the next item has not yet been completed at the prior station. Maintaining a constant flow is most important at the "bottleneck" workstation. Define the relative work load for station k as $C_k = \sum_{i \in S_k} t_i$. The bottleneck station k_b is the station with maximum total work (or equivalently average work per cycle), that is, $k^b = \mathrm{argmax}_k C_k$. If a partial sequence overloads this workstation with respect to average cycle time C, subsequent stations are starved. If a partial sequence underloads this station, the initial output rate from the line will be too high, causing inventory to accumulate. Likewise, later in the sequence this station will be overloaded, thus starving its successors. It therefore is desirable to try to pace the lines' production rate by keeping this station on an even keel. Rocking this boat either way as we sail through the cycle can sink our overall objective.

We now develop a model for determining the dispatch sequence. Let X_{jn} be 1 if item type j is placed in the nth position and 0 otherwise. Likewise $j(n)$ will denote the type of item placed nth. Our approach becomes selecting the nth item to be entered to the line to optimize the following problem:

$$\text{minimize } \underset{1 \le n \le N}{\text{maximum}} \sum_{j=1}^{n} \left| \sum_{i \in S_{k^b}} t_{i,j(n)} - n C_{k^b} \right|$$

subject to

$$\sum_{n=1}^{N} X_{jn} = N_j \qquad j = 1, \ldots, P \tag{2.7}$$

$$\frac{nN_j}{N} - s_1 \le \sum_{b=1}^{n} X_{jb} \le \frac{nN_j}{N} + s_1 \qquad n = 1, \ldots, N \qquad j = 1, \ldots, P \tag{2.8}$$

$$\sum_{b=1}^{n} \sum_{j=1}^{P} \sum_{i \in S_k} t_{ij} X_{jb} \le (n + s_2) C_k \qquad n = 1, \ldots, N \qquad k = 1, \ldots, K \tag{2.9}$$

$$X_{jn} \quad 0 \quad \text{or} \quad 1$$

The objective function minimizes the maximum deviation from assigning average work load to the bottleneck station at any point during the period. Constraints 2.7 ensure that all items are produced during the cycle. Constraints 2.8 restrict the production rate of each product to be within s_1 of its average rate at all times. This controls production rate to suitably match utilization. Last, constraints 2.9 limit maximum overutilization at all times. These constraints attempt to restrict unplanned station idle time due to starving. While the formulation appears complex, use of a greedy[2] approach to the objective yields a straightforward sequencing heuristic.

---

[2]Recall that a greedy heuristic has the characteristics (1) decision variables are ordered and (2) values are assigned to these decision variables one by one using a (typically) simple computational rule that does not look ahead. For our model sequencing problem, if we let decision variable Y_n be the model placed nth in the production order, our heuristic satisfies the greediness definition.

Sequencing Heuristic

STEP 0. Initialization. Create a list of all products to be assigned during the cycle. Call this list A.

STEP 1. Assign a product. For $n = 1, \ldots, N$ from list A, create a list B of all product types that could be assigned without violating any constraint. From list B select the product type (j^*) that minimizes

$$\sum_{j=1}^{n} \left| \sum_{i \in S_k b} t_{i,j(n)} - nC_{k^b} \right| \tag{2.10}$$

Add product type j^* to the nth position. Remove a product type j^* from A and if $n < N$, go to 1.

The approach used in the heuristic should seem familiar. The creation of the list of unassigned products, which is then reduced first to a list of feasible products and then to the single best feasible product, is conceptually similar to the line-balancing procedures studied previously.

We have basically assumed that operators can intermingle to a small degree to keep the line moving even if a station is temporarily overloaded. This is often a reasonable practice if the line is designed accordingly, products are not too different, and the line is designed to accommodate occasional overlap. The issue of line length to permit such mobility by the workers is discussed more fully by Dar-El and Cother [1975].

Often a basic product family is produced with special custom options constituting the individual products. In this environment it may be easiest to produce the basic model on a single product line and then employ utility operators at the end of the line to perform customizing. Nevertheless, with the development of integrated information and handling systems, it is now feasible to directly deliver customized kits for each individual product unit to the proper workstation and to build the custom units on-line. This is often done for large product assembly such as automobiles and heavy equipment.

Small item assembly and packaging lines are normally characterized by medium batch sizes (100 to 5000), cycle times under a minute, low skill level (no learning), general purpose machines, and less than 100 operations and 15 workstations. Here it is generally preferable to use multiple product lines, periodically stopping the line to change over from one product to another. The line may be rebalanced and workers reassigned for each run.

EXAMPLE 2.4

Of course, we would be foolish to market just one toy car. Every child wants to "collect the whole set." Suppose the times used to balance the line are actually averages over all the models in this product line. Through knowledge of current children's television programs, advertising programs, and past experience, we estimate sales by model to be as shown in Table 2.6. The reader may recall that the bottleneck (highest average use) station was station 2 with assigned workload of 68 seconds per cycle. Actual workload by model type for this station is provided in the table.

Table 2.6	Estimated Weekly Sales by Model Type		
Model	Sales	Percentage	Station 2 Time
Red Z	250	16.7	72
Blue Q	250	16.7	68
Black R	500	33.3	68
RWB American	500	33.3	66

Solution

To sequence the mixed line we need to find the required production of each model in a cycle. From the table we see the need for 1 Red, 1 Blue, 2 Black, and 2 RWB per cycle, that is, $N_{red} = 1, N_{blue} = 1, N_{black} = 2, N_{rwb} = 2$, and $N = 6$. For our heuristic, we will set $s_1 = s_2 = 0.9$. Initially, all models are eligible. Assignment of the blue or black models will maintain the lower bound of 0 for the objective. We choose to break the tie by assigning black to minimize the maximum deviation of actual to desired production for any assignable product. The rationale in assigning black puts us 2/3 item ahead of schedule for black and 1/6 behind for blue. Assigning blue puts us 5/6 ahead for blue and 1/3 behind for black. Values for (1) deviation from smooth production rate $[(nN_j/N) - \sum_{r=1}^{n-1} X_{jr}]$ as required in equation 2.8 and (2) objective 2.10 are shown in Table 2.7 for each stage. For assignment at $n = 2$ black is not eligible. Its assignment would place it $4/3 > s_1$ ahead of schedule. Consequently, we assign blue to stay on time schedule. The reader who believes that we should assign RWB is implicitly stating a preference for the inventory/demand constraint over smoothing station usage. Blue is now complete for the cycle. We next assign RWB. The bottleneck workstation is now underloaded by two seconds, but failure to assign RWB violates the demand constraint. At stage four, we are indifferent between Red and Black. In either event we will remain two seconds off the desired station utilization and the worst demand constraint is violated by 2/3. We arbitrarily choose Red. Fifth we choose RWB again to bring workstation time variance back to zero; last, another Black. This cycle, Black-Blue-RWB-Red-RWB-Black, can then be repeated indefinitely. ∎

Table 2.7	Production Shortage and Objective by Stage				
Stage	Red Z	Blue Q	Black R	RWB American	Assigned
1	1/6, 4	1/6, 0	1/3, 0	1/3, 2	Black
2	1/3, 4	1/3, 0	−1/3, 0	2/3, 2	Blue
3	1/2, 4	—	0, 0	1, 2	RWB
4	2/3, 2	—	1/3, 2	1/3, 4	Red
5	—	—	2/3, 2	2/3, 0	RWB
6	—	—	1, 0	—	Black

2.6
UNPACED LINES

In a paced line with K stations and cycle time C, each item spends KC time units in the system and the production rate is C^{-1} units per time. Unpaced lines require a little extra scrutiny.

Let s_k be the sum of the task times for tasks assigned to station k ($s_k = \sum_{i \in S_k} t_i$). Let us assume for simplicity that task times are all deterministic. The bottleneck workstation k_b is the slowest workstation. Capacity and output rate are determined by this workstation, that is, production rate is $s_{k_b}^{-1}$. Buffers are of no use in this reliable, deterministic world. Indeed, if we set $C = s_{k_b}$, the paced and unpaced lines have the same production rate.

The time in the system can differ from the paced line, however. To see how this can be, divide the line into two sublines, stations 1 to k_b and $k_b + 1$ to K. (If multiple bottlenecks exist we break the line at the lowest numbered bottleneck station.) First, consider line 1. Stations 1 through $k_b - 1$ work faster than station k_b. By definition of the unpaced line, station 1 will obtain a new item and begin processing as soon as it passes on its just completed item. Since station 1 (likewise every other station) can outproduce k_b at the end of the line, all stations will fill up, and as soon as k_b completes an item, all stations will pass on their item. Hence, each item will spend s_{k_b} time units in each workstation. The item in station k is blocked for the last $s_{k_b} - s_k$ time units.

Next consider line 2. Items enter the line every s_{k_b} time units. Since each station's required production time is at most s_{k_b}, each station has time to complete the item before the next arrives. Hence, items do not become blocked. Line 2 throughput time is then just the sum of the station times. We can briefly illustrate this result. Consider a three-station line. Items enter every 5 time units. Let $s_1 = 2, s_2 = 4, s_3 = 3$. Table 2.8 shows the sequence of event times as items begin to go through the line. All items are in the system for $s_1 + s_2 + s_3 = 9$ time units and blocking never occurs. Once loaded, each station k will have a slack time (time between passing on an item and receiving the next) equal to $s_{k_b} - s_k$. Combining results for the two lines, production time in system is $k_b s_{k_b} + \sum_{k=k_b+1}^{K} s_k$. This interesting result says that time in the line and, hence, work in process is smaller for unpaced lines. The advantage of unpaced lines is maximized when the slowest station is placed first.

An interesting problem is the estimation of performance of general serial systems. Workstation loads may be slightly unbalanced because of the natural processing time requirements on various machines, and processing times may be random. Work-in-process inventory buffers are often provided to cushion the effects of processing-time variability. The next chapter will explore such systems in greater detail.

Table 2.8   Illustration of Item Progession

Item	Enter 1	Leave 1	Enter 2	Leave 2	Enter 3	Leave 3	Flow Time
1	0	2	2	6	6	9	9
2	5	7	7	11	11	14	9
3	10	12	12	16	16	19	9
4	15	17	17	21	21	24	9
5	20	22	22	26	26	29	9

2.7
SUMMARY

Assembly lines have greatly enhanced production because of their constant concentration on the objective: producing good product. Many final products are produced by assembly systems where feeder lines of components and subassemblies gradually are combined via a network of assembly lines and inventory stock points. An important problem is the balancing of the assembly lines to match production rates for each line in the network and each station in the line. The objective is minimization of idle time and is obtained by designing lines with constant and equivalent cycle time and then assigning an equivalent amount of work to each workstation. Simple heuristics have often been used to find good solutions. Mathematical programming formulations of the assembly-line balancing problem exist. Advances in computational speed and algorithm design are beginning to make possible the finding of optimal solutions for many problems. An efficient sequence tree exploration procedure is one such implicit enumeration approach.

Many extra concerns must be considered in balancing assembly lines. Lines often must handle mixed models. Such cases can be handled by unpaced lines, which are designed for the statistically average product, and then controlling the model input sequence. In general unpaced lines have the advantage of allowing some variability in station assembly times while minimizing flow time. Paced lines, however, can avoid the need to remove and replace the product unit on the transport mechanism at each station and can help motivate workers to maintain a pace.

Despite the appearance of literally hundreds of papers on assembly lines, little work has been done on modeling the full range of practical considerations in designing assembly lines. Many of these concerns are discussed in Gunther and Peterson [1983]. Baybars [1986] reviews exact algorithms for the "simple" (deterministic, single product) line-balancing problem. Kottas and Lau [1981] and Smunt and Perkins [1985] discuss stochastic problems. Ghosh and Gagnon [1989] provide a thorough review of current capabilities and how these match up to the real-world considerations. Stochastic and deterministic, single and multiple product cases are covered. Pinto et al. [1981] consider the case where parallel workstations are allowed to facilitate balancing and handling long tasks. Bard [1989] uses dynamic programming to investigate the use of parallel workstations when deadtime exists at the start of each work cycle.

REFERENCES

Arcus, A. L. (1966), "COMSOAL: A COmputer Method for Sequencing Operations for Assembly Lines," *International Journal of Production Research*, 4(4), 259–277.

Bard, Jonathan F. (1989), "Assembly Line Balancing with Parallel Work Stations and Dead Time," *International Journal of Production Research*, 27(6), 1005–1018.

Baybars, I. (1986), "A Survey of Exact Algorithms for the Simple Assembly Line Balancing Problem," *Management Science*, 32, 909–932.

Chakravarty, Amiya K. and Avraham Shtub (1985), "Balancing Mixed Model Lines with In-Process Inventories," *Management Science*, 31(9), 1161–1174.

Dar-El, E. M., and R. F. Cother (1975), "Assembly Line Sequencing for Model Mix," *International Journal of Production Research*, 13(5), 463–477.

Dar-El, E. M., and S. Cucky (1977), "Optimal Mixed-Model Sequencing for Balanced Assembly Lines," *Omega*, 5, 333–342.

Ghosh, Soumen, and Roger J. Gagnon (1989), "A Comprehensive Literature Review and Analysis of the Design, Balancing and Scheduling of Assembly Systems," *International Journal of Production Research*, 27(4), 637–670.

Gunther, R. E., and R. S. Peterson (1983),"Currently Practiced Formulations for the Assembly Line Balancing Problem," *Journal of Operations Management,* 3(4), 209–221.

Inman, Robert R., and Robert L. Bulfin (1991), "Sequencing JIT Mixed-Model Assembly Lines," *Management Science,* 37(7), 901–904.

Johnson, R.V. (1988), "Optimally Balancing Large Assembly Lines with 'FABLE'", *Management Science,* 34(2), 240–253.

Kottas, J. F., and H. S. Lau (1981), "A Stochastic Line Balancing Procedure," *International Journal of Production Research,* 19(2), 177–193.

Kubiak, W., and S. Sethi (1991), "Level Schedules for Mixed Model Assembly Lines in JIT Production Systems," *Management Science,* 37(1), 121–122.

Monden, Yasuhiro (1984), *Toyota Production System,* Industrial Engineering and Management Press, Norcross, GA.

Muth, Eginhard J. (1973), "The Production Rate of a Series of Work Stations with Variable Service Times," *International Journal of Production Research,* 11(2), 155–169.

Pinto, P., D. G. Dannenbring, and B. M. Khumawala (1981), "Branch and Bound Heuristic Procedures for Assembly Line Balancing with Paralleling of Stations," *International Journal of Production Research,* 19, 565–576.

Prenting, Theodore O., and Nicholas T. Thomopoulos (1974), *Humanism and Technology in Assembly Line Systems,* Hayden Book Company, Rochelle Park, NJ.

Smunt, T. L., and W. C. Perkins (1985), "Stochastic Unpaced Line Design: Review and Further Experimental Results," *Journal of Operations Management,* 5(3), 351–373.

Talbot, F. B., and J. H. Patterson (1984), "An Integer Programming Algorithm with Network Cuts for Solving the Assembly Line Balancing Problem," *Management Science,* 30(1), 85–99.

PROBLEMS

2.1. Define the Principle of Interchangeability and the concept of Division of Labor and explain their significance to assembly lines.

2.2. What are the key factors in deciding whether to have one assembly line with many workers and a small cycle time or multiple short lines each with a longer cycle time?

2.3. A manufacturer of communications equipment is constructing a line to assemble several similar models of speaker phones. An industrial engineer has divided assembly of each model into elemental tasks. Phones require about 30 assembly, test, and packing operations, which will be performed on the line. Task times vary in duration from 5 seconds to 36 seconds. Skill level is minimal, as is equipment cost except for one moderately priced test machine. Determine the appropriate cycle time if demand requires producing 750 phones per shift. Each shift has 8 productive hours.

2.4. Suppose in Problem 2.3 that only an estimate of demand is known. The line must be capable of producing to daily demand of between 500 and 1000 phones. Overtime is possible up to 2 hours per day, but workers must be paid 1.5 times their normal rate for overtime. Alternatively, a second shift of 4 to 8 hours could be added at the same wage rate as the first shift. Storage space for several thousand phones is also available. Discuss what effect this additional information would have on your choice of a cycle time.

2.5. Consider using the integer programming formulation of Section 2.2 to solve the toy car problem of Example 2.1. Let $K = 4$ workstations.

  a. Determine the number of decision variables X_{ik}.

  b. Determine the number of constraints needed of type 2.2, 2.3, and 2.4.

2.6. Write out the integer programming formulation of Section 2.2 for the toy car problem of Example 2.1. Find an integer programming software package and solve for an optimal line balance. Note the amount of computer time needed to solve this problem.

2.7. Write out the complete binary integer programming formulation for the following line balancing problem. Let $C = 100$.

Task	Time	Immediate Predecessors
a	40	—
b	75	a
c	50	a
d	35	c
e	80	d

2.8. Write a flowchart for COMSOAL using the decision rule that all feasible tasks are equally likely to be selected.

2.9. Write a flowchart for COMSOAL using the decision rule that feasible tasks are selected with probability proportional to their positional weight.

2.10. The assembly activities of a new product must follow the precedence restrictions of Figure 2.11. Task times (minutes) are shown above each task node. Cycle time has been set at 40 minutes.
　　a. Find a lower bound on the number of workstations required.
　　b. Balance the line using the Ranked Positional Weight technique.
　　c. Is your solution optimal?

2.11. Suppose a second product with the same precedence structure as Figure 2.11 is to be balanced. The new product will have random assembly times. Expected task times are as follows:

Task	a	b	c	d	e	f
E (time)	15	8	10	14	6	14

Task	g	h	i	j	k
E (time)	3	3	4	20	12

Task times are normally distributed with a coefficient of variation (standard deviation/mean) of 0.05. Management policy dictates that the line should permit each worker to complete their assigned tasks in at least 99 percent of cycles. Balance the line using Ranked Positional Weight for this new product. Use $C = 40$.

2.12. Suppose a line is to be developed to produce the products of Problems 2.10 and 2.11 simultaneously. The line must be able to work on both parts at the same time, that is, at any point in time some stations will have the first product and other stations the second product. Due to equipment requirements, stations must be assigned the same set of tasks for both products.
　　a. Balance the line with $C = 40$.
　　b. If demand is equal for both products, determine the line's long-range balance delay.

2.13. The assembly of a product has been divided into elemental tasks suitable for assignment to unskilled workers. Task times and constraints are given in Table 2.9.

Table 2.9	Data for Problem 2.13	
Task	Time	Immediate Predecessors
a	20	—
b	18	—
c	6	a
d	10	a
e	6	b
f	7	c,d
g	6	e,f
h	14	g

　　a. Draw the precedence network.
　　b. Suppose cycle time $C = 30$. Find a lower bound on the number of workstations required.
　　c. Suppose $C = 30$; find an upper bound on the location of each task if a three-workstation solution is desired.
　　d. Consider the pair of tasks d and e. Does either task "dominate" the other in the sense

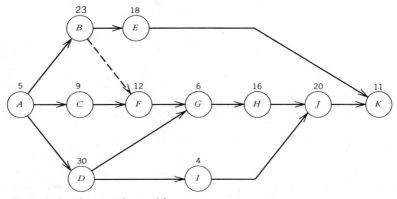

Figure 2.11　Precedence diagram for Problem 2.10.

that it may help to fathom solutions in a branch and bound algorithm?

e. Balance the line (you may use the Ranked Positional Weight technique).

2.14. Determine all task dominance pairs for the data given in Problem 2.13.

2.15. Use the implicit enumeration scheme of Section 2.3.3 to find an optimal solution to the data in Problem 2.13.

2.16. Suppose the cycle time for Problem 2.13 was 20 instead of 30.

a. Find a lower bound on the number of workstations required.

b. Find the Ranked Positional Weight solution.

c. Give an optimal solution.

2.17. Consider the five assembly tasks given in Problem 2.7.

a. Construct the complete tree of precedence-feasible task sequences.

b. Explicitly evaluate each sequence in the tree to determine the number of workstations required. List all optimal sequences.

2.18. For each of the terms listed below, give a definition of the term including any algebraic relations that apply, and also give an example from the problem described in Table 2.10. $C = 50$ and assume the best known solution uses the four workstations (a, b, e, f), (c, d), (g), and (h, i).

Table 2.10   Data for Problem 2.18

Task	Time	Immediate Predecessors
a	6	—
b	25	a
c	15	a
d	23	b,c
e	12	b
f	5	e
g	36	d
h	16	f,g
i	12	h

a. Lower bound on solution
b. Fittable task
c. Positional weight
d. Upper bound on station for a task
e. Task dominance

2.19. Find the Ranked Positional Weight solution for the problem stated in Problem 2.18. Do you know if this solution is optimal?

2.20. Use the implicit enumeration scheme of Section 2.3.3 to find an optimal solution to the line-balancing problem given in Problem 2.18.

2.21. Resolve Problem 2.19 using a cycle time of 40. Which solution has a lower balance delay?

2.22. Resolve Problem 2.20 using a cycle time of 40. Which solution has a better balance delay?

2.23. An assembly line is being designed. Each station will have one worker. Cycle time for meeting demand is 40 time units. Using the data in Table 2.11 find an optimal (minimum number of workstations) solution. A zoning restriction exists such that tasks d and f must be performed in the same workstation. If your solution has more than 10 percent idle time, suggest alternatives that could be explored to improve efficiency. What additional data would be required to analyze these alternatives?

Table 2.11   Data for Problem 2.23

Operation	Time	Immediate Predecessors
a	3	—
b	5	—
c	10	a,b
d	11	c
e	24	c
f	26	d
g	24	e
h	15	g

2.24. Suppose a second product model is to be simultaneously produced on the same line as that constructed for the product described in Problem 2.18. Demand for the original product is unchanged. Initial demand for the new product is forecasted to be one-fourth of that for the existing product; thus, cycle time will be decreased to 40. The new product has the same set of operations and precedence restrictions, but operation times for tasks a through i are (7, 22, 10, 25, 0, 6, 40, 15, 10), respectively.

a. Balance the line assuming synchronous (paced) transfer. In this instance, all workstations must be able to perform all the assigned tasks for either product within the cycle time.

b. Suppose a line was to be built for just this new product. Find an appropriate line balance with $C = 40$. Compare the balance delay for the separate-line and mixed-line solutions.

2.25. Resolve Problem 2.24 assuming asynchronous transfers (unpaced line). In this case, all workstations must be able to perform the weighted average of assigned task times within the cycle time.

2.26. Reconsider the situation described in Problem 2.12. Instead of a paced line, an asynchronous line will be used wherein each workstation must only be able to complete its average work load within the cycle time. The demand rate for both products is the same. Find the feasible allocation of tasks to workstations that minimizes the number of workstations required. Compare this solution to that of the paced line.

2.27. Resolve Problem 2.23 for the case that processing times are random with a coefficient of variation of 0.1. Each station should have at least a 99 percent probability of completing all its assigned tasks within the cycle time.

2.28. Three products are produced on the same line. Task times are shown in Table 2.12. One half of demand is for product A. The other one half of demand is evenly split between products B and C.

Table 2.12   Task Times for Problem 2.28

Task	Immediate Predecessors	Product A	B	C
a	—	30	25	35
b	a	15	10	10
c	a	5	7	5
d	a	40	40	45
e	d	50	45	45
f	d	30	40	40
g	b,c	15	15	20
h	e,g	65	60	65
i	h	25	25	0
j	f	50	50	60
k	i,j	45	40	40
l	k	5	5	5

a. Find a line balance for which no workstation has an average work requirement in excess of 100 units per product produced.
b. Identify the bottleneck workstation.

c. Find a repeating cycle or product order entry onto the line that smooths the production rate of the bottleneck workstation without building unnecessary inventories or shortages for any product during the cycle.

2.29. Suppose a line is to be designed to produce just product A from Problem 2.28. Demand is such that cycle time must not exceed 50 time units. Construct two alternatives. First, try using two parallel lines each with $C = 100$. Second, use a single line with $C = 50$, but allow parallel workstations to exist along the line. Compare your solutions in terms of balance delay and the cost and difficulties in implementing the solutions.

2.30. Five products are produced on the same assembly line. Table 2.13 contains weekly demand for each product and the time required by workstation 6 per unit. Workstation 6 is more heavily loaded than the other workstations. Find a repeating cycle for entering product onto the mixed model line. The demand rate is constant and continuous for each product and management desires to avoid accumulating excess or shortage inventory of more than two units of any product at any time. Management would like to operate the line with the largest average production rate possible.

Table 2.13   Demand and Bottleneck Processing Times for Problem 2.30

Product	Weekly Demand	Bottleneck Station Processing Time
A	1000	45
B	500	40
C	750	45
D	500	50
E	250	55

2.31. Four products are built on a rotating basis using a multiple product philosophy. The plant sets up the machines to make one product, produces a specified amount, and then halts production while all machines are changed over to produce the next product. The production rate of the line is very high relative to total demand. However, labor cost is high and therefore the line is operated only a small fraction of the time. Using the data in Table 2.14 find the optimal run length for each product. (*Hint:* You should trade off setup cost and inventory holding cost.)

Table 2.14   Product Data for Problem 2.31

Product	Annual Demand	Setup Cost	Inventory Holding Cost per Unit-Year
A	10,000	500	2.50
B	50,000	1500	1.75
C	20,000	1000	4.00
D	25,000	700	3.25

2.32. A part has a completely specified precedence order, that is, operation 1 must come first followed by operation 2 and so on up to operation N. Each operation is performed on a separate machine. Machines are laid out in the operation order. Each task requires a_i concurrent time of the worker and the machine. Additionally, task i requires b_i off-line operator time and t_i unattended machine time. It is assumed that all $a_i + t_i \leq C$, where C is the desired cycle time for the line. Assume also that $b_i \leq t_i$ so that workers may be able to handle more than one machine. The objective is to minimize the number of workers required. Show that the optimal solution can be found by assigning tasks 1 to k^* to worker 1 where $k^* = \text{argmax}\{k : \sum_{i=1}^{k}(a_i + t_i) \leq C\}$ and assignments to subsequent workers are found by deleting machines 1 to k^* and repeating the assignment step.

2.33. Disassemble an old calculator. List the parts, assembly tasks, and precedence constraints for this item. Estimate assembly times. Design a system to manufacture 1000 calculators per hour.

2.34. Disassemble a hand-held stapler. List the set of parts and propose a set of elemental tasks that would allow reassembly of the stapler. Indicate any precedence restrictions. Can you suggest any design changes that would simplify the assembly process?

2.35. An assembly system is to be developed to produce 10,000 units per month of product A from Problem 2.28. The plant will operate one 8-hour shift an average of 22 days per month. Task times in Table 2.12 are in seconds. Workers are paid $12 per hour. Tasks b, e, and g require a V25 machine, which costs $1200 per month. Tasks l and i require a TL 100 machine costing $950 per month. You may purchase as many of these machines as needed. Design an assembly system (number of lines, cycle times, and allocation of tasks to workstations). Determine labor and machine costs per unit.

CHAPTER 3

TRANSFER LINES AND GENERAL SERIAL SYSTEMS

The grim shape
Towered up between me and the stars, and still,
For so it seemed, with purpose of its own
And measured motion like a living thing,
Strode after me.
—William Wordsworth, *The Prelude*

3.1
INTRODUCTION

The assembly lines discussed in the preceding chapter were assumed to be 100 percent reliable (workstations never broke down). Processing times were for the most part assumed to be deterministic. In this chapter we discuss flow-line systems that are subject to breakdowns and processing time variability. Transfer lines are one important class of flow-line systems for which consideration of breakdowns is important. Although the chapter title may imply that the material in this chapter will be specific to transfer lines, in reality the transfer line models apply to any paced, serial system subject to breakdowns of various types. The breakdown could be caused by a fractured tool or jamming of the transport mechanism in a transfer line. However, it might also be due to the temporary unavailability of a worker on an assembly line or an operation that occasionally takes an unusually long time. At the end of the chapter we discuss asynchronous serial systems with random processing times at each workstation. Thus, this chapter covers serial production systems subject to machine failures and random processing times.

A transfer line may be defined as a set of serial, automatic machine and/or inspection stations linked by a common material transfer and control system. As with an assembly line, éach workstation performs its operation on each unit of product. Every C time units the handling system indexes parts on the line ahead to the next workstation. Figure 3.1 illustrates the layout and hardware of a typical transfer line.

The original transfer lines represented "hard automation." In hard automation the line is designed for mass production of a single product. Even minor changes in product design can render the line obsolete. Recent advances in automation and the development of low-cost controllers have resulted in programmable workstations and flexible flow lines. Flexibility is becoming more important as rapid technological innovation and intense competition shorten product life cycles.

67

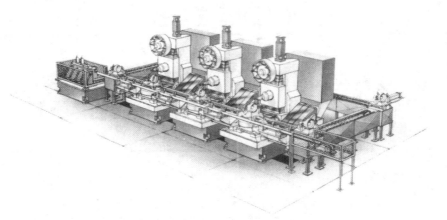

Figure 3.1   A pallet transfer line (courtesy of Toyota Machinery USA).

Transfer lines and automated flexible flow lines are capital intensive and must be kept running to be justifiable. Breakdowns of single workstations or the entire line are particularly important issues in the design of such lines. As the number of stations along a line increases, the probability of all stations being "up" (i.e., operational) decreases. Buffers, though expensive to install and maintain, provide a means for insulating workstations from failures elsewhere in the line, thus improving station utilizations. The objective of this chapter is to determine the effectiveness of a line given both buffer capacities and failure and repair rates for each workstation. Effectiveness is measured by the net production rate. A general schematic representation of a transfer line with several buffers is shown in Figure 3.2.

A station may be "down" for one of four reasons:

1. Station Failure
2. Total Line Failure
3. Station Blocked
4. Station Starved

Station failures are caused by events such as a fractured tool, quality out-of-control signal, missing/defective part program, or jammed mechanism. Although the failed station must stop producing, other stations may continue provided they are fed product and have space for sending completed product. A total line failure exists if all stations are inoperative. A power outage or error in the central line controller would

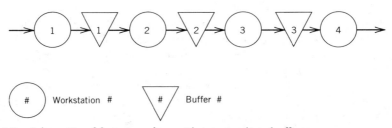

Figure 3.2   Schematic of four-stage line with intermediate buffers.

cause a total line failure. Station i is "blocked" if on completion of a cycle it is unable to pass the part to station $i + 1$. The inability to pass the part may be due to a failure of the handling system, failure of a downstream station prior to the next buffer, or failure of a downstream station with the intermediate buffer between these stations currently being full. If station $i + 1$ is down, and its input buffer is filled, then station i must remain idle while it waits for downstream space for the just completed part. Last, station i is said to be "starved" if an upstream failure has halted the flow of parts into station i. Even if operational, a starved station will sit idle.

Failures may also be classified as time or operation (cycle) dependent failures. Time failures occur with a chronological frequency independent of the number of successful cycles since the last failure. Hence, the span of time between time-dependent failures is measured in units such as hours. Daily maintenance would be an example. Operation-dependent failures, such as tool wear, occur only while the system is running. Time between operation failures is measured in cycles. It has been estimated that approximately 80 percent of transfer line failures are operation-dependent station failures and, hence, we will concentrate on these in the discussion that follows. We begin by examining lines without buffers and then consider the improvements possible from the addition of buffers.

Performance of a line configuration is measured by its effectiveness E (also referred to as availability). We define

$$E = \lim_{t \to \infty} \frac{q(t)}{Q(t)}$$

where $q(t)$ is actual output over time t and $Q(t)$ is theoretical capacity over t in the absence of any work stoppages. As a practical matter, we estimate t by comparing productive cycles (time) to total cycles (time) and

$$E = \frac{E(\text{uptime})}{E(\text{uptime} + \text{downtime})} \tag{3.1}$$

The $E(\cdot)$ operator in equation 3.1 refers to expected value. Uptime refers to the interval during which production is occurring, and downtime refers to the period during which finished product is not leaving the line. Throughout this chapter we assume that durations of uptimes and downtimes are independent and their respective distributions are stationary. Renewal theory can be used to show the validity of expression 3.1 provided that the sequence of uptime + downtime periods are independent (Barlow and Proschan [1975], p. 192).

3.2
PACED LINES WITHOUT BUFFERS

We begin by discussing the basic paced line. As in the previous chapter, the product unit is transferred to the next workstation at the end of every cycle. The difference from the previous chapter comes in the recognition that the transfer mechanism (or workstation) may fail, thus stopping the line.

3.2.1 Operation-Dependent Failures

We first consider the case of operation-dependent station failures on an M-stage transfer line. In doing so, we make the following assumptions.

Assumptions

1. The number of cycles until failure of station i is assumed to be a geometric random variable T with failure rate α_i. Hence, the density function for the number of cycles to failure is

$$f_i(t) = \alpha_i(1 - \alpha_i)^{t-1}$$

Mean cycles to failure (MCTF) is accordingly α_i^{-1}.

2. The number of cycles for repair at station i is geometric with mean b_i^{-1} cycles. We will normally assume $b_i = b, i = 1, \ldots, M$.
3. All uptime and downtime random variables are independent.
4. Idle stations do not fail.
5. Failures occur at the end of a cycle; failures do not destroy the product.
6. At most one station can fail on any cycle.

As with any model, a thorough understanding of the assumptions is necessary for valid use of the model. The motivation for assumptions 1 to 3 is both analytical convenience and the discrete analog of the justification for exponential interarrival times in random processes. These assumptions permit use of a discrete time, discrete state Markov chain model to solve the problem defined above. Exact analysis for other distributions is significantly more difficult and beyond the scope of this book. Assumption 4 defines our distinction between operation-dependent and time-dependent failures. Assumption 5 provides a convenient, intuitive view of the Markov chain and simplifies the counting of actual production. Assumption 6 allows us to ignore low-probability, second-order terms such as $\alpha_i \alpha_j$. This assumption also obviates the issue of the capacity of the repair facility (if multiple stations could fail, we would have to consider the number of stations that can simultaneously be under active repair, i.e., the number of repair stations). Essentially, we require only that failure probabilities be small. In this case, multiple failures are rare and do not seriously affect results. As you read the following model development, try noting where these assumptions are invoked.

Let Q_r be the event that all M stations survived r cycles. The probability that the line first fails at the end of the tth cycle (t good products are produced but the tth cannot be passed on) is given by

$$P(T = t) = \text{Prob}(Q_{t-1}) * [1 - \text{Prob}(Q_t/Q_{t-1})]$$

$$= \left\{ \prod_{i=1}^{M} (1 - \alpha_i) \right\}^{t-1} \left[1 - \prod_{i=1}^{M} (1 - \alpha_i) \right]$$

Commuting terms and expanding the first product by adding and subtracting 1, we obtain

$$P(T = t) = \left[1 - \prod_{i=1}^{M} (1 - \alpha_i) \right] \left\{ 1 - \left[1 - \prod_{i=1}^{M} (1 - \alpha_i) \right] \right\}^{t-1} \tag{3.2}$$

If we define

$$\beta = \left[1 - \prod_{i=1}^{M} (1 - \alpha_i) \right] \tag{3.3}$$

equation 3.2 becomes

$$P(T = t) = \beta(1 - \beta)^{t-1}$$

which is once again a geometric distribution, but now with parameter β. Thus, the M station line behaves like a single station but with failure parameter β replacing α_i.

The reader may have noted one inconsistency in the foregoing derivation. We used the expression $[1 - \Pi_{i=1}^{M}(1 - \alpha_i)]$ for the probability of not surviving on the tth cycle. This is true, except it states that one or more stations fail. Assumption 6 allows only one failure per cycle. The difference between our stated assumption and result to this point can be seen by expanding the terms in β. We obtain

$$\beta = 1 - (1 - \sum_i \alpha_i + \sum_{i>j} \alpha_i \alpha_j - \cdots + \sum_{i>j>\cdots>M} \alpha_i \alpha_j \cdots \alpha_M)$$

Noting that α_i are small, higher order terms approach zero, and we have

$$\beta \approx \sum_{i=1}^{M} \alpha_i \tag{3.4}$$

The higher order terms just eliminated account for the events of more than one failure in a cycle. In most cases the choice of equation 3.3 or equation 3.4 is unimportant. For instance, if $\alpha_i = 0.001$ and $M = 20$, the expressions yield $\beta = 0.0198$ and $\beta = 0.0200$, respectively. The accuracy of our data (α_i, β) and assumptions would normally be a more limiting factor in model accuracy than this discrepancy.

Line effectiveness can now be calculated by the ratio of expected productive cycles between failures (uptime) divided by expected total cycles between failures (uptime + downtime). Using equation 3.4 and the repair time distribution assumption, we find that

$$E = \frac{\beta^{-1}}{\beta^{-1} + b^{-1}} = \frac{1}{1 + \beta b^{-1}} \tag{3.5}$$

EXAMPLE 3.1

Consider a two-station line where the first station fails on the average every 10 cycles and the second station averages a failure every 15 cycles. Average repair time is two cycles. Find line availability.

Solution

The notation above yields $\alpha_1 = \frac{1}{10}$, $\alpha_2 = \frac{1}{15}$, and $b = \frac{1}{2}$. We then have

$$E = \frac{1}{1 + \left(\frac{1}{10} + \frac{1}{15}\right)2} = 0.75$$

To illustrate, suppose by chance a sequence of random events occurred with each failure and repair event realizing its mean time. Figure 3.3 shows the event sequence assuming both machines are newly operational at time 0. After 10 cycles station 1 fails and is down for 2 cycles. Five cycles later, station 2 fails for 2 cycles. Station 1 has now been running for 5 cycles and, hence, it performs for another 5 cycles before failing. On repair, both stations are up for another 10 cycles. At time 36 both stations want to fail. Our assumption of one failure at a time translates to the need to repair station 1 for 2 cycles, whereupon station 2 suddenly fails and requires another 2 repair cycles. Thus, after 40 cycles, 30 productive cycles are observed and the process is ready to restart.

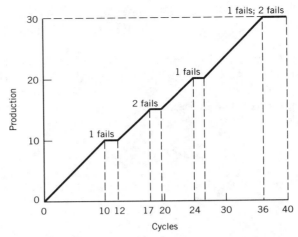

Figure 3.3   Production path for deterministic (average outcomes) sequence.

The reader is asked in the chapter problems to show that if we had used the exact expression for β we would have obtained $E = 0.7575$. This effectiveness corresponds to a rotation every 39.6 cycles. Intuitively, of the six cycles during which station 1 was down, α_2 or one-fifteenth of this time station 2 would have also been under repair had either joint downtimes or two repairpersons been allowed; $6/15 = 0.4$, precisely the difference between the actual and approximate model for how long it takes to make 30 good units and return to starting status of two newly repaired stations. ∎

Equation 3.5 can also be obtained from the following intuitive argument. For every T cycles operated (good cycles), the expected number of station i failures is $T\alpha_i$. This is a simple binomial model. If failure times and repair times are independent, expected repair cycles may be obtained by the product of expected number of failures and expected repair time per failure, that is, $T\alpha_i/b$. We then find, as above, that

$$E = \frac{T}{T + \sum_{i=1}^{M} T\frac{\alpha_i}{b}}$$

Dividing by T yields the above result. Note that we may generalize this result by letting b_i be the average repair cycles for station i.

Finally, we note that from the viewpoint of output produced, individual station failures and total line failures are equivalent. In either case, the entire line is idled. We can thus model total line failures by use of a "pseudo workstation 0" with cycle failure and repair rates α_0 and b_0, respectively. The lower limit on our summation in equation 3.4 becomes 0. The M station line with station failures and separate line failures can be modeled as an $M + 1$ station line with only station failures. The line's failures are those of the added station.

3.2.2 Time-Dependent Failures

We now consider the case where failures occur without regard to operating cycles. Instead, failures occur continuously through time with the exception that a failed station cannot fail again until it is repaired. (The student is asked to consider the effect of this timekeeping assumption in the end-of-chapter exercises.) Thus, time to failure is measured as clock time since the station was last repaired. Completion of a station repair initializes a renewal of the failure process. We will assume that on repair completion, time to failure for station i is exponential with rate parameter α_i, that is, mean time to failure is $1/\alpha_i$. Repair times are exponential with parameter b_i. As before, total line failures have the same effect as station failures; thus we use the pseudo workstation 0 to represent the line. The expected number of failures during an operating time of t is $\alpha_i t$ for station i. Expected repair time is $\alpha_i t / b_i$. Thus, for station i alone, its effectiveness is

$$E_i = \lim_{t \to \infty} \frac{E(\text{uptime})}{E(\text{uptime}) + E(\text{downtime})} = \frac{t}{t + \dfrac{\alpha_i t}{b_i}} = \frac{1}{1 + \dfrac{\alpha_i}{b_i}}$$

for station i. In cycle-dependent failures, time was suspended when any station failed. This is not the case with time-dependent failures. As stations are independent,[1]

$$E = \prod_{i=0}^{M} \left[1 + \frac{\alpha_i}{b_i} \right]^{-1} \tag{3.6}$$

EXAMPLE 3.2

Suppose the line in Example 3.1 is subjected instead to time-dependent failures. How is the effectiveness affected?

Solution

For the case $\alpha_1 = \frac{1}{10}$, $\alpha_2 = \frac{1}{15}$, and $b = \frac{1}{2}$ we obtain

$$E = \left(1 + \frac{2}{10} \right)^{-1} \left(1 + \frac{2}{15} \right)^{-1} = 0.735$$

Effectiveness has been reduced from 0.75 to 0.735 since stations 1 and 2 can now continue the aging process while idle as a result of a failure at the other station. ∎

3.3
TWO-STAGE PACED LINES WITH BUFFER

We now consider two serial stages separated by an inventory buffer. The buffer reduces the dependence between stations, that is, if the buffer contains parts when station 1 fails, then station 2 continues operating. Likewise, if station 2 fails, 1 continues operating until the buffer is full. We will concentrate on operation-dependent failures, the most common mode. We let Z be the buffer capacity.

---

[1]With operation-dependent failures, at most one station could be down at a time. However, with time-dependent failures, several stations could be down at the same time. One requirement for independence is that the repair facility can work on all failed stations simultaneously.

3.3.1 Operation-Dependent Failures

Except for the addition of buffers, our assumptions are as before. We can describe the environment with a Markov chain. The states of the Markov chain are (S_1, S_2, z) where S_i is the status for station i and z is the number of items in the buffer. Status for station i is denoted by W for working (operational) condition and R for repair required. Note that if station 1 is "working" but 2 is under repair, station 1 may still be idle because of a full buffer. If 1 is under repair and 2 is working, an empty buffer will starve station 2.

We adopt the following convention. We view the system at the start of a cycle. Failures and repairs occur at the end of a cycle; hence, if a station begins a cycle under repair, it cannot produce an item during this cycle. Likewise, if the station starts a cycle operational, it does produce an item during the cycle unless it is starved. When a cycle starts, if both stations are working, station 2 receives its next part from station 1. If station 2 is working but 1 is down, station 2 attempts to remove a part from the buffer. Unless the buffer is empty, a part is removed from the buffer and processed in station 2. If station 1 is working but 2 is down, station 1 attempts to place its finished part in the buffer and continue. If the buffer is full, station 1 becomes blocked. By this convention, buffer size is unaffected when both stations start a cycle in the same state, either R or W. The buffer level will decrease by 1 during the cycle when station 2 is active and 1 is under repair ($RWz, z > 0$). The buffer level will increase by 1 each cycle when station 1 is active but 2 is under repair ($WRz, z < Z$). System transitions are shown in Table 3.1. Column 1 gives the possible inital state. Columns 2 and 3 indicate what, if any, changes occur in the indicated stations at the end of the cycle. The probability of these changes yields the probability column. The final column indicates the ending state based on the initial state and the associated events. We continue to assume multiple events, such as both stations failing in the same cycle, have zero probability. This assumption manifests itself in the use of the approximations in the probability column of Table 3.1. All resultant possibilities are listed.

Let S be the set of states of the system. In accord with the above system description, we can define the steady-state balance equations by applying the Chapman-Kolmogorov result:

$$P(s_1) = \sum_{s \in S} P(s)p(s, s_1) \tag{3.7}$$

where $P(s)$ is the probability of being in state s and $p(u, v)$ is the transition probability for ending in state v given that we began the cycle in state u. To develop the steady-state equations we employ the relationships in Table 3.1. To enter a resultant state from the table, we must have been in the initial state and have made the transition with the corresponding probability. Hence, we group rows of Table 3.1 with similar resultant states to obtain steady-state equations.

Consider the state $WW0$. When we observe the system, there are three possible ways to end up in this state. We could have begun in state $WW0$ and have had no failures. Thus, an item transferred from station 1 to 2 at the start of the cycle and was produced in station 2. Meanwhile, station 1 produced a new unit, which is ready to be transferred at the end of the cycle. Now, at the end of the cycle, neither station fails and we remain in state $WW0$, consuming station 2's output. Second, we could have begun in state $RW0$ and have had a repair to station 1. Beginning in state $RW0$ implies no production by either station for this cycle. Ending the cycle with a repair, however, moves us to $WW0$. Third, we could have begun in state $RW1$ and have ended the cycle with a repair. Entering this cycle in $RW1$ implies no movement of

Table 3.1   Transitions for Two-Stage Line with Buffer

Initial State	Station 1	Station 2	Probability	Resultant State
$WWx, \ 0 \le x \le Z$	Up	Up	$(1 - \alpha_1)(1 - \alpha_2) \approx 1 - \alpha_1 - \alpha_2$	WWx
	Up	Fail	$(1 - \alpha_1)\alpha_2 \approx \alpha_2$	WRx
	Fail	Up	$\alpha_1(1 - \alpha_2) \approx \alpha_1$	RWx
$RW\,0$	Repaired	Idle	b_1	$WW0$
	Down	Idle	$1 - b_1$	$RW0$
$RWx, \ 0 < x \le Z$	Fixed	Up	$b_1(1 - \alpha_2) \approx b_1$	$WW\,x - 1$
	Down	Up	$(1 - b_1)(1 - \alpha_2) \approx 1 - b_1 - \alpha_2$	$RW\,x - 1$
	Down	Fail	$(1 - b_1)\alpha_2 \approx \alpha_2$	$RR\,x - 1$
$WRx, \ 0 \le x < Z$	Up	Down	$(1 - \alpha_1)(1 - b_2) \approx 1 - \alpha_1 - b_2$	$WR\,x + 1$
	Up	Repaired	$(1 - \alpha_1)b_2 \approx b_2$	$WW\,x + 1$
	Fail	Down	$\alpha_1(1 - b_2) \approx \alpha_1$	$RR\,x + 1$
WRZ	Idle	Down	$1 - b_2$	WRZ
	Idle	Repaired	b_2	WWZ
$RRx, \ 0 \le x \le Z$	Down	Down	$(1 - b_1)(1 - b_2) \approx 1 - b_1 - b_2$	RRx
	Down	Fixed	$(1 - b_1)b_2 \approx b_2$	RWx
	Fixed	Down	$b_1(1 - b_2) \approx b_1$	WRx

product from station 1 to either station 2 or the buffer, but the part previously in the buffer is completed at station 2 and shipped out. Thus, we are about to end the cycle in state $RW0$, but then the repair event occurs at station 1. The item that was left stranded at station 1 when its failure occurred is now ready to be transferred at the start of the next cycle. Combining these possibilities, expression 3.7 becomes

$$P(WW0) = (1 - \alpha_1 - \alpha_2)P(WW0) + b_1 P(RW0) + b_1 P(RW1)$$

The remaining equations are as follows:

$$P(WWx) = (1 - \alpha_1 - \alpha_2)P(WWx) + bP(RW, x + 1) + b_2 P(WR, x - 1)$$
$$0 < x < Z$$
$$P(WWZ) = (1 - \alpha_1 - \alpha_2)P(WWZ) + b_2 P(WR, Z - 1) + b_2 P(WRZ)$$
$$P(RW0) = \alpha_1 P(WW0) + (1 - b_1)P(RW0) + (1 - b_1 - b_2)P(RW1) + b_2 P(RR0)$$
$$P(RWx) = \alpha_1 P(WWx) + (1 - b_1 - \alpha_2)P(RW, x + 1) + b_2 P(RRx) \qquad 0 < x < Z$$
$$P(RWZ) = \alpha_1 P(WWZ) + b_2 P(RRZ)$$
$$P(WR0) = \alpha_2 P(WW0) + b_1 P(RR0)$$
$$P(WRx) = \alpha_2 P(WWx) + (1 - \alpha_1 - b_2)P(WR, x - 1) + b_1 P(RRx) \qquad 0 < x < Z$$
$$P(WRZ) = \alpha_2 P(WWZ) + (1 - \alpha_1 - b_2)P(WR, Z - 1) + (1 - b_2)P(WRZ)$$
$$\qquad + b_1 P(RRZ)$$
$$P(RR0) = \alpha_2 P(RW1) + (1 - b_1 - b_2)P(RR0)$$
$$P(RRx) = \alpha_2 P(RW, x + 1) + \alpha_1 P(WR, x - 1) + (1 - b_1 - b_2)P(RRx)$$
$$0 < x < Z$$
$$P(RRZ) = \alpha_1 P(WR, Z - 1) + (1 - b_1 - b_2)P(RRZ) \qquad (3.8)$$

If we let π be the row vector of steady-state probabilities, and $P = [p(i, j)]$, the balance equations are $\pi P = \pi$ or $\pi(P - I) = 0$. P is a singular, stochastic matrix. However, states are mutually exclusive and exhaustive. Thus, any of the linearly dependent equations can be replaced by the additional constraint

$$\sum_{s \epsilon S} P(s) = 1$$

to form an independent system. We then find the state probabilities by $\pi = 0_1(P - I)^{-1}$. 0_1 is a vector of 0's except for a 1 in the location of the substituted equation.

Our primary objective is the determination of system effectiveness, that is, production rate. Production during the cycle is determined by the beginning state. During a cycle, an item is produced whenever both stations are working, or the second station only is working, but at least one part rests in the buffer. System effectiveness for a buffer of maximum size Z, E_Z, can thus be measured:

$$E_Z = \sum_{x=0}^{Z} P(WWx) + \sum_{x=1}^{Z} P(RWx) \tag{3.9}$$

Expression 3.9 could be used regardless of whether the low-probability events, such as two station failures on the same cycle, are excluded from the Markov chain model. However, for the case of their exclusion, Buzacott [1971] presents a closed-form expression for the effectiveness of this model. Let $x_i = \alpha_i / b_i$ be the ratio of average repair time to uptime. Then define $s = x_2 / x_1$, $r = \alpha_2 / \alpha_1$, and

$$C = \frac{(\alpha_1 + \alpha_2)(b_1 + b_2) - \alpha_1 b_2(\alpha_1 + \alpha_2 + b_1 + b_2)}{(\alpha_1 + \alpha_2)(b_1 + b_2) - \alpha_2 b_1(\alpha_1 + \alpha_2 + b_1 + b_2)}$$

It can then be shown that

$$E_z \begin{cases} \dfrac{1 - sC^Z}{1 + x_1 - (1 + x_2)sC^Z} & s \neq 1 \\[2ex] \dfrac{1 + r - b_2(1 + x) + Zb_2(1 + x)}{(1 + 2x)[1 + r - b_2(1 + x)] + Zb_2(1 + x)^2} & s = 1 \end{cases} \tag{3.10}$$

There are several meaningful measures of buffer usefulness. E_Z, the effectiveness with a buffer of size Z, is immediately found from expression 3.10. The gain due to the buffer G_Z is found from $G_Z = E_Z - E_0$. The reader should note that E_0 reduces to our nonbuffer expression obtained in the last section. Likewise, limiting effectiveness is found from $\lim_{Z \to \infty} E_Z = 1/(1 + x^*)$ where $x^* = \max(x_1, x_2)$. In other words, *as the buffer capacity is increased, asymptotic effectiveness approaches the capacity of the least effective station.*

EXAMPLE 3.3

Find the availability of the line in Example 3.1 if a buffer of four spaces is placed between the workstations. Also, find the availability if unlimited buffer space is available between the workstations.

Solution

The first step is to compute the model parameters used in expression 3.10:

$$x_1 = \frac{\alpha_1}{b_1} = \frac{0.1}{0.5} = 0.2$$

Likewise,

$$x_2 = \frac{\frac{1}{15}}{0.5} = 0.1333$$

Since $x_2 < x_1$, station 2 has greater natural availability. The ratio $s = x_2/x_1 = \frac{2}{3}$. Since $s \neq 1$, we must compute C:

$$C = \frac{(\alpha_1 + \alpha_2)(b_1 + b_2) - \alpha_1 b_2(\alpha_1 + \alpha_2 + b_1 + b_2)}{(\alpha_1 + \alpha_2)(b_1 + b_2) - \alpha_2 b_1(\alpha_1 + \alpha_2 + b_1 + b_2)}$$

$$= \frac{0.108335}{0.127778} = 0.847823$$

Finally, from expression 3.10,

$$E_4 = \frac{1 - sC^4}{1 + x_1 - (1 + x_2)sC^4} = 0.81$$

Thus, the addition of buffer space for four units has increased availability from 0.75 to 0.81. In every 100 cycles we will average six additional good units.

To find the maximum possible production rate, we use the result

$$E_\infty = \frac{1}{1 + \max(0.2, 0.13333)} = \frac{1}{1.2} = 0.833$$

∎

EXAMPLE 3.4

Consider a line with $\alpha_1 = \alpha_2 = 0.01$ and $b_1 = b_2 = 0.1$. Initially assume $Z = 2$. Solve for line effectiveness using both the steady-state equations and equation 3.10.

Solution

The $4(Z+1)$ steady-state equations become

$$P(WW0) = 0.98P(WW0) + 0.1P(RW0) + 0.1P(RW1)$$

$$P(WW1) = 0.98P(WW1) + 0.1P(RW2) + 0.1P(WR0)$$

$$P(WW2) = 0.98P(WW2) + 0.1P(WR1) + 0.1P(WR2)$$

$$P(RW0) = 0.01P(WW0) + 0.9P(RW0) + 0.89P(RW1) + 0.1P(RR0)$$

$$P(RW1) = 0.01P(WW1) + 0.89P(RW2) + 0.1P(RR1)$$

$$P(RW2) = 0.01P(WW2) + 0.1P(RR2)$$

$$P(WR0) = 0.01P(WW0) + 0.1P(RR0)$$

$$P(WR1) = 0.01P(WW1) + 0.89P(WR0) + 0.1P(RR1)$$

$$P(WR2) = 0.01P(WW2) + 0.89P(WR1) + 0.9P(WR2) + 0.1P(RR2)$$

$$P(RR0) = 0.01P(RW1) + 0.8P(RR0)$$

$$P(RR1) = 0.01P(RW2) + 0.01P(WR0) + 0.8P(RR1)$$

$$P(RR2) = 0.01P(WR1) + 0.8P(RR2)$$

Replacing the last dependent equation with the normalizing constraint $\mathbf{1}'\boldsymbol{\pi} = 1$ and solving the system of linear equations to obtain the state probabilities yields

$$\boldsymbol{\pi} = \begin{bmatrix} 0.3964 \\ 0.0398 \\ 0.3964 \\ 0.0753 \\ 0.0040 \\ 0.0040 \\ 0.0040 \\ 0.0040 \\ 0.0753 \\ 0.0002 \\ 0.0004 \\ 0.0002 \end{bmatrix}$$

Using equation 3.9, E_2 is

$$E_2 = P(WW0) + P(WW1) + P(WW2) + P(RW1) + P(RW2)$$

$$= 0.3964 + 0.0398 + 3964 + 0.0040 + 0.0040 = 0.841$$

Of course, we could have obtained this result with less effort by using the closed-form expression 3.10. This approach would yield $s = \alpha_2 b_1 / \alpha_1 b_2 = 1$. The average repair time to uptime ratio is $x = \alpha_i / b_i = 0.1$, and $r = \alpha_2 / \alpha_1 = 1$. Hence,

$$E_2 = \frac{1 + r - b_2(1 + x) + Zb_2(1 + x)}{(1 + 2x)[1 + r - b_2(1 + x)] + Zb_2(1 + x)^2}$$

$$= \frac{1 + 1 - 0.1(1.1) + 2(0.1)(1.1)}{1.2[1 + 1 - 0.1(1.1)] + 2(0.1)(1.1)^2} = 0.841 \qquad \blacksquare$$

3.3.2 Deterministic Failures and Repairs

To this point we have only considered exponential failures and repairs. Suppose times between failures and repair times are deterministic. To simplify the explanation, we will also assume identical stations. In this system, failures will alternate between stations. When station 1 fails, station 2 will continue to operate from the buffer until 1 is repaired or the buffer empties. If $Z \geq b^{-1}$, the cycles to repair, then station 1 worked through station 2's previous repair time, at least enough to ensure a buffer of b^{-1} parts. Accordingly, when station 1 subsequently fails, station 2 is not starved. It does however empty the buffer by b^{-1} units so that when station 2 next fails there is room in the buffer for station 1 to continue operating. Hence, no starving or blocking occurs and $E_Z = 1/(1 + x)$ for $Z \geq b^{-1}$. Figure 3.4 illustrates the status of the system through time. In the figure, we arbitrarily set $b^{-1} = 5$ and $\alpha_i^{-1} = 20$. Station 1 fails 10 cycles after station 2 the first (and each successive) time. The system repeats itself every 25 ($b^{-1} + \alpha_i^{-1}$) cycles. This observation and other empirical results have led to the general rule that buffers should normally be large enough to accommodate at least average repair time production.

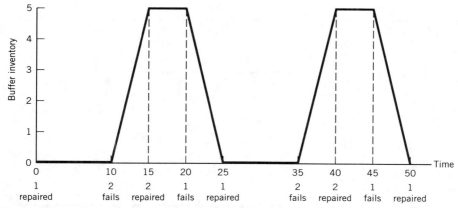

Figure 3.4   Buffer levels for deterministic up- and downtimes.

3.3.3 System Reduction

It may appear that the foregoing model is rather limited in that most systems contain more than two workstations. However, the important aspect is the existence of only one buffer. A set of stations that must be jointly active or idle can be aggregated into a single station providing they have a common repair rate. This logic holds whether the stations are all in series or if certain stations act as feeder stations for the main line. The aggregated failure rate is obtained by summing the individual failure rates, that is, solving an unbuffered line problem as seen by equation 3.4. This result is summarized by the following rule.

Rule 1: Any set of stations $i = 1, \ldots, m$ with no intermittent buffers, with failure rates α_i and repair rates $b_i = b$, can be replaced with the single station j with failure rate $\alpha'_j = \sum_{i=1}^{m} \alpha_i$ and $b_j = b$ provided that all stations must stop if any individual station fails.

Figure 3.5 shows several equivalent system designs assuming all stations have a common repair rate.

EXAMPLE 3.5

An engineer is trying to decide where to place a buffer in a four-stage production process. Each stage has the same fixed cycle time. Uptimes are unpredictable but the mean cycles between failures is about 250 for each station. Repair times average 10 cycles. The buffer will have a capacity of 20 units, since most repairs can be completed in 20 cycles.

Solution

Three possible locations exist for the buffer, namely following station 1, 2, or 3. In all cases, the repair time parameter is $b = 0.1$. Individual workstations have $\alpha_i = 0.004$. The possible buffer placements and corresponding availabilities are shown in Table 3.2. We do best by placing the buffer between stations 2 and 3. This is the middle of a four-station line.   ■

The example illustrates two important results.

Rule 2: Median Buffer Location. If only one buffer is to be inserted, it should be placed in the "middle" of the line.

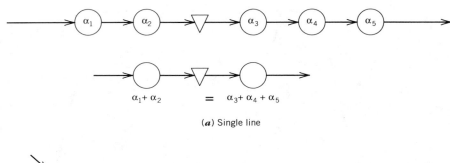

(a) Single line

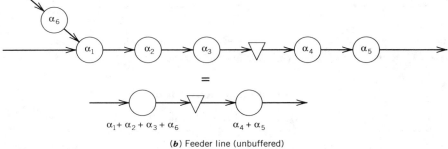

(b) Feeder line (unbuffered)

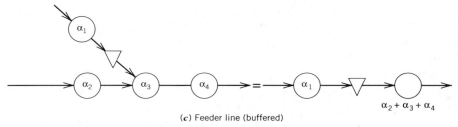

(c) Feeder line (buffered)

Figure 3.5   Equivalent transfer lines for analysis.

Consider an M-stage line with failure rates α_i and equal repair rates b. The upper bound on availability is given by the

$$\min_{1 \leq i \leq M} \left(1 + \frac{\alpha_i}{b}\right)^{-1} \tag{3.11}$$

This bound is reached when an infinite-capacity buffer is placed between each pair of workstations. The bounds are shown in the last column of Table 3.2 for the two-stage model of aggregated workstations.

Table 3.2   Possible Buffer Locations for Example 3.5

Buffer after WS	α'_1	α'_2	E_{20}	E_∞
1	0.004	0.012	0.885	0.893
2	0.008	0.008	0.895	0.926
3	0.012	0.004	0.885	0.893

Suppose we insert just one buffer and place it after workstation r. The two-stage, single buffer model of this line has $\alpha_1' = \sum_{i=1}^{r} \alpha_i$ and $\alpha_2' = \sum_{i=r+1}^{M} \alpha_i$. The upper bound $(1 + b^{-1} \max\{\alpha_1', \alpha_2'\})^{-1}$ is determined by the larger of α_1' and α_2'. This upper bound and, hence, potential output is maximized by moving the buffer to the **median** location with respect to the failure rates of the actual workstations. A median location is one for which, if r is the last station before the buffer,

$$\max\left\{\sum_{i=1}^{r} \alpha_i, \sum_{r+1}^{M} \alpha_i\right\} \le \max\left\{\sum_{i=1}^{r+1} \alpha_i, \sum_{i=r+2}^{M} \alpha_i\right\} \tag{3.12a}$$

and

$$\max\left\{\sum_{i=1}^{r} \alpha_i, \sum_{r+1}^{M} \alpha_i\right\} \le \max\left\{\sum_{i=1}^{r-1} \alpha_i, \sum_{i=r}^{M} \alpha_i\right\} \tag{3.12b}$$

Relation 3.12a states that if the buffer is moved forward one station and placed after workstation $r + 1$, then the upper bound on availability of the two-stage model, evaluated via expression 3.11, either stays the same or is reduced. Relation 3.12b gives a similar result if the buffer is placed earlier in the line.

Rule 3: Reversibility. If the direction of flow is reversed in a serial line, production rate stays the same.

This rule can reduce the number of possible line designs we need to investigate. In Table 3.2 we see that the two lines with $(\alpha_1, Z, \alpha_2) = (0.004, 20, 0.012)$ and $(\alpha_1, Z, \alpha_2) = (0.012, 20, 0.004)$ have the same availability. In terms of physical flow, each line is the reverse of the other.

3.4*
APPROXIMATE THREE-STAGE MODEL

In theory, systems can be modeled via Markov chains for any number of stages. In practice, the state space quickly becomes prohibitively large. For instance, M stages with intermediate buffers of size Z require $2^M (Z + 1)^{M-1}$ states. Several approximate approaches have been proposed for long lines. Most of these heuristics are based on the use of a sequence of two-stage models to progressively build up estimates for the entire line. We will look briefly at an approach useful for three-stage lines (two buffers).

We make the simplifying approximation that at most one station is down at a time. The probability that station i is down is given by

$$P_i = \frac{x_i}{1 + \sum_{m=1}^{M} x_m} \tag{3.13}$$

where, as before, $x_i = \alpha_i / b_i$. Equation 3.13 is most easily seen for a line without buffers. For every unit produced, that is, a good cycle, station i is down for x_i cycles. The term x_i reflects the station failure probability and repair rate. Possible system states are producing (all stations operational) and station i down, for $i = 1, \ldots, M$. Accordingly, we note that $E_0 + \sum_{i=1}^{M} P_i = 1$.

One way to determine effectiveness is by noting that production must be the same in the long run at all three stations. Consider station 2. This station produces in three types of states. First, station 2 produces when all stations are up. Station 2 also produces when station 1 is down but the first buffer allows station 2 to operate. Third, we could have station 3 down but space available in the second buffer. If we define $b_{ij}(Z_1, Z_2)$ as the proportion of time station i operates when i is under repair for the specified buffer limits, then

$$E_{Z_1 Z_2} = E_{00} + P_1 b_{12}(Z_1, Z_2) + P_3 b_{32}(Z_1, Z_2) \tag{3.14}$$

Consider the viewpoint of an observer at buffer 1 peering down the line. Upstream we see station 1. Looking downstream, the line is forced down when either station 2 fails (α_2), or 3 fails and the second buffer is full $\{\alpha_3[1 - b_{32}(Z_1, Z_2)]\}$. Now, if we had an estimate of $b_{32}(Z_1, Z_2)$, we could combine stations 2 and 3 into the pseudo station seen by the buffer and solve a two-stage line problem. We model the three-stage line as a two-stage line where stations 2 and 3 and their connecting buffer are replaced by a pseudo station 2, say $2'$, whose failure rate is

$$\alpha_{2'} = \alpha_2 + \alpha_3[1 - b_{32}(Z_2)] \tag{3.15}$$

Our reason for omitting reference to the first buffer state in equation 3.15 will become apparent shortly. Basically, our estimate of $b_{32}()$ will not depend on the first buffer.

Now, for a two-stage line we may write

$$\begin{aligned} E_Z &= E_0 + P_1 b_{12}(Z) \\ &= E_0 + P_2 b_{21}(Z) \end{aligned} \tag{3.16}$$

P_i refers to the probability that station i is down as given in expression 3.13. For our line consisting of stations 1 and $2'$, $b_{12}(Z)$ in equation 3.16 actually represents $b_{12}(Z_1, Z_2)$. Thus, given $b_{32}(Z_1, Z_2)$, we could solve the two pseudo-station line by using the methods of the previous section, and then evaluate expression 3.14 to obtain an estimate of effectiveness for the three-stage line. Of course, the problem is that we do not know $b_{32}(Z_1, Z_2)$.

Our decision to aggregate stations 2 and 3 into $2'$ was arbitrary. We could just as easily have combined stations 1 and 2 into station $1'$. Buffer 2 sees the line in front of it stop when 2 fails or 1 fails and buffer 1 is empty. Thus, for this alternative aggregation we would use

$$\alpha_{1'} = \alpha_1[1 - b_{12}(Z_1, Z_2)] + \alpha_2 \tag{3.17}$$

The second stage in this modified line has α_3 failure rate. We may solve this two-stage pseudo line, and in this instance $b_{21}(Z)$ of expression 3.16 estimates $b_{32}(Z_1, Z_2)$. Now an approach begins to crystallize. Solving either aggregation provides an estimate of the $b_{ij}()$ factor required as input to the other aggregation. By successively utilizing these two model aggregations, we can estimate performance. Eventually our estimates will solidify.

Solution Procedure

1. Initialize $b_{12}(Z_1, Z_2)$ at, say, 0.5. Denote stages 1 and 2 in any psuedo two-stage approximation as $1'$ and $2'$, respectively. Compute E_{00}, the effectiveness for the unbuffered line.
2. Solve the two-stage line with $\alpha_{1'}$ given by equation 3.17. Estimate $b_{32}(Z_1, Z_2) = b_{2'1'}(Z)$ from equation 3.16. $\alpha_{2'} = \alpha_3$.
3. Solve the two-stage line with $\alpha_{2'}$ given by equation 3.15. Estimate $b_{12}(Z_1, Z_2) = b_{1'2'}(Z)$ from equation 3.16. $\alpha_{1'} = \alpha_1$.

If suitable convergence criteria are satisfied, go to 4, otherwise go to 2.

4. Estimate effectiveness for the three-station line by

$$E_{Z_1 Z_2} = E_{00} + P_1 b_{12}(Z_1, Z_2) + P_3 b_{32}(Z_1, Z_2)$$

EXAMPLE 3.6

A 20-stage transfer line with two buffers is being considered. Tentative plans place buffers of size 15 after workstations 10 and 15. The first 10 workstations have a cumulative failure rate of $\alpha = 0.005$. Workstations 11 through 15 have a cumulative failure rate of $\alpha = 0.01$ and workstations 16 through 20 together yield an $\alpha = 0.005$. Repair of any station would average 10 cycles in length. Estimate the effectiveness of this line design.

Solution

STEP 1.
$$E_{00} = \frac{1}{1 + b^{-1} \sum\limits_{i=1}^{3} \alpha_i} = \frac{1}{1 + 10(0.005 + 0.01 + 0.005)} = 0.83333$$

Set $b_{12}(15, 15) = 0.5$.

STEP 2. Combine stations 1 and 2.

$\alpha_{1'} = \alpha_1[1 - b_{12}(15, 15)] + \alpha_2 = 0.005[.5] + 0.01 = 0.0125$

$\alpha_{2'} \equiv \alpha_3 = 0.005$. Solving we find $x_{1'} = 0.125$, $x_{2'} = 0.05$, $s = x_{2'}/x_{1'} = 0.4$. Using expression 3.10 with $s \neq 1$, we find $C = 0.951898$ and $E_{15} = 0.8751$. Then employing equation 3.16 with $P_2 = x_{2'}/(1 + x_{1'} + x_{2'})$, we find

$$b_{32}(15, 15) \approx \frac{E_{15} - E_0}{P_2} = \frac{0.8751 - 0.8511}{0.04255} = 0.564$$

STEP 3. Combine stations 2 and 3. $\alpha_{1'} \equiv \alpha_1 = 0.005$.

$\alpha_{2'} = \alpha_2 + \alpha_3[1 - b_{32}(15, 15)] = 0.01 + 0.005(0.436) = 0.01218$

$x_{1'} = \alpha_{1'}/b_{1'} = 0.05$, $x_{2'} = \alpha_{2'}/b_{2'} = 0.1218$, and $s = x_{2'}/x_{1'} = 2.436$. Solving we obtain $C = 1.04916$ and from equation 3.10 $E_{15} = 0.087740$. Now, use the result $E_{15} = E_0 + P_1 b_{12}$ to estimate $b_{12}(15, 15)$. First, $P_1 = x_{1'}/(1 + x_{1'} + x_{2'}) = 0.04267$, then $b_{12}(15, 15) \approx (E_{15} - E_0)/P_1 = 0.563$.

As our new estimate of 0.563 differs from our initial guess of 0.5, we return to step 2.

STEP 2. Combine stations 1 and 2.
Now, with

$$\alpha_{1'} = \alpha_1[1 - 0.563] + \alpha_2 = 0.01219$$

and $\alpha_{2'} = 0.005$ (α_3), we obtain $E_0 = 0.8533$, $E_{15} = 0.8773$, and with $P_2 = x_{2'}/(1 + x_{1'} + x_{2'}) = 0.042667$, $b_{32}(15, 15) \approx 0.563$. As this differs from the previous estimate of 0.564, we continue the process.

STEP 3. Combine stations 2 and 3.

Using the new estimates we have parameters $\alpha_{1'} = 0.005$ and $\alpha_{2'} = 0.01218$, which yield $E_0 = 0.8534$, $E_{15} = 0.8774$, and our updated estimate $b_{12}(15, 15) = 0.563$. As we seem to have converged, we will move to step 4.

STEP 4. Estimate 3 stage effectiveness.

$$E_{15\,15} = E_{00} + P_1 b_{12}(15, 15) + P_3 b_{32}(15, 15)$$

$$\approx 0.8333 + \frac{0.05}{1 + 0.05 + 0.1 + 0.05}(0.563)$$

$$+ \frac{0.05}{1 + 0.05 + 0.1 + 0.05}(0.563) = 0.88 \qquad \blacksquare$$

The solution of longer lines is beyond the scope of this book. Simulation models are often built for analyzing such systems. However, accurate analytical approximations are available. Suri and Diehl [1986] and Gershwin [1987] provide procedures. These approximations basically operate by decomposing the line into a set of two-workstation, one-buffer sublines. Gershwin, for instance, considers each of the $k - 1$ buffers of a k-workstation line. Each buffer is surrounded by a pseudo workstation on each side. This produces $k - 1$ two-workstation, one-buffer lines. The trick is to estimate failure rate and repair rate parameters for the upstream and downstream workstation in each two-stage problem so that these workstations behave like the entire portion of the real line upstream (or downstream) of the buffer being modeled. The two-stage models are tied together by conservation, that is, each two-stage model must have the same production rate. An iterative procedure is used to find appropriate failure and repair rate parameters. The production rate (availability) for the line is that of each of the two-stage subproblems, and the buffer utilizations of the two-stage subproblems are used as actual buffer utilizations for the corresponding real transfer line.

3.5
UNPACED LINES

The transfer lines discussed to this point were mechanized with all stations indexing parts at the same time. Many serial production systems involve workstations with random processing times. A variety of jobs may be produced, each with its own requirements at each workstation. Even if the line is balanced in terms of average work load being nearly equal for each workstation, individual jobs may have grossly inequitable time requirements from station to station. In the unpaced (asynchronous) line, each workstation acts independently. As soon as its assigned tasks are completed, the workstation attempts to pass its part along to the next workstation. If the successor station is idle or buffer space exists, the part is passed along; otherwise, the station becomes blocked. Once the part is passed, the workstation checks its input buffer. If a part is available, it begins working; otherwise, it is starved.

Conway et al. [1988] examined serial production systems. The results stated in this section rely heavily on that study. Several scenarios were investigated. Simulation (see Chapter 12) was used except for a few simple cases that could be solved analytically. We first look at cases where workstations do not fail. To put this in perspective, this corresponds to the section on unpaced lines in Chapter 2. We then consider unreliable workstations. A number of other authors have also examined serial lines with buffers. Many of these models assume serial systems with random workstation processing times. Buzacott and Hanifin [1978] review early results. Gershwin and Berman [1981] allow workstation failure probability to depend on service time. Altiok and Stidham [1983] consider allocation of buffer capacity to systems with more general service time distributions. Altiok and Perros [1986] consider splitting and merging into parallel stations as units pass down the line. These models lead us to the more general queueing networks to be discussed in Chapter 11.

3.5.1 Identical Workstations, Random Processing Times, No Buffers, No Breakdowns

We start by considering a serial production system with M identical stages. By identical we mean that all workstations have the same processing time distribution. The processing times for each stage of each job are assumed to be independent (and identically distributed). Workstations do not break down and no buffer space exists. Thus, workstations are starved until the upstream station finishes its operation. Likewise, workstations are blocked until the downstream station finishes and is able to pass on its job. An important determinant of throughput in this situation is the coefficient of variation of processing time (cv = standard deviation/mean). Throughput decreases as the number of stations increases, but levels off quickly. For instance, with $cv = 0.3$ a two-stage line has only 85 percent of the throughput of a single-stage line, but even at eight or more stages throughput is maintained at 75 percent of the single stage. Beyond five or six stages, further increases in the number of stages result in only minor reductions in throughput. For $cv = 0.1$, throughput levels off at 90 percent of single-stage production. For variability levels of $cv = 0.5$ and 1.0 (exponential), throughput levels off at about 65 percent and 45 percent of single-stage output respectively. These findings (see Conway et al. [1988]) are illustrated in Figure 3.6.

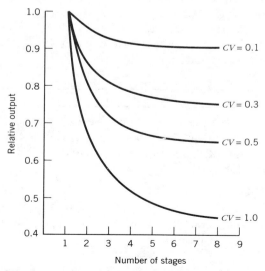

Figure 3.6   Effect of random processing time in balanced, unbuffered lines.

EXAMPLE 3.7

A proposed serial processing system will have five stages. Each station will require an exponentially distributed service time with mean 10 minutes. If no buffers are used, find the production rate for the line. Supposing service time variability could be reduced to a standard deviation of one minute (nonexponential), what would then be the production rate?

Solution

The exponential distribution has a coefficient of variation of 1.0. Using Figure 3.6 with five workstations, we find a value of approximately 0.5. Thus, instead of one unit being produced every 10 minutes, for a production rate of 0.1 per minute, we have production rate equals 0.1(0.5) or 0.05 units completed per minute, or 3 per hour.

If the process could be improved such that $cv = \frac{1}{10}$, then Figure 3.6 predicts a factor of about 0.9. Thus, production rate is 0.1(0.9) or 0.09 units completed per minute. Output is 5.4 units per hour, a significant increase.   ■

3.5.2 Identical Workstations, Random Service, Equal Buffers, No Breakdowns

Suppose buffers of the same capacity are placed between each pair of successive workstations. Since the first workstation is never starved, it will tend to keep the first buffer full. Since the last workstation is never blocked, it will tend to keep the last buffer empty. In general, buffer utilization decreases from the front to the rear of the line with the middle buffer being about half full on the average.

Throughput is dependent on the ratio of buffer capacity (Z for each buffer) to processing time cv. The percentage of the capacity lost in the unbuffered line that is recovered by adding buffers is only marginally dependent on line length. Figure 3.7 gives the average recovery proportion. For $Z/cv = 10$, 80 percent of capacity lost because of variability of processing times is recovered. For $Z/cv = 20$, about 90 percent is recovered. Thus, given cv we can find the capacity lost because of variablility using Figure 3.6. Then, given Z, we use Figure 3.7 to find the portion of this loss that is recovered by buffering.

EXAMPLE 3.8

A serial system has six balanced stages, but whereas mean job processing time is 15 minutes, times vary from 5 to 25 minutes in each workstation. Estimate throughput for the line and determine the increase possible from inserting buffers between workstations.

Solution

Assuming times are uniformly distributed between 5 and 25 minutes, the standard deviation of processing time is

$$\sigma = \frac{(\text{max} - \text{min})}{\sqrt{12}} = \frac{25 - 5}{\sqrt{12}} = 5.77 \text{ minutes}$$

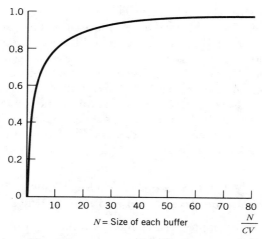

Figure 3.7   Proportion of lost output recovered by buffering in balanced lines.

This gives a coefficient of variation of $cv = \sigma/\mu = 5.77/15 = 0.38$. Interpolating between the results for a cv of 0.3 and 0.5, we estimate availability to be about 70 percent of that for a single station line. At 1 job per 15 minutes, the single station line would average 4 jobs per hour. We therefore estimate $4 \cdot 0.70 = 2.8$ jobs per hour for the six-stage line without buffers.

Buffers of size $10 \cdot cv = 3.8 \approx 4$ will recover 80 percent to 85 percent of the 1.2 jobs lost per hour. This would give an increase of about 1 job per hour. This increases to about 1.1 job per hour extra (3.9 jobs per hour total) for buffers of size $20 \cdot cv = 20(0.38) \approx 8$.   ∎

Blumenfeld [1990] combined the results of prior theoretical and simulation studies into an approximate expression for the production rate of serial lines with identical random processing time distributions at each stage and identical buffer capacities between each pair of stages. Let T be the mean service time at each workstation. The production rate is given by

$$X = T^{-1}\left[1 + \frac{1.67(M-1)cv}{1 + M + 0.31cv + 1.67MZ/(2cv)}\right]^{-1} \qquad (3.18)$$

As an example, with $Z = 4$ in Example 3.8 we obtain

$$X = 4\left[1 + \frac{1.67(5)(0.38)}{1 + 6 + 0.31(0.38) + 1.67(6)(4)/[2(0.38)]}\right]^{-1} = 3.80 \text{ jobs/hour}$$

This agrees with our earlier estimate. Likewise, equation 3.18 estimates 2.8 jobs per hour for $Z = 0$ and 3.9 jobs per hour for $Z = 8$.

Expression 3.18 assumes $M - 1$ buffers, each of size Z. Earlier in the chapter we noted that if we can only place one buffer, the optimal location is the center of the line such that the aggregation of upstream workstations has the same availability as the downstream aggregation. Suppose it is possible to insert more than one buffer, but total buffer capacity is fixed. Are we better off with one large or several small buffers? It turns out that for lines with identical workstations, the best allocation is many buffers of nearly equal size. The largest buffers should be in the middle, but the difference in size between the largest and smallest buffers should be no more than one slot. Buffer sizes should be symmetric from the center moving to the front and rear of the line. For instance, if we have a four-station line (three buffers) and 10 buffer slots, then the first and third buffers should have a capacity of three and the middle buffer should have a capacity of four. As stations become unequal, the less reliable stations should have larger input and output buffers. As you might suspect, buffers are less useful in unbalanced lines. Slow, bottleneck workstations rarely are starved or blocked precisely because the surrounding workstations outperform them. However, if breakdowns or high processing-time variability occurs at a workstation other than the bottleneck, then input and output buffers at the bottleneck can be important for maintaining its utilization.

3.5.3 Constant Processing Times, Random Breakdown and Repair Times

Even in a balanced, constant-processing-time environment, it may be worthwhile to employ asynchronous part transfer. Because of station breakdowns and random repair times, stages can lose synchronization. Buffers allow workstations to start production cycles independently. In Section 3.3 we noted that buffers should be at least large enough to hold average repair-time production. Once again, assume identical workstations and buffers between every pair of adjacent stations. Conway et al. discovered that the increase in throughput is largely determined by the ratio $Z \cdot b / (1 + cv_R^2)$, where cv_R is the coefficient of variation of the repair time distribution and b^{-1} is average repair time measured in processing time cycles. It should not be surprising that the term Zb is important; this is the size of the buffer measured in multiples of average repair time. Table 3.3 summarizes the approximate relationship. Improvement is measured by proportion of possible improvement gained from the buffer, that is, $(E_Z - E_0)/(E_\infty - E_0)$.

EXAMPLE 3.9

A 10-workstation line (9 buffers) must produce 900 parts per 16-hour day. Processing times are constant at 1 minute per station. The predominant cause of station failures is tool breakage, which happens about once every 1000 cycles for each station. It takes 12 minutes to replace a broken tool. Find the required buffer sizes.

Solution

Our basic time unit is a 1-minute cycle. For each station $\alpha_i = 0.001$ and $b_i = \frac{1}{12}$ First, let us check to see if the requirement is feasible. With infinite buffers, $E_\infty = 1/(1 + \alpha b^{-1}) = 0.988$. At 16 hours of 1-minute cycles (960 cycles per day), this would yield 949 good parts per day. Since this exceeds the required 900 parts per day, a feasible design will exist.

Perhaps buffers are not needed at all. Let's check output for the unbuffered line. From equation 3.5, $E_0 = 1/[1 + 10(0.001)12] = 0.893$, which translates to $0.893 \cdot 960 = 857$ good parts per day.

We must determine the proportion of the gap between zero and infinite buffer output that is needed. This proportion is $(900 - 857)/(949 - 857) = 0.47$. From Table 3.3 we see that a buffer satisfying $Z \cdot b / (1 + cv_R^2) = 1$ supplies 50 percent of possible increase and will suffice (0.50 is close but conservative compared to the 0.47 required) Thus, $Z = b^{-1}(1 + cv_R^2) = 12(1 + 0) = 12$. ∎

**Table 3.3   Effect of Buffers with
Random Failures and Repair Times**

$\dfrac{Z \cdot b}{1 + cv_R^2}$	$\dfrac{E_Z - E_0}{E_\infty - E_0}$
0	0.0
0.25	0.25
0.5	0.35
1	0.5
2	0.7
4	0.8
8	0.9

The impact of cv_R is important. An exponential repair distribution has $cv_R = 1$. Compared to deterministic repair times, buffers must be twice as large to obtain the same increase in output!

3.5.4 Buffers and Production Control

Let us summarize what we have learned thus far about inventory levels in buffers. If the portion of the line in front of a buffer has significantly higher availability than the portion of the line behind the buffer, then the buffer will tend to be full. If the front portion of the line has significantly lower availability, the buffer will tend to be empty. The larger the discrepancy in availabilities, the closer the average buffer level is to being full or empty. As the size of the discrepancy between upstream and downstream availabilities diminishes, the average buffer level moves toward one-half buffer capacity. Now a buffer's being full or empty does not make it useless; it may be important to ensure that the bottleneck resource (the section of the line with smallest capacity) will not be starved or blocked on the rare occasions that it gets ahead of the rest of the line. A frequent practice has been to design serial production lines to have equal capacity at each workstation. However, if the cost of capacity differs between workstations, and throughput time or WIP levels are important, it is possible that an unbalanced line is preferable! Allocating extra capacity to less expensive workstations reduces the system WIP levels needed to obtain a given production rate.

Just-in-time (JIT) production control systems have received much credit recently for reducing inventory levels as compared with materials-requirements-planning (MRP) systems. JIT is often referred to as a pull system. Workstations have a limit on the size of their output buffers. If the buffer is full, the workstation ceases work (as we would say, the workstation is blocked). Pull refers to the fact that workstations react to the pulling of stock by the successor workstation. Several authors (Wang and Wang [1990] and Mitra and Mitrani [1990], for instance) have modeled JIT systems as a series of finite buffer queues.

MRP is considered a push system. A higher level planning model determines how much each workstation should produce during each period. Unfortunately, the MRP system does not react in real time when downstream stations break down. Instead, upstream stations continue to operate according to the plan for that period, and inventory levels (buffer stocks) can accumulate. If we can sell all the product we can make, then MRP's implicitly unlimited buffer capacity is good for the bottleneck station, but still wasteful for other stations. If sales are not limited by capacity, allowing large buffer stocks makes virtually no sense. As we know, large buffers increase throughput time, space requirements, material handling costs, and scrap/rework costs. The latter costs increase since quality problems are temporarily hidden in idle inventory, allowing out-of-control processes to continue. In light of our findings in this section, however, we should note that an MRP system with short planning periods and real-time knowledge of workstation operational and inventory status can effectively control the amount of work released to the shop floor and function as well as or better than a JIT system. MRP has the advantage of looking ahead and planning for changes in demand. If the MRP system also has limited buffers such that workstations can be blocked, then the MRP push system will behave like the JIT system for limiting inventory accumulations. Much of the observed improvements from using JIT systems come about from the efforts usually made to reduce setup time. We will examine this effect more closely in Chapter 6, but suffice it for now to say that lower setup times permit smaller batch sizes. Smaller batch sizes mean smaller processing times. Smaller processing times translate directly into smaller waiting times at machines. Since throughput time drops, the value of WIP must drop as well.

3.6
SUMMARY

Transfer lines are serial production systems that are usually capital intensive and subject to random failures. Failures can be caused by time or effective production cycles and can affect the entire line or just a single station. Basic reliability models can be used to estimate system effectiveness when buffers are not provided. Buffers allow partial independence between stages in the line, thus improving line effectiveness against station failures. Markov chain models can be used to determine the increase in output from a single buffer. Accurate determination of output for a general line with many buffers is a more difficult problem.

Empirical results have been compiled to aid in the design of serial lines. For random processing times but reliable workstations, knowledge of the coefficient of variation allows estimation of the production rate. If workstations can fail, average repair time and the coefficient of variation for repair time are needed to estimate the effect of buffers.

Many real systems are of an assemblylike nature with feeder lines combining to final assembly. Such systems are harder to evaluate because of the rapid growth in the size of the state space of the discrete-time Markov chain. Simplifying assumptions and the use of continuous supplementary variables can, however, lead to performance estimates.

REFERENCES

Altiok, Tayfur, and H. G. Perros (1986), "Open Networks of Queues with Blocking: Split and Merge Configurations," *IIE Transactions*, 18(3), 251–261.

Altiok, Tayfur, and Shaler Stidham, Jr. (1983), "The Allocation of Interstage Buffer Capacities in Production Lines," *IIE Transactions*, 15(4), 292–299.

Anderson, D. R., and C. Moodie (1969), "Optimal Buffer Storage Capacity in Production Line Systems," *International Journal of Production Research*, 7(3), 233–240.

Barlow, Richard E., and Frank Proschan (1975), *Statistical Theory of Reliability and Life Testing*, Holt, Rinehart and Winston, New York.

Blumenfeld, D. E. (1990), "A Simple Formula for Estimating Throughput of Serial Production Lines with Variable Processing Times and Limited Buffer Capacity," *International Journal of Production Research*, 28(6), 1163–1182.

Buzacott, J. (1967), "Automatic Transfer Lines with Buffer Stacks," *International Journal of Production Research*, 5(3), 183–200.

Buzacott, J. (1971), "Methods of Reliability Analysis of Production Systems Subject to Breakdowns," in *Operations Research and Reliability*, Daniel Grouchko, ed., Gordon and Breach Science Publishers, New York, 211–217, 222–232.

Buzacott, J., and L. Hanifin (1978), "Models of Automatic Transfer Lines with Inventory Banks: A Review and Comparison," *AIIE Transactions*, 10, 197–207.

Conway, R., W. Maxwell, J. O. McClain, and L. J. Thomas (1988), "The Role of Work-in-Process Inventory in Serial Production Lines," *Operations Research*, 36(2), 229–241.

Gershwin, S. B. (1987), "An Efficient Decomposition Method for the Approximate Evaluation of Tandem Queues with Finite Storage Space and Blocking," *Operations Research*, 35(2), 291–305.

Gershwin, S. B., and O. Berman (1981), "Analysis of Transfer Lines Consisting of Two Unreliable Machines with Random Processing Times and Finite Storage Buffers," *AIIE Transactions*, 13(1), 2–11.

Mitra, Debasis, and Isi Mitrani (1990), "Analysis of a Kanban Discipline for Cell Coordination in Production Lines," *Management Science*, 36(12), 1548–1566.

Muth, Eginhard (1973), "The Production Rate of a Series of Work Stations with Variable Processing Times," *International Journal of Production Research*, 11(2), 155–169.

Ohmi, Takayoshi (1981), "An Approximation for the Production Efficiency of Automatic Transfer Lines with In-Process Storage," *AIIE Transactions*, 13(1), 22–28.

Simon, J. T., and W. J. Hopp (1991), "Availability and Average Inventory of Balanced Assembly-Like Flow Systems," *IIE Transactions*, 23(2), 161–168.

Suri, R., and G. Diehl (1986), "A Variable Buffer-Size Model and Its Use in Analyzing Closed Queueing Networks with Blocking," *Management Science*, 32(2), 206–224.

Wang, Hunglin, and Hsu-Pin Wang (1990), "Determining the Number of Kanbans: A Step Toward Non-Stock Production," *International Journal of Production Research*, 28(11), 2101–2115.

PROBLEMS

3.1. A four-station transfer line is to be designed without buffers. Demand dictates that 100 products must be produced every hour the line is in operation. A cycle time of 0.5 minute is being considered. Experience with similar lines has indicated that each station will fail at random about once every 500 cycles. Downtime will average about 6 minutes per station failure. Station failures are independent. The line must also be down for an average of 5 minutes every hour of scheduled time for maintenance, loading, and adjustment. Determine the reliability of this transfer line. Will the goal of 100 units per hour be met on the average?

3.2. A new transfer line is being planned. From experience, the line designer believes workstation reliabilities can be divided into three types of stations. All failures are operation-dependent station failures. Type 1 stations fail about every 1000 cycles and average 10 minutes for repair. Type 2 stations fail every 500 cycles on average and take 5 minutes to repair. Type 3 stations fail only once in about 100,000 cycles but take on the order of 1 hour to repair. The current line design has 6 type 1 stations, 5 type 2 stations, and 20 type 3 stations. No buffers are planned. Cycle time for this design is 0.5 minute. Estimate the line's effectiveness. State your assumptions.

3.3. Consider a two-stage transfer line with intermediate buffer. Let failure and repair rates be geometric. However, assume that only one repairperson exists. Thus if both stations are down, only one can be worked on at a time. Assuming that buffer capacity size is 1 and the worker always repairs station 1 if it is broken (station 1 has repair priority), give the steady-state balance equations. In practice, what rule would you suggest be used to assign repair priorities when both stations are broken? Justify your rule as best you can.

3.4. A three-stage transfer line has a buffer with capacity of 5 located after station 2. Station failures are independent. Each station fails on the average once every 100 cycles and each station averages 5 cycles for repair. Find the effectiveness of this transfer line. How much will output increase if the capacity of the buffer is doubled?

3.5. Consider an eight-stage transfer line. Each stage fails about once every 1000 cycles. Repair times average 12 cycles. A feeder line of subassemblies merges into the main transfer line at station 5. The feeder line has six stations and these stations individually fail about once every 600 cycles. Repairs also average 12 cycles for the feeder line.

a. Find the effectiveness of this line.

b. What is the maximum effectiveness that could be obtained by adding buffers between every stage in both the main and feeder lines?

c. What is the maximum effectiveness that could be obtained by adding a buffer between the main and feeder line?

d. Suppose a buffer of size 5 was provided between the main and feeder lines. Find the line effectiveness in this case. What proportion of the time that the main line is down will the feeder line continue to produce?

3.6. A five-station transfer line is being considered. All failures are expected to occur at workstations and be operation dependent. Average repair time will be 5 cycles for each station. Average failure rates are estimated to be 0.01, 0.02, 0.02, 0.03, and 0.02, respectively.

a. Compute the effectiveness of the line if no buffers are used.

b. Suppose one buffer of size 10 is to be added. Where should it be placed?

c. Compute the effectiveness of the line with the buffer included.

3.7. Consider a three-stage transfer line with buffers between each pair of stages. Stage i has failure rate α_i and repair rate b_i. The maximum buffer sizes are Z_1 and Z_2, respectively. You may assume geometric failure and repair rates and ample repair workers.

a. How many states are there for the system?

b. Consider state $(RWWz_10)$ where $0 < z_1 < Z_1$. Write the balance equation for this state.

c. Give an expression for the effectiveness of the system in terms of state probabilities.

3.8. An engineer is designing a transfer line. The line has 15 stations. Stations 1 through 10 should have the same reliability. Each station is expected to operate approximately 1000 cycles between failures. The last 5 stations are expected to break down about once every 600 cycles. Repair times will vary but should average about the equivalent of 12 cycles in duration.

a. Find the availability of the line.

b. Suppose you could add one buffer of size 20. Where would you place this buffer? Find the increase in availability.

3.9. Using the state probabilities derive an expression for the average number of parts in the buffer for a two-stage transfer line separated by a buffer. Use the notation given in the chapter.

3.10. Equation 3.5 gives the probability that all workstations are operational. If sufficient repair stations are available and buffer size is 0 in equation 3.10, then these two equations should agree. Show that this is the case.

3.11. Destructive Failures. Suppose that when a failure occurs at a workstation, the product is destroyed with probablility D. The part remains on the line but is rejected when it exits the final workstation.

a. Derive an appropriate alternative to equation 3.5 for line availability in this case.

b. Describe how the state space of two-stage lines with buffers (Section 3.3.1) would be changed in the presence of destructive failures.

3.12. Show that equation 3.5 still holds for availability even if repair times are dependent on uptimes. Availability is defined as

$$E = \frac{E(\text{uptime})}{E(\text{uptime} + \text{downtime})}$$

3.13. Suppose that data on station failures consists of clock times at which failures occurred and a notation as to which station failed. Uptimes are found by subtracting consecutive failure times. These uptimes include the time to repair the station. Actual repair times are unknown, but the repair crew thinks that they can estimate average repair time (b^{-1}). Assume all failures are time dependent.

a. Suppose that the line contains only one workstation. Justify the use of the single-station effectiveness measure $E = (1 - \alpha/b)$ for this case, that is, show that

$$\frac{1}{1 + \dfrac{\alpha'}{b}} = 1 - \frac{\alpha}{b}$$

where $1/\alpha$ is average time between failures including repair time, and $1/\alpha'$ is average time between failures excluding repair time. [*Hint:* $1/\alpha = 1/\alpha' + 1/b$.]

b. Give an appropriate expression for the effectiveness of an M station line. Given the et of failure times, failed station indicators, and average repair time, explain how you would estimate the parameters in your expression.

Table 3.4   Historical Failure and Repair Records (Cycles)

WS1		WS2		WS3	
Up	Repair	Up	Repair	Up	Repair
178, 56	12, 13	133, 120	45, 30	467, 30	60, 10
34, 1198	100, 4	127, 973	12, 5	1231, 238	5, 12
119, 23	12, 5	326, 146	45, 7		
21, 54	4, 43	30, 13	9, 14		
113, 26	32, 2	98	43		
2, 67	21, 1				
74, 1	18, 4				

3.14. Table 3.4 contains historical records on cycles to failure and repair times for a three-station transfer line. The data represent one week of activity. The three workstations incurred 14, 9, and 4 failures, respectively.

a. Construct a histogram for each set of data. Do the assumptions of geometric failure and repair rate models appear reasonable?

b. Estimate line effectiveness. (No buffers are used.)

3.15. A processing line has 10 workstations. Each station has a unit processing time of C time units. Workstations fail once every 1000 cycles on the average. Repairs take about 5 cycles.

a. Estimate the availability of the line if no buffers are used.

b. Estimate line availability if a buffer of size 10 is placed after station 5.

c. Estimate line availability if a buffer of size 10 is placed after station 2.

3.16. Repairmen can fix a failed workstation in 10 minutes on the average. Cycle time is 20 seconds. Stations seem to operate 250 cycles on average before failing. Estimate the daily (8-hour shift) production for a six-station line. What is the average daily workload of the repairman?

3.17. An engineer is designing an automated, paced assembly system. The line will have four workstations. Mean cycles to failure are estimated to be 100, 200, 100, and 50 cycles, respectively. Repair times should average 8 cycles.

a. Assuming no buffers, find line availability.

b. A buffer of size 5 would be profitable if availability increased by at least 0.04. The buffer could be located after station 2 or 3. Which buffer location is best? Should the buffer be included?

3.18. Suppose that workstation repair rates vary from station to station. Develop a definition of a median location for buffer placement that does not require $b_i = b$ for all i.

3.19. An assembly line consists of 10 reliable stations. However, unit processing times vary randomly at each workstation with a standard deviation equal to 20 percent of average cycle time. If the line is operated asynchronously, what proportion of output is lost due to processing time variability?

3.20. A paced assembly line has a cycle time of 3.0 minutes. The line has eight workstations. Each station has a 1 percent chance of breaking down in any cycle. Repairs average 12 minutes.

a. Estimate the number of good parts made per hour.

b. If a buffer of 10 units is provided between the fourth and fifth station, how many extra parts will be made per hour (on the average)?

3.21. Using the approximate relationships of Section 3.5, show that buffers must be twice as large in the presence of exponential repair times as compared to constant repair times to achieve the same throughput.

3.22. A transfer line is composed of workstations that fail once every 5000 cycles on average and take an average of 20 cycles to be repaired. Construct a graph showing line availability as a function of the number of workstations. No buffers are to be used.

3.23. A serial production line has 20 workstations. Cycle time is constant, but stations occasionally fail. Repairs take an average of 10 cycles. Construct a graph of line availability as a function of α. Let α vary from 0.1 to 1.0 and assume the failure rate is the same for all stations.

3.24. Consider a two-stage transfer line. Stations are identical with $\alpha = 0.001$ and $b = 0.1$. The current buffer size is 0. For the same cost, the system can be modified to reduce α by 0.0001, or an additional buffer capacity of 10 can be installed. Which do you prefer? Suppose that after the first improvement, the same option was again available for continued improvement. Would your decision change? If the process continued, would your decision ever change?

3.25. A processing line is being designed. There will be five stages with deterministic processing times of (5.0, 4.0, 3.5, 6.0, 4.5). Stages are expected to be 100 percent available, that is, no breakdowns. No precedence relations exist to constrain the order in which stages are sequenced.

    a. Show that the production rate is unaffected by the ordering of process stages.

    b. Show that the time spent on the line by a part is minimized by placing the slowest station first.

    c. Is there any advantage to adding buffers? Why or why not?

3.26. A six-stage, serial processing system has reliable workstations but random processing times. The unpaced line has mean processing time of 30 minutes at each workstation. The standard deviation of processing times is about 9 minutes in each station. Find the production rate of the line while it is operating.

3.27. Four automatic insertion machines are set up in series, without intermediate buffers, to add components to printed circuit boards. Each machine inserts 20 different component types. Due to board design and component requirements, automatic setup and insertion times for machines vary from about 2 minutes to 6 minutes per workstation. Assuming that machines do not fail, estimate the number of boards produced by this line per hour while the machines are all operational.

3.28. Suppose in Problem 3.27 that a buffer space for up to two boards is added between each pair of

insertion machines. Estimate the production rate per hour.

3.29. A 25-station automated assembly system is being designed to produce the frame of a toy truck. The operations are mainly bending, punching, and spot welding. Cycle time is 10 seconds. Average cycles to failure for workstations is 2000. Workstation repair times are exponential with mean of 5 minutes.

    a. Find the production rate of the line if no buffers are added.

    b. Find the production rate of the line if unlimited buffers are placed between each pair of workstations.

    c. Find the production rate if a buffer of capacity 20 is placed between each pair of workstations. What effect will the presence of such buffers have on throughput time for truck frames?

3.30. Suppose in Problem 3.29 that we decided to use 4 buffers of size 120 instead of 24 buffers of size 20. Where should the buffers be placed? What will be the production rate?

3.31. A mixed-model assembly line has eight stations. Average cycle time is 15 minutes, but times vary due to model type and natural randomness. The estimated standard deviation of processing time in a station is 4.5 minutes. Estimate the production rate for the asynchronous line if no buffer is allowed.

3.32. A serial assembly process has 10 workstations. The line is paced and stations fail an average of once every 10,000 cycles. Repair times average 25 cycles.

    a. Estimate system availability. You may assume that operational and repair times are exponentially distributed.

    b. Estimate system availability if a buffer of capacity 10 is inserted after the fifth workstation.

    c. Suppose a buffer of capacity 50 units is placed between every pair of workstations. Estimate system availability.

CHAPTER 4

SHOP SCHEDULING WITH MANY PRODUCTS

*"I mean that part where you say how the First Industrial
Revolution devalued muscle work, then the second one
devalued routine mental work.... That would be the third
revolution, I guess—machines that devaluate human
thinking."*
—Kurt Vonnegut, Jr., *Player Piano*

4.1
INTRODUCTION

In the two previous chapters we were concerned with serial or assembly flow systems. All products followed the same route. In addition, production was repetitive with a small number of similar products being manufactured in high volume. In this chapter, we examine systems that manufacture a variety of different products. Products may or may not follow the same route, but relative processing times at workstations vary significantly. Products are released to the production system in **batches** of one or more parts. Batches will also be referred to as jobs. For each job, the production sequence and processing time at each workstation in the sequence are assumed known.

If all batches visit the same sequence of workstations, the system is called a **flow shop.** In a **job shop,** on the other hand, each part type has its own route. These individual routes may be carefully planned by an experienced process planner; however, the total mix of part batches results in a virtually structureless overall material flow pattern.

Because of the variety of product requirements, job shops must be designed for maximum flexibility. Batches must be free to move between any pair of workstations and, normally, to be processed in any order. Individual workstations must be capable of performing a wide variety of tasks. Expertise tends to be process related as opposed to product related, since products may have little or no relationship to one another. Accordingly, job shops tend to be physically and administratively organized by processing function. Milling machines, for instance, form one department and lathes another. A product batch flows between departments, visiting perhaps only one machining center in each department along its route. Throughout most of this book we argue against such configurations. Indeed, perhaps the single most agreed-on principle of manufacturing is that process layouts are less efficient than product layouts. One widely observed phenomenon is that while in a job shop, jobs spend 95 percent of their time in nonproductive activity. Much of this time is spent waiting in queue. The remaining 5 percent is split between lot setup and processing. Unfortunately, many facilities do not produce high enough volumes of specific parts to justify product layouts. Random batch arrival rates and processing times mean variability in resource requirements through time. If we wish to keep down capital investment and maintain high levels of machine utilization, queueing theory leaves no doubt that long average waiting times for resources will occur. The complexities of scheduling and engineering change orders in this environment exacerbate the problem. Chapter 5 discusses the use of group technology to classify product similarities that will allow medium volume/variety facilities to implement multiple product layouts. However, even this is of little help to the pure custom job shop, where all jobs differ significantly. Customers may demand specific features and practices that prevent exploitation of job similarities. Thus, environments exist in which job shops are necessary.

Throughout time, the time between when a job is released to the shop and when it is completed and ready for delivery, is an important measure for production facilities. From Little's Law we know that WIP inventory is proportional to throughput time with the production rate of jobs being the proportionality constant. Since WIP costs money and occupies space and other resources, we desire just enough WIP to allow workstations to be scheduled effectively. The term "effective" here refers to avoiding machine starvation that results in late delivery of product. Thus our objective should be to find the minimum WIP plan that permits on-time deliveries.

Throughput time is composed of processing time, setup time, move (material handling) time, plus waiting time. Waiting time includes the time spent waiting for machines, waiting to be moved, and waiting for parts that are also needed for the next operation. Processing time per unit is normally fixed by machine capabilities. However, smaller setup times allow smaller economic batch sizes and thus shorter batch processing times. The majority of throughput time tends to be waiting time. Continuous or high-frequency material handling can reduce move wait time. Small batch sizes (high batch processing rates) result in short wait times at machines. In general, actions that reduce the variability in the system allow resources to maintain effective utilization levels without excessive inventory. Variability in workcenter queue lengths, machine availability, and process yield all require that slack time be added to lead time estimates which, in turn, increase WIP levels. Reliable, paced assembly lines do not require inventory between workstations. Even in less repetitive situations, work should be released to the shop to level work loads through time.

Recognizing the importance of throughput time, queueing theory justifies process layouts for a nonrepetitive, highly variable environment. When interarrival and processing times vary, a single group of multiple servers, corresponding to a process layout with machines of a common type grouped together, is more efficient than parallel servers, each with its own queue. The idea of parallel servers with individual queues is similar to having separate serial production lines for different types of products. Another choice may be between general purpose machines, capable of performing many or all of the required operations on a part, and use of a serial line of special purpose machines, each machine intended to perform a subset of the part operations. We will address these issues in more detail in Chapter 11 but, for now, consider the two configurations shown in Figure 4.1. To gain insight into system behavior, we will assume a steady-state environment where jobs arrive randomly (exponential interarrival times) at a rate of 25 per day. Processing times are also exponentially distributed with a mean of 0.1 machine days. With arrival rate $\lambda = 25$, service rate $\mu = 10$, and $c = 4$ servers, Poisson queueing theory (an $M/M/c/\infty$ model, see Table 11.1 for details) tells us that job shop configuration a will have an average of 3.0 jobs in process and job throughput time will average 0.12 days. If, instead, we assign one-fourth of the workload to each machine and use the serial configuration b, we will see in Chapter 11 that the system behaves like a serial network of $GI/G/1$ queues; we will have 3.5 jobs in the system on average, and throughput times will average 0.14 days. In both instances machines are busy five-eighths of the time, but the use of general machines performing the complete job is clearly preferred. The difference would be much greater if job requirements could not be divided into four equal subtasks. Why then were assembly/transfer lines so useful in the last two chapters? Remember, in those chapters processing times were basically constant as opposed to the exponential assumption used here. If jobs arrived precisely every twenty-fifth of a day and could be divided into four equal subtasks, throughput times would drop to 0.10 days as queueing disappeared in configuration b. As utilizations increase, the differences between the operating systems become more exaggerated. Suppose $\lambda = 38$. With constant interarrival and processing times, the serial line still manages fine, since cycle time permits 40 jobs per day. Jobs never wait and throughput time is still 0.1 days. However, suppose the time between job arrivals is exponentially distributed, as is their processing time. The four parallel servers of configuration a will yield an average throughput time of 0.55 days. The serial line of configuration b will yield an average throughput time of 0.72 days. For repetitive manufacturing, the reader should recall other advantages as well, such as lower skill requirements, faster learning, and setup elimination.

Now that we have seen that job shops have a place in manufacturing, the remainder of this chapter contains some simple deterministic models and basic scheduling concepts. We begin with the issue of when to release individual orders to the production facility. We then look at how to sequence jobs at a single workstation and eventually consider scheduling jobs through the entire facility.

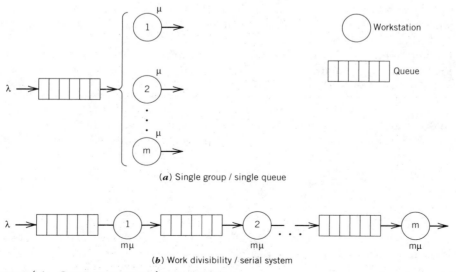

(*a*) Single group / single queue

(*b*) Work divisibility / serial system

Figure 4.1   Group versus serial organization.

4.2
ORDER RELEASE

Suppose we are assigned the task of releasing jobs to the shop. From a list of pending orders we must select the time to begin processing. The shop manager's objective is to keep all machines busy. The sales department wants to ensure that all customer due dates are met. In either case, it is helpful to know the rate at which released jobs will progress through the shop. The simplest approach is to use average workstation delay times. We will describe this simple approximation by an example.

EXAMPLE 4.1

Consider the product shown in Table 4.1.

Table 4.1   Process Plan for Example 4.1

Operation	Workcenter	Setup Time (days)	Processing Time (days)
1	A	0.5	1.2
2	C	0.1	0.6
3	F	1.0	3.4
4	D	0.3	1.2

Waiting times at workcenters (A, B, C, D, E, F) historically average (3.0, 4.0, 2.5, 1.0, 6.5, 3.5) days, respectively. After completion at a workcenter, jobs usually sit for about 0.1 days before being moved to their next workcenter. A job waiting to be released is due on day 21. When should it be released?

Solution

Using the average times we can project the flow of a batch of this product. On release, the batch enters queue A. It remains there for 3.0 days and then spends 1.7 days being setup and run. After 4.7 days the job is ready to move to C. Taking 0.1 days for transit, the job arrives at C after 4.8 days, at which time it enters the queue. Proceeding with this loading, we obtain the time-phased loading shown in Figure 4.2. The sum of processing, setup, transit, and wait times is 18.6 days. Thus the job should be released at time $21.0 - 18.6 = 2.4$. ∎

In summary, let p_{ij} be the processing time for job i on machine j and w_j be the average waiting time in queue at j. If $S\{i\}$ is the set of workstations visited by i and m_j is the time to collect and move part i after it finishes at j, then we can estimate throughput time as

$$T = \sum_{j \in S\{i\}} (p_{ij} + w_j + m_j)$$

We release job i at time T before its due date. In our example we assumed jobs left the system immediately after completing the last operation, and thus the final m_j term was discarded.

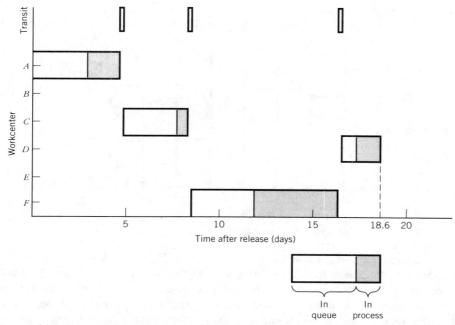

Figure 4.2   Deterministic projection of lead time for Example 4.1.

Unfortunately, this simple approach is valid only under stable conditions without infrequent events of lasting impact. Experience informs us that (1) queues do vary through time and (2) although preventive maintenance is predictable and can be factored into the above scheme, nevertheless, machines do break down at random. Since this is just an average estimate, the prudent manager would release the job earlier to avoid the 50 percent chance that the job takes longer than average and finishes late. The result of such "insurance" is to simply add inventory to the system, which must result in increased waiting times (Little's Law). The increased inventory level causes the variability of queueing time to increase, and hence the manager decides to add even more "insurance." The outcome: output does not increase, but we develop a chaotic mess of excess WIP!

Instead, temporally varying work loads can be partially stabilized by order release strategies that dampen the effects of demand variability. If workstation load predictions are available, dynamic queue averages can be used to improve lead time estimates. Preventive maintenance, process design improvements, and standardized procedures should be used to reduce variability induced by unpredictable events such as machine breakdowns.

Load reports are a common tool for controlling work loads. Finite-loading production planning systems create reports such as Figure 4.3. Assume that it is currently time 0. The report indicates available time (P_t) for each time period t. P_t would reflect any planned maintenance, overtime, or reduction in assigned workers due to schedule changes. Also shown are jobs currently in the queue (W_0), work load scheduled to arrive in period t from jobs already released to the shop (C_t), and work load expected to arrive from jobs scheduled to be released (E_t). Starting with the first operation, the job is tentatively scheduled into the first available time slot after it arrives at the workstation. If a rule other than "first come, first served" (FCFS) is used for selecting the batch to process, the batch can be penciled in where

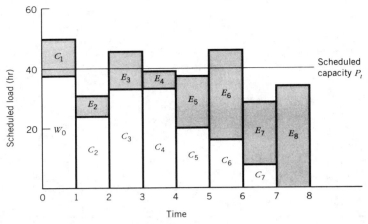

Figure 4.3   Sample workcenter load report.

appropriate. However, this causes all displaced jobs to be rescheduled, both here and at subsequent workstations, creating quite a disruption in the schedule and extensive replanning. In this case it would be preferable to assign only the period in which an operation will be performed. Effectively, each operation is assumed to be performed at the end of the period and unavailable at the next workstation until the next period. In such a system, the minimum flow time for a job is the number of periods equal to its number of operations.

EXAMPLE 4.2

A large job shop is operated with a FCFS rule at workstations. A planner is trying to determine the proper release time for a job requiring 2.5 hours at workstation L followed by 6.3 hours at workstation CG. All workstations are scheduled for 40 hours of productive time per week. The job is due in 3 weeks. Current load information is shown in Figure 4.4. When should the job be released?

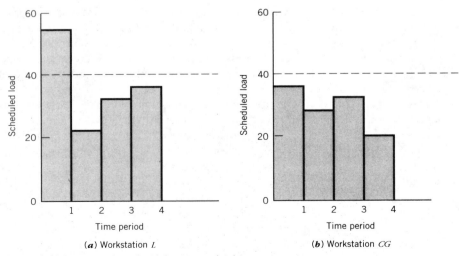

(**a**) Workstation L (**b**) Workstation CG

Figure 4.4   Current shop load for Example 4.2.

Solution

From the figure we see that there is little point in releasing the job now. Workstation L is already overloaded. We can, however, release it in period 2. Including the 12 hours of work carried over from the first period, this will increase the workload at L in the second period to an acceptable 36.5 hours. Workstation CG's load will be increased in period 3 to 38.3 hours, and thus we should be able to complete the job on time. ∎

In preparing load reports it is convenient to make use of **load profiles** for each part. The load profile is a time-phased listing of the resource requirement on each workcenter to produce a single part unit. For instance, suppose the standard batch size of the part described in Example 4.1 is 10 units and each workcenter has a one-period lead time. Then the load profile states that in the first period after release, each unit of this part requires 0.17 days at workcenter A. In the second period, 0.07 days are needed at workcenter C. In the third period, 0.44 days are used at F, and in the fourth period, 0.15 days at workcenter D. Times are one-tenth of the normal batch time.

Jobs are released so products can be sold, not to occupy space with inventory. Consequently, we should follow two basic rules.

First, if you can't sell it, don't release it!

Second, if you can't make it now, don't release it now!

The rules carry the understanding that normal lead times should be taken into account when asking whether we can make and sell the item. For instance, if the first machine in the production sequence currently has a two-week supply of jobs in its input queue, there is no point in releasing a new job. However, if normal throughput time is two weeks for a part and the two-week queue is at the last machine in its sequence with no other currently released jobs heading for that machine, then it may be time to release the job.

Most manufacturing systems involve assembly of component parts into products prior to shipping. Assembly cannot occur unless all parts are available. In this case, part releases should be coordinated so that the components arrive simultaneously at assembly. Suppose component A and component B are combined to make a product. Component A has a lead time of one week to arrive at assembly, whereas component B takes two weeks. Order releases should be offset such that B is released a week ahead of A. This practice is common in the assembly of complex products such as automobiles. The engine and other subassemblies are fed into the final assembly line with precise timing and synchronization to produce a custom-ordered vehicle. In batch manufacturing, MRP (Materials Requirements Planning) systems attempt to do the same and can be successful if valid part lead times are known. Unfortunately, the attempt to schedule compound assemblies is hindered by the system reliability law encountered in Chapter 1. Suppose parts are released such that each part has a 90 percent chance of being available when it is needed for assembly. If we have 10 parts, then the probability that we can complete the assembly as scheduled is $0.9^{10} = 0.35$. If the assembly has 100 parts, this probability is only 0.00003!

4.2.1 Bottleneck Scheduling

In both job shops and flow shops, for a specified part mix and routings there is a bottleneck process. We define the bottleneck as the workcenter with the highest

utilization, that is, the largest ratio of required processing time to time available.[1]
Bottlenecks are easy to identify. By multiplying the load profile by demand rate
we obtain the total load on each resource by part type. Summing over part types
gives total load per resource. Dividing by time available for each resource we obtain
utilization. Thus, utilization for workcenter m is

$$u_m = \frac{\sum_i p_{im} D_i}{P_m}$$

where p_{im} is processing time for part type i at workcenter m, P_m is available (schedulable) time for m, and D_i is demand for i. By our definition, the largest utilizations are
the bottlenecks. Any workcenter with a utilization greater than one limits sales, frus-
trating customers, marketing and manufacturing personnel, and stockholders alike.
If all workstation utilizations are less than one, then marketing is the bottleneck. It
is also possible that material availability may limit production and effectively be the
bottleneck.

Scheduling decisions should key on the bottleneck resources as these facilities
determine the output rate. Commercial scheduling systems are available that maximize the productive utilization of the bottlenecks and work outward from that point.
Predecessor workstations are scheduled to ensure available input into the bottleneck.
Orders are released in accord with these schedules. There is simply no reason to
release work faster than the bottleneck can process it. Successor workstations are
scheduled to handle output of the bottlenecks.

It may be desirable to accumulate significant levels of WIP in front of the bottleneck. Jobs can then be sequenced so as to minimize the time needed to change
machines over from one order to the next. Some jobs may be able to share partial or
full machine setups. This increases the net processing time available and effectively
increases system capacity. Following the objective of keeping the bottleneck busy
in productive activity, batch sizes should be large. However, large batch sizes mean
long batch processing times and large WIP levels. To avoid this problem while still
maintaining efficient usage of the bottleneck, a smaller **transfer batch** can be used.
The transfer batch is the number of units transported between workstations. As soon
as a number of parts equal to the transfer batch size are completed, the parts are
transported to the next workstation. For nonbottleneck stations, the transfer batch
can serve as the production batch. Thus, by the time the full production batch is
completed at the bottleneck, many of the parts in this batch will have moved downstream and may have already completed subsequent operations. Batch processing
times, WIP levels, and throughput times may therefore be small at the nonbottleneck
stations.

When the facility produces only a few part types on a repetitive basis, order releases
must adhere to the part demand rates. Batch sizes should be determined by standard
economic order quantity and setup-time considerations. The basic expression is

$$Q = \sqrt{\frac{2AD}{h}}$$

where Q is the batch size, A is setup cost, D is average demand rate, and h is inventory
holding cost per time (generally a percentage of inventory value). In using economic

---

[1]Other definitions of bottleneck occasionally appear. If batch processing times are variable, then a workstation with multiple, parallel servers will have a smaller average queue length than a single server station
with the same utilization. Hence, average queue length or time could be used to define the bottleneck.

order quantity formulas, setup cost should reflect "the bottom line." For instance, if workers are dedicated to a department that is not a bottleneck and hence has extra capacity, the labor cost component of setup is effectively zero. If not used for setup, the workers are paid to be idle. On the other hand, for a bottleneck with capacity less than customer demand, setup cost should include the cost of lost contribution to profit and overhead while the resource is down for setup. Setups reduce sales in this case.

Han and McGinnis [1988] discuss the use of linear programming for determining the best timing of releases when machines are subject to breakdown. The program compromises between the requisite long-term relative production rates dictated by demand and the desire to keep the bottleneck busy to avoid future resource shortages in case of a breakdown.

4.3

FLOW SHOP SEQUENCING

Sequencing is the process of defining the order in which jobs are to be run on a machine. **Scheduling** is the process of adding start and finish time information to the job order dictated by the sequence. Essentially, the sequence determines the schedule, since we will assume each job is started on a machine as early as possible, that is, as soon as both the job has finished all predecessor operations and the machine has completed all earlier jobs in its sequence.[2] This is referred to as a **semiactive schedule**, and it can be shown that it is an optimal policy for minimizing what are called **regular measures of performance**. Measures (criteria) in this class include average and maximum completion time, flow time, lateness, and tardiness. A measure is considered "regular" if it is nondecreasing in job completion times, that is, if any job is made to finish later, the measure will stay the same or increase.

Before proceeding, we need a few more definitions.

Problem Variables

N —the number of jobs to be scheduled

M —the number of machines; each job is assumed to visit each machine once

d_i —due date of job i

p_{ij} —setup and processing time of job i on machine j

Solution-Dependent Measures

C_i —time at which job i is completed

F_i —the length of time job i is in the shop (flow time)

L_i —lateness ($C_i - d_i$)

T_i —tardiness (max$\{0, L_i\}$, i.e., positive lateness values)

We will be assuming that all jobs are in the shop and ready for processing at time 0, and hence flow time and completion time are the same. Typical objectives are minimizing average flow time ($\overline{F} = N^{-1} \sum_i F_i$), minimizing the time required to complete all jobs ($C_{\max}$, also referred to as **makespan**), minimizing average

---

[2]This assumption is common in scheduling literature. However, in light of the validity of the just-in-time production philosophy and the recognition that many world-class systems plan for slack capacity, the efficacy of this assumption deserves consideration.

tardiness ($\overline{T} = N^{-1} \sum_i T_i$), minimizing maximum tardiness (T_{max}), and minimizing the number of tardy jobs ($\sum_i \delta_i$, where δ_i is 1 if $T_i > 0$ and 0 otherwise).

Scheduling problems are often denoted by $N/M/A/B$, indicating N jobs, M machines, job flow pattern A, and performance measure B, which is to be appropriately minimized or maximized. Minimizing average flow time ($\overline{F}$) in a general job shop with arbitrary flow pattern (G) would be classified as $N/M/G/\overline{F}$.

In a flow shop, all N jobs must visit machines in the same sequence. It is convenient in this case to number machines in accord with the processing order such that jobs visit machine 1 first and M last. Suppose we also restricted our solution to have the property that all machines process jobs in the same order. This is referred to as a **permutation** schedule, and although the optimal solution for flow shops with more than three machines need not be a permutation schedule, there is usually a permutation schedule that is near optimal.[3] Given the sequence, scheduling is easy. At time 0 the first job is started on machine 1. As soon as this operation is completed, the first job begins on machine 2 and the second job begins on machine 1. This process is continued until the last job finishes on machine M. The good news is that we have only to consider $N!$ sequences instead of $(N!)^M$. The bad news is that $N!$ still grows quickly in N.

EXAMPLE 4.3

Consider the set of jobs and processing times shown in Table 4.2. Generate the schedule assuming jobs are processed in the order {2, 4, 1, 3}.

Table 4.2   Flow Shop Processing Times for Example 4.2

Job	Machine			
	1	2	3	4
1	2.0	3.5	1.5	2.0
2	4.5	3.0	2.5	1.0
3	1.5	1.5	5.0	0.5
4	4.0	1.0	2.5	0.5

Solution

The final solution is summarized in the Gantt chart of Figure 4.5. We start by assigning job 2 to machine 1 at time 0. Since $p_{21} = 4.5$, the operation lasts until time 4.5. Since all jobs must go to machine 1 first, the other machines are idle and the other jobs are queued. At time 4.5, job 2 is loaded onto machine 2 and machine 1 starts on job 4, the second job in the sequence. Machine 2 finishes job 2, $p_{22} = 3.0$ time units later (time is now $t = 7.5$). Machine 1 is still busy with job 4; thus, while job 2 is begun on machine 3, machine 2 is idle, waiting for job 4. The remainder of the schedule is shown in the figure. Note that the schedule reflects the rule that machine j starts job i only when job i is finished on machine $j - 1$ *and* all jobs with earlier locations in the schedule have finished with machine j.   ■

---

[3]See Theorems 5.1 and 5.2 of French [1982] to see that an optimal permutation schedule exists for minimizing makespan in two- and three-machine flow shops.

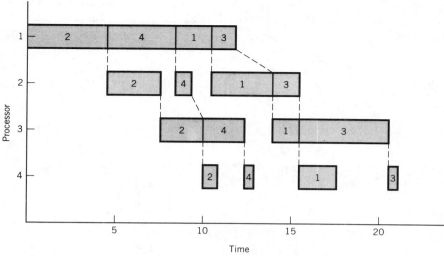

Figure 4.5   Gantt chart for Example 4.3.

The first thing that should strike you about the figure is that even for this rel-atively well-behaved processing system, machine utilization is not very high. The makespan is 21 time units. During that time the machines are busy 12.0, 9.0, 10.5, and 4.0 time units, respectively. Makespan is a good measure if the facility is truly devoted to this set of jobs during their manufacture. In reality, however, the next schedule of jobs may be assigned to machine 1 starting at any time after 12, and machine 4 could be busy completing the previous schedule until time 10 without conflict. Alternatively, we might have workers assigned to these machines only from the time of their first job arrival until last job completion, time 4.5 to 15.5 for ma-chine 2, instead of the entire 21 time units. Nevertheless, makespan does indicate the length of time for which the facility will be occupied by one or more of the cur-rent jobs.

Putting aside for a moment the question of the appropriateness of makespan as a criterion, we may ask whether this is the best makespan schedule. Idle time clearly exists in the schedule, leading us to question its efficiency. It turns out that this is often a difficult question to answer. Nonetheless, we can, at least, try by finding a lower bound on schedule makespan. Each machine supplies a lower bound. The makespan must accommodate (1) the delay before the machine can begin process-ing, (2) total processing time on the machine, and (3) the remaining processing time for the last job after it leaves the machine. Thus, a lower bound based on machine j is

$$LB_j = \min_i \left\{ \sum_{r=1}^{j-1} p_{ir} \right\} + \sum_{i=1}^{N} p_{ij} + \min_i \left\{ \sum_{r=j+1}^{M} p_{ir} \right\} \qquad (4.1)$$

Summations from 1 to 0 and $M + 1$ to M are defined to be 0. The three terms in equation 4.1 correspond to a lower bound on when machine j can start, processing time on j, and a lower bound on how soon j must finish before total schedule completion. Since each machine provides a valid lower bound, the largest of these is also a lower bound.

EXAMPLE 4.4

For the flow shop scheduling problem of Example 4.3, find a lower bound on makespan.

Solution

Consider machine 1.

$$LB_1 = 0 + \{2.0 + 4.5 + 1.5 + 4.0\}$$
$$+ \min\{3.5 + 1.5 + 2.0, \ 3.0 + 2.5 + 1.0,$$
$$1.5 + 5.0 + 0.5, \ 1.0 + 2.5 + 0.5\}$$
$$= 0 + 12.0 + \min\{7.0, \ 6.5, \ 7.0, \ 4.0\} = 16.0$$

For machine 2,

$$LB_2 = \min\{2.0, \ 4.5, \ 1.5, \ 4.0\} + (3.5 + 3.0 + 1.5 + 1.0)$$
$$+ \min\{3.5, \ 3.5, \ 5.5, \ 3.0\} = 13.5$$

For machine 3,

$$LB_3 = \min\{5.5, \ 7.5, \ 3.0, \ 5.0\} + (1.5 + 2.5 + 5.0 + 2.5)$$
$$+ \min\{2.0, \ 1.0, \ 0.5, \ 0.5\} = 15.0$$

Last, for machine 4,

$$LB_4 = \min\{7.0, \ 10.0, \ 8.0, \ 7.5\} + (2.0 + 1.0 + 0.5 + 0.5) + 0 = 11.0$$

The largest (tightest) lower bound is 16.0. We do not know if 16 is obtainable, but we may suspect we can do better than 21. If we examine the LB_1 equation we see that this lower bound came from putting job 4 last. Job 4 went last because it can proceed most quickly through machines 2, 3, and 4. Let's try the schedule (2, 1, 3, 4) formed by putting job 4 last. The Gantt chart is shown in Figure 4.6a. We save a modest 0.5 on makespan. It now appears that the problem with the schedule is that it takes too long to start machines 2 through 4 into operation. Suppose we place job 3 first; it has the smallest time requirement on machine 1 and will let us initiate the other machines sooner. Sequence (3, 2, 1, 4) is shown in Figure 4.6b. Makespan is now 17, just one more than our lower bound of 16. To summarize, we have improved by placing jobs with small p_{i1} early and small $\sum_{j=2}^{M} p_{ij}$ late in the sequence. Can you see another switch we could make by using these guidelines? Try it and see what happens! ∎

4.3.1 Single-Machine Scheduling

Before looking further at the general scheduling problem, let's try some simpler problems. Assume $M = 1$. This may seem like a trivial problem, but there are systems that are either scheduled as a single facility (only one product is manufactured at a time) or for which the schedule of a single bottleneck resource forms the basis for the entire facility.

Several useful results exist for single-machine scheduling. Suppose, for instance, that our objective is minimization of average job flow time. This is equivalent to minimizing average WIP. Since flow times and completion times are the same, we

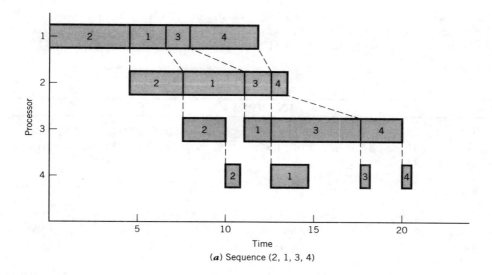

(*a*) Sequence (2, 1, 3, 4)

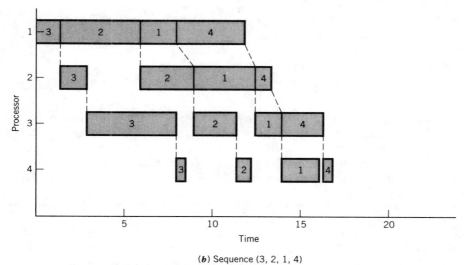

(*b*) Sequence (3, 2, 1, 4)

Figure 4.6   Gantt charts for alternate sequences.

can write average flow time as

$$\overline{F} = \frac{\sum\limits_{i=1}^{N} C_i}{N} \tag{4.2}$$

Let $[i]$ represent the ith ordered job in the production sequence. $C_{[i]} = \sum_{r=1}^{i} p_{[r]1}$. Thus, expression 4.2 becomes

$$\overline{F} = \frac{\sum\limits_{i=1}^{N} \sum\limits_{r=1}^{i} p_{[r]1}}{N}$$

$$= N^{-1} \left[N p_{[1]1} + (N-1)p_{[2]1} + (N-2)p_{[3]1} + \cdots + 1 \cdot p_{[N]1} \right] \tag{4.3}$$

Since each job must fit into a slot $[1]$ to $[N]$, the best we can do is to put the job with smallest p_i in first position, the job with second smallest p_i second, and so on.[4] Thus, we have the result (Conway et al. [1967], p. 27):

Fact: Any sequence with $p_{[1]} \leq p_{[2]} \leq \cdots \leq p_{[N]}$ minimizes average flow time.

EXAMPLE 4.5

Consider the set of six jobs with processing times $(21.4, 5.7, 16.2, 8.4, 9.0, 11.2)$. Sequence these jobs to minimize the average number of jobs in the shop (or equivalently $\overline{F}$). Find $\overline{F}$.

Solution

Simply placing the jobs in order based on their processing times, we have the sequence $(2, 4, 5, 6, 3, 1)$. Considering jobs in this order, completion times are $(5.7, 14.1, 23.1, 34.3, 50.5, 71.9)$. We can find average flow time from equation 4.2 by

$$\overline{F} = \frac{5.7 + 14.1 + 23.1 + 34.3 + 50.5 + 71.9}{6} = 33.27$$

Equivalently, from (4.3)

$$\overline{F} = \frac{6(5.7) + 5(8.4) + 4(9.0) + 3(11.2) + 2(16.2) + 21.4}{6} = 33.27 \qquad (4.4)$$

We could also estimate the average number of jobs at the machine $\overline{N}$. The numerator of equation 4.4 gives the product of the number of jobs in process and corresponding duration of that WIP level over the makespan. This is the total WIP and is equivalent to the integral over the makespan of the number of jobs in process. If we divide by the makespan (71.9) instead of N, we find $\overline{N} = 2.78$ jobs. ■

Sequencing jobs such that processing time is nondecreasing is referred to as **shortest processing time** (SPT) scheduling. SPT is a widely used rule that has been found to perform reasonably well on a number of performance measures in a variety of shop environments. Unfortunately, SPT ignores due dates.

As an alternative to SPT, we might sequence jobs in order of their due dates. The job with the **earliest due date** goes first. Earliest due date (EDD) scheduling is also very popular and useful. Suppose we wished to minimize maximum lateness for any job. Let S be a sequence that is not in EDD order. There must be some job $[k]$ in S for which $d_{[k]} > d_{[k+1]}$. Now, $L_{[k]} = C_{[k]} - d_{[k]}$ and $L_{[k+1]} = C_{[k+1]} - d_{[k+1]}$ where $C_{[k]} < C_{[k+1]}$. What happens if we switch jobs $[k]$ and $[k + 1]$? The completion times of the other $N - 2$ jobs, and hence their lateness, is unchanged. The relationships between the completion times and due dates for $[k]$ and $[k + 1]$ tell us that $L_{[k+1]} > L_{[k]}$ in S. Switching order reduces $C_{[k+1]}$ without affecting $d_{[k+1]}$. Hence, $L_{[k+1]}$ decreases. Unless the new lateness for $[k]$ is larger than the previous lateness for $[k + 1]$, we must have at least as good a solution as before. In fact, the new solution is better if $[k + 1]$ was the latest job in S. Let's examine the revised lateness for $[k]$.

---

[4]For the remainder of this single-machine section, we delete the machine subscript and use p_i.

After switching,

$$L_{[k]} = C_{[k+1]} - d_{[k]}$$

Since $d_{[k]} > d_{[k+1]}$, $[k]$'s new lateness must be less than $[k + 1]$'s old lateness. We have shown that if any sequence violates the EDD rule, then switching a violating job will either lower maximum lateness or leave it unchanged. Thus, (Conway et al. [1967], p. 31):

Fact: A sequence with $d_{[1]} \leq d_{[2]} \leq \cdots \leq d_{[N]}$ minimizes maximum lateness.

As a final note, since EDD minimizes $L_{\max}$, it also minimizes $T_{\max}$, maximum tardiness.

EXAMPLE 4.6

Let the due dates for the six jobs in the previous example be (25., 34., 82., 54., 70., 45.), respectively. Sequence the jobs to minimize maximum tardiness.

Solution

Placing the jobs in EDD order we have the sequence (1, 2, 6, 4, 5, 3). The associated completion times are (21.4, 27.1, 38.3, 46.7, 55.7, 71.9). With completion times and due dates known, we have lateness values $(-3.6, -6.9, -6.7, -7.3, -14.3, -10.1)$. For example, $L_6 = C_6 - d_6 = 38.3 - 45.0 = -6.7$. The corresponding tardiness values are all 0. ∎

It would be useful to be able to find the sequence that satisfied due dates but also minimized average flow time among such sequences. Fortunately, a single-pass construction algorithm was presented by Smith [1956] to do just that. Our problem is

$$\text{minimize } \overline{F}$$

subject to:

$$T_{\max} = 0$$

Let the sequence be denoted $S = \{s_{[1]}, \ldots, s_{[N]}\}$. Working from the end of the schedule, we will add a new job to the sequence at each step. Consider the task of selecting a job to go last. Due to our delivery constraint, only those jobs k for which $d_k \geq \sum_i p_i$ are admissible, since the last job will have completion time $C_{[N]} = \sum_i p_i$. We know that SPT minimizes $\overline{F}$, and thus long jobs should be at the end of the schedule. This is accomplished by selecting the admissable k with maximum processing time and placing this job last. We now have $N - 1$ jobs to schedule. The process is repeated to select the job that goes $N - 1$st in sequence. The process continues until all jobs are sequenced. If at some point no admissable jobs are found, then no sequence exists with $T_{\max} = 0$. In this instance our best option is to change the constraint to $T_{\max} = \delta$ for some small δ and try to resolve the problem. The same algorithm is applied by adding δ to all due dates.

Flow Time-Due Date Algorithm

STEP 0. Initialize. Set $k = N + 1$, $A = \{1, \ldots, N\}$, $S = \phi$, $\tau = \sum_{i=1}^{N} p_i$.

STEP 1. Select job. Set $k = k - 1$. Find $s_{[k]} = \arg \max_{i \in A}\{p_i : d_i \geq \tau\}$.

STEP 2. Assign job. Update the list of unassigned jobs $A = A - s_{[k]}$, $\tau = \tau - p_{s_{[k]}}$. If $A = \phi$ stop; otherwise, go to 1.

EXAMPLE 4.7

Consider the single-machine scheduling problem of Table 4.3. The supervisor realizes that all jobs cannot be delivered on time. Is there a sequence for which no job is more than two days late? If so, find it. If options exist, the supervisor would like to minimize the number of jobs at the machine as well.

Table 4.3   Problem Data for Example 4.7

Job	Processing Time (days)	Due Date (day)
1	2.4	11
2	5.0	8
3	4.2	5
4	1.5	12
5	0.9	13

Solution

STEP 0. Initialize. Total processing time is $\tau = 2.4 + 5.0 + 4.2 + 1.5 + 0.9 = 14.0$. Due dates are adjusted to (13., 10., 7., 14., 15.), respectively, by adding the 2-day allowance.

STEP 1. Select job. Admissable jobs are those with due dates of 14 or later. This includes jobs 4 and 5. Since $p_4 > p_5$, $s_5 =$ job 4.

STEP 2. Update. $S = \{,,,4\}$, $A = \{1, 2, 3, 5\}$, and $\tau = \tau - 1.5 = 12.5$.

STEP 1. Select job. Admissable jobs with due dates of 12.5 or greater are 1 and 5. Since $p_1 > p_5$, $s_4 = 1$.

STEP 2. Update. $S = \{,, 1, 4\}$, $A = \{2, 3, 5\}$, and $\tau = 10.4$.

STEP 1. Select job. Only job 5 is admissable.

STEP 2. Update. $S = \{, 5, 1, 4\}$, $A = \{2, 3\}$, and $\tau = 9.5$.

STEP 1. Select job. Job 2 is now the only admissable job.

STEP 2. Update. $S = \{, 2, 5, 1, 4\}$, $A = \{3\}$, $\tau = 4.5$.

STEP 1. Select job. Job 3 is now admissable.

STEP 2. Final schedule is $S = \{3, 2, 5, 1, 4\}$.

The final schedule is shown in Figure 4.7. Average flow time is

$$\overline{F} = \frac{4.2 + 9.2 + 10.1 + 12.5 + 14.0}{5} = 10.0$$

Job 2 is 1.2 days late, job 1 is 1.5 days late, and job 4 is 2 days late. As compared with an EDD solution, job 5 was moved up because of its short processing time. The EDD solution $\{3, 2, 1, 4, 5\}$ has $\overline{F} = 10.4$. ∎

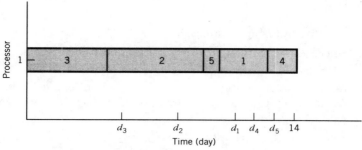

Figure 4.7   Satisfying due-date schedule of Example 4.7.

4.3.2 Two-Machine Flow Shops

Beyond single-machine scheduling, most problems are difficult to solve. However, it is easy to find the minimum makespan schedule for a two-machine flow shop. In fact, we have already encountered the logic for such an algorithm; jobs with short processing times on machine 1 should go early in the schedule and jobs with short processing time on machine 2 should go late in the schedule. We will construct the optimal sequence by adding one job at a time. The selected job goes at either the earliest available slot in the sequence or the latest. Job assignments are permanent. For example, once a job is assigned to be first, it stays in that position and subsequent assignments fall into line behind it.

Johnson's Algorithm

STEP 0. Initialize. $A = \{1, 2, \ldots, N\}$.

STEP 1. Select job. If $A = \phi$, stop; otherwise, find $k = \arg\ \min_{k \in A}\{p_{i1}$ or $p_{i2}\}$.

STEP 2. Assign job k. If minimum is on machine 1, place k as early as possible in sequence S. If minimum is on machine 2, place k as late as possible in S. $A = A - k$. Go to 1.

EXAMPLE 4.8

A plant produces six families of products. Production consists of parts fabrication followed by assembly. Only one family can be fabricated or assembled at a time. Given the data in Table 4.4, find the minimium makespan schedule for the plant.

Table 4.4   Family Setup and Processing Days for Example 4.8

Family	Parts Fabrication	Assembly
1	6	4
2	10	8
3	4	9
4	7	2
5	6	3
6	5	6

Solution

STEP 0.   Initialize. All families are available: $A = \{1, 2, 3, 4, 5, 6\}$.

STEP 1.   Select job. The minimum processing time is $p_{42} = 2$; so select job 4.

STEP 2.   Assign job. Since the selected time is on machine 2, the selected family is placed last. The optimal partial sequence is $S = \{ , , , , , 4\}.A = \{1, 2, 3, 5, 6\}$.

STEP 1.   Select job. The smallest time in A is $p_{52} = 3$.

STEP 2.   Assign job. Family 5 is placed as late as possible, that is, into position 5. $S = \{ , , , 5, 4\}$, and $A = \{1, 2, 3, 6\}$.

Continuing in this fashion we select family 1 and place it in position 4 (ties are broken arbitrarily), then we select family 3 for position 1, then family 6 for position 2, and last family 2 goes to position 3. The final sequence is $S = \{3, 6, 2, 1, 5, 4\}$.   ∎

Johnson's algorithm can be used as a heuristic when $M > 2$. The M machines are split into two pseudo machines. The pseudo machines have processing times equal to the sum of the processing times on the real machines assigned to them. Johnson's algorithm can then be applied to the N-job, two-pseudo-machine problem to obtain a job sequence. The only question is how to split the M machines into the two groups. One well-known scheduling rule is that for the three-machine flow shop, if either

$$\min p_{i1} \geq \max p_{i2} \quad \text{or} \quad \min p_{i3} \geq \max p_{i2}$$

then the optimal sequence is found from groups $\{1, 2\}$ and $\{2, 3\}$. Thus for the pseudo machines a and b, we use processing times

$$p_{ia} = p_{i1} + p_{i2} \qquad p_{ib} = p_{i2} + p_{i3}$$

Campbell et al. [1970] have suggested solving $M - 1$ two-machine problems in general and selecting the best. The $M - 1$ problems come from forming the groups

$\{1\}, \{M\}$

$\{1, 2\}, \{M - 1, M\}$

$\{1, 2, 3\}, \{M - 2, M - 1, M\}$

$\{1, 2, 3, \ldots, M - 1\}, \{2, 3, \ldots, M\}$

EXAMPLE 4.9

Consider the four-machine flow problem of Table 4.5. Find a low makespan schedule.

Solution

We must solve three different two-machine flow shop problems. The first problem uses machines 1 and 4 only. The second problem combines processing times on machines 1 with 2 and 3 with 4. The third problem combines 1, 2, and 3 for the first pseudo machine and 2, 3, and 4 for the second. Resulting pseudo processing times are shown in Table 4.6.

Table 4.5   Data for Example 4.9

Job	Processing Time on Machine			
	1	2	3	4
1	2	3	1	5
2	4	2	6	9
3	1	4	8	5
4	6	3	5	2
5	4	2	3	7

Table 4.6   Combined Processing Times for Example 4.9

Job			Machine Combination			
	1	4	1, 2	3, 4	1, 2, 3	2, 3, 4
1	2	5	5	6	6	9
2	4	9	6	15	12	17
3	1	5	5	13	13	17
4	6	2	9	7	14	10
5	4	7	6	10	9	12

The sequences generated by the algorithm for the three problems are $S_1 = \{3, 1, 2, 5, 4\}$, $S_2 = \{1, 3, 2, 5, 4\}$, and $S_3 = \{1, 5, 2, 3, 4\}$. The corresponding schedules are shown in Figure 4.8 on page 114. The best makespan of 34 days results from schedule $S_3 = \{1, 5, 2, 3, 4\}$. Is this schedule optimal? Here's a hint: how soon can machine 4 begin working? ∎

4.4
JOB SHOP SCHEDULING

The general job shop problem is to schedule production times for N jobs on M machines. At time 0, we have a set of N jobs. For each job we have knowledge of the sequence of machines required by the job and the processing time on each of those machines. Due dates may also be known. The objective may be to minimize the makespan for completion of all jobs, minimizing the number of tardy jobs or average tardiness, minimizing the average flow time, or achieving some weighted combination of these criteria. This problem is notoriously difficult to solve. On each of the M machines there are $N!$ possible job orderings making a total of $(N!)^M$ possible solutions. For just 10 jobs on five machines that gives over 6×10^{31} choices. Despite considerable effort by researchers, and simplistic abstractions of reality, most scheduling problems are not even close to being solved. Rather than diving into the mathematical quagmire engulfing job shop scheduling, we will skim the top and look at some popular, simple rules and some basic results.

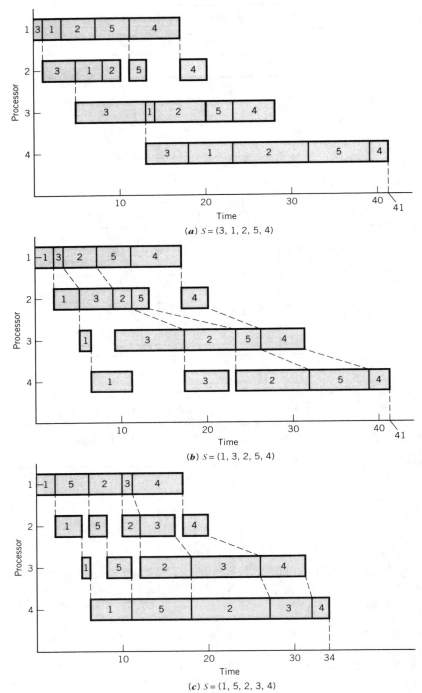

Figure 4.8 Gantt charts for Example 4.9 candidate V sequences.

4.4.1 Dispatching Rules

When a processor becomes available, a job must be selected from its input queue for immediate setup and processing. We refer to this as **dispatching**. Simple dispatching rules are often used in shop scheduling. Table 4.7 lists a dozen of the more popular rules. Rules can be classified as being **static** or **dynamic**. Static rules have priority indices that stay constant as jobs travel through the plant, whereas dynamic rules change with time and queue characteristics. LTWK and EDD (assuming due dates are fixed) are static rules. LWKR is dynamic, since the remaining processing time decreases as the job progresses through the shop, that is, through time. **Slack-based** rules are also dynamic. Slack is defined as slack = due date − current time − remaining work, that is, the time until the due date minus remaining processing time. Due dates may also change while the job order is open; however, Berry et al. [1984] concluded that for reorder point inventory systems, use of static due dates based on average demand rates is preferable to the use of dynamic due dates, which constantly reflect random fluctuations in demand. Rules can also be classified as **myopic** or **global**. Myopic rules look only at the individual machine, whereas global rules look at the entire shop. SPT is myopic but WINQ is global.

Table 4.7   Standard Dispatching Rules

Name	Description
SPT	(Shortest Processing Time) Select a job with minimum processing time.
EDD	(Earliest Due Date) Select a job due first.
FCFS	(First Come, First Served) Select a job that has been in the workstation's queue the longest.
FISFS	(First in System, First Served) Select a job that has been on the shop floor the longest.
S/RO	(Slack per Remaining Operation) Select a job with the smallest ratio of slack to operations remaining to be performed.
Covert	Order jobs based on ratio of slack-based priority to processing time.
LTWK	(Least Total Work) Select a job with smallest total processing time.
LWKR	(Least Work Remaining) Select a job with smallest total processing time for unfinished operations.
MOPNR	(Most Operations Remaining) Select a job with the most operations remaining in its processing sequence.
MWKR	(Most Work Remaining) Select a job with the most total processing time remaining.
RANDOM	(Random) Select a job at random.
WINQ	(Work in Next Queue) Select a job whose subsequent machine currently has the shortest queue.

EXAMPLE 4.10

Current time is 10. Machine *B* has just finished a job, and it is time to select its next job. Table 4.8 provides information on the four jobs available. For each of the dispatching rules above, determine the corresponding sequence.

Table 4.8　Available Jobs for Example 4.10

Job	Arrival to System	Arrival at B	Due Date	Operation (machine, p_{ij})		
				1	2	3
1	10	10	30	$(B, 5)$	$(A, 1)$	$(D, 6)$
2	0	5	20	$(A, 5)$	$(B, 3)$	$(C, 2)$
3	0	9	10	$(C, 3)$	$(D, 2)$	$(B, 2)$
4	0	8	25	$(E, 6)$	$(B, 4)$	$(C, 4)$

Solution

SPT: Jobs (1, 2, 3, 4) have processing times of (5, 3, 2, 4) on machine B. Placing jobs in increasing order of processing time yields the job sequence {3, 2, 4, 1}. Thus, load job 3 onto machine B.

EDD: Jobs (1, 2, 3, 4) have due dates (30, 20, 10, 25), respectively. Ordering by due date we have job sequence {3, 2, 4, 1}.

FCFS: Jobs arrived at B at times (10, 5, 9, 8). Placing earliest arrivals first, we obtain the job sequence {2, 4, 3, 1}.

FISFS: Ordering jobs by system arrival, job 1 goes last. The other three jobs can be sequenced arbitrarily. One such sequence is {2, 3, 4, 1}.

S/RO: Job slack is due date−current time−remaining processing time. For jobs 1 through 4 this gives (30 − 10 − 5 − 1 − 6 = 8, 20 − 10 − 3 − 2 = 5, 10−10−2 = −2, 25−10−4−4 = 7). Dividing by the number of remaining operations in each case, S/RO ratios are (2.67, 2.50, −2.00, 3.50). Placing jobs in increasing order of S/RO produces the sequence {3, 2, 1, 4}.

Covert:[5] The Covert rule uses the measure (delay cost)/(operation processing time).

$$\text{Delay cost} = \begin{cases} 1 & \text{if slack } < 0 \\ 0 & \text{if slack } > E \text{ (wait time)} \\ E(\text{wait time}) - \dfrac{\text{slack}}{E(\text{wait time})} & \text{otherwise} \end{cases}$$

The three categories in delay cost correspond to jobs that are late, jobs for which there is no urgency, and jobs that must proceed at least at the normal pace to be completed on time. E (wait time) is often taken as a multiple of the number of remaining operations. Let us assume a value of 0.5 per remaining operation as normal waiting time. The E (wait time) for our jobs are (1.5, 1.0, 0.5, 1.0). Except for job 3, these times are less than the job's slack, and their delay costs are 0. Delay cost for job 3 is 1 since that job has negative slack. The Covert job measures are (0, 0, 1/2, 0). Job 3 has the highest value and goes first. The rest are tied and can be sequenced arbitrarily. Let's choose sequence {3, 1, 2, 4}. In empirical testing the Covert rule (likewise Critical Ratio) has performed well.

LTWK: Total work values are (12, 10, 7, 14). Thus, our sequence for jobs is {3, 2, 1, 4}.

---

[5]The Covert rule is closely related to critical ratio priority scheduling. The Critical Ratio rule usually defines *delay cost* as due date − current time, thus basing priority on the ratio of (remaining time)/(remaining work).

LWKR: Remaining processing time until job completion is (12, 5, 2, 8). The corresponding LWKR sequence is {3, 2, 4, 1}.

MOPNR: The number of remaining operations by job is (3, 2, 1, 2). The MOPNR sequence is {1, 2, 4, 3}.

MWKR: Remaining work loads were found in LWKR. Reversing that order we have sequence {1, 4, 2, 3}.

RANDOM: Using a random number table the author obtained the sequence {3, 1, 2, 4}.

WINQ: We require information on the length of the queue at machines A and C. Suppose these values are 10 and 4. Job 3 goes first since it has no next queue. Jobs 2 and 4 come next since they are headed for machine C, which has less work in its queue than A. Our sequence is thus {3, 2, 4, 1}. ∎

4.4.2 Schedule Generation

We have already argued the futility of trying to generate optimal schedules in a random environment. Nevertheless, techniques have been developed using optimization ideas, such as dynamic programming and branch and bound, to do just that. We will not discuss these techniques, but it is useful to consider what is known about the optimal schedule. Semiactive schedules were defined earlier. Not all semiactive schedules must be examined. The optimal schedule must be **fully active**. Active schedules are those that never make a job wait in queue when it can be completely processed before the next job is scheduled to start. Active scheduling recognizes that by going ahead and producing the job, nothing is delayed. Time that would otherwise be wasted is used productively. Consider the four-machine, four-job scheduling problem shown in Table 4.9.

Table 4.9 Sample Scheduling Problem

Job	Operation (machine, time)			
	1	2	3	4
1	$(A, 5)$	$(B, 3)$	$(C, 7)$	$(D, 1)$
2	$(B, 8)$	$(D, 6)$	$(C, 2)$	$(A, 5)$
3	$(B, 3)$	$(D, 6)$	$(C, 4)$	$(A, 4)$
4	$(A, 2)$	$(C, 2)$	$(B, 7)$	$(D, 3)$

Job 1 has high priority, and it has been decided that this job should not be delayed if possible. Adhering to this request, consider the semiactive permutation schedule {1, 2, 3, 4}. This schedule is shown in Figure 4.9a. Adhering to the technological ordering of each job's operations and the permutation requirement for each machine, every operation is scheduled as early as possible. The permutation schedule is not active. Job 3 could be started at time 0 on machine B without delaying job 1. Job 4 could also start on machine A before jobs 2 and 3 are ready for machine A. Moving job 3 forward first and then job 4, we obtain the active schedule shown in Figure 4.9b. Note the significant reduction in makespan.

The next step in schedule elimination is to consider only **nondelay** schedules. Nondelay schedules are such that a machine is never idle when its queue

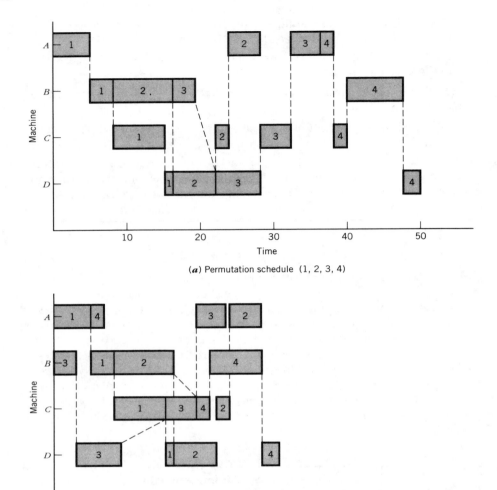

(*a*) Permutation schedule (1, 2, 3, 4)

(*b*) Active schedule

Figure 4.9   Four-job, four-machine schedules.

is nonempty. In active scheduling we started a job if it could be completed before the next job was scheduled to start. In nondelay scheduling we start a job if it can be started before the next job is scheduled to start. This is a logical policy but not necessarily optimal. It may have been preferable to wait for a specific job that is important to the grand scheme. The active schedule of Figure 4.9*b* is not a nondelay schedule. Job 4 could begin on machine *C* at time 7. However, it cannot finish on machine *C* until after the currently scheduled start time for job 1. Nevertheless, a nondelay rule would require job 4 to be processed on *C* from time 7 to 9, thus delaying job 1. This move may be undesirable if job 1 truly dominates in importance.

The process of generating all the nondelay schedules (or the active schedules for that matter) is straightforward. The number of such schedules can be computationally prohibitive, however. We will present a procedure that will allow generation of as many of the nondelay schedules as desired. Basically, we construct a schedule by scheduling one operation at a time. At any point in time each job is either finished or has one operation that can be scheduled. We will simply move forward in time and whenever one or more operations is ready to be scheduled and its desired machine is available, we schedule an operation. If we select one of the available operations at each point and continue until all MN operations are assigned, we have one complete nondelay schedule. If instead we form a tree of all possible decisions at each decision point, all possible nondelay schedules are formed.

At stage t, let

S_t be the partial schedule of $(t - 1)$ scheduled operations.

A_t be the set of operations schedulable at stage t, that is, all predecessor operations are in S_t.

e_k be the earliest time that operation $k \in A_t$ can be scheduled, that is, predecessors are completed and machine is available. Then:

Nondelay Schedule Generation

STEP 0. Initialize. Let $t = 1, S_1 = \phi$. A_1 contains the first operation of each ready job.

STEP 1. Select operation. Find $e^* = \min_{k \in A_t} e_k$. If several e^* exist, choose arbitrarily. Let m^* be the machine needed by e^*. Choose any $k \in A_t$ that requires m^* and has $e_k = e^*$. (If all nondelay schedules are to be created, choose all such k and create a new partial schedule for each.)

STEP 2. Increment. Add the selected operation k to S_t to create S_{t+1}. Remove k from A_t and add the next operation for its job unless that job is completed; this creates A_{t+1}. Set $t = t + 1$. If $t = MN$ stop; otherwise go to 1.

An example will illustrate the procedure. Let o_{ir} denote the rth operation of job i.

EXAMPLE 4.11

Generate a nondelay schedule for the four-job, four-machine problem data of Table 4.9. When choices between operations exist, give highest priority to job 1. Otherwise, choose based on SPT.

Solution

STEP 0. Initialize. $t = 1, S_1 = \phi$, and $A_1 = \{o_{11}, o_{21}, o_{31}, o_{41}\}$.

STEP 1. Select operation. The early start time of all first operations is 0. Job 1 has priority, so select operation o_{11}, which implies $m^* = A$. Operations o_{11} and o_{41} are eligible for selection. We will choose o_{11} due to job 1's priority. (If all nondelay schedules were to be generated, we would have to start two different schedules at this point and continue each through 16 operation assignments.)

STEP 2. Increment. $t = 2, S_2 = \{o_{11}\}, A_2 = \{o_{12}, o_{21}, o_{31}, o_{41}\}$. Early start times for the four operations in A_2 are now (5, 0, 0, 5). Note that operation o_{41} has been moved back since we decided to place operation o_{11} ahead of it.

STEP 1. Select operation. We must choose between operations o_{21} and o_{31}, since they have the earliest start time. Choose o_{21}, which yields $m^* = B$. Both operations are eligible since both require B. Neither operation is from job 1. Invoking the SPT tie-breaker, we choose o_{31} and place this job first on machine B.

The process continues until all 16 operations are assigned. Steps are summarized in Table 4.10, and the final schedule is shown in Figure 4.10. In each row of the table, m^* was selected to be one of the machines used by an element of A_t with the minimum earliest start time. We have a new best makespan of 28. ∎

Table 4.10   Construction of Nondelay Schedule for Example 4.11

Step t	A_t	e_k	m^*	k
1	$o_{11}, o_{21}, o_{31}, o_{41}$	0, 0, 0, 0	A	o_{11}
2	$o_{12}, o_{21}, o_{31}, o_{41}$	5, 0, 0, 5	B	o_{31}
3	$o_{12}, o_{21}, o_{32}, o_{41}$	5, 3, 3, 5	B	o_{21}
4	$o_{12}, o_{22}, o_{32}, o_{41}$	11, 11, 3, 5	D	o_{32}
5	$o_{12}, o_{22}, o_{33}, o_{41}$	11, 11, 9, 5	A	o_{41}
6	$o_{12}, o_{22}, o_{33}, o_{42}$	11, 11, 9, 7	C	o_{42}
7	$o_{12}, o_{22}, o_{33}, o_{43}$	11, 11, 9, 11	C	o_{33}
8	$o_{12}, o_{22}, o_{34}, o_{43}$	11, 11, 13, 11	B	o_{12}
9	$o_{13}, o_{22}, o_{34}, o_{43}$	14, 11, 13, 14	D	o_{22}
10	$o_{13}, o_{23}, o_{34}, o_{43}$	14, 17, 13, 14	C	o_{34}
11	$o_{13}, o_{23}, -, o_{43}$	14, 17, −, 14	C	o_{13}
12	$o_{14}, o_{23}, -, o_{43}$	21, 21, −, 14	B	o_{43}
13	$o_{14}, o_{23}, -, o_{44}$	21, 21, −, 21	D	o_{14}
14	$-, o_{23}, -, o_{44}$	−, 21, −, 22	C	o_{23}
15	$-, o_{24}, -, o_{44}$	−, 23, −, 22	D	o_{44}
16	$-, o_{24}, -, -$	−, 23, −, −	A	o_{24}

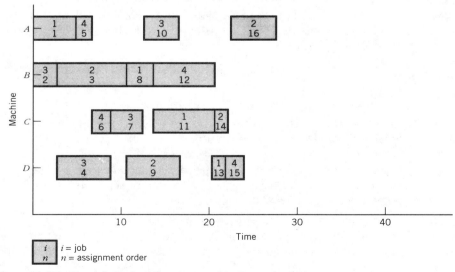

Figure 4.10   Nondelay schedule generated in Example 4.11.

4.5
SUMMARY

Job shops appear inefficient largely because of the production demands placed on them. Order release strategies should be carefully planned to meet due dates while attempting to smooth resource utilization through time.

A few scheduling problems such as minimizing mean flow time or maximum lateness for a single machine are easy to solve. However, general problems such as minimizing average tardiness or scheduling in dynamic, multimachine job shops are notoriously difficult. Even most oversimplified academic problems, which assume perfect knowledge and a deterministic world, are hard. The dynamic nature of actual shops requires real-time schedule updating. For this reason, heuristic dispatching rules are widely used in job shops. SPT performs well in minimizing average flow time and normally makespan as well. EDD minimizes maximum lateness or tardiness.

Permutation schedules work well in flow shops but not necessarily in job shops. If a job has increasing processing times from its first to last operation, it should most likely be early in the production schedule of machines. Likewise, jobs with decreasing processing times should be scheduled later.

For regular measures of performance, there is an active schedule that is optimal. The best nondelay schedule may not be optimal but will be close in most cases. Nondelay schedules also reduce the number of schedules that must be examined and follow the oft-imposed organizational restriction that machines not sit idle while work sits in their input queue.

McKay et al. [1988] defined the real scheduling problem as "how to schedule and dispatch work in such a way that many stated and unstated conflicting goals are satisfied using hard and soft information that is possibly incomplete, biased, outdated and erroneous." Expert systems (Fox and Smith [1984]) have the ability to capture experience and make decisions based on complex interactions of current system status data and objectives. As such, they may be potentially useful in complex scheduling environments.

REFERENCES

Baker, Kenneth R. (1974), *Introduction to Sequencing and Scheduling*, John Wiley & Sons, New York.

Berry, William L., Richard J. Penlesky, and Thomas E. Vollmann (1984), "Critical Ratio Scheduling: Dynamic Due-Date Procedures under Demand Uncertainty," *IIE Transactions*, 16(1), 81–89.

Blackstone, John H. Jr., Don T. Phillips, and Gary L. Hogg (1982), "A State of the Art Survey of Dispatching Rules for Manufacturing Job Shop Operation," *International Journal of Production Research*, 20(1), 27–45.

Campbell, Herbert G., Richard A. Dudek, and Milton L. Smith (1970), "A Heuristic for the *N* Job, *M* Machine Sequencing Problem," *Management Science*, 16(10), 630–637.

Conway, Richard W., William L. Maxwell, and Louis W. Miller (1967), *Theory of Scheduling*, Addison-Wesley, Reading, Mass.

Fox, Mark S., and Stephen F. Smith (1984), "ISIS — A Knowledge-Based System for Factory Scheduling," *Expert Systems*, 1(1), 25–49.

French, Simon (1982), *Sequencing and Scheduling: An Introduction to the Mathematics of the Job Shop*, John Wiley & Sons, New York.

Goldratt, Eliyahu M., and Jeff Cox (1984), *The Goal: Excellence in Manufacturing*, North River Press, New York.

Han, Min-Hong, and Leon F. McGinnis (1988) "Throughput Rate Maximization in Flexible Manufacturing Cells," *IIE Transactions*, 20(4), 409–417.

Hopp, Wallace J., Mark L. Spearman, and David L. Woodruff (1990), "Practical Strategies for Lead Time Reduction," *Manufacturing Review*, 3(2), 78–84.

Johnson, S. M. (1954), "Optimal Two and Three Stage Production Schedules with Setup Times

Included," *Naval Research Logistics Quarterly*, 1(1).

McKay, K. N., F. R. Safayeni, and J. A. Buzacott, (1988), "Job Shop Scheduling Theory: What Is Relevant?" *Interfaces*, 18(4), 84–90.

Moore, J. M. (1968), "Scheduling *n* Jobs on One Machine to Minimize the Number of Tardy Jobs," *Management Science*, 17(1).

Smith, W. E. (1956), "Various Optimizers for Single-Stage Production," *Naval Research Logistics Quarterly*, 3(1).

Stecke, Kathryn E., and James J. Solberg (1985), "The Optimality of Unbalancing Both Workloads and Machine Group Sizes in Closed Queueing Networks of Multiserver Queues," *Operations Research*, 33(4), 882–910.

Shanthikumar, J. G., and J. A. Buzacott (1981), "Open Queueing Network Models of Dynamic Job Shops," *International Journal of Production Research*, 19(3), 255–266.

Wilkerson, L. J., and J. D. Irwin (1971), "An Improved Algorithm for Scheduling Independent Tasks," *AIIE Transactions*, 3(3).

PROBLEMS

4.1. Describe the conditions under which process-oriented facility organization is appropriate.

4.2. A welding department has three workers. On average 15 jobs arrive each week. About 30 percent of these jobs are difficult. Currently, jobs are handled FCFS and the job at the head of the queue is assigned to the first available worker. A worker can complete an average of six jobs per week. To improve quality, the supervisor is considering assigning the difficult jobs and only the difficult jobs to the most experienced worker. The other two workers would continue to handle the remaining 70 percent of jobs. Assuming Poisson interarrival and exponential processing times, how will this change effect worker utilizations, WIP levels, and job throughput times?

4.3. Consider the process plan shown in Table 4.11 for a part. A batch of 25 units of this part is due in 21 days. When should the part be released? Each workcenter has an average waiting time of 2 days.

Table 4.11   Part Process Plan for Problem 4.3

Operation	Workcenter	Setup Time (days)	Unit Processing Time (days)
1	A	0.2	0.01
2	C	0.3	0.04
3	G	0.2	0.08

4.4. Resolve Problem 4.3 assuming a batch of 100 parts due in 25 days.

4.5. Assuming a batch of 25 units and periods of one week (seven days), construct a load profile for the part described in Table 4.11.

4.6. What is different about determining the cost of a setup for a bottleneck and a nonbottleneck resource? Do you think a typical accounting system would reflect this difference?

4.7. A nonbottleneck workstation requires 30 minutes of worker time to set up a part. The worker is paid $25/hour. The machine has a fixed overhead charge of $10/hour and costs $15/hour to operate. The part has a demand rate of 10,000 units per year, and holding cost per unit is $3.50/year.

  a. Determine the setup cost.

  b. Find the economic order quantity.

4.8. Suppose in Problem 4.7 that the workstation was limiting production and the plant makes $450/hour when producing output. Find the setup cost and economic order quantity.

4.9. What is the difference between semiactive, active, and nondelay schedules?

4.10. Three jobs are to be scheduled in a five-machine flow shop. Job data are summarized in Table 4.12.

  a. For each machine, find a lower bound on makespan.

  b. Construct a Gantt chart for job processing sequence 1, 2, 3.

Table 4.12   Job Processing Times for Problem 4.10

Job	Machine				
	A	B	C	D	E
1	2.0	4.5	1.0	3.2	4.1
2	0.5	2.4	0.2	0.1	2.8
3	1.2	0.3	0.5	1.4	8.2

4.11. Enumerating all possible orderings, find an optimal permutation schedule for the jobs in Table 4.12 through a flow shop. The objective is minimizing makespan.

4.12. Apply the Campbell procedure to generate a low makespan schedule for the flowshop problem detailed in Table 4.12.

4.13. Consider the parts summarized in Table 4.13. The shop is a flowshop.
  a. Find a lower bound on makespan.
  b. Generate a permutation schedule using Campbell's procedure.

Table 4.13   Flowshop Processing Times for Problem 4.13

	Machine			
Job	A	B	C	D
1	1	4	2	5
2	4	2	9	11
3	1	6	9	3
4	3	4	8	2
5	7	1	2	5
6	3	2	1	8

4.14. Seven jobs are available to be produced on a single machine. The jobs have processing times (2, 1, 5, 3, 9, 13, 24) and due dates (12, 34, 8, 42, 65, 51, 29), respectively.
  a. Find the schedule that minimizes average flow time. What is the maximum tardiness of this schedule?
  b. Find the schedule that minimizes maximum tardiness. What is the average flow time of this schedule?

4.15. A steel-rolling mill has eight outstanding orders. Processing time in hours for the jobs are (13.4, 2.4, 1.8, 7.9, 12.3, 10.5, 2.4, 8.1), respectively. Each day has 24 working hours. Jobs 1 and 2 are due by the end of today (24 hours from now). Jobs

3 through 5 are due tomorrow, and jobs 6 through 8 are due in 2 days.
  a. Schedule the mill to minimize average completion time.
  b. Schedule the mill to minimize maximum lateness.

4.16. Schedule the orders in Problem 4.15 to minimize average flow time without letting any jobs be tardy.

4.17. In a paint shop, parts are first coated and then baked. The jobs shown in Table 4.14 are currently awaiting processing. Schedule these jobs to minimize makespan.

Table 4.14   Processing Times for Problem 4.17

Job	Coat	Bake
1	4	2
2	5	3
3	12	1
4	8	7
5	6	5
6	4	1
7	8	4

4.18. Parts go through a turning operation and are then inspected. Six parts are available with turning times of (2.4, 1.7, 0.4, 0.1, 3.5, 2.8) hours, respectively. Inspection times are (0.1, 0.4, 0.8, 1.1, 0.4, 0.2) hours. Schedule jobs to minimize the makespan.

4.19. Six jobs are waiting at a milling station. Job data are provided in Table 4.15. Current queue lengths at Inspection, Grinding, Turning, and Drilling are 12.0 hours, 4.5 hours, 3.9 hours, and 0.0 hours, respectively. However, on the average each station requires a 5-hour wait. For each of the standard dispatching rules in Table 4.7, find the corresponding sequence. Current time is 20.

Table 4.15   Available Milling Jobs for Problem 4.19

Job	Arrival to System	Arrival at Mill	Due Date	Operation (machine, p_{ij}) 1	2	3
1	10	10	30	(Mill, 5)	(Turn, 3)	(Drill, 5)
2	0	0	20	(Mill, 3)	(Grind, 4)	(Inspect, 1)
3	5	12	35	(Drill, 4)	(Mill, 6)	
4	7	18	26	(Turn, 3)	(Mill, 7)	(Inspect, 3)
5	12	12	45	(Mill, 10)	(Grind, 4)	(Inspect, 3)
6	2	17	29	(Turn, 4)	(Mill, 8)	

Table 4.16 Job Data for Problem 4.20

Job	Arrival to Plant	Arrival at WS	Due Date	Operation Time (hr)			
				AI	WS	HI	CC
1	12	34	72	1.3	0.10	0.3	0.2
2	34	41	46	1.2	0.12	2.1	0.9
3	25	31	49	5.2	0.20	1.4	0.8
4	29	42	53	2.1	0.30	12.4	1.2
5	36	43	58	3.1	1.05	4.2	1.3
6	18	24	64	8.5	1.35	6.3	2.1
7	24	38	52	5.3	0.85	5.1	1.0

4.20. In a circuit card assembly plant, seven jobs are waiting to be wave soldered. Jobs pass through Auto Insert, Wave Solder, Hand Insert, and Conformal Coating. Job data are given in Table 4.16. Find the wave solder schedules for each of the rules described in Table 4.7. It is currently time 46. Normal waiting time is 2 hours at each workcenter.

4.21. Consider the set of jobs shown in Table 4.17. This is a five-job, four-machine general job shop problem. Generate a nondelay schedule. Compute makespan and percentage utilization for each machine.

Table 4.17 Job Data for Problem 4.21

Job	Operation (machine, processing time)			
	1	2	3	4
1	$(A, 4)$	$(C, 12)$	$(B, 3)$	$(D, 7)$
2	$(C, 11)$	$(D, 5)$	$(A, 3)$	$(B, 4)$
3	$(D, 1)$	$(C, 4)$	$(A, 7)$	$(B, 12)$
4	$(A, 4)$	$(C, 8)$	$(B, 9)$	$(D, 6)$
5	$(D, 7)$	$(C, 14)$	$(A, 3)$	$(B, 8)$

4.22. Prove that there must exist an optimal schedule for a regular measure that is an active schedule.

4.23. Suppose operations could be processed in parts. Total processing time for an operation was fixed but it could be loaded, partially processed, unloaded, queued, loaded, partially processed, unloaded, and so on until completed. If no time was lost in loading and unloading, would it still be true that the optimal schedule need not be a nondelay schedule? What are the advantages and disadvantages of using such an operating policy in practice?

4.24. Johnson's algorithm was described in Section 4.3.2 for two-machine flow shops. This procedure can also be used in two-machine job shops. Suppose jobs may follow any of the sequences $\{a, b\}, \{a\}, \{b\}$, or $\{b, a\}$. Explain how you could use Johnson's algorithm in this case to minimize makespan.

4.25. Modify the nondelay schedule-generation algorithm to generate an active schedule. Use this algorithm to generate an active schedule from Table 4.15.

4.26. Modify the nondelay schedule-generation algorithm to generate all active schedules. (*Hint:* Let f_k be the earliest finish time of operations in A_t given the partial schedule S_t. Modify step 1 to accommodate the definition of an active schedule.)

4.27. Suppose that we have N jobs and job i takes p_i time to setup and run. There are M identical, parallel machines. Each job must be assigned to exactly one of the machines. The objective is to minimize the total makespan. The makespan for any machine is simply the sum of the processing times for the jobs assigned to that machine. The total makespan is the maximum of the makespans for the individual machines.

  a. State a lower bound on total makespan as a function of the p_i.

  b. Show that the schedule formed by ordering the jobs from largest to smallest p_i, and then sequentially assigning jobs to the first available machine, will minimize total makespan. This is referred to as the LPT (Longest Processing Time) rule. Either show this yourself, find a proof in the literature, or disprove by counterexample.

  c. Given the assignment of jobs to machines, how would you sequence jobs on machines to minimize mean flow time?

CHAPTER 5

FLEXIBLE MANUFACTURING SYSTEMS

*It is well to observe the force and virtue and consequences of
discoveries, and these are to be seen nowhere more
conspicuously than in those three which were unknown to
the ancients, and of which the origin, though recent, is
obscure and inglorious; namely, printing, gunpowder and
the magnet. For these three have changed the whole face and
state of things throughout the world.*
—Francis Bacon, translated from *Novum Oganum*

5.1
INTRODUCTION

Flexibility measures the ability to adapt "to a wide range of possible environments" (Sethi and Sethi [1990]). The dynamic, probabilistic, and individualistic world in which we compete makes flexibility a requirement for long-term survival.

The term **flexible manufacturing system**, or FMS, refers to a set of computer numerically controlled (CNC) machine tools and supporting workstations that are connected by an automated material handling system and are controlled by a central computer. FMS technology represents an evolutionary step beyond transfer lines and offers one means by which manufacturing can address the growing customer demand for quick delivery of customized products. The preceding definition includes the key elements of an FMS: (1) automatically reprogrammable machines, (2) automated tool delivery and changing, (3) automated material handling both for transferring parts between machines and for loading/unloading parts at machines, and (4) coordinated control. Many part types can be simultaneously loaded onto the system because machines have the tooling and processing information to work on any part. Thus, parts can arrive at individual machines in any sequence. By reading a code on the part or following supervisory instructions, the part type can be identified (or verified) and the proper processing sequence can be retrieved from the machine's computer memory. The system may include as many as 20 machines. Small systems of one or two machines are normally referred to as a flexible cell.

Sethi and Sethi [1990] categorize 11 types of manufacturing flexibility. Basic flexibilities include Machine (range of operations a machine can perform with minor setup, it is enhanced by generic fixtures, large tool magazines, and automatic tool changers), Material Handling (ability to move various parts between machines and storage areas and maintain proper orientation), and Operation (based on part design, this refers to the ability to use different processing operations to produce part features). These yield system flexibilities for Process (the variety of parts that can be produced with the same setup), Routing (ability to use different machines or operations to produce parts under the same setup, routing flexibility ensures against breakdowns and temporary bottlenecks but requires redundant machines and tooling and sophisticated control capability), Product (ease of changing over the system to produce a new set of parts), Volume (insensitivity of profit to production level), and Expansion (ease of adding additional capacity). Finally, the system flexibilities and control system combine to determine the aggregated flexibility measures Program (ability to run unattended for a long period), Production (range of part types that could be produced without major capital expense), and Market (combination of product, process, volume, and expansion). Microprocessor technology has been the prime factor behind the development of equipment and information technologies that enable flexibility.

Although flexibility does not remove variability from the system, it does enable the system to perform effectively in the presence of wide and varying requirements. The important issue is whether a system can be designed with usefulness over a sufficient time horizon and part mix and with sufficiently small changeover times that it can offer an economical alternative for simultaneous production of multiple, medium-volume part types. The part types assigned to the FMS should have sufficient production volumes to make automation attractive but insufficient to justify dedicated production lines.

The beginnings of the FMS lie in the "link-lines" that began appearing in the 1960s. Link-lines consisted of numerically controlled (NC) machines linked by conveyors. These systems were designed for batch processing, unlike the mass producing transfer lines. The NC machines ran on tapes and required changeover time in switching from one part type batch to another. Many of the machines did have automatic tool changers, however, to permit multiple operations per part.

A modern flexible cell is shown in Figure 5.1, and a full FMS is shown in Figure 5.2. As direct numerical control (DNC) and computer numerical control (CNC) technologies developed for machine tools, and robots and AGVSs (Automated Guided Vehicle Systems) appeared for part handling, the concept of flexible systems became technologically feasible. The loading and unloading of parts onto a machine and the rotation of tools are automatic, taking only seconds and offering the advantage of economic batch quantities of size one.

FMSs are expensive to implement but yield significant savings. Equipment utilization normally runs at most at 30 percent in conventional systems but may be at 85 percent or higher in an FMS. Machines can be kept running for three shifts with lower crew levels, an important consideration as the supply of quality machinists dwindles and resistance to late-shift employment grows. Fewer machines are required, albeit more expensive machines. Talavage and Hannam ([1988], p. 65) give several specific examples of machine savings, including one system where a 10-machine FMS replaced 25 CNC machines at 70 percent of the cost. Floor space requirements are often reduced by one third. The ability to change plans instantaneously suits the trend toward just-in-time manufacturing. Detailed production sequences are not needed well in advance. Use of batches of size 1 reduce work-in-process and throughput time (we will discuss this phenomenon in more detail in the next chapter). Reductions in direct labor cost also accrue with an FMS. An individual worker may monitor several machines, being responsible mainly for maintenance. The only full-time manned workstation is generally the part loading and unloading station.

Figure 5.1   A flexible manufacturing cell. (Photo courtesy of KT-Swasey, Milwaukee, WI)

Many benefits are intangible. Reduced variable cost and throughput time greatly enhance manufacturing competitiveness. Although difficult to quantify, these advantages can and often do mean the difference between survival and dissolution as competitors lower their prices and improve quality. In regard to quality, CNC machines allow on-line inspection and control for improving quality. The constant use of machines also tends to standardize performance by eliminating startup cycles.[1] Another advantage is that FMSs usually have a modular design, which allows phased implementation and expansion as financial and market forces allow. Whereas transfer lines must be economically justified over the life of the intended product, the FMS capability can easily be turned over to a new set of products as the current life cycles end.

A 1985 survey (see Smith et al. in Stecke and Suri [1986]) of 22 U.S. FMSs found that whereas 22 percent of FMSs produce less than 10 part types, 36 percent produce more than 100 part types. Systems were equally divided between those that must periodically batch part types and retool machines and those with sufficient tool storage capacity to run any part. The majority of those systems that operated with batches changed over daily. AGVs, tow carts, conveyors, and robots were each used in approximately the same number of systems for material handling. Three-fourths of the systems were designed for material cutting, and the others were designed for forming or assembly. Utilizations ranged from 90 percent to 30 percent. The low utilizations seemed to be due, at least in part, to changes in business conditions instead of system availability. Meeting due dates was the most important scheduling objective for most systems. The second priority was to maximize machine utilization. This latter objective perhaps derives from the manner in which accounting systems allocate overhead, and the concern of finance managers in justifying these large capital expenditures.

---

[1]Many machines perform significantly differently when hot than when cold. This adversely affects quality and reduces capacity.

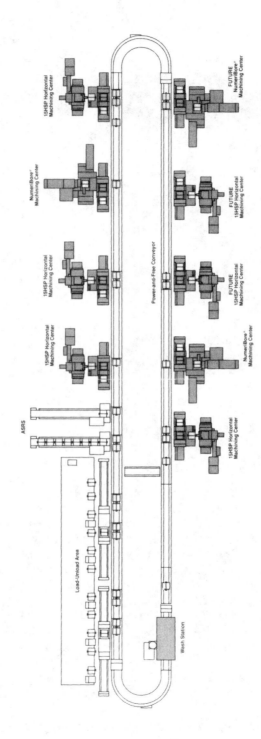

Figure 5.2 A flexible manufacturing system for producing truck brake system components. (Photograph courtesy of Giddings & Lewis, Inc.)

From both the shop floor and strategic perspective it would seem more prudent to concentrate on minimizing throughput times (and equivalently work-in-process levels) once the system is in place and part types are assigned. Users indicated a frequent use of routing flexibility in the systems to adjust when machines went down.

5.2
SYSTEM COMPONENTS

In this section we describe the basic components of a typical FMS. In the next section we consider planning and control algorithms.

5.2.1 Machines

The backbone of any FMS is the set of machines used for value-added operations. Many existing FMSs have been installed for the production of either **prismatic** or **rotational** parts. Prismatic parts, such as engine blocks or pump housings, have no consistent shape. They are often castings or forgings, which are then machined. Rotational parts are those with a symmetrical quality to their cross section and are produced partially with turning operations. A rotational part can be held in place by the lathe's chuck during machining. Robots can grasp the rotational part for loading and unloading onto a machine. Prismatic parts rest in a **fixture** that is specially designed for the part. The fixture has a generic exterior shape that allows it to be located and held in the proper position on the machine. While loading the part onto the FMS, it is placed in the fixture. The part remains in the fixture until it has completed all processing operations and is unloaded from the system. Probes and stops are used each time a fixture is placed on a machine to determine its exact position.

Horizontal machining centers (HMCs) and Head Indexers (HIs) are typically used for machining prismatic parts. Figure 5.3 shows an HMC. The HMC allows spindle access to several sides of the workpiece with the use of a rotary, indexing work-table. Machines hold up to 100 tools. Tool magazines and automatic tool changers at the machines allow many operations to be performed on a part each time it is

Figure 5.3   A horizontal machining center. (Photograph courtesy of Giddings & Lewis, Inc.)

loaded onto the machine. Likewise, many parts can be machined without tooling changeovers. Head indexers are used for machining regular patterns of features. An HI is capable of storing several machining heads, each head containing multiple spindles. With the multiple spindles, each machine head can efficiently produce a regular pattern of holes or bores by simultaneously drilling with each spindle. In addition to increased efficiency, multiple spindles improve dimensional tolerance between hole positions, since the spindles are prepositioned. The actual tools used by the machines are an important aspect of an FMS. Gray et al. [1988] survey issues related to tool management and discuss control policies.

5.2.2 Part Movement System

Automated handling of workpieces is another integral aspect of an FMS. The handling system is designed to transport parts on pallets between workstations. Conveyors, tow carts, rail carts, and AGVSs have all been used in the past, with AGVSs being the dominant approach in the 1980s. Tow carts are simple platforms on wheels that can be engaged by drive chains in the floor and carried along to the desired destination. Rail carts are used when all workstations lie on a straight line. The carts can be self-powered and, unlike the other systems, can move in either direction. This bidirectional movement is possible only when the number of carts is small. With more than one or two, cart interference becomes too frequent. An AGV is a self-powered vehicle, as shown in Figure 5.4. In FMS application, the vehicle generally transports a pallet containing one or more fixtured parts. Most systems follow a wire-guided path in the floor, but self-guided systems are under development. The vehicles generally move in only one direction around the path circuit. An example of this kind of a path is shown in Figure 5.2. Instructions can also be sent through underground wires on vehicle-specific frequencies. In practice, the need for wires embedded in the floor has not been a major problem as the path can be occasionally modified, and, moreover, wireless systems also exist. It seems to be more difficult to design the control system to avoid deadlocking (vehicles blocking each other so none can move) and to decide what the path should be. We will discuss these problems in Chapter 9.

Figure 5.4 An automated guided vehicle.

A mechanism is required for connecting the part movement system, such as the AGVS, to the machine. Parts and their fixtures must be off-loaded from the transport system when they arrive at the destination. A shuttle receives the loaded pallet and stores it until the machine is ready to receive the next part. On completing a part, the part and its fixture are off-loaded to a pallet on the shuttle to await the arrival of an empty vehicle. The new part and its fixture can then be loaded onto the machine for processing. In designing the FMS, it is important to allow sufficient pallet locations in the shuttle to avoid blocking. Blocking would occur, for instance, if all shuttle locations were full of completed parts. New parts could not be loaded onto the shuttle from the part handling system nor could finished parts be removed from the machine. Shuttles can be linear, with input and output to the handling system on different sides of the machine, or have a rotary indexing form.

5.2.3 Supporting Workstations

A number of supporting workstations can also be found in FMSs. The previously mentioned Load/Unload station is used to enter and remove parts from the system. Automatic washers are used to clean parts for machining.

A centralized storage area for pallets may be included. A third, unmanned shift could be accommodated by loading raw parts onto the system before this shift and then unloading the completed parts after the shift. The area can also be used to absorb backlogs at machining centers due to machine breakdowns or other changes in plans.

Coordinate measuring machines (CMMs) have become commonplace for inspecting parts. Dimensions and locations of features can be accurately measured. Each part type has a specific inspection plan. A probe on the machine locates various points on the surface of the machined part. Statistical quality control is applied to ensure the compliance of manufactured product with specifications. The use of CMMs opens up interesting standards questions. Standards are the precise definitions and procedures used to facilitate unambiguous communication. In the past, gages were used for inspection. A round part either did or did not fit through a standard ring with known inside diameter. However, CMMs measure only coordinates. If all you know are three points on the outer surface of the part, and you cannot assume the surface is a perfect circle, then how do you determine the diameter?

5.2.4 System Controller

The brain of the FMS is the system controller. The typical controller is a computer with an attendant worker who keeps track of performance and intercedes when necessary to change priorities or solve problems. The controller must be capable of keeping track of system status. System status involves the location of all parts, tools, and carts, including those waiting to be loaded, and the operational status of each machine. Based on current status and production plans, the controller downloads commands to the individual system components. The components acknowledge receipt of the command and later respond that the command has been executed (or failed to execute). Unless individual machine computers have sufficient storage capability to maintain all part plans, the controller may store part programs, which are downloaded to individual machines as required. The controller must also decide when and how parts should be moved between machines and when parts must be loaded.

In a sophisticated system, the controller is thinking ahead. Predicted part-completion times could be used to send empty carts to a pickup site in advance. Instead of having a predetermined machine sequence, changed only in case of breakdowns, parts can be dynamically routed to the closest available machine with the necessary tooling. Either the controller or the CNC machine itself can keep track of cutting time for each tool and know when it is time to change the tool to prevent quality problems. Travel times of AGVS carts are monitored to indicate when a battery recharge is needed.

The vast number of routine decisions and possible system states makes the development of controller software very complex. Each system has its own characteristics, requiring customized logic. The difficulty of achieving near-optimal control of complex systems has led to the use of small FMSs.

5.3
PLANNING AND CONTROL HIERARCHY

The presence of a system controller naturally implies a hierarchy of planning and control decisions for an FMS. In this section we are primarily concerned with the process of making decisions, that is, the determination of *what* should be communicated. In practice, we must also consider the mechanics of *how* system components communicate.

The National Institute of Standards and Technology (NIST) has been working on a model of "how" (Jones and McLean [1986]) at their Automated Manufacturing Research Facility (AMRF). The AMRF operates a test bed for manufacturing automation and integration. The manufacturing facility is envisioned as consisting of five levels: facility, shop, cell, workstation, and equipment. The FMS would, in most instances, constitute a shop. A hypothetical implementation of this structure is shown in Figure 5.5. Information flows only between adjacent layers. The milling workstation controller can speak to its cell controller and its robot, milling machine, and part buffer, but it cannot communicate directly with the cell 2 controller or other workstations. Planning is performed at each level but the planning horizon differs. While the facility is attempting to meet monthly or quarterly production goals, the cell is planning to finish its current batch of parts today, and the robot is only looking at how it should proceed to the next point on its path. Table 5.1 summarizes the planning horizon and decision types at each level.

In order to accomplish planning, execution, and feedback at each level, a generic control structure is used. This structure is shown in Figure 5.6. Commands are received from the next higher level. The task is broken down into subtasks. Subtasks are assigned to components at the next lower level. A schedule is maintained and lower level resources are allocated to the task. Commands are then sent to the next level for execution. Subtask monitoring is performed through the receipt of status feedback from the lower level. Likewise, task status information is relayed to the next higher level. Each controller has a production manager that receives commands, plans, and schedules subtasks. A queue manager is maintained for each lower level component to manage its assigned subtasks. A dispatch manager receives orders for dispatching and manages subtask execution for each queue manager.

Several hierarchies have been proposed for deciding "what" to communicate in FMSs. We will discuss a popular model based on Stecke [1983]. One should keep

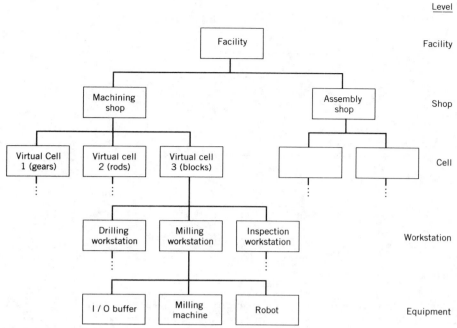

Figure 5.5   Five-level control hierarchy.

Table 5.1 Overview of Facility Control Hierarchy		
Level	Planning Horizon	Decision Types
Facility	Months to years	Manufacturing engineering (CAD, process planning) Administrative functions (accounting, purchasing) Production management (capacity planning, aggregate production planning, quality planning)
Shop	Weeks to months	Order grouping and scheduling Dynamic cell formation Preventive maintenance Inventory control
Cell	Hours to weeks	Sequencing job batches Job routing
Workstation	Minutes to hours	Job setup and fixturing Inspection and cleanup
Equipment	Milliseconds to minutes	Machine level control Sensor data acquisition

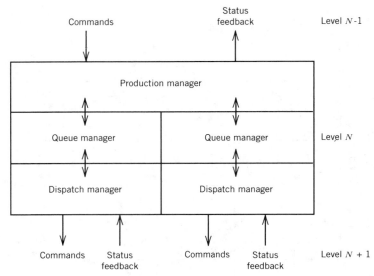

Figure 5.6   Generic control model.

in mind, however, that each system is different. For instance, in one system the key decision may be selection of the part types to load onto the system for simultaneous production during the upcoming week. Another system may use a continuous retooling approach whereby tools are automatically sent to machines from an on-line tool crib as required. Capacity of the tool delivery system and the number of tools of each type required may be the key issues. A third system may have adequate tool capacity to avoid all retooling. Machines are capable of holding all of the tools ever used, and are changed only because of wear. This latter design allows truly flexible scheduling.

The decision hierarchy can be divided into three basic steps: system design, medium-range planning, and short-term operation. System design is concerned with selecting the part types to be assigned to the FMS and the equipment required to manufacture those parts. Medium-range planning revolves around the weekly or daily decisions of what to load onto the system—both part types and tooling. Short-term operation includes scheduling and control of machines and material handling devices. We will examine each of these topics in more detail.

5.3.1 System Design

Several basic questions must be asked when designing an FMS. These include strategic questions concerning why an FMS is being considered. Implementing a high-technology process sends a message of progressiveness to customers. This may be important for several product lines independent of current financial justification. The FMS may complement corporate strategy for increased responsiveness to market and engineering changes. Whatever the reason, it must be decided which part types will be made on the FMS and what operation sequences will be used for these parts. It is best to answer this question initially with a range of part sizes, materials, and machining operations that the system will accommodate. By the nature of flexible manufacturing, we expect the specific parts to change over the life of the system. But, if the system is designed for tractor engine blocks, we clearly will not be using it for key chains. However, if demand drops and engine blocks do not utilize all the capacity, we might move transmission housings to the FMS.

Given the scope of the system, specific hardware and software must be selected. Often this is done in concert with a vendor to assure compatibility. Designing the system is a complex process. Experienced vendors have a better understanding of the possible and desirable features along with their costs, and it is best to involve these system integrators early in the design process. System hardware includes machining centers, tool changers, pallet loaders, cutting tools, pallets, fixtures, system storage, machine buffers, material handling carts, and system controller. Machining centers have their own controllers and features, such as tool-wear sensors and on-line inspection devices. The desired data to be acquired automatically for feedback and feedforward control of operations must be stated and corresponding data-acquisiton devices must be selected. This is true for both machining centers and the material handling system.

Once the system size is determined and the hardware is selected, specific parts can be picked for current assignment to the FMS. Parts designed to be produced by manned workstations may need modification to facilitate FMS manufacture. Ideally a part can have all of its operations performed while positioned in the same fixture, but this is not always possible. Much has been written about the importance of design for (automated) manufacturing, and hence we shall not pursue this topic further here.

We will consider an economic criterion for selecting part types for the FMS. Strategic considerations can then adjust the solution. For each part type being considered for assignment to the FMS, we should determine its variable manufacturing cost on the FMS and on the existing alternative process (in-house conventional machines, unlinked in-house NC machines, or subcontracting, for instance). Fixed cost for conversion to the FMS must also be determined. This would include any necessary redesign and fixtures. Fixed costs can be apportioned over the expected future demand and allocated to the variable cost. Any part with lower estimated production cost for the FMS is a candidate. For a given set of FMS resources, the problem of selecting parts becomes a knapsack problem: Load the FMS to maximize the savings subject to FMS capacity. After first stating the problem mathematically, we will present a simple heuristic and then solve the problem optimally.

We assume all or none of a part type will be made on the FMS. Let P be the productive time per period available on the key, bottleneck FMS resource; p_i be the time per period (time per unit $\times$ units per period) required by part type i on this resource; and s_i be the savings per period (savings per unit $\times$ units per period) if part type i is added to the FMS. With binary decision variables X_i indicating whether i is assigned to the FMS, we have the knapsack problem

$$\text{maximize} \sum_{i=1}^{N} s_i X_i \tag{5.1}$$

subject to

$$\sum_{i=1}^{N} p_i X_i \leq P \tag{5.2}$$
$$X_i \in \{0, 1\}$$

The objective 5.1 accumulates the savings resulting from the assignment of part types to the FMS. We would like to assign all part types with positive s_i, but we are constrained by the key resource, as shown in equation 5.2. The resource can be thought of as a knapsack. We wish to fill the knapsack with the most valuable (largest s_i) items without violating its size P. The relative desirability of including a part type is given by the savings per resource consumed. Thus it is informative to place part types in nonincreasing order of s_i/p_i. Our greedy heuristic is as follows:

Greedy Knapsack Heuristic

STEP 1. Order part types $[1]$ to $[N]$ such that

$$\frac{s_{[1]}}{p_{[1]}} \geq \frac{s_{[2]}}{p_{[2]}} \geq \cdots \geq \frac{s_{[N]}}{p_{[N]}}$$

STEP 2. For $i = 1$ to N: select part type $[i]$ if $s_i > 0$ and inclusion is feasible.

The heuristic simply checks each part type in turn and assigns it to the FMS if savings are positive and sufficient capacity exists.

EXAMPLE 5.1

Eight parts are being considered for an automated workcell. These parts are currently purchased from a vendor. The workcell will be available 250 hours per period. The upper portion of Table 5.2 gives current unit cost for each part type, raw material cost, demand rate, and unit production time. The cell is charged at the rate of $50/hour, which includes all related expenses. Which parts should be added to the new cell?

Table 5.2   Part Type Data for Example 5.1

	\multicolumn{8}{c}{Part Type}							
	1	2	3	4	5	6	7	8
Unit purchase price	200	155	300	125	300	86	93	165
Material cost	45	35	124	50	120	34	36	114
Demand per period	100	50	50	75	60	30	50	600
Hours per unit	1.0	2.0	4.0	1.0	2.0	1.0	1.0	0.5
Unit savings	105	20	−24	25	60	2	7	26
Hours per period	100	100	200	75	120	30	50	300
Savings per hour	105	10	—	25	30	2	7	—

Solution

As a first step we might check to see which products are desirable and feasible. A part is desirable if conversion to the FMS will save money. Savings per unit are given by purchase cost−[material cost + processing cost]. Processing cost is determined by production time and the cell overhead rate. For part type 1 we obtain savings per unit = $200. - [45. + (50.) \cdot 1.0] = 105$. The remaining unit savings values are found in the same manner and displayed in Table 5.2. Part type 3 is immediately eliminated; it is cheaper to purchase this part. The part would be manufactured in-house only if justifications such as quality or local control overrode economics. The s_i values are found by multiplying unit savings by demand per period. For instance, $s_1 = (105)(100) = 10,500$. Next, we check for resource usage. For part type 1, $p_1 = (1.)100 = 100$ hours per period. Other p_i values are also shown in the Hours per Period row of the table. Part type 8 is eliminated, since it requires more than the allotted 250 hours per period. Examining the data we note that part 8 is a high-volume, low-value-added part that might justify its own dedicated cell. Finally, it is interesting to order remaining parts based on savings per machine hour. These values form the last row in Table 5.2. Part type 1 appears the best candidate for the FMS. Parts 5 and 4 are distant second and third choices, with parts 2, 7, and 6 being even less desirable. Let us try applying the heuristic to determine which parts it recommends.

STEP 1. Ordering. From the Savings per Hour row of Table 5.2, the order of part types in step 1 is 1, 5, 4, 2, 7, 6.

STEP 2. Assignment.

First, part 1 is assigned to the FMS, setting resource usage to 100 hours.

Second, part 5 is assigned, setting resource usage to $100 + 120 = 220$ hours.

Third, part 4 is passed over, since it requires 75 hours and only 30 of the 250 remain.

Fourth, part 2 is passed over, since it requires 100 hours.

Fifth, part 7 is passed over, since it requires 50 hours.

Sixth, part 6 is assigned to the FMS, since the required 30 hours are available.

The result is the assignment of parts 1, 5, and 6 to the FMS, using all 250 hours per period and saving $105(100) + 60(60) + 30(2) = 14,160$. ∎

We will now examine how this knapsack problem can be solved optimally. The discrete nature of resource consumption complicates the solution, but for a single key resource the optimal assignment can readily be found by dynamic programming. In dynamic programming we make a series of decisions. Each decision is called a **stage** and is based on some input **state.** The state corresponds to the amount of resource available. Decisions yield a return but consume resources, thus changing the state for the next stage. A **recursive equation** computes the return and ties the state variable together between stages. Stages and decisions must be picked to satisfy the **Principle of Optimality**. This principle states that for any initial stage, state, and decision, subsequent decisions must be optimal for the remainder of the problem that results from this initial decision.

For our knapsack problem each decision variable is a stage. The decision is whether or not to include that variable (part type) in the knapsack (FMS). The state is the remaining space in the knapsack (time on the FMS). Define $f_i(\rho)$ as the cost savings for the optimal decision regarding part types 1 to i if they are allowed to occupy ρ time per period on the FMS. Then, assuming $s_i \geq 0$,

$$f_1(\rho) = \begin{cases} s_1 & \rho \geq p_1 \\ 0 & \rho < p_1 \end{cases} \tag{5.3}$$

Equation 5.3 acknowledges that if we were considering only part type 1, we would assign it to the FMS provided that this saved money and sufficient time was available on the FMS.

For $2 \leq i \leq N$,

$$f_i(\rho) = \begin{cases} \max_{X_i = 0,1} [s_i X_i + f_{i-1}(\rho - p_i X_i)] & \rho \geq p_i \\ f_{i-1}(\rho) & \rho < p_i \end{cases} \tag{5.4}$$

Equation 5.3 starts the process. Recursive equation 5.4 controls the transitions between stages. The problem is scaled such that all p_i are integers. Then, f_1 is found for all integers $\rho \leq P$. Storing these results, $f_2(\rho)$ values are found for all integers $\rho \leq P$ using equation 5.4. With $f_2(\rho)$ values known, equation 5.4 can be applied to find $f_3(\rho)$. The process continues until $f_N(P)$ is found. This is the maximal savings, and we trace back through the stored stage solutions to find the optimal solution.

EXAMPLE 5.2

Let's now solve the problem presented in Example 5.1 optimally.

Solution

Note that although we must find $f_i(\rho)$ values for all values of $1 \leq \rho \leq 250$, in reality, we require at least $\rho = 30$ hours before any part can be assigned to the FMS. Moreover, we really need to check for changes in $f_i(\rho)$ values only when ρ increases sufficiently to allow a different option for the parts assigned to the FMS. In other words, if $\rho < 30$, no parts can be assigned. At $\rho = 30$, part type 6 becomes eligible for assignment. Once we allow 30 hours of resource usage, the solution cannot change until ρ is increased to at least 50 hours. At this point we may choose between parts 6 or 7. Our choices do not change again until $\rho = 75$, at which point part 4 can be considered. The next change occurs at $\rho = 80$; we now have the option of choosing both 6 and 7. We will make use of this discrete nature of the problem to reduce our calculations.

STAGE 1. Assume only part type 1 exists. For $\rho < 100$ the part cannot be selected, because $X_1 = 0$ and $f_1(\rho) = 0$. For $\rho \geq 100$, we select $X_1 = 1$ and obtain $f_1(\rho) = 100(105) = 10,500$.

STAGE 2. Solve the two-stage problem for part types 1 and 2. If $\rho < 100$, neither part is feasible. If $\rho = 100$ we can select either part. Our choices are $X_2 = 0$ or 1 corresponding to $0 + f_1(100) = 10,500$ or $50(20) + f_1(0) = 1000$, respectively. The best choice is $X_2 = 0$. On reaching $\rho = 200$ we may also select part type 2 for a total return of $f_2(200) = 50(20) + f_1(100) = 11,500$.

The process continues up through stage 6. The recursive function 5.4 is implicitly evaluated $6 \cdot 250$ times, once for each ρ value at each stage. A summary of outcomes showing break points in decisions is shown in Table 5.3.

As an example of the computations, consider finding $f_5(195)$. We have two choices. If $X_5 = 0$, we obtain $f_5(195) = 0 + f_4(195) = 12,375$. If we set $X_5 = 1$, we obtain $f_5(195) = 60(60) + f_4(195 - 120) = 5475$. (The 120 hours subtracted in the state variable are used for part type 5 and, hence, are not available for parts 1 to 4.) Clearly, we stick with $X_5 = 0$. The optimal solution is to load parts 1, 5, and 6 for a return of 14,160. ∎

Table 5.3   Dynamic Programming Computation Outcomes

State	Stage					
	1	2	3	4	5	6
ρ	$f_1(\rho)$	$f_2(\rho)$	$f_3(\rho)$	$f_4(\rho)$	$f_5(\rho)$	$f_6(\rho)$
0	0	0	0	0	0	0
30	0	0	0	0	60	60
50	0	0	0	0	60	350
75	0	0	1,875	1,875	1,875	1,875
100	10,500	10,500	10,500	10,500	10,500	10,500
130	10,500	10,500	10,500	10,500	10,560	10,560
150	10,500	10,500	10,500	10,500	10,560	10,850
175	10,500	10,500	12,375	12,375	12,375	12,375
200	10,500	11,500	12,375	12,375	12,375	12,375
205	10,500	11,500	12,375	12,375	12,435	12,435
220	10,500	11,500	12,375	14,100	14,100	14,100
250	10,500	11,500	12,375	14,100	14,160	14,160

The last aspect of system design is layout. Machines must be located, machine buffers and central storage planned, and the material handling path laid out. The techniques of Chapters 7 and 9 are applicable with the added reminder that flexibility must be a key consideration for the FMS. In planning the layout, flexibility must be designed in for product **routings, demand,** and **mix**. Routing here refers both to the use of tools (operations) and the movement of parts between machines. Future possibilities should be considered as well, since system life should extend beyond the current conditions.

In this section we have listed the basic sequence of decisions necessary in designing an FMS. These can be summarized as (1) set strategic objectives and budget; (2) set capability of system (operations, work envelopes); (3) select machines and outline control strategies; (4) plan and select parts based on capacity and economics; and (5) finalize layout. To aid in this endeavor, models were provided for part selection. Other chapters in this text cover layout and material handling systems. During the design stage, multiple alternative part and machine selections should be coupled with potential layouts and should be modeled as a system. Only then can the optimal system be determined. Approaches for building and evaluating such models will be discussed in Chapters 11 and 12.

5.3.2 System Setup

Now that the system has been brought on-line and its mission defined, we must operate the system. The main tactical issue concerns the assignment of operations, and the accompanying tooling, to machines. Two basic environments are possible. First, the complete set of tools used by the parts assigned to the FMS is larger than the storage capacity on the machines. Thus, only a subset of the part types may be run on the system at any time. In the second environment, machine tools can hold all the tools required. In the first case, parts must be combined into batches. A batch consists of a specific number of parts of each type to be produced. If the part type is not to be included in this batch, it is assigned 0 parts. Batches are produced sequentially. The system is set up to produce a batch, and when the planned number of each part type is completed, machines are retooled and the next batch started. We refer to batch formation as the **part selection** problem. Keeping due dates in mind, batches may be formed to balance work loads on machine centers and to utilize a fixed period length, generally one day to one week. Whitney and Gaul [1985] discuss the use of batches. Stecke and Kim [1988] note that it may be advantageous to operate on a continuous basis. As soon as any part type in the batch is completed, a new part type is started in its place. Any required tools for the new part are loaded at this time.

In all of these operating environments, parts scheduled to enter the system must be sequenced and routed. We refer to this as the **loading** problem. Loading includes the assignment of tools and part operations to specific machines. The allocation of pallets to part types is also included.

Part Selection Problem

The objective of part selection is to place required parts into compatible batches. Each batch should use all machines, require a limited number of tools on each machine, and have similar due dates for parts in the batch. A simple, greedy heuristic for forming batches would be to first arrange part orders based on due dates. Then, part orders are sequentially added to the current batch provided no constraint is violated. We are not yet concerned with placing specific tools on specific machines, but only with the set of tasks to be assigned to each type of machine, such as the set of horizontal machining centers. Thus, constraints include available tool space and time allotted to each machine type. If an entire job cannot be accepted because of a lack of machine time, it may be desirable to break up the job and to produce as many parts as possible in this shift (period) and save the remainder of the job for the next shift. Once the batch is formed, we solve a loading problem to assign tools to specific machines and to route parts. The parts are then run, and the process repeats next period.

EXAMPLE 5.3

Consider the set of parts shown in Table 5.4. (Note that two orders exist for part type *a* with different due dates.) Three machines of type *A* and one of type *B* are available. Most part types require processing on both machine types. To make a unit of part *a*, 0.1 hours are required on an *A* machine *and* 0.3 hours are required on a *B* machine. Machines are set up once each day and are normally available for 12 hours per day. Both machine types are capable of holding two tools. The initial letter in the tool name indicates the machine required. Our objective is to select the parts to be produced today.

Table 5.4   Part Data for Example 5.3

Part Type	Order Size	Due Date	Unit Processing Time (hrs) Machine Type *A*	Machine Type *B*	Tools
a	5	0	0.1	0.3	$A1, B2$
b	10	1	1.2	0.0	$A2$
c	25	1	0.7	0.4	$A3, B4$
d	10	1	0.1	0.2	$A1, B2$
e	4	2	0.3	0.2	$A5, B3$
a	10	4	0.3	0.2	$A1, B2$

Solution

Parts are already ordered by due dates. We will begin to select parts. Table 5.5 tracks results at each iteration. We are allowed 36 hours and six tools on machine type A. Machine type B permits 12 hours and two tools.

Table 5.5   Iterative Selection for Example 5.3

Step	Assigned Parts	Time Assigned A	Time Assigned B	Tools Assigned A	Tools Assigned B
1	a	0.5	1.5	A1	B2
2	a, b	12.5	1.5	A1, A2	B2
3	a, b, c	30.0	11.5	A1, A2, A3	B2, B4
4	$a, b, c, d(2/10)$	30.2	11.9	A1, A2, A3	B2, B4
5	$a, b, c, d(2/10)$	30.2	11.9	A1, A2, A3	B2, B4
6	$a, b, c, d(2/10)$	30.2	11.9	A1, A2, A3	B2, B4

STEP 1. Check order a. Order a places a load of 0.5 hours on machines A and 1.5 hours on machine B. Sufficient tool space exists; thus, order a is assigned to the first batch.

STEP 2. Check order b. Order b adds an additional 12 hours to machines A and requires one new tool space. This load is feasible, and thus order b is added.

STEP 3. Check order c. Order c adds 17.5 hours to machines A bringing the total to 30.0 hours. This is feasible for three machines. Space also exists for tool $A3$. Utilization of machine B is increased to 11.5 hours, and the tool space is filled.

STEP 4. Check order d. Order d must be assigned to batch 2. Insufficient time exists for machine B. Note that this means order d will be completed late. We could make 2 units of order d this period and carry 8 units forward. Since the order is due today, let us suppose the planner decides to make the 2 units today.

STEP 5. Check order e. Order e must also be assigned to batch 2 because of the insufficient time and tool space of machine B.

STEP 6. Check order a. Order a must be assigned to a later batch. No new tools are required but the required 3 hours of time on machine B are unavailable. ∎

The part selection problem can be formulated as a mixed-integer program. We assume that a time-phased set of part orders, D_{it} for part i in period t, is given along with process plans for each part type. Machine type j has P_j hours available per period and each unit of i produced requires p_{ij} hours. Let x_{it} be the number of part type i made in period t. A period could be anywhere between a shift and a week, probably a day. y_{ljt} will be 1 if tool type l is assigned to machine type j in period t and 0 otherwise. Machines of type j have K_j tool slots, and tool l requires k_{lj} tool slots on machine j. This notation allows tools of different sizes to be modeled. We let $l\,j(i)$ denote the set of tools l required on machine j to produce part type i. Our objective will be to minimize inventory costs while meeting due dates. Holding cost is h_i per period for part i. Our formulation is

$$\text{minimize} \sum_{i=1}^{N} \sum_{t=1}^{T} h_i \sum_{r=1}^{t} (x_{ir} - D_{ir}) \tag{5.5}$$

subject to

$$\sum_{r=1}^{t} x_{ir} \geq \sum_{r=1}^{t} D_{ir} \quad \text{for all } i, t \tag{5.6}$$

$$\sum_{i=1}^{N} p_{ij} x_{it} \leq P_j \quad \text{for all } j, t \tag{5.7}$$

$$x_{it} \leq M \cdot y_{ljt} \quad \text{for all } l\, j(i), t \tag{5.8}$$

$$\sum_{l=1}^{L} k_{lj} y_{ljt} \leq K_j \quad \text{for all } j, t \tag{5.9}$$

$$0 \leq x_{it} \quad y_{ljt} \quad 0 \text{ or } 1 \tag{5.10}$$

The objective 5.5 accumulates holding costs for each period based on the cumulative excess of production over demand. Shortages are prevented by constraints 5.6. For each part type in each period of the planning horizon, cumulative production must at least equal cumulative demand. Production time is treated in constraints 5.7 to avoid overloading any machine type in any period. Constraints 5.8 ensure that all required tools are assigned to machines before any production is scheduled. (M is the notorious Big-M and should be, at least, as large as cumulative demand.) Constraints 5.9 restrict the assignment of tools to the tool storage space available on each machine type in each period.

The difficulty with this formulation lies in the large number of binary variables required for tooling decisions. However, if capacity in certain periods is the major concern and sufficient tool space exists on machines for most desired part mixes, the tooling variables y_{lkt} and constraints 5.8 and 5.9 can be dropped. The remaining linear program is easily solved.

The reader may have noticed that whereas the heuristic piled as much work as possible into the first period, this formulation will try to delay work as long as possible without missing due dates. Which approach is right? The answer again depends on the situation. If complete information on orders is available well in advance, it makes sense to delay production until parts are needed. However, if we frequently are besieged with rush orders, it makes more sense to take care of the demand we know about in order to maintain slack capacity for the rushes. In either formulation compromises could be added. For instance, the heuristic might add a rule saying never produce orders more than X days in advance. In the mathematical formulation we could either add safety lead time by subtracting X periods from the real due date before entering the data or add safety stock by artificially increasing the first order for each part type by the desired safety stock.

EXAMPLE 5.4

We can list the mathematical program for the data of Example 5.3. Unfortunately, we note at the start that an order for five units of part type a is already tardy.

We could wait until the model returns an answer of "no feasible solution" before recognizing this problem; instead, let's assume that the problem has been recognized and that due dates have been modified such that the first three orders are due on day 1 and the second three are due on day 2.

Instead of giving the entire formulation, we will focus on certain aspects.

a. How many continuous and binary variables are needed?

Solution

Ten x_{it} variables corresponding to $i = a, b, c, d, e$ and $t = 1, 2$ are required. Fourteen binary variables are needed corresponding to tools $A1$, $A2$, $A3$, $A5$ on machine type A and tools $B1$, $B2$, $B3$ on machine type B in both periods.

b. State the objective function.

$$\text{minimize } h_a[(x_{a1} - 5) + (x_{a1} + x_{a2} - 15)]$$
$$+ h_b[(x_{b1} - 10) + (x_{b1} + x_{b2} - 10)]$$
$$+ h_c[(x_{c1} - 25) + (x_{c1} + x_{c2} - 25)]$$
$$+ h_d[(x_{d1} - 0) + (x_{d1} + x_{d2} - 10)]$$
$$+ h_e[(x_{e1} - 0) + (x_{e1} + x_{e2} - 4)]$$

c. State the number of constraints of type 5.6, 5.7, 5.8, 5.9, and 5.10, respectively.

There are 10 constraints of form 5.6, one for each part type in each period. There are 4 constraints of the form 5.7 and 4 of the form 5.9. In both instances, there is a constraint for machine type A and one for B in both periods. Eighteen constraints are needed for form 5.8. The five part types use a total of nine tools. Thus, we need 9 constraints in each period. As an example, part type a in period 1 requires

$$x_{a1} - M y_{A1,A,1} \leq 0 \qquad x_{a1} - M y_{B2,B,1} \leq 0$$

Each variable has a constraint of the form 5.10, making 24 constraints.   ∎

Incremental Part Selection

In using discrete (quantity and time) production batches, a major tooling change-over is required at the end of each period. The machines are idle while the setup workers, who may have been idle during much of the previous production period, now rush to change tools. Also, as we approach the end of a period, the shortage of remaining work may lead to low utilization levels for certain machines. Third, batch time normally corresponds to an organizationally meaningful period such as a shift or a day. There is no reason to believe that this will conveniently tie in to production requirements for part lots. Thus, whereas fixed-length production batches may be easiest to implement, Stecke and Kim [1988] found a "flexible" incremental approach to be preferable from a productivity viewpoint. As before, several part types are in process at any time. However, the system operates on more of a continuous basis. Whenever the demand requirements for any part type are completed, the part selection and loading problem is resolved with the completed part type removed. The system may be constrained to continue producing the part types still in process. In addition, one or more new part types, and the required tools, may be added.

Let us assume that each operation requires a specific machine type and that due dates are not relevant. Our objective is to minimize makespan to complete all currently available part orders. As a surrogate, we will minimize idle time by balancing work loads subject to part demand and tool magazine capacity. We let x_i be the relative production rate for part i. The t subscript is dropped since we plan only the immediate policy. The objective is to balance work load, and thus we define c_{j1} and c_{j2} as penalties per time for over- and underuse of machine j. If one or more machines are known to be bottlenecks, they should be assigned large undertime penalty costs. The formulation is then

$$\text{minimize} \sum_j c_{j1} w_{j1}^+ + \sum_j c_{j2} w_{j2}^- \tag{5.11}$$

subject to

$$\sum_i p_{ij} x_i = P_j + w_j^+ - w_j^- \qquad \text{for all } j \tag{5.12}$$

$$x_i \leq M \cdot y_{lj} \qquad \text{for all } l\, j(i) \tag{5.13}$$

$$\sum_{l=1}^{L} k_{lj} y_{lj} \leq K_j \qquad \text{for all } j \tag{5.14}$$

$$0 \leq x_i \leq f_i \qquad x_i \text{ integer} \tag{5.15}$$

$$y_{lj}\, 0 \text{ or } 1 \qquad w_j^+, w_j^- \geq 0 \tag{5.16}$$

Equations 5.12 record the overtime (w_j^+) and undertime (w_j^-) assigned to machine type j relative to the available time P_j. Since no fixed duration periods are planned, P_j can be selected arbitrarily. Small P_j values are advisable since this will force relative production ratios x_i to be small—a desirable result. Moreover, large P_j ignore the reality that the system will soon complete the requirements for a part type and need to be reconfigured. However, P_j must be large enough to allow integer amounts of all part types to be scheduled. If all part types have similar throughput times, then one approach would be to set P_j equal to mean throughput time. In this case, x_i represents the number of part type i in process, which is limited in expression 5.15 by the number of fixtures f_i. Equations 5.13 ensure that all tools necessary to produce the selected parts are loaded. Constraints 5.14 limit the number of tools loaded onto the system.

After solving this formulation, the system is set up and is operated until all demand is met for some part type. The model is then resolved with the remaining parts plus any new part orders that may have arrived. To ensure continuity, we could constrain x_i to be at least one for part types currently loaded onto the system but not yet completed. Alternatively, we could add a cost for any tools not currently loaded. This cost term would be $\sum_{lj \in \theta} c_{lj} y_{lj}$ where θ is the set of tools currently not loaded.

If only one machine of each type exists, or K_j is tool magazine capacity per machine, then the formulation of equations 5.11 through 5.16 determines tool loading as well as production rates. (This is also true for formulations 5.5 through 5.10.) If K_j is the total tool space for several machines of type j, then we must add a second step to determine on which machine(s) each tool is loaded. We discuss this issue next.

Loading Problem

We now assume that the batch of parts to be produced this period is known. Objectives and active constraints for the loading problem will differ from installation to installation. Objectives that may be important include minimizing work-in-process inventory, cost of tools, or variable machining cost. On the other hand, batches were formed by a deterministic model. We may prefer a loading solution that is robust and flexible, that is, one that continues to function acceptably in the face of random events such as machine breakdowns and changes in plans. Even within a given installation, the importance of the various objectives changes with time as organizational finances and demand vary. Accordingly, no single problem statement can be made. We will, however, provide a basic mathematical programming formulation of a typical loading problem and then discuss heuristics approaches.

Letting index i range over all part operations for any part included in the current batch, we define the following decision variables:

$$x_{ij} = \text{proportion of operations } i \text{ assigned to machine } j$$

$$y_{lj} = \begin{cases} 1 & \text{if tool l is assigned to individual machine } j \\ 0 & \text{otherwise} \end{cases}$$

Other parameters of relevance are

c_{ij} the cost to perform operation i (all parts) on machine j

$l(i)$ the tool required for operation i

n_l the number of type l tools available

Part routings are comprised of operations. Each operation requires a specific tool and has a set of feasible machines to which it may be assigned. The coefficients c_{ij} reflect the fact that although several machines may be capable of performing a given operation, certain machines may be more efficient than others. As an extreme example, a pattern of holes may be drilled sequentially on a machining center or simultaneously by a multi-spindle drill head. A cost minimization model is

$$\text{minimize} \sum_{i=1}^{I} \sum_{j=1}^{J} c_{ij} x_{ij} \tag{5.17}$$

subject to

$$\sum_{j=1}^{J} x_{ij} = 1 \qquad \text{for all } i \tag{5.18}$$

$$\sum_{i=1}^{I} p_{ij} x_{ij} \le P_j \qquad \text{for all } j \tag{5.19}$$

$$\sum_{l=1}^{L} k_{lj} y_{lj} \le K_j \qquad \text{for all } j \tag{5.20}$$

$$x_{ij} - y_{l(i),j} \le 0 \qquad \text{for all } i, j \tag{5.21}$$

$$\sum_{j=1}^{J} y_{lj} \le n_l \qquad \text{for all } l \tag{5.22}$$

$$0 \le x_{ij} \le 1 \qquad y_{lj} = 0 \text{ or } 1 \tag{5.23}$$

The objective function 5.17 indicates the objective of minimizing variable production cost. Constraints 5.18 ensure that each operation i is assigned to one or more machines. By restricting x_{ij} to be integers, we could force each operation to be assigned to a unique machine. Constraints 5.19 restrict the amount of processing time assigned to each machine to be no more than the time available during the period. Constraints 5.20 ensure sufficient space in the tool magazine to hold those tools assigned to machine j. The capacity constraints 5.19 and 5.20 have the standard form that the resource usage, expressed on the left-hand side of the constraint as a function of the decision variables, does not exceed the amount of resource available, as given on the right-hand side. To assure that tools are actually mounted on the necessary machines, we add constraints 5.21. Finally, constraints 5.22 recognize the limit on the number of tools available for each tool type.

One advantage of such a formulation is the clear specification of the objective and constraints. Another advantage is the flexibility to add restrictions that can be expressed algebraically. For example, if L_l is the maximum allowable usage per period for tool l (tools can be replaced or reground between periods), we could add

$$\sum_{l(i)=l} p_{ij}x_{ij} \leq L_l \qquad \text{for all } l$$

On the other hand, this formulation does not readily handle concerns about part movement within the system. It would be nice to keep parts on one machine for as many consecutive operations as possible. We might consider aggregating consecutive operations into a pseudo-operation. When the pseudo-operation is assigned to a machine, this implies that all of the component operations are assigned to that machine. Unfortunately, this significantly complicates specification of the tool assignment (feasibility) constraints if several operations on different parts use the same tool. We will look at heuristics that consider part movement shortly.

Numerous formulations of the loading problem have appeared in print. The large number attests to the realization that the problem is important; each installation faces a slightly different problem, and a general solution is beyond our reach. O'Grady and Menon [1987] applied goal programming to handle the multiple objectives of the loading problem. Computationally, an average of 25 minutes were required to solve a 6-machine, 14-part type, 311-tool problem on a VAX11/780 computer. Sarin and Chen [1987], however, solved a standard cost-minimization integer-programming formulation of a 4-machine, 10-operation problem in under 1 minute. It seems reasonable to conclude that small problems can be solved optimally *if* the relevant data can be automatically assembled into the proper input format for the solution algorithm. Nevertheless, there still seems to be a need for simple heuristics that can generate solutions that satisfy multiple criteria and complex constraints.

Heuristic rules can be useful for achieving solutions that score highly on multiple criteria of interest without requiring prohibitive computational effort. The loading problem requires assigning tools and production operations to machines. We have mentioned the objectives of balancing work loads, minimizing intermachine moves, providing routing flexibility, and minimizing tooling investment. Let us see how we might apply heuristics to the loading problem. Our first objective was to equalize work loads. We know from scheduling theory that placing tasks in nonincreasing order of processing time and then assigning tasks sequentially to the first available machine tends to balance work loads.[2] We could pretend that all operations are

---

[2]This is the LPT (Longest Processing Time) rule. The LPT rule minimizes the makespan when scheduling identical, parallel processors.

separate tasks and assign them accordingly to machines. Suppose we were instead interested in minimizing intermachine moves. It seems clear that a good strategy would be to sequence the operations on each part so that long strings of operations could be performed on the same machine. Likewise, we could use the rule that once a part was assigned to a machine, as many operations as possible would be completed before removing the part. Let us next consider routing flexibility, the objective being to make as many machines as possible capable of performing an operation. We could divide machines of each type into the fewest number of groups possible and assign the same set of tools to each machine in a group. Machines within a group are thus interchangeable. A lower limit on the number of groups is given by the number of machines needed to store the different types of tools required. If tooling investment dominates, simply limit the number of tools of each type allowed, that is, some machines may have empty tool slots. Each of these rules is simple to implement and useful for accomplishing our objectives. We will link these rules together to sketch a heuristic loading procedure. This is only one of many possible heuristics. The reader faced with an actual loading problem is encouraged to first consider what their objectives are, the relative importance of these objectives, and whether the solution frequency and problem size are amenable to optimization before opting for such an approach.

The loading heuristic will be divided into two phases. Phase I assigns operations to machine types. This phase is required only when operations can be performed on more than one machine type. We utilize this latitude to equalize work load. Operations are ordered based on the number of different machine types to which they may be assigned. From the operations with the fewest choices, we select the longest operation (total batch processing time) and assign it to the machine type that will end up with the lowest utilization. In this way we hope to generate a feasible assignment with nearly equal work loads on each machine type. If we fail to find a feasible assignment, we must backtrack and reassign operations. (This latter step is left out of the description that follows but could be implemented similarly to the backtracking scheme discussed in Section 2.3.)

Phase II involves three steps. Operations are combined to reduce material handling transfers between machines, machine groups are formed, and operations (with their tools) are assigned to groups. We call a set of sequential operations on a part that are combined into a single operation, a **cluster**. Thereafter a cluster is treated as a single operation requiring all of the tools needed by any operation in the cluster. In this fashion, problem size is reduced. Also, parts cannot be moved between machines except when passing from one operation cluster to another. This reduces part movement. In the second step, machine groups are formed by identically tooling machines of the same type. Groups provide routing flexibility but obviously increase tooling cost. Assuming routing flexibility to be more important, we form large groups. The number of groups is determined by the number of tool slots needed for the operations assigned to this machine type. Last, operations are assigned to machine groups within each machine type to equalize work load.[3] Since multiple operations (clusters) may require the same tool, routing flexibility may be further enhanced as operations are assigned. This sequence of decisions provides a mechanism for dividing the hard loading problem into a set of simpler problems that can be solved sequentially.

---

[3]If group sizes are unequal, the modeler interested in minimizing throughput times instead of maximizing capacity may wish to purposely unbalance work loads, allocating a higher average work load to machines in the larger groups.

Our added notation is

m_j = the number of machines of type j in the FMS

k_{ij} = tool space required if operation cluster i is assigned to machine type j

κ_j = unassigned available tool slots per machine type j per period

τ = the maximum operation time allowed for a cluster

ψ_j = unassigned available time per machine type j per period

τ is a control parameter for the heuristic. When forming clusters, we use τ to limit the operation aggregation process. As τ increases, fewer but longer operations will be fed to the solution procedure. Material handling should decline but work - load balancing may become more difficult. We will use the term Δk_{ij} as a dynamic variable to indicate the number of tools that must be added to machine type j to perform operation cluster i. This term is dynamic in the sense that it depends on which tools have already been assigned to j. If two operations use the same tool, then once the first operation is assigned to type j machines, no additional tool slots are used by assigning the second operation. Such considerations are relatively easy to handle in the coding of heuristics as compared with their inclusion in mathematical programming models.

Loading Heuristic

Phase I: Machine Type Selection

STEP 0. Initialize. $\psi_j = P_j$. $\kappa_j = K_j$.

STEP 1. For each unassigned operation of each part, determine the number of feasible machine types. Type j is feasible iff $\psi_j \geq p_{ij}/m_j$ and $\kappa_j \geq \Delta k_{ij}/m_j$. (If the number of feasible machine types = 0 for any operation, backtrack.)

STEP 2. From the set of operations with the fewest number of feasible machine types, select $i^* = \text{argmax}_i\{\min_j p_{ij}\}$. Assign operation i^* to $j^* = \text{argmax} \{\psi_j - p_{i^*j}/m_j\}$. Update $\psi_{j^*} = \psi_j - p_{i^*j}/m_j, \kappa_j = \kappa_j - k_{i^*j}/m_j$. If any operations are unassigned, go to 1.

Phase II: Tool and Operation Assignment (for Each Machine Type j)

STEP 1. Clustering. For part type $n = 1$ to N: Starting at the first operation and continuing to the last, add the next operation to the current cluster (combine this operation with the last); unless a new machine type is required, cluster time would exceed τ or cluster tool space would exceed K_j. *Note:* Clusters now replace operations.

STEP 2. Form groups. Let σ_j be the number of tool slots required by the set of tools needed to produce all operations assigned to machine type j. Find the minimum number of feasible groups $g_j = \lceil \sigma_j/K_j \rceil$. Sequentially assign machines to groups to equalize group size, that is, group 1 receives machines $1, g_j + 1, 2g_j + 1, \ldots$.

STEP 3. Assign operations. Order operations for machine type j in nonincreasing order of processing time, that is, $p_{[1]j} \geq p_{[2]j} \geq \cdots \geq p_{[n]j}$. For operation [1] to [$n$], consider the machine groups with sufficient unfilled tool slots: Assign operations to the group with smallest work load/machine after assignment (LPT).

EXAMPLE 5.5

A machine group is to be loaded to produce three part types. Relevant data are shown in Table 5.6. Certain operations can be performed by more than one machine type, as shown by the processing times in the table. Only one machine type should be selected for each operation, however. The machine group has two machines of type A, two of type B, and one type C machine. Machines of type $A, B,$ and C can hold three, one and four tools, respectively. Each machine is expected to be available 800 minutes during this period.

Table 5.6   Batch Data for Example 5.5

Part	Demand	Operation	Unit Processing Time (min)			Tool
			A	B	C	
		1	12	11	10	a
1	40	2	13	15	∞	b
		3	14	14	∞	c
2	100	1	2	4	∞	a
		2	2	6	6	c
		1	4	∞	∞	d
3	100	2	5	∞	8	e
		3	∞	∞	4	f

Find a good assignment of tools and operations to machines. Relevant objectives are routing flexibility, minimizing material handling, and balancing work load.

Solution

We will denote operations by their part and operation number, that is, 31 indicates the first operation on part type 3.

Phase I

STEP 0. Initialize. $\psi_A = \psi_B = \psi_C = 800$. $\kappa_A = 3, \kappa_B = 1, \kappa_C = 4$.

STEP 1. We must execute step 1 eight times, each time assigning a new operation to a machine type. Each machine has sufficient capacity to receive any operation with a finite processing time. Accordingly, operations 11 and 22 have three feasible machine types. Operations 12, 13, and 21 can be assigned to A or B. Operation 32 can be assigned to A or C. Operations 31 and 33 have only one feasible machine type.

Results of each iteration through step 1 are given in Table 5.7. We could assign 31 or 33 first; we arbitrarily select 33.[4] This assignment is made by using 400 minutes and 1 ($\frac{1}{4}$ per machine) tool slot. We must update machine C feasibility. Operation 11

---

[4]We actually selected 33 because there is only one machine C. The stated procedure does not have any such tiebreaker since both operations require 4 minutes. Add your own tiebreaker if you prefer.

will still fit, since the required 400 minutes and a tool slot are still available. However, operation 22 would require 600 minutes and, since only 400 minutes remain available, we remove C as an option here. Operation 32 is also excluded. These updates can be seen in the second iteration row of Table 5.7.

Operations 31 and 32 now have only one choice. Since 32 has a larger p_{ij}, 5 versus 4, we assign 32 according to our procedure. After updating, we obtain the first portion of the iteration 3 row in Table 5.7. Operation 31 is the only operation with one choice and, hence, it is assigned.

In iteration 4 we choose between 12, 13, 21, and 22; all have two choices. Since 13 has the largest minimal p_{ij} (14 in this instance) among its feasible machines, we select operation 13. If we assign 13 to machine type A, we are left with only 70 minutes ($350 - 14 \cdot 40/2$) of time per machine. If we select machine type B, we are left with 520 minutes per machine ($800 - 40 \cdot 14/2$). We thus select machine type B, producing the results shown on the right side of iteration 4 in the table. The rest of the iterations are conducted in the same manner. The final assignment has operations 21, 22, 31, and 32 on machines A with tools a, c, d, e. Operations 12 and 13 are sent to machines B and take tools b and c. Operations 11 and 33 are assigned to C along with tools a and f.

Phase II

STEP 1. Clustering machine type A. We will let $\tau = 800$ minutes, essentially one machine. Operations 21 and 22 are combined into one cluster. When attempting to cluster operations for part 3, however, we fail since operations 31 and 32 would take 900 minutes and this exceeds τ.

STEP 2. Form groups. $\sigma_A = 4$ tools, $a, c, d,$ and e. With $K_A = 3$ tools per machine, $g_A = [4/3]^+ = 2$. Since two groups are needed to support the tools, each machine of type A must be its own group.

STEP 3. Assign operations. Our operation clusters are now {21, 22}, 31, and 32, requiring 400, 400, and 500 minutes, respectively. We assign by LPT. Thus 32 goes to the first machine A with tool e. Next cluster {21, 22} is assigned to the second machine A along with tools a and c. Last, operation 31 must also be assigned to the second machine since its utilization is lower. Tool d follows its operation.

STEP 1. Clustering machine type B. Operations 12 and 13 cannot be clustered because machines B can hold only one tool and the operations use different tools.

STEP 2. Form groups. We have $g_B = [\sigma_B/K_B]^+ = [2/1]^+ = 2$ groups. Each group must be one machine.

STEP 3. Assign operations. Using LPT, we first assign the 600 minutes of operation 12 to the first machine B and then assign the 560 minutes of operation 13 to the other machine B.

STEP 1. Clustering machine type C. Since we only have one machine of type C, both its operations and tools a and f are assigned to that machine.   ∎

The reader may (should) be asking questions such as "Why not aggregate operations before assigning to machine types; this would even further reduce the potential number of material handling transfers?" As long as the adjacent operations can be

Table 5.7 Iterative Operation Assignments to Machine Types

	Feasible Machines by Operation								Selected Operation	Selected Machine Type	Remaining per Machine	
Iteration	11	12	13	21	22	31	32	33			Time	Tools
1	A, B, C	A, B	A, B	A, B	A, B, C	A	A, C	C	33	C	400	3.0
2	A, B, C	A, B	A, B	A, B	A, B	A	A	—	32	A	550	2.5
3	A, B, C	A, B	A, B	A, B	A, B	A	—	—	31	A	350	2.0
4	A, B, C	A, B	A, B	A, B	A, B	—	—	—	13	B	520	1.0
5	A, B, C	A, B	—	A, B	A, B	—	—	—	12	B	260	0.0
6	A, B	—	—	A	A	—	—	—	21	A	250	1.5
7	A, C	—	—	—	A	—	—	—	22	A	150	1.0
8	C	—	—	—	—	—	—	—	11	C	0	2.0

performed on the same machine types and material handling reduction is a priority, this is a good suggestion. The performance of heuristics is often problem specific, so if you are confronted with a loading problem, go ahead and try it; you might like it.

5.3.3 Scheduling and Control

At the bottom of the decision hierarchy are the scheduling and control issues. We can define three basic problem areas. The first problem concerns the sequencing and timing of part releases to the system. When should a new part be dispatched to the system and which part type should be entered? The second issue relates to the setting of internal priorities in the system. When a machine becomes available, which part should be placed on the machine? Job shop priority scheduling rules may be used here except that the problem also requires control of the material handling system. Parts must be transferred between machining centers, storage, load/unload stations, and washers. Scheduling was discussed in the previous chapter; many of those ideas are equally applicable here. Internal dispatching rules must be specified for assigning priorities to parts waiting for both machines and transporters. In addition to assigning specific transport devices to transport demands, we must control the transporters to avoid collisions and deadlocking. Additional issues are covered in Chapter 9, on material handling systems. The third issue relates to the ability of the system to take corrective action when system components fail. We will not attempt to solve these three problems, but simply make a few pertinent comments.

Parts should be dispatched to the system to match demand while keeping in mind system capacity. When the system requires frequent changeovers, the dispatching problem dominates. As space becomes available, we select a remaining (not yet entered) part from the batch. The part should be selected to smooth the entry rate of all part types in the batch and to add work to machines with a relatively low load level based on parts currently in the FMS. Internal scheduling remains important, the objective being to keep machines busy while matching part production rates to relative batch sizes. The second case covers systems that are capable of simultaneous production of all part types. The goal should be to meet demand with minimal inventory, essentially operating as a just-in-time system. Finding optimal strategies is a difficult problem. Kimemia and Gershwin [1983] apply optimal control theory to this problem for the case of possible machine breakdowns. Seidmann and Schweitzer [1984] proposed a Markov decision approach. Policy iteration was used to minimize the expected penalty cost in matching output to demand. In both instances, a policy is developed that describes the optimal operating strategy for any possible state. The system state is the status of machines (up or down) and the level of inventory of each part type.

As with general job shops, general results on internal scheduling and control are hard to come by. Job shop scheduling rules can be applied for selecting between parts waiting at a machining center or for a transporter. Often in an FMS environment we have a choice of machines for performing an operation. Wilhelm and Shin [1985] find that mean flow time can be substantially reduced by selecting an alternate, compatibly tooled machine for part operations when the primary machine is busy. Nasr and Elsayed [1990] propose solving a series of simple assignment problems (see Chapter 8) at the conclusion of each operation to select the next machine for a part. Normally, we think of moving parts between machines. The versatility of machining centers opens new possibilities, however. Han et al. [1989] explore the option of moving tools instead of parts. Whenever possible, parts are kept on the same workcenter and any required tools are brought to the machine. This approach has the advantages of avoiding part repositioning (a potential source of reduced positional accuracy) and lower WIP. Once a part is released to the system, a workcenter is dedicated. Adopting such a strategy requires initial grouping of parts for machine assignment based on a combination of tool similarity and total machine utilization. Tools must be simultaneously assigned a home machine in conjunction with part assignments.

Control actions must be taken when reality deviates from plans. How should parts be rerouted when a machine fails? Efficient system operation requires constant data acquisition to keep track of system status and preplanning of alternative process plans for each part. Preventive maintenance and safe machining speeds may at first appear to reduce capacity, but in the long run such strategies often increase machine availability and hence output. Planned maintenance is far less disruptive than failures from riding the machine too hard. Such failures, by the nature of their causes, tend to occur at the least opportune times. Sodhi et al. [1991] discuss methods for allocating tools to machines and selecting routes so as to minimize the adverse effect of disruptions.

5.4

FLEXIBLE ASSEMBLY SYSTEMS

To this point in the chapter we have concentrated on flexible machining systems. Flexible automation is also possible for assembly. Whereas flexible machining systems generally remove material to convert the raw material into components with desired geometric features, flexible assembly systems (FASs) are concerned with combining raw material and components into products with functional characteristics. Although flexible assembly is generally associated with smaller parts than flexible machining, many of the principles are the same. The multiproduct assembly lines discussed in Chapter 2 are flexible assembly systems. The machining centers of an FMS are replaced by assemblers. In an automated system the assemblers are robots, whereas humans perform the operations in a manned system. Unlike the serial system envisioned when assembly lines are mentioned, an FAS may have parallel and serial workstations. As in the FMS, material may circulate, allowing material transfer between any pair of workstations.

The earliest FASs were designed for activities such as assembling circuit boards. Parts could be inserted and even wave soldering machines could be integrated into the systems. More recently, systems have been designed for mechanical assemblies. Hall and Stecke in Stecke and Suri [1986] describe an FAS used for assembling nine products using press operations, vision, screw-driving, and grease and loctite operations. Part kits are manually constructed and routed to workstations on AGVs.

Readily reprogrammable operation, computer control, high-speed operation, and high quality are musts in today's world marketplace. For FAS, quality is often determined by positioning accuracy, repeatability, and part verification. Often we are dealing with large volumes of small components; this makes part orientation important, especially in automated assembly. Vibratory bowl feeders and vision systems are often employed to obtain the correct part orientation. Vibratory bowl feeders start with a bin of parts randomly oriented. Parts are set into random motion. Parts are then sent individually through a narrow channel, where they are forced to pass a series of mechanical or computer vision checkpoints. Only those with the proper orientation pass through. The other parts are sent back to the starting bin. The feeder can be designed with features to help parts find the correct position. Once obtained, this orientation information is carefully maintained.

An important aspect of FAS implementation involves design for assembly. Most parts are not totally symmetric functionally; hence, they must be designed with easily detected asymmetric geometric features to facilitate orientation. Likewise, product designers should make use of easy assembly characteristics. Robots can "snap" parts into place more easily than "screwing" fasteners. Robots can also easily dispense adhesives to proper locations for fastening parts. This generally has the added advantage of lower material cost as well when compared with the use of metal screws. The IBM Proprinter has been widely touted as an example of how assembly can be simplified and automated through design.

The topic of design for assembly is discussed in Boothroyd et al. [1982] and Andreasen [1987]. Freeman [1990] uses an actual case history for illustrating the technique of designing for assembly and converting a manual assembly system to an automated flexible system. Basic principles for product and component design include the following:

1. Minimize the number of components to be handled.
2. Design a base component (assembly) that can be fixtured into a stable position and assembled into.
3. Design product variations to allow common handling and assembly techniques.
4. Minimize the assembly directions; in particular, attempt to use only top-down, vertical assembly motions.
5. Use guides to facilitate accurate positioning. Parts should be tapered and chamfers should be added to openings.
6. Minimize the number of fastener types and sizes and avoid difficult fastening operations.
7. Design components to avoid tangling when feeding.
8. Use symmetric components when possible; if not possible, exaggerate asymmetrical features to facilitate identification and orientation.
9. Design parts such that only the proper assembly method can physically combine parts. This helps to avoid incorrect assembly.

Operation times tend to be much shorter in FASs than in FMSs. Machining operations often are measured in minutes, whereas pick and place, adhesive application and insertion operations, which characterize assembly, are measured in seconds. This has the effect of placing a higher demand on the material handling system.

We mentioned kitting and bowl feeders earlier. A kit contains all of the necessary components for a product assembly and is sent along with the assembly. Feeders deliver a continual stream of oriented parts. Other options exist as well. Magazines of individual part types can be kept at each workstation to be dispensed as needed. Magazines are replaced as they become empty or new part types are needed. Alternatively, components could be palletized and sent on the material handling system to each workstation as required. However, in some instances, this approach would overwhelm the handling system. Typically, small, widely used parts are stored at workstations, whereas kits of product-specific parts travel with the part to the workstation. In addition to the need for component parts at the assembly workstation, various tools may also be needed. A robot, for instance, may need access to several different grippers to handle its range of assembly operations. Thus, computer-directed, automated tool changing is important for carrying out assembly plans. Several analytical problems related to the layout and operation of FAS cells will be covered in

Chapter 8. These include the location of feeders/magazines. Those interested in an actual case study of the process and the advantages of design for (automated) assembly are referred to Elmaraghy and Knoll [1989]. The authors describe development of a flexible assembly cell for a family of DC motors.

5.5
SUMMARY

A flexible manufacturing system is a group of automated, multipurpose machines linked by a material handling system and all under central control. Flexible manufacturing systems offer an opportunity to meet customer demand for product variety in a timely fashion and inexpensively (at least, at low variable cost). The full capability of CNC machines can be used to ensure system reliability and product quality.

In deciding to install an FMS, strategic considerations may be as important as economics. Market requirements may necessitate the flexibility and efficiency inherent in these systems.

Basic FMS decisions can be divided into system design and long-term planning, medium- to short-term setup, and real-time scheduling and control. First comes system design. The relative lack of industrial experience with the complex issues of system coordination and control suggest that the prudent approach may be to implement, at least initially, a small system. Systems are modular and with a little forethought can be easily designed to facilitate expansion. From an economic perspective, part types can be ranked based on the potential benefit of assigning them to the FMS. The FMS can be loaded with parts showing an FMS preference.

The second basic problem relates to the construction of part batches for simultaneous production on the system and the loading of machines. Tools must be assigned to machines and parts routed to machines with the required tooling, power, and tolerance capability. Routing should be planned to minimize the load on the material handling system. Tools should be assigned to allow routing flexibility in the event of machine breakdowns.

The third problem concerns real-time scheduling of the system. Rules are needed for selecting among available transporters and properly tooled machines when part transfers are needed. Transporters must also be controlled to avoid collisions and deadlocks.

Flexible assembly systems have some of the same problems as flexible manufacturing systems. However, they are generally designed for higher volume production on smaller products. Robots are often used in place of machining centers. Important considerations include the design of products for easy assembly and the layout of the cell for rapid, flexible operation.

REFERENCES

Andreasen, M. M. (1987), "Design for Flexible Automated Assembly," in *Computer Integrated Manufacturing: Proceedings of the 3rd CIM Europe Conference*, K. Rathmill and P. MacConaill, eds., Springer-Verlag, New York, 213–225.

Askin, Ronald G., Manbir Sodhi, and Suvrajeet Sen (1991), "An Hierarchical Model for Flexible Manufacturing Systems," in *Factory Automation and Information Management*, M. M. Ahmad and W. G. Sullivan, eds., CRC Press, Boca Raton, FL, 934–943.

Boothroyd, Geoffrey, Corrado Poli, and Laurence E. Murch (1982), *Automatic Assembly*, Marcel Dekker, New York.

Chen, Yung-Jung, and Ronald G. Askin (1990), "A Multiobjective Evaluation of Flexible Manufacturing System Loading Heuristics," *International Journal of Production Research*, 28(5), 895–911.

Elmaraghy, Hoda A., and Larry Knoll (1989), "Flexible Assembly of a Family of DC Motors," *Manufacturing Review*, 2(4), 250–256.

Freeman, Brad (1990), "The HP Deskjet: Flexible Assembly and Design for Manufacturability," *CIM Review*, 7(1), 50–55.

Gray, Ann E., Abraham Seidmann, and Kathryn E. Stecke (1988), "Tool Management in Automated Manufacturing: Operational Issues and Decision Problems," Working Paper 88-03, Center for Manufacturing and Operations Management, University of Rochester, Rochester, NY.

Han, Min-Hong, Yoon K. Na, and Gary L. Hogg (1989), "Real-Time Tool Control and Job Dispatching in Flexible Manufacturing Systems," *International Journal of Production Research*, 27(8), 1257–1267.

Jaikumar, R., and L. N. Van Wassenhove (1989), "Production Planning in Flexible Manufacturing Systems," *Journal of Manufacturing and Operations Management*, 2, 52–78.

Jones, Albert T., and Charles R. McLean (1986), "A Proposed Hierarchical Control Model for Automated Manufacturing Systems," *Journal of Manufacturing Systems*, 5(1), 15–25.

Kimemia, J. G., and S. B. Gershwin (1983), "An Algorithm for the Computer Control of Production in a Flexible Manufacturing System," *IIE Transactions*, 15(4), 353–362.

Kiran, Ali S., and Barbaros C. Tansel (1985), "A Framework for Flexible Manufacturing Systems," *Proceedings of the Annual International Industrial Engineering Conference*, 446–450.

Kusiak, Andrew (1983), "Loading Models in FMS," *Proceedings of the 7th International Conference on Production Research*, Windsor, Ontario, 641–647.

Kusiak, Andrew, ed. (1986), *Modelling and Design of Flexible Manufacturing Systems*, Elsevier Science Publishers, Amsterdam.

Menon, U., and P. J. O'Grady (1984), "A Flexible Multiobjective Production Planning Framework for Automated Manufacturing Systems," *Engineering Costs and Production Economics*, 8, 189–200.

Nasr, Nabil, and E. A. Elsayed (1990), "Job Shop Scheduling with Alternative Machines," *International Journal of Production Research*, 28(9), 1595–1609.

O'Grady, P. J., and U. Menon (1987), "Loading a Flexible Manufacturing System," *International Journal of Production Research*, 25(7), 1053–1068.

Owen, A. E. (1984), *Flexible Assembly Systems*, Plenum Press, New York.

Purdom, P. B., and T. Palazzo (1982), "The Citroen (CCM) Flexible Manufacturing Cell," *Proceedings of the First International Conference on FMS*, Brighton, UK, 151–169.

Ranky, Paul (1983), *The Design and Operation of FMS*, IFS Publications, Bedford, UK.

Sarin, S. C., and C. S. Chen (1987), "The Machine Loading and Tool Allocation Problem in an FMS," *International Journal of Production Research*, 25(7), 1081–1094.

Seidmann, Abraham, and Paul J. Schweitzer (1984), "Part Selection Policy for a Flexible Manufacturing Cell Feeding Several Production Lines," *IIE Transactions*, 16(4), 355–362.

Sethi, Andrea Krasa, and Suresh, Pal Sethi (1990), "Flexibility in Manufacturing: A Survey," *International Journal of Flexible Manufacturing Systems*, 2, 289–328.

Sodhi, Manbir, Alessandro Agnetis, and Ronald G. Askin (1991), "Tool Addition Strategies for Flexible Manufacturing Systems," *Technical Report #91-022*, Department of Systems & Industrial Engineering, The University of Arizona, Tucson, AZ.

Stecke, Kathryn E. (1983), "Formulation and Solution of Nonlinear Integer Production Planning Problems for Flexible Manufacturing Systems," *Management Science*, 29(3), 273–288.

Stecke, Kathryn E., and Ilyong Kim (1988), "A Study of FMS Part Type Selection Approaches for Short-Term Production Planning," *International Journal of Flexible Manufacturing Systems*, 1(1), 7–29.

Stecke, Kathryn E., and Rajan Suri, eds. (1986), *Proceedings of the 2nd ORSA/TIMS Conference on Flexible Manufacturing Systems*, Elsevier, New York.

Talavage, Joseph, and Roger G. Hannam (1988), *Flexible Manufacturing Systems in Practice: Applications, Design and Simulation*, Marcel-Dekker, New York.

Whitney, Cynthia K., and Thomas S. Gaul (1985), "Sequential Decision Processes for Batching and Balancing in FMS," *Annals of Operations Research*, 3, 301–316.

Whitney, Cynthia K., and Rajan Suri (1985), "Algorithms for Part and Machine Selection in Flexible Manufacturing Systems," *Annals of Operations Research*, 3, 239–261.

Wilhelm, W. E., and H. Shin (1985), "Effectiveness of Alternative Operations in a FMS," *International Journal of Production Research*, 23(1), 65–79.

PROBLEMS

5.1. Discuss the strategic implications of installing an FMS. How would you use it to enhance your marketing position?

5.2. Describe the advantages of having *flexible* manufacturing processes. What direct and indirect cost savings might accrue?

5.3. List the hardware and software components of an FMS designed to manufacture large, prismatic parts. What information must be communicated between components?

5.4. What are the basic levels in the planning and control of an FMS? For each level, list the basic questions that must be answered.

5.5. A plant operates five days a week. An average of three days' worth of production is kept at each machine. This average is necessary to ensure that input queues only rarely become empty for the heavily used machines. Parts are produced in batches of size 20 taking about one half day per batch on each machine. The company is considering installing an FMS with general-purpose machines. Batch sizes would be one and parts would only need to visit two machines. Compare the throughput times for the two alternatives for a part that currently visits 10 special-purpose machines. Assume that total processing time per part is the same for both systems.

5.6. A manufacturing system uses four special purpose machines to make a set of parts. Total utilization of these machines, including setup time, is (0.50, 0.35, 0.20, 0.65) for a two-shift operation. Each machine is manned separately at a cost of $30,000 per machine per shift-year. Setup constitutes one third of utilization. Machines cost $10,000 per year plus $8 per hour to operate for utilities and maintenance. Flexible machines could be obtained at a cost of $18,000 per year. Each machine could perform all four operations and operate unattended for up to 2 hours at a time. While attended, one operator can oversee two machines. Setup would be reduced by 75 percent since tool magazines would hold most tools needed with automatic tool exchangers on the machines. Should the flexible machines be purchased?

5.7. Repeat problem 5.6 by using the additional information that current machines occupy 80 square feet each and the new machines require 125 square feet. Space is changed at $20 per square foot per year.

5.8. What are the five levels of control in the NIST hierarchical model of automated manufacturing? Why do you suppose a strict hierarchical model is used?

5.9. Consider a five-level hierarchical model of a facility with four shops, four cells per shop, four workstations per cell, and four pieces of equipment per workstation. How many two-way hierarchical links are required? Suppose all pieces of equipment were linked to each other directly. How many links would be required at just the bottom level, that is, between pieces of equipment?

5.10. An existing FMS has 20 hours per week of unused capacity. Six part types are being considered for FMS production. These parts are currently purchased at costs of (100, 50, 75, 84, 95, 165) dollars, respectively. Each product is expected to be in production two more years, with annual demand rates of (100, 300, 1000, 500, 500, 250). Horizontal machining centers are the most limited resource on the FMS. HMC hours per unit for the contemplated part types are (2.0, 3.5, 2.5, 1.0, 1.4, 2.2), respectively. If produced on the FMS, parts are expected to cost (95, 40, 60, 70, 75, 125) dollars.

   a. Suppose the FMS is to be fully utilized. Parts may be partially made in-house and the remaining units purchased. Which parts should be added to the FMS? What is the cost savings?

   b. Suppose parts must be either have all units purchased or made on the FMS. Which parts should be added to the FMS?

5.11. An FMS is being planned. The system will include one turning center and two machining centers. The turning center is not expected to be as heavily utilized as the machining centers. The system will be run 16 hours per day, 6 days per week, and machines are expected to be available 90 percent of this time. Machines will cost $45 per hour

Table 5.8   Data for FMS Design in Problem 5.11

Part Family	Weekly Demand	Machining Center Time per Unit	Turning Time per Unit	Material Cost per Unit	Subcontracting Cost per Unit
1	400	0.15	0.00	10.50	25.00
2	800	0.05	0.05	5.75	20.00
3	400	0.35	0.00	12.25	35.00
4	400	0.30	0.00	18.55	29.50
5	800	0.50	0.01	21.45	62.35
6	200	0.10	0.02	9.55	18.50
7	400	0.20	0.00	6.45	24.35
8	400	0.05	0.02	7.35	10.40
9	800	0.15	0.00	15.45	21.95
10	200	0.25	0.00	3.45	19.95

to operate. Using the data in Table 5.8, apply the greedy heuristic of Section 5.3.1 to determine the set of part families to produce on the FMS.

5.12. Solve Problem 5.11 optimally using dynamic programming.

5.13. Suppose in Problem 5.11 that an extra turning center could be purchased for $1,000 per month and/or extra machining center could be purchased for $1250 per month. Would either acquisition be cost effective? What part types would be added to the system?

5.14. Ten part types are being considered for production on a new machine. These parts are currently purchased. For each part type a decision will be made to either make all its units on the new machine or to continue to purchase all units. This is necessary to ensure product consistency. Total estimated savings per month to make the product internally are (100, 50, 75, 67, 213, 96, 88, 432, 21, 123), respectively. The hours required per week to produce the product's demand are (12, 9, 21, 8, 34, 16, 17, 64, 1.5, 18), respectively. If the machine is available 80 hours per week, what products should be made internally?

5.15. Suppose overtime hours could be arranged at a cost of $10 per hour for up to 18 hours per week, four weeks per month in Problem 5.14. How would this change your solution?

5.16. Two team members are having a disagreement over the design of a new FMS. Both members agree on a set of similar size and material part types to be made on the FMS. However, these parts do not fully utilize all FMS machine capability during the year. One member claims that it is important to maximize utilization on all equipment for financial and organizational purposes. As such, any additional parts that can feasibly be made, both

technologically and time-wise, on the FMS should be produced on the FMS. The second member claims that all decisions should be based on economics and customer service. This member believes that some planned idle time on the average will allow for variability in part demands and machine availabilities. The second member is willing to charge only the variable cost of FMS production to these "extra" part types, but believes the savings must *more* than cover this cost to justify the operational loss of "flexible" idle time. Which argument do you agree with? Defend your stand.

5.17. Several jobs are waiting to be produced in a manufacturing cell. Job characteristics are shown in Table 5.9. The cell has two identical machines. All tooling is standard, and thus no tools must be added or removed from the machines except for normal regrinding after excessive tool wear. Find the set of jobs to produce in the next 8-hour shift. Jobs must be produced completely during the shift in which they are scheduled. It is currently the morning of day 103.

Table 5.9   Job Data for Problem 5.17

Job	Due Date	Order Size (units)	Unit Processing Time (hrs)
1	103	34	0.005
2	103	10	0.010
3	103	25	0.230
4	104	125	0.120
5	104	14	0.205
6	105	100	0.071
7	105	25	0.200
8	105	250	0.023

Table 5.10   Assembly Job Data for Problem 5.20

Job	Due Date	Processing Time (standard hours)		
		Kitting	Assembly	Inspection
1	1	10.5	2.4	0.5
2	1	3.5	1.0	0.6
3	1	5.7	0.6	0.6
4	1	2.3	0.3	0.4
5	2	23.9	4.2	3.5
6	2	4.5	1.2	1.5
7	3	2.3	1.1	0.8
8	3	4.1	0.6	3.5
9	3	12.4	4.6	7.2
10	4	10.4	2.4	0.1
11	4	18.9	5.4	4.5

5.18. Suppose in Problem 5.17 that jobs could be accepted for partial completion during the shift. Resolve the problem.

5.19. Let h_i be the daily inventory holding cost for a part from job i. Give a mathematical programming formulation for determining the optimal schedule in Problem 5.17. Your schedule should cover all three days, one shift per day. Assume that partial production of a job in a shift is permissible. Given h_i values, can you find an easy method to solve this problem?

5.20. An assembly system uses kitting, assembly, and inspection stations. There are three kitters, one automatic assembler, and one inspector. Each worker is available 7.5 hours per day. Table 5.10 contains the set of jobs ready to be produced. To avoid leaving kits exposed, a job must be completed the same day it is started. Which kits should be made today?

5.21. Write out a mathematical progamming formulation to schedule orders in Problem 5.20 for the next four shifts.

5.22. A drill press and three CNC milling machines have been combined into a flexible system. The drill press can store seven bits on a rotating head. The milling machines can each hold six tools. The operating policy is to set up machines once per day, then produce for 10 hours. The system makes eight part types. Additional data are contained in Table 5.11. Devise a rotating plan (the plan will repeat every T days) to meet demand.

5.23. Using your rotating production schedule for Problem 5.22, find an assignment of tools to machines in each day. Attempt to maximize routing flexibility whenever possible.

5.24. A three-machine cell has four part types to produce in the next period. Machine A costs 50

Table 5.11   Data for Problem 5.22

Part Type	Daily Demand	Unit Processing Hours		Tools Required	
		Drill Press	Milling	Drill Press	Milling
1	35	0.02	0.23	3, 5, 6, 9, 10	5, 6, 9, 11
2	15	0.04	0.18	3, 6, 9	2, 4, 11, 12
3	13	0.04	0.05	1, 4, 6, 13	3, 7, 12
4	18	0.01	0.26	2, 4	5, 11, 12
5	30	0.02	0.17	4, 7, 9, 10, 12	4, 6, 7
6	5	0.03	0.05	1, 2, 6	7, 8, 13
7	20	0.15	0.20	1, 3, 4, 5	1, 2, 3, 4, 5
8	10	0.20	0.30	1, 5, 6, 9	2, 4, 5, 8

Table 5.12   Data for Loading Problem 5.24

Part	Operation	Machine A	Machine B	Machine C	Tool Number
		Unit Processing Time (hrs)			
a	1	0.2	0.3	0.3	3
	2	0.4	0.6	0.7	2
	3	0.1	0.1	0.1	1
b	1	0.5	0.6	0.7	1
	2	0.3	0.5	0.5	2
c	1	0.2	0.2	0.3	3
d	1	0.3	0.4	0.4	1
	2	0.1	0.2	0.2	2

dollars per hour to operate, whereas B and C are charged at the rate of $30 per hour. Any machine can be used to produce any part; however, machine A has better precision, which is important for part type b. Demand for parts $a, b, c,$ and d is (20, 40, 40, 30), respectively. Each machine can hold three tools and has 60 hours available. Tools will be replaced as they wear out, and so you need not consider this in your formulation. Given the data in Table 5.12,

  a. Formulate the loading problem as a mathematical program.

  b. Solve the mathematical program on a software package.

5.25. Suppose in Problem 5.24 that we were told that operation 2 on part a could be avoided if operation 1 was assigned to machine A and the machine was run at one half the initial feed rate. (Operation 2 is a finishing cut, which is no longer necessary if operation 1 is modified.) How would you incorporate this into your model?

5.26. A new flexible manufacturing cell is being designed. Your objective is to select the set of part types to be produced in the cell. All parts are

currently purchased. The cell will perform only boring and turning operations. Tools will be kept at the machines. A robot will be used for changing tools and the loading and unloading of parts. Machining operations are time consuming compared with tool changes and part moves; hence, the changing of tools and utilization of the robot should not be a problem. All units of a part must be acquired by the same alternative. You cannot, for instance, make some part a's and purchase the others. The set of potential part types is shown in Table 5.13.

  a. Suppose boring machines are inexpensive, and we can buy as many as we need. The turning center, however, is expensive, and only 600 time units are available per period. Order parts based on their savings per unit of resource consumed. Using a greedy heuristic, recommend a set of parts to manufacture in the cell. What is the cost savings per period?

  b. Suppose both machine types are expensive and only 600 time units are available on each per period. Now find the best set of parts to manufacture in the cell.

Table 5.13   Part Data for Problem 5.26

Part	Demand/Period	Turning (time/part)	Boring (time/part)	Purchase (cost/part)	Variable Machining (cost/part)
a	4	20	15	100	35
b	3	35	10	125	45
c	2	40	50	250	90
d	5	10	60	165	70
e	2	55	25	110	80
f	5	75	65	300	140
g	8	15	55	215	70

5.27. The following jobs are available for manufacture on an FMS. The FMS has one HMC and one HI machine. The HMC can hold many tools, but the HI can hold only three tool heads. All parts that start with the same letter use the same tool head. For instance, all $A\#\#$ parts could be done on the HI and require only one tool head. Machines are available 14 hours per day. Select the jobs to be run today.

Part Type	No. of Units	Due Date	Unit Processing Hours	
			HMC	HI
A23	10	1	0.3	0.05
A23	5	3	1.0	0.05
A99	20	2	0.4	0.10
C45	10	2	0.8	0.20
D13	20	2	0.5	—
D13	10	3	0.5	—
E11	50	1	0.2	0.1
E4	50	2	0.4	0.05
F21	10	1	0.1	0.1

5.28. Formulate the problem of assigning part types and tools to machines for the strategy of Han et al., whereby parts stay on machines but tools move. Tools have an initial machine, but can be moved if necessary. Your objective should be to minimize the number of tool movements. Define all notation. (*Hint:* Let x_{it} be 1 if tool t is assigned to machine i and 0 otherwise. y_{ij} is 1 if part j is assigned to machine i and 0 otherwise.)

5.29. Suggest a solution for Problem 5.28. The large number of binary variables and nonlinear objectives suggest the need for a heuristic. (*Hint:* What type of problem results if the x_{it} variables are fixed? What if the y_{ij} variables are fixed?)

5.30. If tools are moved between machines and parts are kept on the same machine, two basic tool-control strategies are possible. Strategy 1 says always return a tool to its home machine after it has been borrowed for use on another machine. Strategy 2 says to leave tools where they are until requested elsewhere. Discuss the relative merits of these two strategies. In real-time control, what additional information would help you to choose between these two strategies?

5.31. An FMS has two type A machines and three type B machines. Five part types are made on the

FMS. Table 5.14 contains a set of jobs currently waiting to be completed. A job consists of a specified number of units of a single part type. Operations requiring machine type A have a tool named $A\#$ and similarly for type B machines. Machines are available 500 minutes a shift (all times are in minutes). Each machine can hold three tools. It is the start of shift one. Company policy is to make all parts due in the current shift and any additional complete jobs that can be accommodated with the same tooling setup.

Table 5.14a   Process Plans for Problem 5.31

Part	Process Plans Operation (unit time, tool)		
	1	2	3
a	(30, A1)	(45, B3)	(15, A2)
b	(15, A4)	(20, A6)	(30, B4)
c	(10, B1)	(25, B2)	(15, A5)
d	(20, B3)	(5, A1)	—
e	(50, B2)	(15, A4)	(20, B1)

Table 5.14b   Available Jobs for Problem 5.31

Part Type	Quantity	Due Date (shift)
a	10	1
b	10	1
e	5	1
a	10	2
c	5	2
d	20	2

a. Select the part types and quantities to be produced during this shift. Justify your decisions.

b. For each machine type assign individual machines to groups.

c. Assign operations to machine groups. State your objective and explain your procedure. Assume that operations can be performed in any order.

5.32. Suppose in Problem 5.31 that operations had to be performed in the order shown in the process plan. How would that change your clustering of operations to reduce material handling?

Table 5.15   Part Data for Problem 5.33

Part	Operation	Machine Type	Total Machine Hours	Tools
31E245	1	A	1.5	A1
	2	A	2.4	A2
	3	A	1.2	A3
	4	B	12.6	B2
20E139	1	B	7.1	B1
	2	B	1.3	B5
	3	A	1.6	A5
	4	B	4.5	B3
	5	A	2.5	A7
10F865	1	A	1.4	A6
	2	B	3.9	B3
	3	A	2.4	A5
	4	A	1.6	A1
	5	B	2.8	B2
24F621	1	A	2.4	A7
	2	A	1.5	A6
	3	B	4.8	B6
	4	B	3.3	B9

5.33. Table 5.15 contains the parts to be made today. There are three machines of type A and six of type B. All machines can hold three tools. Each machine is available 7 hours. Tabulated time values are aggregated over the part batch. Operations can be performed in any order, and part jobs can be subdivided since each job consists of multiple pallets of material. Load part operations and tools onto specific machines. Attempt to maximize routing flexibility.

5.34. Resolve Problem 5.33 assuming that operations must be performed in the order listed in the table and that you wish to minimize material handling transfers between machines as well as maximizing routing flexibility.

5.35. Consider the loading objective 5.17. Modify this objective to reflect the fact that by placing a tool type on more than one machine we enhance routing flexibility.

5.36. Consider the data in Table 5.15. The data represent a demand of 100 units of each part type. Assume that an incremental approach is to be used to select parts and to load machines. Furthermore, all machines of the same type must carry the same tools. No more than six parts of any type can be in process at one time. Normal throughput time is 2 hours for a part.

    a. Write out the mathematical program to select parts.

    b. If software is available, solve the integer program.

CHAPTER 6

GROUP TECHNOLOGY

"Small Is Beautiful"
—E. F. Schumacher

6.1
INTRODUCTION

In 1925, R. E. Flanders described the use of product-oriented departments to manufacture standardized products with minimal transportation at a machine company. As pointed out in the history chronicled by Snead [1989], this could be considered the start of group technology. In 1937, A. P. Sokolovski proposed that parts be classified and parts with similar features be manufactured together with standardized processes. This theme was subsequently advanced by his fellow countryman S. P. Mitrofanov. Working in England in the 1960s, J. L. Burbidge developed a systematic planning approach based on this concept. In the ensuing 20 years, this philosophy has spread throughout the manufacturing world. Today, small "focused factories" are being created as independent operating units within large facilities. This effort is largely based on the concept of Group Technology (GT)—a theory of management based on the principle that *similar things should be done similarly*. For our purposes, the nondescript "things" include product design, process planning, fabrication, assembly, and production control. However, GT may be applied to all activities, including administrative functions. In this chapter we study this management philosophy and examine how GT can be applied to manufacturing systems.

One tenet of group technology entails dividing the manufacturing facility into small groups or cells of machines, each cell being dedicated to a specified set of part types. The term **cellular manufacturing** is often used in this regard. The connotation of a cell is a small group of perhaps only one or two machines but seldom more than five. A typical cell might contain a machining center, on-machine inspection and monitoring devices, tool and part storage, a robot for part handling, and the associated control hardware. In this chapter, however, we will also be concerned with larger groups, for instance, a department, possibly composed of multiple automated cells or a dozen manned machines of various types. Configuring machines with different capabilities into a cohesive group is an alternative to process layout. This group configuration is most appropriate for medium-variety, medium-volume environments. If volumes are very large, pure item flow lines are possible; if volumes are small, and parts are varied to the point of only slight similarities between jobs, then there is less to be gained by grouping. Nevertheless, GT can produce highly significant improvement where it is appropriate and presents important lessons that should be utilized in all manufacturing environments.

The manufacturing engineer can view GT as an attempt to obtain the advantages of flow line systems in environments previously ruled by job shop procedures. Instead of a large process layout with each job being designed in a way that is independent of previous designs, we aim toward a product-type layout within each group. The resultant groups are each dedicated to a family of parts. Whenever possible, new parts are designed to be compatible with the processes and tooling of an existing part family. Experience is quickly amassed on part families, and standard process plans and tooling can be developed for this restricted part set. Standard tooling quickens part changeovers. Fast changeovers in turn allow short cycle, just-in-time, production.

The design engineer can view GT as an attempt to standardize products and process plans. Items with similar geometric features should have similar designs. The design of a new part is initiated by retrieving the design for a similar, existing part. This plan is then modified as necessary for the new part. We may even eliminate the need for the new part if an existing part will suffice. If the part is truly needed, the new plan can be developed more quickly by relying on decisions and documentation previously made for similar parts. The resulting plan will match current manufacturing procedures. In this fashion the potential of CAD systems for reducing document preparation time can be realized. The design engineer is freed to concentrate on optimal design. Often, a manufacturing firm designs many low- to medium-volume parts for which time-consuming design and process planning optimization is not warranted. If this knowledge can be transferred, however, to future parts, the effort is now justifiable and all parts benefit. It is economically feasible to spend resources finding the optimal plan for a part of this type since this effort will have recurring benefits.

One approach to GT for design is the use of composite part families. An example of a composite part is shown in Figure 6.1. All current and future parts that can be defined by setting the parameter values for the features of this composite part within allowable ranges comprise a single family. Each part in the family requires the same set of machines and tools. For our example part this would include internal boring, face milling, drilling, and so on. Fixtures can be designed for the composite parts that are capable of supporting all the actual realizations of the composite. Raw material should be reasonably consistent as well as geometric entities. Plastic and metallic parts require different manufacturing operations, for instance, and should not be in the same family. The definition of composite parts simplifies the identification of groups and also provides a basis for the design of group tooling. Standard machine setups are often possible with little or no changeover required between parts within the composite family. To accomplish this, all members of a composite part family would normally need to be of the same material and fixtured by the same method as well as being similar in size and requiring the same tools and machines. For example, all parts in a composite part family should fit in the same collet or chuck for turning and use the same cutting tools, angles, fluids, and chip-removal system.

The facility layout aspect of GT is reflected in Figure 6.2. In the functional process layout, all parts travel through the entire shop. Scheduling and material control are complicated. Job priorities are difficult to set, and foremen have traditionally aimed for large work-in-process inventories to assure ample work. With GT, each part type flows only through its specific group area. Houtzeel and Brown [1984] describe a study in which 150 similar parts were placed into a group of 8 dedicated machines. Previously, the same parts had been made on 51 different machines with 87 routings. Workers may be cross-trained on all machines within the group and follow the job from start to finish. Machine scheduling is simplified as we need only look behind us to see what is coming. The reduced setup time allows faster adjustment to changing

Operations   Example parts

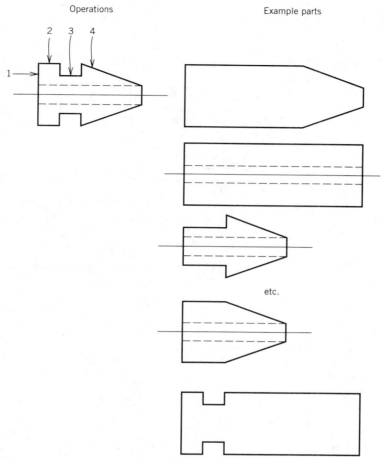

Figure 6.1 Composite Group Technology part.

conditions—an essential requirement to compete successfully in today's dynamic world markets.

It may be necessary to include several part families in a machine group to justify machine utilization. Even within a family, parts may skip certain operations or require different machine sequences. Three types of group layout thus arise, as shown in Figure 6.3. The GT **flow line** is used when all parts assigned to the group follow the same machine sequence and require relatively proportional time requirements on each machine. The GT flow line operates as a mixed-product assembly line system. In some cases, automated transfer mechanisms can be used within the group for handling parts. The GT **cell** allows parts to move from any machine to any other machine. Although flow is not unidirectional, machines are located in close proximity. The GT **center** is a logical arrangement. Machines may be located as in a process layout by using functional departments, but each machine is dedicated to producing only certain part families. The tooling and control advantages of GT can be achieved despite the increased material handling. GT centers are appropriate when large machines have already been located and cannot be moved, or when product mix and part families are dynamic and would require frequent relayout.

GT offers numerous benefits. In a 1989 survey of 32 firms operating cells, Wemmerlov and Hyer found the benefits shown in Table 6.1. We should remember, how-

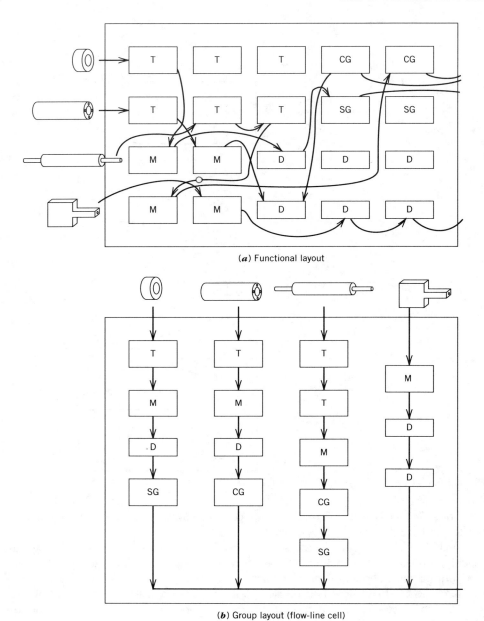

(*a*) Functional layout

(*b*) Group layout (flow-line cell)

Figure 6.2   Facility layout for Group Technology. (*Source:* N. Hyer and U. Wemmerlov, 1982, *Decision Sciences,* published with permission from Decision Sciences Institute.)

ever, that the advantages to design engineering, such as ease of design retrieval, design standardization, and reduction in part proliferation, and the corresponding increase in designer productivity, are just as significant. In general, GT simplifies and standardizes. As any good athletic coach knows, the approach of simplify, standardize, and internalize through repetition produces the consistent proficiency needed to obtain winning performance.

Setup time reduction is an important aspect of GT. A workcenter will work only on a family (or families) of similar parts. This allows the development of generic fixtures

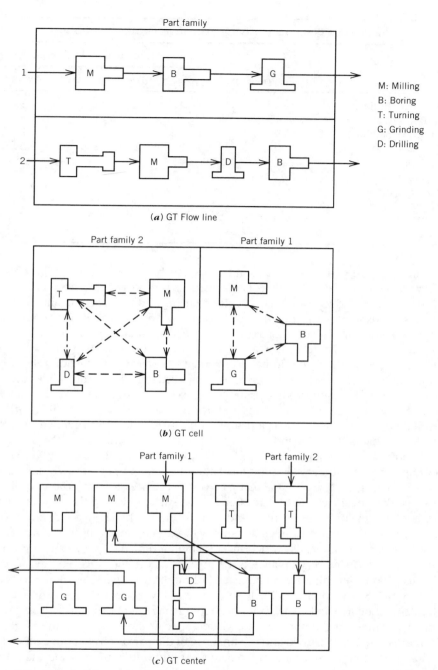

Figure 6.3   Types of Group Technology layout.

for that family. Tooling can be stored locally since a part will always be processed through a given machine: not only today's batch but also next month's. A turret lathe may be loaded with the tools required to perform all operations of all parts in a family. Tool changes may be required due to tool wear only, not part changeovers. A press may have a generic fixture that can hold all the parts in a family simply by changing a movable stop or part-specific insert secured by a single screw. In general,

Table 6.1   Percentage Savings Through Group Technology

Category	No. Respondents	Improvement Percentage		
		Average	Minimum	Maximum
Throughput time	25	46	5	90
WIP inventory	23	41	8	80
Materials handling	26	39	10	83
Job satisfaction	16	34	15	50
Fixtures	9	33	10	85
Setup time	23	32	2	95
Space needs	9	31	1	85
Quality	26	30	5	90
Finished goods	14	29	10	75
Labor cost	15	26	5	75

component shape and processing similarities can be exploited to design common tooling. Tooling cost is reduced, as is changeover time between parts.

The advantages of setup reduction extend far beyond the machine and labor time saved. As an example consider the inventory cost and storage space requirements for WIP. If setup time is cut in half, then for the same total labor and machine time, we can set up each product twice as often. Let's examine what happens when batch sizes are correspondingly cut in half to maintain a constant production rate. If batch interarrival and service times are exponentially distributed, we can model the work cell as an $M/M/1$[1] queue with arrival rate λ, service rate μ, and utilization $\rho = \lambda/\mu$. In this system we know that the time in system is given by $W = \rho/[\lambda(1 - \rho)]$. Our modified GT system, denoted by a prime (') has twice the arrival rate and twice the service rate, that is, $\lambda' = 2\lambda$ and $\mu' = 2\mu$. Utilization is unchanged ($\rho' = \rho$). Waiting time is now $W' = \rho/[\lambda'(1 - \rho)] = \frac{1}{2}W$. This is an astonishing result! We have cut throughput time for the system, not just setup time, by 50 percent. WIP inventory is likewise cut in half. Not only do many of our headaches for material coordination and space disappear but the comptroller will be sure to pick up the next dinner tab.

In addition to lower setup time and WIP inventory, finished goods inventory is reduced. Instead of make-to-stock systems with parts either being run at long, fixed intervals or random intervals (which make planning difficult), parts can be produced either just in time in very small lots or at fixed short intervals. For instance, the ability to produce each part weekly instead of monthly reduces finished goods cycle inventory by 75 percent and is likely to reduce safety stock by 50 percent.[2]

Implementing focused GT factories changes management orientation. In a process-based department, the blueprint and written specifications become the target. The customer is seldom consulted and the final use of the part is rarely understood. There are simply too many different parts to take the time to fully understand the purpose and problems of each. With small groups, a smaller set of suppliers, producers, and

---

[1]The $M/M/1$ notation is standard in queueing theory. The first M indicates that the time between successive arrivals of customers is exponentially distributed. The second M indicates that customers require an exponentially distributed service time. The 1 indicates a single server for the system.

[2]Cycle inventory is the average on-hand inventory due to production batches of more than one unit. Average cycle inventory is one half of the batch size. Safety stock is often taken as a multiple of the standard deviation of demand between replenishments. Assuming independent weekly demands, monthly demand variance would be four (weeks per month) times weekly variance; thus, monthly standard error is twice weekly standard error.

customers (yes, that does spell SPC) are involved. Communication can be verbal, and it is not uncommon for producers to visit suppliers and customers to better understand requirements—both written and unwritten.

Three basic steps comprise group technology planning. These steps are coding, classification, and layout. Coding involves the specification of knowledge concerning the similarities between parts. Coding often involves the assignment of a symbolic or numerical description to parts based on their design and manufacturing characteristics. However, in the general sense we use the term coding here, it may also refer to simply listing the set of machines used by each part. Section 6.3 provides an overview of the issues and approaches to coding. Classification (also called group formation) refers to the use of part codes and other information to assign parts to families. Part families are assigned to groups along with the machines required to produce the parts. A variety of models for forming part-machine groups are described in Sections 6.4 and 6.6. To group machines, part routings must be known. Section 6.5 presents a method for clustering part operations onto specific machines to provide this routing information. Layout is concerned with the physical placement of productive facilities (machines) on the shop floor. Within a group this becomes the choice of flow line, center, or cell as discussed above. The relative location of groups throughout the facility is discussed in Chapter 7. Once the groups are formed, an effort should be made to develop standard tooling for groups, which allows quick changeovers. The advantages of family based tooling have already been described. Before attacking the stages of the group planning problem, we will try to paint a more vivid picture of what typical groups look like and the principles required for effective use of GT in manufacturing.

6.2
PRINCIPLES OF GROUPS

Through survey results and the reported experiences of GT implementors, a picture of the standard cell has begun to take form. Any such statements are by necessity transitory in nature, however. Current movement seems to be in the direction of small automated cells; nevertheless, some of the discussion in this section is based on manned groups with human supervision.[3]

J. L. Burbidge has noted seven characteristics of successful groups. These are summarized in Table 6.2. These characteristics should be embedded into the group structure. Size is important. The group must be small enough to act as a close-knit team with a common goal, yet be large enough to contain all necessary resources for achieving that goal. Although groups as large as 30 workers have been successfully implemented, social scientists have learned that a range of 6 to 12 workers is best (the magic number 7 is, of course, optimal). As group size grows beyond a dozen, it becomes increasingly difficult to maintain sufficient camaraderie, level of communication, and cohesiveness toward a common goal. It is likewise important to remember to level the long-run work load (utilization) in each group and to match this to production plans. Failure to balance work load creates a strong temptation to route extra work to the cell or to off-load jobs to other cells, thus destroying the concept of GT. To facilitate the organizational authorization process, it may be

---

[3]The push in recent years to hierarchical control, simplicity, and multipurpose machines has led to minicells within a larger group. The discussion of flexible manufacturing systems in Chapter 5 is a case in point. The FMS may represent one group with its assigned parts. Each machining center with its peripherals such as tool storage, part storage, and robot may constitute a cell.

Table 6.2	Characteristics of Successful Groups
Characteristic	Description
Team	Specified team of dedicated workers
Products	Specified set of products, and no others
Facilities	Specified set of (mainly) dedicated machines/equipment
Group layout	Dedicated contiguous space for specified facilities
Target	Common group goal, established at start of each period
Independence	Buffers between groups; groups can reach goals independently
Size	Preferably 6–15 workers; more can be accommodated, however

advisable to maximize utilization of existing machines and to minimize investment requirements.

Groups should be designed with safety in mind. Painting may be required after welding, but placing these functions too close may not be the spark of enlightenment for which you are searching. Likewise, machines and processes that are incompatible for other reasons, such as tolerance capability or environmental factors, should not be grouped together.

A survey in 1976, reported by Pullen [1984], produced the following typical group. The group was physically separated from all other groups with its own input and output point. The group employed 11 workers who had access to 14 machines. These machines were purposefully laid out to minimize material movement. The basic assigned work load was a set of similar parts. Batches averaged 330 parts and spent 2.4 weeks in the group. Approximately 2 extra weeks' worth of work was released for manufacture and a 5-week supply of work was queued up in front of the group. Management occasionally moved workers about in the group and even changed those assigned to the group. Although most parts were produced totally within the group, several parts did require other processes. In slack times other work was moved in from elsewhere in the shop. Group tooling resided locally. All relevant material and drawings were supplied to the group when production orders arrived. Design and process planning was done externally and the information supplied, but all inspection and testing was done within the group. When the group was designed, no consideration was given to the workers that would be initially assigned, and these workers were not provided any special training. Workers clocked in and out, and management set production goals and staffing levels. Management also appointed the group leader and retained responsibility for staffing and discipline. Workers were granted greater autonomy and responsibility than previously permitted, however. Fortunately, although not yet fully achieved, the importance of reducing the input queue and maintaining the integrity of the group and its goals tended to be recognized by the organization.

A discussion of group technology principles would not be complete without some comment on the implications for organizational structure. The organization should be structured around groups. Each group performs functions that in many cases were previously attributed to different functional departments. Labor reporting and the entire accounting system should be based on group activity. For instance, in most situations employee bonuses should be based on group performance. If skill or experience levels vary significantly, it is possible to assign shares of the bonus based on relevant, quantified measures, but the entire group should be rewarded from a common evaluation.

Worker empowerment is a key aspect of manned cells. The group must contain individuals who are dedicated to team success and who can work with one another, exchanging ideas and work load. Many groups are allocated the responsibility for hiring, firing, and work assignments. To encourage the group, it is important that the group be free to determine its destiny by setting its own policies and not be constrained by outside entities. Thus, it is unfair to fault a group for not achieving its production goal if raw material was unavailable. In addition to having cross-training of technical skills, so that at least two workers can perform each task and all workers can perform multiple tasks, compatibility of personalities and willingness to both share ideas and to consider the ideas of co-workers are important traits for group members.

In a truly enlightened environment, the group should be an independent profit center. The group members purchase raw material from supplier groups and sell their product to the company or customer groups. Decisions on purchasing new equipment or changes in staffing levels could be made by group members. Group members should visit with their "supplier" and "customer" groups to iron out quality wrinkles and to understand the uses of their product. In general, the group should retain responsibility for its performance and authority to effect that performance. When reporting to authority outside the group, all members should share a common destiny. Once the group has assumed responsibility for acquiring resources, no individual can claim that incoming product was poor or someone else made the mistake. The group is a single entity and must act together to resolve problems. If adequate freedom and breadth is assigned to the group, then when the group points a finger, the tip curls right back on itself.

6.3
CODING SCHEMES

The part-coding scheme forms the basis for GT. The intent of the part code is to compactly describe those part characteristics that will facilitate determination and retrieval of similar parts. Ideally, the code contains the information necessary to guide the entire planning process successfully. Codes should reflect how activities could or should be performed. When constructing a code system, we must be careful to avoid institutionalizing existing practices regardless of their necessity and desirability. The code should inform us as to what machines or processes could be used, not which are currently used. Clearly, it is self-defeating to constrain subsequent decisions to continue any poor practices.

Four major issues guide the construction of a coding system. These issues are part (component) population, code detail, code structure, and digital representation. We discuss each of these in greater detail and then complete this section with an example of Opitz, one of the most widely used coding systems. The intent is to give the reader an idea of how a code might be constructed. Numerous codes now exist, including Brisch-Birn, MULTICLASS, and KK-3; several of these are referenced at the end of the chapter. Many firms customize existing coding systems to their specific needs.

The coding system should be *inclusive*, that is, designed to cover the entire class or **population** of parts to be coded. The scope of part types to be included must be known. For instance, are the parts rotational, prismatic, sheet metal, or some other substance? The code should be sufficiently *flexible* to handle future as well as current parts. To be useful, the code must *discriminate* between parts with different values for key attributes. Among the class of parts, what characteristics differ and may affect the choice of manufacturing processes, machines, and tooling? As an example, material may be an important characteristic. Iron castings require different cutting tools, speeds, and forces than aluminum castings. Part sizes may vary to the degree that different machine tools are required. Some parts may require threading or turning operations, whereas others do not. Tolerances may differ, requiring grinding for some finishes and not others or reaming for holes on certain parts.

The choice of **code detail** is crucial to the success of the coding project. Efficiency is important; use of extra detail results in cumbersome codes and the waste of resources in data collection. Too few details and the code becomes useless. It would be nice to have a short code that could uniquely identify each part and fully describe the part from design and manufacturing viewpoints, that is, a unique descriptive code value for every part to be manufactured. This would normally require a code too long and complex to be practical, however. Indeed, we would basically have to repeat the entire part record kept in the data base. A more reasonable alternative is the composite part discussed above. Each small family of parts may have a unique code. At least for the purposes of routing in process planning, all parts in this family may be identical. Only this set of parts would share a common code value. This greatly facilitates the development and retrieval of process plans. As a general rule all information necessary for grouping the part for manufacture should be included whenever possible in the code. For design, if a feature is such that the design for existing parts with this feature forms a useful starting point, then the feature should be included in the code. Features like outside shape, end shape, internal shape, protrusions, and holes in addition to a main internal shaft are typically included in the coding scheme. The scale of major part dimensions is clearly relevant. The next question is whether secondary shape information such as chamfers and threads should be included. The code developer must consider the trade-off of having this information in the code and the cost and time of collecting this information for each part. Information likely to change with time is normally not included in the primary code, even if it may be significant. Part demand can be important, but it is fleeting. Whether a part is standard or special custom could be included, though. Production rate and lifetime can certainly influence tooling, fixture, and routing designs.

The third issue is **code structure.** Codes are generally classified as hierarchical (also called monocode), chain (also called polycode), or hybrid. The basic structure of these codes is shown in Figure 6.4. A hierarchical code structure exists when the meaning of a digit in the code depends on the values of preceding digits. The value of 3 in the third place of the code value "1232" may indicate the existence of internal threads in a rotational part. The same 3 in the code value "4532" may indicate a smooth internal feature. Hierarchical codes are efficient in that only relevant information need be considered at each digit. The values for each digit can be used to convey meaningful data in all instances. However, hierarchical codes are difficult to learn because of the large number of conditional inferences.

Chain codes are easier to learn but less efficient. In a chain code each value for each digit of the code has a consistent meaning. The value 3 in the third place of the codes above would have to mean the same thing for all parts. As a result, certain digits may be almost meaningless for some parts. Rotational and nonrotational parts may require different knowledge regarding shape elements.

Since both hierarchical and chain codes have advantages, many commercial codes are of a hybrid nature, exhibiting aspects of each. A hybrid code will typically utilize a section of code that is chain in nature and then switch to several hierarchical digits to further detail the specified characteristics. Several such chain/hierarchical sections may exist. Opitz is one example of a hybrid code.

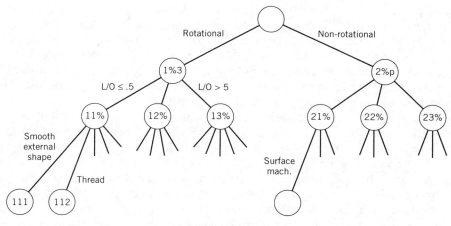

(*a*) Hierachical structure

Code Digit Feature	1 Outside shape	2 Inside shape	3 Holes	4 Surface Machining	• • •
Value 1	None	None	No	None	
2	Smooth	Smooth	Smooth axial	External groove	
3	Stepped ends	Stepped ends	Smooth radial	External spline	
4	Stepped and threads	Stepped and threads	Axial and radial	Internal curved	
⋮					

(*b*) Chain structure

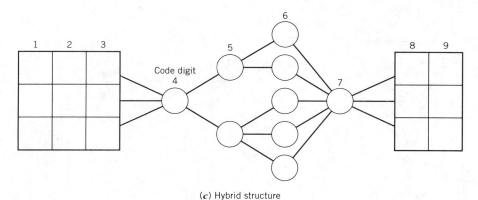

(*c*) Hybrid structure

Figure 6.4   Code structures.

The last issue concerns **code representation.** Computers speak binary best, and the many parts coded justifies consideration of storage and retrieval efficiency. Humans, however, are more comfortable with alphanumeric characters, especially if characters convey meaning such as use of "S" for smooth or "T" for thread. Individual digits should be alphabetic or numeric. A code that always produces *numeric/alphabetic/numeric* values such as $3B2$ and $5A6$ is easier to remember and verify than one that can produce $3B2$ and $A56$. The selection of a binary, octal, alphanumeric, or other similar code also depends on the desired number of categories for each digit. In dividing the coded information into digits and in defining the descriptors for each code digit, consideration should be given to standard industrial terminology and groupings. Adhering to this convention will facilitate comprehension of the code and interpersonal communication. We are left with the conclusion that the proper decision process is one that involves the design engineer, manufacturing engineer, and computer scientist working together as a team.

To illustrate the ideas of code design, we will briefly look at the Opitz coding system. The code consists of a five-digit "geometric form code" followed by a four-digit "supplementary code." This may be followed by a company-specific four-digit "secondary code." The first digit of the form code details whether the part is rotational and also the basic dimension ratio (length/diameter if rotational, length/width if nonrotational). Digit two specifies main external shape and is partly dependent on the first digit. Digit three describes internal shape entities. Digit four describes the machining requirements for plane surfaces. Auxiliary features are described in digit five. The supplementary code begins with a value for main dimension. Combined with the first digit, the part size can be determined. The supplemental code continues by describing the work material, the original shape of the raw material, and the required accuracy. The installation-specific secondary code is intended for describing production operations and sequencing. Figure 6.5 provides an overview of the Opitz code. The form code for certain rotational parts is further defined in Figure 6.6. An example of a coded part is shown in Figure 6.7.

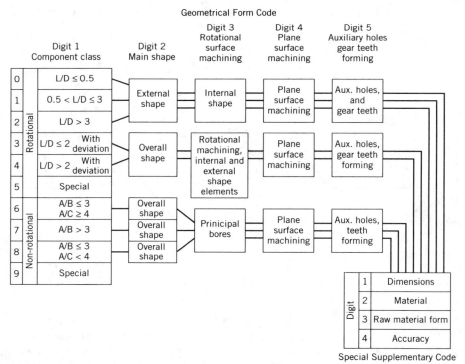

Figure 6.5 Overview of Opitz code. (Reprinted from Rembold, Blume, and Dillmann, 1985, *Computer Integrated Manufacturing Technology and Systems*, p. 305, by courtesy of Marcel Dekker Inc.)

Code	Digit 1 Part class	Digit 2 External shape, external shape elements	Digit 3 Internal shape, internal shape elements	Digit 4 Plane surface machining	Digit 5 Auxiliary holes and gear teeth
0	$L/D \leq 0.5$ (Rotational parts)	Smooth, no shape elements	No hole, no breakthrough	No surface machining	No auxiliary hole (No gear teeth)
1	$0.5 < L/D < 3$ (Rotational parts)	No shape elements (Stepped to one end/or smooth)	No shape elements (Smooth or stepped to one end)	Surface plane and/or curved in one direction, external	Axial, not on pitch circle diameter (No gear teeth)
2	$L/D \geq 3$ (Rotational parts)	Thread (Stepped to one end/or smooth)	Thread (Smooth or stepped to one end)	External plane surface related by graduation around a circle	Axial, on pitch circle diameter (No gear teeth)
3	(Rotational parts)	Functional groove (Stepped to one end/or smooth)	Functional groove (Smooth or stepped to one end)	External groove and/or slot	Radial, not on pitch circle diameter (No gear teeth)
4	(Rotational parts)	No shape elements (Stepped to both ends)	No shape elements (Stepped to both ends)	External spline (polygon)	Axial and/or radial and/or other direction (No gear teeth)
5	(Rotational parts)	Thread (Stepped to both ends)	Thread (Stepped to both ends)	External plane surface and/or slot, external spline	Axial and/or radial on PCD and/or other directions (No gear teeth)
6	(Nonrotational parts)	Functional groove (Stepped to both ends)	Functional groove (Stepped to both ends)	Internal plane surface and/or slot	Spur gear teeth (With gear teeth)
7	(Nonrotational parts)	Functional cone	Functional cone	Internal spline (polygon)	Bevel gear teeth (With gear teeth)
8	(Nonrotational parts)	Operating thread	Operating thread	Internal and external polygon, groove and/or slot	Other gear teeth (With gear teeth)
9	(Nonrotational parts)	All others	All others	All others	All others (With gear teeth)

Figure 6.6 Opitz detail for rotational parts. (Reprinted from Rembold, Blume, and Dillmann, 1985, *Computer Integrated Manufacturing Technology and Systems*, pp. 305–306, by courtesy of Marcel Dekker Inc.)

175

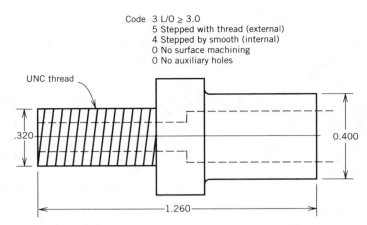

Code 3 L/O ≥ 3.0
          5 Stepped with thread (external)
          4 Stepped by smooth (internal)
          0 No surface machining
          0 No auxiliary holes

Figure 6.7   A sample coded part.

Part coding aids in design and group formation, but can be a formidable task for existing facilities. In fact, the major disadvantage of GT is the time and cost involved in collecting data, determining part families, and rearranging facilities. For designing new facilities and product lines, these costs are not as overwhelming. Parts must be identified and designed, and facilities must be constructed. The extra effort to plan under a GT framework is marginal, and the framework allows improved standardization and operation thereafter. Indeed, GT is a logical approach to product and facility planning. Formalization of the scope and objectives of the system (volumes and variety) is a necessary step in the detailed design of any facility. For existing facilities, GT centers can be used to avoid relayout costs. Although not the best choice, existing part routings can be used as the code. In this case, the code may be represented as a binary string indicating, for each machine type, whether it is needed or not. The disadvantage of this approach is that we are restricted to continuing past routings. The more informative codes described above allow intelligent, dynamic (time-varying), selection of machine routings based on machine availability and capability.

6.4

ASSIGNING MACHINES TO GROUPS

It would be convenient if each part could justify its own group. We could skip GT and return to the serial flow systems of earlier chapters. However, this is not the case, and once parts have been coded, the next step is to form groups. A group is a collection of part types along with the machines, workers, and tooling required to produce them. To the extent possible, each part will be made totally within its group. The machines that constitute the group will be laid out to facilitate material flow—a GT flow line if possible, a GT cell otherwise.

We now look toward finding good solutions. Kusiak and Chow [1988] present a comprehensive survey of approaches for configuring groups. We discuss several approaches in some detail. Others may be found in the references at the end of the chapter. Pay attention to each model. Certain assumptions may be modified and desirable traits may be sacrificed or ignored at times. Many solution approaches, for instance, assume that a machine type can be placed in at most one group. We consider this antithetic to the concept of GT. Instead, Section 6.5 discusses preassignment of parts to specific machines. Each machine becomes a type, and groups may be formed without concern for the assumption of a unique group for each machine type. In the future, part design, coding, and grouping decisions will most likely be integrated through sophisticated data-base information systems (see Ham et al. [1988], for instance).

We assume that the coding step has led to a process plan for each part. This process plan is optimal in the sense that it will produce the highest quality, most reliable parts at the lowest price. Costs here are necessarily based on either historical machine usage for overhead or the assumption of fully utilized machines. Actual costs may differ as we organize groups and determine machine utilizations. In practice, parts that turn out to be troublesome or exceptional, in that they do not fit conveniently into any group, should have their alternative routings considered for primary use.

6.4.1 Production Flow Analysis

Burbidge [1977] originally developed a procedure called *Production Flow Analysis* (PFA) for using part routing information to form machine groups. We assume knowledge of the sequence of machines visited by each part type from the process plan. PFA requires manual decision activity but, unlike some purely automated techniques, it allows a machine type to be placed in more than one group when appropriate.

In a large manufacturing facility previously organized around a process concept, PFA can be carried out in a hierarchical fashion. First Factory Flow Analysis (FFA) is applied to link together processes (machining, pressing, welding) or subprocesses (turning, milling, boring) used by a significant number of parts. Large departments are formed by combining all of the related processes. Process types could appear in more than one department. Large departments are essentially independent plants that manufacture dissimilar products. Exceptional parts that do not fit any of these standard process route requirements can be redesigned, rerouted, subcontracted, or allowed to drag the extra required equipment into a department. Departments are independent in the sense that they use physically distinct processes. The departments are likely to be larger than desired for administrative and control purposes. We must then divide these into smaller groups of the desired size. This step is sometimes referred to as **group analysis** and is the subject of this section. The objective here is to assign machines to groups so as to minimize the amount of material flow between these groups. Small, inexpensive machines such as washers and bench-top presses are not considered in this step, as we can replicate and locate these as needed.

Group analysis consists of three basic steps. The first step involves constructing a list of the part types that require each machine type. The machine with the fewest part types is called the "key" machine. A subgroup is formed from all the parts that visit this key machine along with all machines required by these part types. Step two is a check to see if, except for the key machine, the machines in the subgroup fall into two or more disjoint sets with respect to the parts they service. If disjoint subsets of the subgroup exist, the subgroup is again subdivided into multiple subgroups. Likewise, if any machine is included in the subgroup due to just one part type, then this machine is termed exceptional and removed. The first two steps are repeated until all parts and machines are assigned to subgroups. The final step involves combining subgroups into groups of the desired size. Subgroups with the greatest number of common machine types are combined. This combination rule reduces the number of extra machines that will be needed and makes it easier to balance machine work loads. Ultimately, each group must be assigned sufficient machines and staff to complete its assigned parts.

EXAMPLE 6.1

Consider the set of process plans shown in Table 6.3a. Each column represents a part, each row a machine. A "1" indicates that the part visits the corresponding machine. Part 1, for instance, requires machines A and B. Parts 1, 2, and 3 visit machine A. While part 1 is seen to visit A and B, no information on operation order or precedence relationships is assumed, the order is arbitrary. Form groups. (For ease, the data has been ordered using a method to be discussed in the next section.)

Table 6.3a   Process Plans for Example 6.1

	Part							
Machine	1	2	3	4	5	6	7	8
A	1	1	1					
B	1	1	1					
C			1	1	1	1		
D				1	1	1	1	
E							1	1
F							1	1

Solution

STEP 1. Identify a key machine. Machines E and F receive the fewest components. Arbitrarily let E be the key machine. To create a subgroup we take parts 7 and 8, which visit E. These parts require machines D, E, and F, thus forming a subgroup.

STEP 2. Check for subgroup division. Ignoring machine E, all parts (7 and 8) visit machine F, and so the subgroup cannot be further subdivided. However, machine D is used only for part 7. D is considered to be an exception for this subgroup and is removed.

STEP 1. Identify a new key machine. Six parts remain. All machines receive at least three parts. A receives only three parts and we will call it the key machine. Parts 1, 2, and 3 form the subgroup along with machines A, B, and C.

STEP 2. Subgroup division. Removing machine A does not create disjoint subgroups for parts 1, 2, and 3. However, C is used for part 3 only and is thus deemed exceptional for this group and removed.

STEP 1. Identify a new key machine. Only parts 4, 5, and 6 remain. C is the key machine. The subgroup becomes parts 4, 5, and 6 along with machines C and D.

STEP 2. Divide subgroup. This third subgroup is "dense" in the sense that all parts use all the machines. This is an ideal situation for GT. No further subdivision is possible.

STEP 3. Aggregation. The designer is now free to attempt to recombine the three subgroups into a set of workable groups of desired size.

Human judgment comes into play at this point to decide how best to combine the atomic subgroups. The final solution must provide adequate machine resources in each group for the assigned parts or a plan to transport some parts between groups to visit the required machines. This would occur, for instance, if the subgroups were kept separate and an extra machine of type C was not provided to the second group.   ■

6.4.2 Binary Ordering Algorithm

The data format shown in Table 6.3a is referred to as a machine–part indicator matrix. The machine–part matrix forms the basis of many procedures for group formation. To see why this is true, suppose the data had been presented as shown in Table 6.3b. Observe that both tables contain identical information—only the row and column ordering differs.

We would like to identify dense blocks of 1's in the matrix where adjacent parts use the same machines. These form natural machine–part groups. Table 6.3a permits recognition of these groups much more readily than Table 6.3b. Fortunately, it turns out that efficient algorithms exist for reordering rows and columns and converting a matrix such as Table 6.3b into that of Table 6.3a. We now discuss such a procedure. Before proceeding, glance back at the previous paragraph on group analysis. Note that the rows of the machine–part matrix summarize the results of data collection in step 1 of group analysis.

The binary ordering algorithm provides an efficient routine for taking an arbitrary 0–1 machine–part matrix and reordering the machine rows and part columns to obtain a nearly block diagonal structure. Block diagonal means that we can partition the matrix such that the "boxes" on the main diagonal contain 0's and 1's but the off-diagonal boxes contain only 0's. In other words, given a matrix, draw $G - 1$ horizontal lines and $G - 1$ vertical lines. We obtain a $G \times G$ partitioned matrix as shown in Figure 6.8. A block diagonal matrix has some 1's in the row i column i partitions for $i = 1, \ldots, G$ but only 0's in the other partitioned boxes. These on-diagonal boxes are then the natural groups for our plant. Ideally we would like as many small groups as possible. However, groups may not be totally independent. For instance, suppose in our example that we drew lines after parts 3 and 6 (see Table 6.3a) and after machines B and D. Elements $C3$ and $D7$ fall outside the diagonal blocks. These are our exceptional elements. They must be handled by one of four

Table 6.3b   Unordered Process Plans for Example 6.1

Machine	Part							
	1	3	5	7	2	4	6	8
A	1	1			1			
E			1					1
C		1	1			1	1	
F				1				1
D			1	1	1	1	1	
B	1	1			1			

Machine	13	2	8	6	11	5	1	10	7	4	3	15	9	12	14
B	▮	▮	▮	▮											
D	▮		▮	▮	▮										
A						▮	▮	▮	▮	▮					
H						▮	▮		▮	▮	▮				
I						▮	▮		▮		▮				
E						▮			▮	▮	▮				
C												▮	▮	▮	
G												▮	▮		▮
F												▮		▮	▮

$G = 3$ groups

Figure 6.8 Example block diagonal ordered matrix.

alternatives. First, we could try to redesign parts 3 and 7 so as not to require machines C and D, respectively. Second, we could acquire extra machines of type C and D, placing a C in groups 1 and 2 and a D in groups 2 and 3. Economics and utilization would be considered in evaluating this possibility. Third, we could route parts 3 and 7 to two groups. Fourth, we could combine these nearly independent groups into one large group.

To illustrate the binary ordering algorithm, we first present a simple manual approach. We will then discuss how the procedure might be implemented on a computer for large problems. Think of the rows and columns as binary strings. Each row (column) consists of a string of 0's and 1's. We will eventually operate on both rows and columns, but for now suppose we want to reorder rows to locate similar rows together. If the row is envisioned as a binary number, then similar rows are those with similar values. Our task is then one of sorting rows by nonincreasing order.[4] We reorder rows in descending order of their binary value. After reordering rows, we switch to columns. The process continues until the ordering does not change. Unfortunately, the ending orderings are not unique for a given data set. Different starting orderings may yield different ending orderings. Moreover, we are still unsure how to treat the exceptional parts, which cause only "nearly" block diagonal structures. Let us not be too critical, however. Binary ordering provides a very simple and efficient way to bring order to chaos. It provides a starting point for most of the grouping procedures used. Ordering also helps reduce problem size. The grouping problem can be addressed separately for each diagonal block in the ordered matrix. There is nothing to gain by combining parts or machines from different blocks. Generally speaking, it is easier to solve r problems of size s then one problem of size rs. This is particularly true of the combinatorial problems faced in manufacturing. We now state the algorithm assuming N parts and M machines.

Binary Ordering Algorithm

1. Order rows: Assign value 2^{N-k} to column k. Evaluate each row. Sort rows in nonincreasing order. If rows were previously ordered, and no change just occurred, stop; otherwise go to 2.

---

[4]As an alternative we could compare pairs of rows, counting the number of columns that differ between each row pair. Identical pairs should be adjacent. Syntactic pattern recognition approaches operate in this fashion.

2. Order columns: Assign value 2^{M-k} to row k. Evaluate each column. Sort columns in nonincreasing order. If no change, stop; otherwise, go to 1.

EXAMPLE 6.2

Consider the $N = 8$ parts and $M = 6$ machines problem of Table 6.3b. Order the part–machine matrix.

Solution

STEP 1. We assign column 8 place value 1, column 7 place value 2, column 6 place value 4, and so on. Row A receives a value of $128 + 64 + 8 = 200$ for its 1's in the first, second, and fifth columns. Evaluating all rows produces the values shown.

Step 1									
	Part								
Machine	1	3	4	7	2	5	6	8	Value
A	1	1			1				200
E				1				1	17
C		1	1			1	1		102
F				1				1	17
D			1	1		1	1		54
B	1	1			1				200
2^{N-k}	128	64	32	16	8	4	2	1	

The rows are reordered to A, B, C, D, E, F. We now proceed to Step 2 and order the columns. This produces the following result:

Step 2									
	Part								
Machine	1	3	4	7	2	5	6	8	2^{M-k}
A	1	1			1				32
B	1	1		1					16
C		1	1			1	1		8
D			1	1		1	1		4
E				1				1	2
F				1				1	1
Value	48	56	12	7	48	12	12	3	

Based on the column values, the new ordering is 3, 1, 2, 4, 5, 6, 7, 8. We return to Step 1 and repeat row ordering.

				Step 1					
				Part					
Machine	3	1	2	4	5	6	7	8	Value
A	1	1	1						224
B	1	1	1						224
C	1			1	1	1			156
D				1	1	1	1		29
E							1	1	3
F							1	1	3
2^{N-k}	128	64	32	16	8	4	2	1	

The row ordering for this step is unchanged and we stop. Apart from the placement of part 3 first, we have achieved the block diagonal structure magically alluded to in Table 6.3a. The algorithm places part 3 first to accentuate its extra machine visit. This is perhaps one drawback of the ordering procedure: it accentuates exceptional elements by moving them as high in the matrix as possible instead of keeping them close to their near neighbors. In Table 6.3a we placed 3 after parts 1 and 2 to indicate the smooth transition between nearly diagonal groups. ∎

Suppose we had an industrial problem with 200 machines and 10,000 parts. Most parts visit less than 20 machines, making the machine–part matrix sparse. Computer scientists have developed more efficient techniques for the storage of these matrices. We could store for each row (column) the list of columns (rows) with a one. Alternatively, we could store the set of row and column combinations that house a one. Along with each two-tuple we might like to include pointers to the next element in its row and column. Adding two "hash" tables for quick access to the first elements in each row or column gives us complete, easy-to-retrieve data. Procedures exist to sort n elements (rows or columns) in a computational effort of $O(n \log n)$, making binary ordering possible for large problems.

6.4.3 Single-Pass Heuristic

Although producing a good starting solution, binary ordering ignores machine utilizations, group sizes, and exceptional elements. In this section we propose a fairly simple, single-pass heuristic for determining groups. At each step the next ordered part is assigned to a group along with the required machines. The process will constrain groups to the desired number of machines. The procedure is relatively quick. What is sacrificed is that the procedure may request extra machines. However, we can easily compute lower bounds on the number of machines of each type that are needed for any feasible solution and compare these to those suggested by this heuristic. If the heuristic finds a solution that does not require any additional machines, we have an optimal solution. In general, the heuristic forms a starting point for subsequent manual modification of the groups. The heuristic is designed for the case where it is desired to totally eliminate all intergroup moves, and an upper bound on the size of any group exists. (Clearly, we assume that no part uses more machines than the maximum group size.)

The first step involves replacing 1's in the machine–part matrix with actual machine utilizations. We assume part batch sizes Q_i are known. Let D_i be period demand for part i and s_{im} be required setup time per batch. Setup time is found from summing part i setup time and its apportioned share of family setup. For D_i/Q_i setups per period, total setup time attributable to part i is then $f_{im} = s_{im}D_i/Q_i$. We can measure in terms of machines needed by dividing by R_m, the time available per machine per period. Variable processing time for part i on machine type m per period (v_{im}) is found by multiplying processing time per unit (t_{im}) by demand, that is, $v_{im} = t_{im}D_i$. Utilizations, measured in machines, are then $u_{im} = (f_{im}+v_{im})/R_m$. A sample utilization matrix is given in Table 6.4. Summing across columns we find the total utilization values for each machine row. Rounding up we find a lower bound on the number of machines of each type.

Using the part ordering from the binary ordering algorithm, iteratively assign parts (and any required machines) to groups. We will assume that the organization has placed an upper limit of m_u on the number of machines in any group. G will represent the most recent group formed. Our decision rule is to assign the next part to the first group $(g \leq G)$ that has sufficient capacity on already allocated machines. If no group has sufficient capacity, we try to add machines to group G to handle this part. If this would violate the limit of m_u machines, we open a new group. To the new group we allocate all of the machines needed for the part. We also start a new group if the new part has no machines in common with the current group.

We can state the algorithm in a slightly more formal manner as follows. A_i will be the group to which part i is assigned. U_{mg} is total work load assigned to machine type m in group g. Group g has n_{mg} type m machines assigned to it. To keep notation as simple as possible, we assume no operation requires more than an entire machine (i.e., $u_{im} \leq 1$), but the procedure is easily extended to this case.

Single-Pass Formation

1. Perform binary ordering.
2. Replace 1's by $u_{im} = [(s_{im}D_i/Q_i) + t_{im}D_i]R_m^{-1}$
   Set $G = 1$, $n_{m1} = 0$, $U_{m1} = 0$.
3. Assign parts: For $i = 1$ to N
   Let $g' = \text{argmin}_{g \leq G} \{U_{mg} + u_{im} \leq n_{mg} \text{ for all } m\}$. M_G is number of machine types with capacity violations at $g = G$.
   (Assign to existing group): If g' exists, set $A_i = g'$, $U_{mg'} = U_{mg'} + u_{im}$ and go to 3.

Table 6.4   Example Utilization Matrix

Machine	1	2	3	4	5	6	7	8	Total Utilization	Minimum Machines
A	0.4	0.6	0.7						1.7	2
B	0.4	0.5	0.2						1.1	2
C			0.5	0.4	0.2	0.4			1.5	2
D				0.3	0.1	0.5	0.1		1.0	1
E							0.5	0.4	0.9	1
F							0.6	0.3	0.9	1

(Add part and machines to current group): If $\sum_m n_{mG} + M_G \leq m_u$, then set $A_i = G$, $U_{mG} = U_{mG} + u_{im}$, and if $U_{mG} + u_{nm} \geq n_{mG}$, $n_{mG} = n_{mG} + 1$ and go to 3.
(Start new group): $G = G + 1$. $A_i = G$, $n_{mG} = 1$ for $u_{nm} > 0$, $U_{mG} = u_{im}$.

On completing this initial group assignment, we check for the total number of machines of each type required. If we are at the lower bound, rejoice. For machine types exceeding this bound consider combining groups. Only group combinations where slack for this machine type exceeds 1.0 need be considered. Small groups may also be combined to approach the m_u limit.

EXAMPLE 6.3

Consider the problem in Table 6.4, and assume $m_u = 4$. Parts and machines are already ordered and utilizations are noted. Use the single-pass heuristic to form tentative groups.

Solution

Updating of results is shown in Table 6.5.

$i = 1$. Part 1 goes to group 1, taking along a machine A and B. This imposes a utilization of 0.4 on both machines, leaving a slack of 0.6 as shown in Table 6.5.

$i = 2$. Part 2 fits into this group with no new machines needed and, hence, is added. Slack time is reduced to 0 and 0.1 for machines A and B.

$i = 3$. Part 3 would require three new machines if it were added to group 1, since neither existing machine can handle the added load. Adding three machines would exceed the allowable group size; hence, a new group is started.

$i = 4$. Part 4 will not fit into either group without adding a machine D. However, there is space in the current group (group 2) for a new machine. We thus add 4 and D to the second group. Machines A and B are unaffected by adding part 4. Available time on machine C is reduced from 0.5 to 0.1.

$i = 5$. Machine C does not have capacity for part 5, and the group is full (in terms of machines), and so we open group 3.

$i = 6$. Part 6 will fit only in group 3 and is added accordingly.

$i = 7$. Part 7 can fit into group 3 by adding machines E and F. This is preferable to opening a new group, since we can utilize the existing machine D. Adding the two machines will not violate the constraint of four machines per group.

$i = 8$. Part 8 is then added to this group, since it requires machines E and F and sufficient slack time exists.

Examining the solution we find that only machine type D exceeds its lower bound. Both groups 2 and 3 have a D allocated, but planned production requires only one machine D for the shop. Our alternatives include combining groups 2 and 3 into one seven-machine group; purchasing the extra machine D;

Table 6.5   Assignments for Heuristic

Iteration	Part Assigned	Group	Machines Added	Resource Update (machine, remaining time)
1	1	1	A, B	$(A,0.6), (B,0.6)$
2	2	1	—	$(A,0.0), (B,0.1)$
3	3	2	A, B, C	$(A,0.3), (B,0.8), (C,0.5)$
4	4	2	D	$(A,0.3), (B,0.8), (C,0.1), (D,0.7)$
5	5	3	C, D	$(C,0.8), (D,0.9)$
6	6	3	—	$(C,0.4), (D,0.4)$
7	7	3	E, F	$(C,0.4), (D,0.3), (E,0.5), (F,0.4)$
8	8	3	—	$(C,0.4), (D,0.3), (E,0.1), (F,0.1)$

removing machine D from group 2 and routing part 4 between the two groups; or redesigning the part or its process plan. On close manual inspection, we notice that part 4 can be moved from group 2 to 3, which does away with the need for a D machine in group 2. Part 3 is its own group. This solution is optimal, since no intergroup material handling is required nor are any extra machines. ∎

6.4.4 Similarity Coefficients

Similarity coefficients (see McAuley [1972], Seifoddini and Wolfe [1986]) provide an alternate means of obtaining nearly independent groups. Emphasis is placed on locating machines with high interaction (material flows) in the same group. Consider a pair of machines i and j. Let n_i be the number of parts that visit machine i and n_{ij} be the number of parts that visit machine i and machine j. We define the similarity coefficient between these two machines as

$$s_{ij} = \max\left(\frac{n_{ij}}{n_i}, \frac{n_{ij}}{n_j}\right) \qquad (6.1)$$

The similarity coefficient indicates the proportion of parts visiting machine i that also visit machine j (or vice versa, whichever is larger). Values are standardized such that values near 1 are important and values near 0 are comparatively unimportant. As $s_{ij} \to 1$, it becomes more desirable to locate machines i and j in the same group. (Note that we could replace the number of parts with the number of handling loads if so desired.)

We present a simple technique called **hierarchical clustering** for determining groups. First, symbolically represent each machine by an icon such as a small circle (node). Connect these nodes with lines (arcs). Arcs are labeled with the similarity coefficient for the two nodes connected by the arc. The resultant graph forms a symbolic model of the problem. Next, eliminate all arcs with $s_{ij} \leq T$ where T is a user-selected threshold value. Eliminating the lower valued arcs allows us to concentrate on the important relationships. All connected nodes (machines) constitute a group. We can trace out the graph as T is varied from 1 to 0. Initially, each machine is a group. As T is reduced, more arcs appear in the graph and groups are combined. The tentative groups at any stage of the procedure are customarily referred to as "clusters." Several methods for updating the similarity coefficients between newly formed clusters and existing clusters have been proposed. One approach uses the maximum s_{ij} for any machine i in the first cluster and any machine j in the second cluster. This approach is called "single linkage," since a single machine pair can cause two clusters to be combined. Eventually, at $T = 0$ all machines are housed in one group. We are left with a range of solutions to select from. We could also modify the procedure to not combine groups once an upper limit on machine size is reached.

Hierarchical clustering has several advantages. First, the important relationships are assured of being satisfied. Second, we can quickly produce a range of solutions. Unfortunately, without the special check mentioned above, we have no control over the size of groups.

Hierarchical Clustering Heuristic

1. Form N initial clusters. Compute s_{ij} from equation 6.1 for all machine pairs.
2. Merge clusters: Let i and j range over clusters. Find i^*, $j^* = \text{argmax}_{i,j} s_{ij}$. Merge clusters i^* and j^*. If more than one cluster remains, go to 3.
3. Update coefficients: Remove rows and columns i^*, j^* from the similarity coefficient matrix. Replace them with row and column k. For all existing clusters r, $s_{rk} = \max(s_{ri^*}, s_{rj^*})$. Go to 2.

As stated, the heuristic produces a range of solutions that can be summarized in a dendogram. This is illustrated in the example that follows. If we redefine the merging rule in step 2 to be: Find i^*, $j^* = \text{argmax}_{i,j} s_{ij}$ such that $N_i + N_j \leq m_u$, where N_i is the number of machines in cluster i, then at the conclusion of the heuristic we are left with one solution that obeys our group size constraint.

EXAMPLE 6.4

Reexamine the six-machine type, eight-part example used above. Assume that one machine of each type is sufficient and apply the hierarchical clustering algorithm.

Solution

The machine–part matrix is as follows:

Machine	Part							
	1	2	3	4	5	6	7	8
A	1	1	1					
B	1	1	1					
C			1	1	1	1		
D				1	1	1	1	
E							1	1
F							1	1

We can compute the symmetric table of similarity coefficients s_{ij}. Consider s_{AC} for example. Machine A serves three parts, machine D serves four parts. Only part type 3 visits both machines. Thus $s_{AC} = \max(\frac{1}{3}, \frac{1}{4}) = \frac{1}{3}$. All values are given in Table 6.6.

Table 6.6   Initial Similarity Coefficients

			Machine			
Machine	A	B	C	D	E	F
A	—	1	0.33	0	0	0
B	1	—	0.33	0	0	0
C	0.33	0.33	—	0.75	0	0
D	0	0	0.75	—	0.50	0.50
E	0	0	0	0.50	—	1
F	0	0·	0	0.50	1	—

We start with six clusters, one for each machine. At $T = 1$ machines A and B are combined into a cluster and so are machines E and F. The table of similarity coefficients is updated using single linkage. The result is shown in Table 6.7. Similarity for any cluster with AB is found from the maximum of the clusters' original coefficient with A or B. EF cluster values are found similarly. For example $s_{AB,C} = \max(s_{AC}, s_{BC}) = 0.33$.

Table 6.7   Updated Similarity Coefficients at $T \leq 1.0$

		Machine		
Machine	AB	C	D	EF
AB	—	0.33	0	0
C	0.33	—	0.75	0
D	0	0.75	—	0.50
EF	0	0	0.50	—

The largest coefficient in the updated matrix of Table 6.7 corresponds to clusters C and D. Accordingly, these machines are grouped at $T = 0.75$. Updated coefficients $s_{AB,CD} = 0.33$ and $s_{CD,EF} = 0.50$ are computed. T is then reduced to 0.50, at which point clusters CD and EF are combined. Updating, $s_{AB,CDEF} = 0.33$ and all machines are grouped for $T \leq 0.33$. The clustering process is graphically summarized in the dendogram of Figure 6.9 on page 188.   ■

6.4.5* Graph Partitioning

Several graph theoretic models of the group configuration problem have been proposed. We define a graph $G = (N, A)$ as a set of nodes or vertices (N) and arcs or edges (A). Each arc connects two nodes. Our arcs are undirected in that neither node is the head nor tail of the arc. For now, it may be helpful to think of nodes as machines and arcs as material flows between machines. Cost c_{ij} may be associated with the level of flow on arc ij. In forming groups we need to partition the graph into a number of subgraphs. Each subgraph represents a group. The nodes of the subgraph are the machines in the group. Arcs connecting nodes in different subgraphs represent flows between groups. These are the flows we wish to minimize. A **cut** is the set of arcs connecting nodes in different subgraphs when we partition the graph

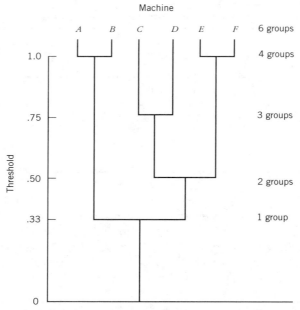

Figure 6.9   Dendogram for a hierarchical clustering with single linkage.

into two subgraphs. The cost of the cut is the sum of the costs associated with the arcs of the cut. From our analogy, it is plain to see that we wish to find the partitions that have at most m_u nodes in a subgraph and minimize cut cost. We let m_l be the minimum number of machines desired in a group. Figure 6.10 depicts the data of Table 6.4 in graph form. We have assumed that part sequences correspond to the alphabetical order of machines. Arc costs should reflect material handling loads; for illustration we simply indicate the part types that flow between machines. In practice, additional information on handling loads would have to be collected.

No clear division of machine types into groups exists. However, suppose that noting the need for two machines of types A, B, and C, we had peeked ahead to Section 6.5 and found the possible assignment of parts to specific machines that has parts 1 and 2 to machines $A1$ and $B1$; part 3 to machines $A2$ and $B2$; part 3 to machine $C1$; and parts 4, 5, and 6 to machine $C2$. Using individual machines we would obtain the graph of Figure 6.11. Now groups are self-evident. In general, even after selecting specific machines, groups will not be so clearly defined, or natural groups may be too large. A procedure for partitioning such graphs is required.

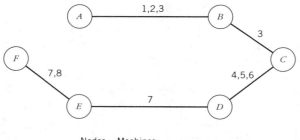

Nodes = Machines
Arc values = Transferred part numbers

Figure 6.10   A graph representation of a part–machine matrix.

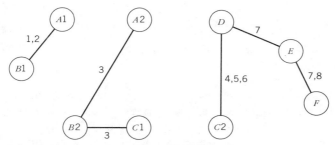

Figure 6.11   A graph representation with replicated machines.

We will first describe the popular Kernighan and Lin [1970] heuristic (K & L) for graph partitioning. This will be followed by its modification and application to our grouping problem. The K & L heuristic is an improvement routine: it tries to find a better partition given a current, feasible partition.

Suppose we have an initial partition and we wish to find one of lower cost. Let the two node sets be called N_1 and N_2 ($N_1 \cup N_2 = N$), each with m nodes. The best partition into two sets of size m must be obtainable by switching nodes n_1 from N_1 with n_2 from N_2. We need only find the best n_1 and n_2 sets. Let us look at how costs will change as we switch nodes. Define $E_i = \sum_{j \in N_2} c_{ij}$ to be the external cost and $I_i = \sum_{j \in N_1} c_{ij}$ the internal cost of node $i \in N_1$. E_i and I_i represent the interaction costs of node i with those outside and inside its current partition, respectively. If node i is moved to partition 2, the gain in cost (cost savings) is

$$G_i = E_i - I_i \qquad (6.2)$$

since all costs in E_i are removed from the cut and all costs in I_i are added. E_j, I_j, and G_j are analogously defined for $j \in N_2$. The heuristic is based on the fact that if we exchange node $i \in N_1$ with node $j \in N_2$, then the gain will be

$$G_{ij} = G_i + G_j - 2c_{ij} \qquad (6.3)$$

The final term reflects the result that ij interaction is unchanged but was already added into G_i and G_j. The K & L procedure first finds the best single node pair to switch, that is, $i \in N_1$ and $j \in N_2$. Next, conditional on i and j being exchanged, we find the additional node pair that should be switched. The search continues up through subsets of size $m - 1$ nodes being switched between initial groups. Whichever switch offers the largest savings is then made, and the process restarts. Thus, we do not officially make any switch until the best $1, 2, \ldots, m - 1$ size switches have all been evaluated and the best of these (conditional) bests is selected. This version of the heuristic assumes that two equal-sized groups exist. Experimentally, it has been found to produce optimal solutions about one half the time on problems with 30 nodes and about one fourth the time on 60-node problems. After stating the procedural steps, we will generalize to our grouping problem, which has multiple groups and variable group sizes. The index "t" will be used to denote the size of the sets n_1 and n_2.

Graph-Partitioning Heuristic

1. Let $n_1 = \phi, n_2 = \phi, t = 0$. Compute all G_i^1, G_j^1 from equation 6.2.
2. Choose best exchange pair: $t = t + 1$. Compute all G_{ij}^t from equation 6.3. Define $i^*, j^* = \text{argmax} G_{ij}^t$. Let $n_1 = n_1 + i^*, n_2 = n_2 + j^*$. $G^t = G_{i^* j^*}^t$.

3. Update exchange costs: $G_i^t = G_i^{t-1} + 2c_{ii^*} - 2c_{ij^*}$ for $i \in N_1 - n_1$ and $G_j^t = G_j^{t-1} + 2c_{jj^*} - 2c_{ji^*}$ for $j \in N_2 - n_2$. If $t = m - 1$ go to 4; otherwise go to 2.

4. Find best size exchange: Find $t^* = \operatorname{argmax}\sum_{t=1}^{t^*} G^t$. If $\sum_{t=1}^{t^*} G^t > 0$ make switch and go to 1. Otherwise stop.

Assuming that specific machine assignments have been made for each operation (we will handle this shortly), we can use graph partitioning to form groups. Each machine is a node. Arc costs between nodes i and j are $c_{ij} = HL_{ij}$, where H is the cost of moving a load between groups and L_{ij} is the number of material handling loads moving directly between i and j based on operation assignments and part routings. We assume that the machine–part matrix has been ordered. This is important for obtaining a good starting solution. We select a target group size as

$$\overline{m} = \lceil \frac{m_l + m_u}{2} \rceil \qquad (6.4)$$

Initial groups are formed by taking $\overline{m}$ machines at a time. We add $m_u - \overline{m}$ dummy nodes with 0 arc costs to each group. Dummy nodes represent vacant machine locations in the group. With this starting solution we proceed to perform Kernighan and Lin improvement procedures on each successive pair of groups, continuing until no changes are made. Improvements need be attempted only for groups that have positive cut costs between them. With the ordered starting solution the number of necessary improvement steps is kept relatively small. A nice aspect of the procedure is that any manually developed formation of groups can be passed to the improvement procedure as well. We simply add the appropriate number of dummy nodes to each group (partition). We will illustrate with an example.

EXAMPLE 6.5

Reconsider the utilization matrix in Figure 6.10. Based on productivity maximization and robot work envelopes, each group is to have between two (m_l) and four (m_u) machines. Assume that the assignment of part operations to machines alluded to earlier has been determined as shown in Table 6.8. (Note the equivalence of Figure 6.11 and Table 6.8.)

Table 6.8   Individual Machine Assignment

				Part				
Machine	1	2	3	4	5	6	7	8
A1	1	1						
B1	1	1						
A2			1					
B2			1					
C1			1					
C2				1	1	1		
D				1	1	1	1	
E							1	1
F							1	1

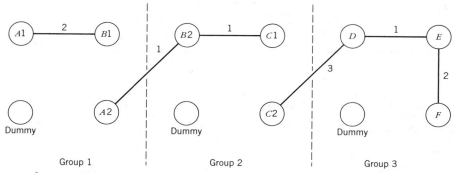

Figure 6.12   Initial groupings for Example 6.5.

Solution

From equation 6.4 we will start with initial groups of $\overline{m} = (2 + 4)/2 = 3$ machines plus $m_u - \overline{m} = 1$ dummy each. This makes the groups ($A1$, $B1$, $A2$, Dummy), ($B2$, $C1$, $C2$, Dummy), and (D, E, F, Dummy). Assuming that all parts have a single unit load, arc costs and the initial solution are as shown in Figure 6.12. Initial cut costs are 1 between groups 1 and 2, 3 between groups 2 and 3, and 0 between groups 1 and 3. We will proceed to improve groups 1 and 2.

   Calculations for the switches are summarized in Table 6.9. We first compute E_i, I_i, and G_i values for each node. Consider node $A1$. External cost is 0, since no arcs leave the group. Internal cost is 2 from the arc connecting $A1$ with $B1$. We next determine the G_{ij}^1 values as $G_{ij}^1 = G_i^1 + G_j^1 - 2c_{ij}$. As an example $G_{A1C1}^1 = G_{A1}^1 + G_{C1}^1 - 0$. Checking the G_{ij}^1 column we find that the best switch is $A2$ with Dummy or $C2$. Either switch has a gain of 1. We arbitrarily select Dummy. Thus, set $n_1 = A2$ and $n_2 = $Dummy. The next step is to update the values to $G_i^2 = G_i^1 + 2c_{iA2} - 2c_{i\,\text{Dummy}}$ and $G_j^2 = G_j^2 + 2c_{j\,\text{Dummy}} - 2c_{jA2}$. These values are also shown in Table 6.9a. Computing G_{ij}^2 using the updated $G_i{}^2$ and G_j^2 values produces the G_{ij}^2 column in Table 6.9b. No switch looks desirable, but Dummy with $C2$ is the least costly. We will consider this switch anyway, that is, $n_1 = \{A2, \text{Dummy}\}$ and $n_2 = \{\text{Dummy}, C2\}$. We can then update to G_i^3, G_j^3, and G_{ij}^3. The switch was considered since a three-node switch may still be found with a positive gain. However, from the G_{ij}^3 column we see that no switch is

Table 6.9a Gains for Considered Node Switches

Group	Node i	E_i	I_i	G_i^1	G_i^2	G_i^3
1	$A\,1$	0	2	-2	-2	-2
	$B\,1$	0	2	-2	-2	-2
	$A\,2$	1	0	1	—	—
	Dummy1	0	0	0	0	—
2	$B\,2$	1	1	0	-2	0
	$C\,1$	0	1	-1	-1	—
	$C\,2$	0	0	0	0	0
	Dummy2	0	0	0	—	—

Table 6.9*b*   **Switch Gains for Considered Switches**

i	j	G_{ij}^1	G_{ij}^2	G_{ij}^3
$A\ 1$	$B\ 2$	-2	-4	-2
	$C\ 1$	-3	-3	$-$
	$C\ 2$	-2	-2	-2
	Dummy2	-2	$-$	$-$
$B\ 1$	$B\ 2$	-2	-4	-2
	$C\ 1$	-3	-3	$-$
	$C\ 2$	-2	-2	-2
	Dummy2	-2	$-$	$-$
$A\ 2$	$B\ 2$	-1	$-$	$-$
	$C\ 1$	0	$-$	$-$
	$C\ 2$	1	$-$	$-$
	Dummy2	1	$-$	$-$
Dummy1	$B\ 2$	0	-2	$-$
	$C\ 1$	-1	-1	$-$
	$C\ 2$	0	0	$-$
	Dummy2	0	$-$	$-$

desirable. We now choose the best switch considered, either one node, two nodes, or three nodes. The best move is thus to switch $A2$ with Dummy for a gain of 1. (In fact, as original cut cost was 1, we could have halted the process at the end of the first stage, recognizing that we had an optimal solution for these two groups.) Our solution for groups 1 and 2 is thus machines ($A1$, $B1$) in group 1 and ($A2$, $B2$, $C1$, and $C2$) in group 2. The "$-$" marks in the table indicate that those switches are no longer considered because one of the machines has already been switched. We now leave it to the reader to show that if we improve groups 2 and 3, we obtain a solution with intragroup flow only. This solution is optimal, since all cut costs are zero.   ∎

Several graph theoretic approaches to cell formation have been proposed in recent years. The reader may wish to consult Rajagopalan and Batra [1975] for one example or Kusiak and Chow [1988] for a summary of several approaches.

6.5*
ASSIGNING PARTS TO MACHINES

Prior to assigning machines to groups, routings of parts to specific machines must be determined. The reader will recall that this information was assumed to be available during graph partitioning. Process planning selects the proper machine type for each operation. If only one machine of this type exists, the routing decision is complete. An important problem develops, however, when any of a number of individual machines could be used for an operation. These machines are referred to as a machine type. The choice of a specific machine for an operation should facilitate within-group production. This implies that if two parts have several machine types in common, then we would like these parts to use the same machines. These machines and parts can then be placed in the same group. In the remainder of this section we show how the graph-partitioning heuristic examined above can also be used for this subproblem.

Assigning operations to specific machines must be performed for every machine type that requires multiple machines. Fortunately, we can view machine types as being independent and solve the problem separately for each type. Let's pick an arbitrary machine type and see what kind of solution we can manufacture.

Notationally, we have I operations to assign to M_m type m machines. As lot sizes are assumed known, we can compute total time required on this machine type per period T. Average utilization of these machines is therefore $\overline{U} = T/M_m$. Work-load balancing is a desirable goal, and so we will aim for a utilization of approximately $\overline{U}$ on each machine. We make use of the binary-ordered part–machine matrix. Since operations belong to parts, operations are implicitly ordered as well as parts. Our solution procedure involves two phases: construction of an initial feasible solution, followed by improvement of this solution.

The initial solution is constructed in a single pass by sequentially assigning the next operation to the first available machine. The operation is assigned to the lowest numbered machine with current utilization not exceeding $\overline{U}$ and sufficient capacity to perform the operation, that is, utilization will not exceed 1.0. New machines are "opened" as needed.

In phase 2 we use the K & L graph partitioning procedure to improve the initial solution. Each operation is a node. Partitions represent the set of operations assigned to a single machine. Taking pairs of machines at a time, we use K & L to switch operation sets between the machines. Each machine can have dummy operations assigned. This will allow improvement by transferring a single real operation. Two issues must be addressed. These issues relate to the proper definition of arc costs and assuring time feasibility of machine assignments.

First, let us consider arc costs. At this stage of group formation, we do not know the exact implications of assigning two operations to the same machine. We do know, however, that parts can share other machines if and only if their routings include additional common machine types. Accordingly, if two parts can share other machines as well, it is advisable to place them on the same machine of the type being considered. The specific machines of each type selected to service these parts can then be placed in a single group. Moreover, the greater the number of material handling loads for a part, the more important it is to manufacture the part totally within one group. With this basis we define arc cost between operations i and j for machine type m as

$$c_{ij} = \sum_{m' \neq m} (L_i + L_j) Z_{ijm'} \tag{6.5}$$

where

$$Z_{ijm'} = \begin{cases} 1 & \text{if } m' \in V\{i\} \text{ and } m' \in V\{j\} \\ 0 & \text{otherwise} \end{cases}$$

indicates whether machine type m' is on the routing ($V\{\ \}$) of both the parts to which operations i and j belong. c_{ij} gives an upper bound on the number of intergroup material handling loads saved if operations i and j are assigned to the same machine of type m and, ultimately, to the same group. If GT advantages are achievable only by combining parts in the same setup family, an additional indicator can be added to equation 6.5 stating whether the two associated parts belong to the same setup family.

The issue of time feasibility for each machine can be easily addressed. Before any pair of nodes is considered for switching, total machine utilization of each machine is checked. If either machine would have assigned period work load in excess of 1.0, we set the gain G_{ij}^t to $-\infty$. Thus, infeasible switches are avoided. As before, assignments are denoted by A_i; t_i is the time required in machine-periods by operation i on machine type m. T_r is the current utilization of machine r, $r = 1, \ldots, M_m$. Briefly stated, the procedure is as follows.

Operation Assignment Procedure

1. Order operations by binary ordering of their parts $T_r = 0$.
2. Initial assignment: For operation $i = 1, \ldots, I$ let $r^* = \text{argmin}_{r \le M_m} T_r \le \overline{U} \cap$ $T_r + t_i \le 1.0$. If no such r^* exists, divide operation i into 2 suboperations and repeat step 2 for i. Set $A_i = r^*, T_{r^*} = T_{r^*} + t_i$.
3. Improvement: For each pair of machines of type m, perform Kernighan and Lin improvements with arc costs c_{ij} (see equation 6.5).

EXAMPLE 6.6

Recall the utilization matrix from Example 6.3. In that example parts 3 and 4 were grouped together. This resulted in either purchasing an extra machine D or routing part 4 to two groups. If part 4 is an important, high-volume part, the latter choice may be undesirable. If machines are expensive, the first option should be avoided. Indeed, part 4 should probably be grouped with parts 5 and 6, which share the same routing. Let us see if our procedure agrees. We will assign parts to the two required machine type C's. Let all $L_i = 1$.

Solution

STEP 1. The ordered part–machine matrix is given in Table 6.4.

STEP 2. Initial assignment. Total utilization is $T = 1.5$, yielding $\overline{U} = 1.50/2 = 0.75$. Part 3 is first assigned to machine $C1$, which sets $T_{C1} = 0.5$. Part 4 likewise goes to $C1$, since $T_{C1} \le \overline{U}$ and total assigned time will only be $0.5 + 0.4 = 0.9$. Since $C1$ has assigned time in excess of $\overline{U}$, part 5 is assigned to machine $C2$, and likewise part 6.

STEP 3. Improvement. We add a dummy part to each machine. Using equation 6.5, c_{ij} values are all zero except $c_{45} = c_{46} = c_{56} = 2$, reflecting the routing of parts 4, 5, and 6 to machine D as well as C and the assumption that all $L_i = 1$. Figure 6.13 shows the result of the initial partitioning. It is clear that when attempting improvement by node exchange the only positive single node exchange is $G_{4\text{Dummy}}^1 = 4$. This switch is borderline feasible as utilization on machine $C2$ goes to 1.0. After this switch all subsequent exchanges are of negative gain. The partitioning heuristic will thus elect to switch nodes 4 and Dummy. Repeating the process will find no new positive gain switches; hence, we stop

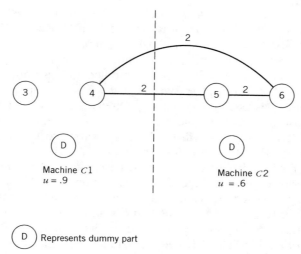

Figure 6.13   The assignment of operations to type C machines.

with the part 3 assigned to $C1$ and parts 4, 5, and 6 assigned to $C2$. This, we saw, will lead us to the optimal solution of three groups that are identified by parts as (1, 2), (3), (4, 5, 6, 7, 8) with machines types (A, B), (A, B, C), and (C, D, E, F), respectively. ∎

6.6
AN ECONOMIC, MATHEMATICAL PROGRAM FOR GROUP FORMATION

The final determination of groups should adhere to the principles of groups described previously in this chapter and should attempt to minimize associated costs. In this section we develop a mathematical statement of the group configuration problem—assigning part operations and machines to groups. Unfortunately, solution of the model is quite difficult except for small problems, and we are forced to resort to heuristic methods similar to those of the two previous sections.

We begin with a formal statement of the problem. The notation is summarized in Table 6.10. Operations are indexed by i, machines by m, families by f, and groups by g. A family is a set of parts S_f that are sufficiently similar that they can share setups. Families should be kept together if possible, and the fewer families per group the better. Machine utilizations may not justify a group for each family, however. It is assumed that the demand, operation sequence, processing times, and handling load size are known for each part. In practice, each part may have several possible plans created during process planning. Alternatives are to allow continuation of production when a primary machine/tool is unavailable. We restrict ourselves to using the primary process routing for each part in our mathematical formulation as even this will be more than we care to attack directly.

Each operation of each part must be assigned to a group. Accordingly, each group must have sufficient machines of each type to perform its assigned operations. To avoid intergroup moves, we would like to assign successive pairs of operations on the same part to the same group (i.e., all of a part's operations to the same group). If a part must move between groups, it incurs a cost of H per load. The number of loads L_i is the ratio of demand to material handling unit load size. Our basic decision variables are whether or not operation i is assigned to group g, X_{ig}, and the number of machines assigned to group g, Y_{mg}. For record-keeping purposes we will use the extra binary variables Z_g and F_{fg}. Z_g will be 1 if we use group g. F_{fg} will be 1 if any part from family f is assigned to group g. Otherwise, these indicators are 0.

With the foregoing notation, we can formulate the economic group configuration problem. The objective function will be the sum of four costs: machines, material movement, group overhead, and tooling. This function is given in equation 6.6. The first two costs represent the basic trade-off to be explored. The importance of the latter two cost terms depends on the specific environment and management's preferences. Variable production costs are fixed owing to set routings, but fixed machine costs must be included and constitute the first cost term. The second cost term is for intergroup material flow. In addition to physical move cost, H should include costs of stockpiling work-in-process and other scheduling difficulties due to intergroup moves. The third cost term allows for the possibility of fixed administrative costs for each group. This is reflected in the cost term C_g, which is charged to each group. Last, if family tooling must be replicated for every group that produces f's parts, we include the cost q_f for each group receiving any of these parts.

Several sets of constraints exist in our decision problem. We earlier discussed limits on the size of a group. These limits are enforced by constraints 6.7. Constraints 6.8 require us to place sufficient machines of each type in each group. The left-hand side of the expression accumulates total processing time required on type m machines in cell g. The right-hand side represents the total available machining time. Our model is a static planning model, and hence we assume that demand is constant. If demand fluctuates, we assume that some combination of overtime and inventory fluctuation is used to meet demand. Naturally, we can plan for growth or random

---

Table 6.10   Summary of Notation for Group Determination

f	Index for part family
g	Index for groups
i	Index for operations on parts
m	Index for machine types
c_m	Fixed period cost per machine type m required
C_g	Fixed period cost per group
D_i	Period demand for product associated with operation i
g_l, g_u	Lower and upper bound on workers per group
H	Intergroup material handling cost per load
L_i	Number of material handling loads following operation i
m_l, m_u	Lower and upper bounds on machines per group
q_f	Cost to assign family f to group
R_m	Regular time available per period for type m machine
t_{im}	Processing time per operation i on machine type m
F_{fg}	Binary indicator of group g containing family f parts
X_{ig}	Binary indicator of operation i being assigned to group g
Y_{mg}	Integer number of machines type m assigned to group g
Z_g	Binary indicator of use of potential group g

fluctuations by the suitable definition of demand D_i in these expressions as well. The constraints 6.9 are required to determine when parts must move between groups. If successive operations on a part are assigned to different groups, v_{ig}^+ will be set to 1, thus incurring the move cost in the objective function. We still need to make sure all operations are assigned to a group. This task is handled by constraints 6.10. The last two constraint sets serve to assure that groups are charged their fixed administrative and family tooling costs. These constraints are not needed if the corresponding cost terms are deleted.

$$\min \text{ cost } = \sum_{g=1}^{G} \sum_{m=1}^{M} c_m Y_{mg} + H \sum_{i=1}^{I} v_{ig}^+ L_i \qquad (6.6)$$

$$+ \sum_{g=1}^{G} C_g Z_g + \sum_{g=1}^{G} \sum_{f=1}^{F} q_f F_{fg}$$

subject to:

$$m_l \le \sum_{m=1}^{M} Y_{mg} \le m_u \qquad \text{for all } g \qquad (6.7)$$

$$\sum_{i=1}^{I} D_i t_{im} X_{ig} \le R_m Y_{mg} \qquad \text{for all } m, g \qquad (6.8)$$

$$X_{ig} - X_{i+1,g} = v_{ig}^+ - v_{ig}^- \qquad \text{for nonterminal operation } i \qquad (6.9)$$

$$\sum_{g=1}^{G} X_{ig} = 1 \qquad \text{for all } i \qquad (6.10)$$

$$\sum_{i=1}^{I} X_{ig} \le I Z_g \qquad \text{for all } g \qquad (6.11)$$

$$\sum_{i \in S_f} X_{ig} \le I F_{fg} \qquad \text{for all } f, g \qquad (6.12)$$

$$Y_{bg} \text{ integer} \qquad X_{ig} \in \{0, 1\} \qquad Z_g \in \{0, 1\} \qquad F_{fg} \in \{0, 1\}$$

The formulation in equations 6.6 to 6.12 is an integer program, far too large to solve for real problems. Several results are apparent, however. First, if we know the assignment of operations to groups, we also know the optimal assignment of machines to groups. Machines cost money ($c_m \ge 0$) and are purchased only as their resource is required. Given X_{ig}, the optimal Y_{mg} are simply the smallest integers that satisfy equations 6.8. If an upper bound on group size is violated in equation 6.7, then the X_{ig} are infeasible. If the lower bound portion of a constraint type 6.7 is violated, based on the formulation, it would be best to add the necessary amount of the cheapest machine type to achieve feasibility. Of course, these machines should not actually be purchased. This phenomenon indicates that we should probably have $m_l = 0$. A second observation concerns the situation when a family can justify its own group. If the fixed group cost and the cost of machines for any family is less than the sum of family tooling cost, then that family should be assigned entirely to one group. This follows, since if the family is split we can improve the cost by creating a new group for just the family. (We assume that the family requires at most m_u machines.) Any intergroup material handling cost being incurred can be added to the family tooling cost in satisfying this condition, since this cost will be eliminated by coalescing the family. This is a rather weak condition in that normal problems will have solutions with entire families within a single group even if this condition is not met.

6.7
SUMMARY

Group technology is a management philosophy. The philosophy espouses the belief that standardizing approaches to similar problems is beneficial. It is preferable to solve a single problem optimally and to apply the learned principles to similar problems than to solve each slightly different problem independently. Taking complete advantage of GT requires first spending the effort to develop a part-coding system that reflects similarities in design and manufacturing. Codes indicate design and manufacturing similarities and should be efficient in size, flexible for expansion, inclusive of entire part population, capable of discriminating between parts based on key attribute values, and compatible with common terminology. Code selection is driven by component population, intended usage, available resources for implementation, and level of detail required. Hierarchical codes are efficient at storing information but more difficult to learn than chain codes. Hybrid codes, such as Opitz, are often the best compromise choice. Coding parts based on similarities also facilitates part-design and process-plan retrieval.

Autonomous, fully equipped cells yield numerous production and organizational behavior advantages. Shorter setups significantly reduce WIP levels and throughput times. Material handling and scheduling are simplified.

Group determination and layout are important aspects of the GT approach. Binary ordering of the machine–part matrix is a good starting point for determining independent groups. Machine costs and utilizations are then factored in for determining the specific assignment of machines to groups and parts to machines. A single-pass heuristic can quickly construct tentative groups. The heuristic solution can be compared to lower bounds. The decision problem can also be modeled as a graph of machine nodes with arc costs reflecting movement between machines. Partitioning the graph forms groups. Within groups, standard facility layout techniques can be used with the aim of achieving advantages of flow lines.

Relationships between machines can be summarized by similarity coefficients. Similarity coefficients permit quick generation of a range of possible grouping decisions. However, use of similarity coefficients sacrifices much important information regarding actual costs.

When several machines of a common type are needed, part operations must be assigned to specific machines prior to (or concurrently with) assigning machines to groups. Graph partitioning can also be used to perform this task.

REFERENCES

Askin, R. G., and S. Subramanian (1987), "A Cost Based Heuristic for Group Technology Configuration," *International Journal of Production Research*, 25(1), 101–114.

Askin, R. G., and K. S. Chiu (1990), "A Graph Partitioning Procedure for Machine Assignment and Cell Formation in Group Technology," *International Journal of Production Research*, 28(8), 1555–1572.

Askin, R. G., and A. Vakharia (1990), "Group Technology Planning and Operation," in *The Automated Factory Handbook*, D. I. Cleland and B. Bidanda, eds., TAB Books, Blue Ridge Summit, PA, 317–366.

Burbidge, J. L. (1977), "A Manual Method of Production Flow Analysis," *The Production Engineer,* 56, 34–38.

Burbidge, J. L. (1975), *The Introduction of Group Technology,* John Wiley & Sons, New York.

Burbidge, J. L. (1989), *Production Flow Analysis for Planning Group Technology,* Clarendon Press, Oxford, England.

Chandrasekharan, M. P., and R. Rajagopalan (1987), "ZODIAC—An Algorithm for Concurrent Formation of Part-Families and Machine Cells," *International Journal of Production Research,* 25(6), 835–850.

Choobineh, F. (1988), "A Framework for the Design of Cellular Manufacturing Systems," *International Journal of Production Research,* 26, 1161–1172.

Gallagher, C. C., and W. A. Knight (1986), *Group Technology Production Methods in Manufacture,* Halsted Press, John Wiley & Sons, New York.

Ham, I., E. V. Goncalves, and C. P. Han (1988), "An Integrated Approach to Group Technology Part Family Data Base Design Based on Artificial Intelligence Techniques," *Annals of the CIRP,* 37(1), 433–437.

Ham, I., K. Hitomi, and T. Yoshida (1985), *Group Technology: Applications to Production Management,* Kluwer-Nijhoff Publishing, Boston, MA.

Houtzeel, A., and C. S. Brown (1984), "A Management Overview of Group Technology," in *Group Technology at Work,* N. L. Hyer, ed., Society of Manufacturing Engineers, Dearborn, MI.

Hyer, N., and U. Wemmerlov (1982), "MRP/GT: A Framework for Production Planning and Control of Cellular Manufacturing," *Decision Sciences,* 13(4), 681–701.

Japan Society for the Promotion of Machine Industry (1980), *Group Technology,* University of Tokyo Press, Tokyo, Japan.

Kernighan, B. W., and S. Lin (1970), "An Efficient Heuristic Procedure for Partitioning Graphs," *The Bell System Technical Journal,* 49, 291–307.

King, J. R. (1980), "Machine-Component Grouping in Production Flow Analysis," *International Journal of Production Research,* 18(2), 213–232.

King, J. R., and V. Nakornchai (1982), "Machine-Component Group Formation in Group Technology: Review and Extension," *International Journal of Production Research,* 20(2), 117–133.

Kumar, K. R., A. Kusiak, and A. Vannelli (1986), "Grouping Parts and Components in Flexible Manufacturing Systems," *European Journal of Operational Research,* 24(3), 387–397.

Kusiak, Andrew, and Wing S. Chow (1988), "Decomposition of Manufacturing Systems," *IEEE Transactions on Robotics and Automation,* 4(5), 457–471.

McAuley, J. (1972), "Machine Grouping for Efficient Production," *Production Engineer,* 51, 53–57.

McCormick, W. T., P. J. Schweitzer, and T. E. White (1972), "Problem Decomposition and Data Reorganization by a Cluster Technique," *Operations Research,* 20, 993–1009.

Opitz, H. (1970), *A Coding System to Describe Workpieces,* Pergamon Press, New York.

Pullen, R. D. (1984), "A Survey of Cellular Manufacturing Cells," in *Group Technology at Work,* N. L. Hyer, ed., Society of Manufacturing Engineers, Dearborn, MI.

Rajagopalan, R., and J. Batra (1975), "Design of Cellular Production Systems—A Graph Theoretic Approach," *International Journal of Production Research,* 13, 56–68.

Rembold, U., C. Blume, and R. Dillmann (1985), *Computer Integrated Manufacturing Technology and Systems,* Marcel Dekker Inc., New York.

Seifoddini, Hamid, and Philip M. Wolfe (1986), "Application of the Similarity Coefficient Method in Group Technology," *IIE Transactions,* 18(3), 271–277.

Shtub, Avraham (1988), "Capacity Allocation and Material Flow in Planning Group Technology Cells," *Engineering Costs and Production Economics,* 13, 217–228.

Snead, Charles S. (1989), *Group Technology: Foundation for Competitive Manufacturing,* Van Nostrand Reinhold, New York.

Tompkins, J. A., and J. A. White (1984), *Facilities Planning,* John Wiley & Sons, New York.

Vakharia, A. J., and U. Wemmerlov (1990), "Designing a Cellular Manufacturing System: A Materials Flow Approach Based on Operation Sequences," *IIE Transactions,* 22, 84–97.

Wemmerlov, Urban, and Nancy Lea Hyer (1989), "Cellular Manufacturing Practices," *Manufacturing Engineering,* 102(3), 79–82.

PROBLEMS

6.1. Describe the typical advantages to be gained by implementing GT.

6.2. What are the major factors to be considered in selecting a coding scheme?

6.3. For a GT flow line, cell, and center organization, discuss the extent to which each of the advantages listed in Table 6.1 might be realized.

6.4. Describe the difference between cellular manufacturing and group technology.

6.5. Suppose, as stated in Table 6.1, that GT succeeds in reducing machine setup times by 32 percent. What is the direct impact of this reduction on throughput time, space needs, and WIP levels? Assume that batch sizes are reduced in proportion to setup time reduction.

6.6. What is meant by a "composite component"?

6.7. Describe the family of parts with Opitz form code 12532.

6.8. Design a coding scheme for the set of parts described below. The code is to be used by process planning and manufacturing. The code is to cover all sheet metal parts. Raw stock ranges between $\frac{1}{32}$ and $\frac{1}{2}$ in. in thickness (light and heavy gauge). Raw sheet alloys of various hardnesses are used. The combination of stock thickness and hardness determine the force needed to perform an operation and, hence, which machines are permissible. Additionally, the demand rate for the part is important because high-volume parts justify specialized multioperation dies and punch presses. For most machine types, a high-power and/or large-dimension part would be made on a different machine than a low-power and/or small part.

Raw sheet is available in many sizes and, hence, larger parts often require only shearing (on a shearing machine) to obtain the individual piece raw material. Other items can have the basic part raw material formed by slitting the sheet stock (on a slitting machine) and then shearing. Parts with irregular shapes undergo trimming early in their production process. Smaller and nonrectangular parts require an initial blanking operation to produce the basic input material for individual pieces.

Most bending operations are performed on punch presses. These presses vary in power and part dimension capabilities. Hole punching machines provide an inexpensive method for individually punching asymmetrical holes on small-volume parts. Press brakes are used for forming long grooves and flanges near the edge of parts.

State any assumptions you make and explain how your code will be useful.

6.9. Consider the 10-part, 12-machine data of Table 6.11. Find any natural groupings of parts and machines. Suggest a set of part/machine groupings and sketch a diagram showing how the groups would be located with respect to one another. Only one machine of each type is to be used.

Table 6.11   Part Routings for Problem 6.9

Part	Sequence of Machines Visited
1	10, 7, 8, 9
2	1, 3, 5, 4, 11
3	6, 12, 2
4	4, 11, 3
5	6, 4, 3, 1
6	9, 8, 7, 12
7	6, 4, 1, 11
8	5, 4, 3, 11
9	10, 9, 8
10	7, 10, 9, 8, 12

6.10. Let c_i be the cost of tooling (fixture) if tooling is designed specifically for part i. An alternative is to use a generic fixture for all parts in a family along with a part-specific insert. Let c_f be the cost of a generic fixture to hold all parts in family f and c_{fi} be the cost of the special insert for part i. Assume $c_{fi} < c_i < c_f$.
a. Indicate graphically the cost of using each alternative as a function of the number of parts in the family. (You may assume that all c_{fi} are the same.)
b. Derive an algebraic expression for determining which tooling approach should be selected as a function of the cost parameters.

6.11. Suppose setup cost is proportional to setup time. Let s_i be the setup time for the part-specific fixture, and s_f and s_{fi} be the setup times for the family fixture and part-specific inserts. Modify your decision model in Problem 6.10b to include setup cost. Assume in either case that each part is produced once a week and that there may be more than one family in the cell (thus, family setups are required once each period). Of course, once the family fixture is set up, all family parts will be run consecutively.

6.12. After examining a sample of several hundred product routings, the six basic routings shown in Table 6.12 have been found to contain 98 percent of products. Reorder machines and products to indicate natural groupings.

Table 6.12   Basic Routings for Problem 6.12

Component–Machine Routing Matrix

Product Routing	Machine								
	A	B	C	D	E	F	G	H	I
1	1			1	1		1		1
2		1	1				1	1	
3			1				1		
4				1	1		1		1
5		1	1			1		1	
6	1					1	1		

Table 6.13   Machine Utilizations for Problem 6.13

Machine Type	Component										
	1	2	3	4	5	6	7	8	9	10	11
A	0.2			0.4		0.2			0.2	0.4	
B		0.7	0.3		0.4	0.2					0.1
C	0.5		0.8		0.2				0.1		
D	0.3			0.2		0.4					
E			0.5	0.3			0.7	0.1			

6.13. Eleven components are to be manufactured on five machine types. Utilizations are given in Table 6.13. Machines are semiautomatic; a worker can operate up to three machines. Determine a set of single-worker manufacturing cells.

6.14. Use the binary ordering algorithm to reorder the seven machines (*A* through *G*) and six components in Table 6.14.

6.15. Form groups for Problem 6.14. What problems arise in trying to lay out the machines within each group? What additional information would you like?

6.16. Consider the product routings of Table 6.15. Reorder the component–machine matrix to suggest natural groups.

Table 6.14   Component Routings for Problem 6.14

Component	Machine Required for Operation			
	1	2	3	4
3B15	B	D	A	E
3B27	C	F	G	—
4C18	F	C	G	—
5CA1	B	A	E	D
1FB1	B	D	C	—
1FB3	B	D	C	E

Table 6.15   Component Routings for Problem 6.16

Component	Machine for Operation			
	1	2	3	4
1	A	C	D	A
2	B	D	E	—
3	C	A	D	—
4	D	C	A	—
5	E	B	F	—
6	F	E	—	—
7	D	C	—	—
8	B	E	—	—
9	C	E	B	F

Table 6.16 Utilizations for Problem 6.17

	Part						
Machine Type	$S1$	$S2$	$S3$	$R4$	$R5$	$R6$	$R7$
A	0.3	0.2					
B	0.2	0.3	0.1	0.2			
C		0.2	0.4	0.3			
D				0.1	0.2	0.4	
E					0.3		0.2

6.17. Part utilizations are given in Table 6.16. Groups may contain up to three machines. Parts belong to either family R or S. Changeovers between parts in the same family require very little time, but changeovers between families is time consuming, and this changeover is not reflected in the utilizations. If both families are assigned to a machine, it is estimated that 20 percent of the machine's capacity will be lost to family changeovers.

a. Construct a set of groups that does not require any intergroup movement.

b. Is it possible to reduce the number of machines by rearranging groups or allowing parts to visit more than one group? If so, suggest an alternate configuration to be investigated. What additional information is needed?

6.18. Consider the binary ordering algorithm of Section 6.4.2.

a. Prove that the algorithm converges to a final solution.

b. Give a counterexample to show that the algorithm does not converge to a unique final ordering.

6.19. Use the data in Table 6.13 and determine grouping alternatives based on similarity coefficients. What information is lost when the groups are formed based solely on similarity coefficients?

6.20. Apply the hierarchical clustering heuristic to determine possible groups for the data in Table 6.15.

6.21. Apply the hierarchical clustering heuristic to suggest groups for the problem described in Table 6.11.

6.22. Construct the dendogram of possible machine groups for the data in Table 6.12.

6.23. Consider the alternate similarity coefficient $s^*_{ij} = n_{ij}/(n_i + n_j - n_{ij})$. Which coefficient do you prefer, s^*_{ij} or s_{ij} from equation 6.1? (*Hint:* Consider the case where machine i manufactures 100 parts and machine j manufactures 5 parts, all of which also use machine i.)

6.24. A colleague has suggested determining machine groups based solely on similarity coefficients. If this suggestion is adopted, what important information is excluded from the decision process?

6.25. A job shop makes four types of products as described in the following discussion. Each product type includes several different stock-keeping items. Each item is produced a number of times each year. After completing an operation on a machine, the batch of product goes to centralized storage. When the queue at its next machine is sufficiently small, the job is transferred to that machine. Currently the shop averages two weeks of work in centralized storage and one day's work waiting at each machine (the shop works 8 hours per day, 5 days per week). The manufacturing manager is wondering if she should convert to a group technology layout utilizing cells. Jobs would enter a cell and not leave until all operations were completed (i.e., no centralized storage, all jobs are transferred to their next machine). The annual inventory-holding cost rate is 0.5 per year based on average inventory level. Jobs of the same product type require no changeover (no setup when switching between them). Otherwise, all setups take 1 hour and machines cost $30 per hour to operate, including labor. Material handling is currently by forklift. The forklift moves an average of 15 jobs per hour and costs $20 per hour. It is estimated that a handling system within a cell could be constructed that would result in a cost of about $.50 per intermachine move. Fixed machine costs are $10,000 per year for machines 1 and 2 and $15,000 for the others. What do you recommend? State operating policies and cost as well as type of layout.

Product	Annual Demand	Raw Material Cost	Value Added
1	1000	$20	$30
2	2000	40	90
3	5000	10	50
4	2000	20	60

The following table of values gives (machine type, unit processing time) for each operation. Processing times are in hours.

Product	Operation Data
1	(1, 0.1), (3, 0.2), (2, 0.5)
2	(4, 0.3), (3, 0.2), (5, 0.05)
3	(5, 0.02), (4, 0.1), (3, 0.07)
4	(2, 0.6), (1, 0.2)

6.26. Resolve Problem 6.14 using the graph partitioning heuristic. Assume that a machine can serve any three parts.

6.27. Resolve Problem 6.17 by using the graph partitioning heuristic.

6.28. It was assumed in the chapter that reducing setup time by 50 percent would reduce lot sizes accordingly. Suppose we use a standard economic order quantity model such as lot size Q, determined by $Q = (2AD/h)^{1/2}$, where A is setup cost, D is demand rate, and h is holding cost. Derive the effect of setup time reduction under this policy. Assume setup cost is proportional to setup time and the new setup time is K times the original setup time.

6.29. Each operation requires a specific machine type. Let u_{im} be the number of type m machines required to perform operation i. Develop a mathematical program for determining the number of machines of each type m to acquire and for assigning all relevant parts to specific machines of this type. Assume that, at most, M_m machines can be purchased. Use the estimate of the number of intergroup moves if parts i and j are assigned to different machines given in equation 6.5. Costs should include fixed machine costs and intergroup movement.

6.30. Assume that each part operation is preassigned to a specific machine and that each part follows a fixed machine sequence. Develop a mathematical program for assigning machines to groups. The objective is to maximize the amount of intragroup material handling (minimize flows between groups). A group may have, at most, m_u machines.

CHAPTER 7

FACILITY LAYOUT

Houses are built to live in and not to look on;
therefore let use be preferred before uniformity,
except where both may be had.
—**Francis Bacon, from** *Of Building*

7.1
INTRODUCTION

A central concern of this book is the task of configuring manufacturing facilities to facilitate material flow and execution of production plans. To this point we have not dealt directly with the problem of arranging the manufacturing departments and support facilities in a large plant. In this chapter we attack this problem directly. We are not concerned with the details, such as the specific position and angular orientation of a worker's bench or where the power outlets should be located. Instead, we concentrate on the relative location of the set of major physical, manufacturing resources with respect to each other. A resource may be a single, large machine for a small problem, but it would more likely be a process department, group, or assembly line for the full plant problem. We hereafter refer to these resources as departments. Questions to consider include: which resources should be adjacent, and how can we arrange the resources to achieve these relationships? Our goal is to produce a **block plan** showing the relative positioning of the departments.

Once a block plan is determined, CAD systems can aid in the production of detailed drawings of facility layouts. Several of these systems include sophisticated space management capabilities. Figure 7.1 shows the type of layout representation possible. Special-purpose facility layout CAD systems exist with options similar to those discussed here for developing block plans as well as detailed layouts.

As a primary criterion for evaluating potential block plans, we select minimization of material handling cost. Direct components of material handling cost include depreciation of material handling equipment, variable operating costs of equipment, and labor expenses for material handlers. Material handling cost is assumed to be an increasing function of the frequency and length of product moves. Although this criteron is important in its own right, it may well be more important as a surrogate measure for other operating costs. Reduced material movement translates into reductions in required aisle space, lower WIP levels and throughput times, less product damage and obsolescence, reduced storage space and utility requirements, simplified material control and scheduling, and less overall congestion in the system. Savings in areas such as space requirements and WIP may be measured quantitatively, but many

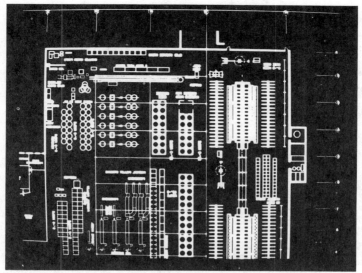

Figure 7.1   CAD layout representation. (Photo courtesy of Auto-trol Technology.)

of the other savings are difficult to estimate. Nonetheless, they are readily observed on the shop floor and find their way into the accountant's bottom line in the long run. Harmon and Peterson [1990] provide a convincing illustration of the impact of handling distance on inventory. If successive processes are immediately adjacent, a single unit is moved at a time, as in an assembly line. If the next process is across the aisle, the handling lot size is a unit load. If the next process is across the plant, the handling lot size is, at least, an hour's supply of product, because more frequent collection is impractical. If the next process is another plant, the handling lot size is at least one day's production. As the WIP between processes will be, at least, one half the handling lot size, we see potential orders-of-magnitude differences in WIP levels based on the layout.

The facility layout problem is often solved with a static, deterministic modeling approach. Material flows and interactions between employees are assumed to be known. Dynamic effects may be taken into account by using a weighted average of expected conditions in future time periods, but the random nature of future demand is not generally directly modeled. Instead, a premium is placed on solutions that embrace flexibility, modularity, maintainability, reliability, and employee morale. Flexibility refers to the ability of an existing system to adjust to minor changes in product design or demand, or in resource availability. Modularity is the ability to easily change the system by adding or subtracting system components to meet major demand changes. Maintainability is concerned with the "wear rate" of the system and the extent to which regular maintenance can keep the system running as it was initially specified. Reliability is a measure of system endurance—how long the system can operate before essential components fail.

The **spine** approach to facility design can aid in modularity, flexibility, and reduction of material flow. The spine refers to a central core or passageway to conduct traffic—material, utilities, information, and people. Figure 7.2 illustrates the spine concept and how it can be useful. Departments expand out from the central core. Secondary aisles can be used to conduct flows into departments. Utilities can be carried overhead to simplify the network of pipes and cables. Material is stored along

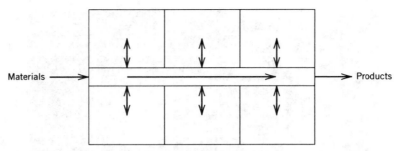

Figure 7.2   Spine approach to material flow design.

the spine. Ideally, each department has an input and output storage area along the spine. This departmental *point of use storage* concept reduces material flow as compared to a system with one centralized warehouse visited by all parts and materials after each operation.

Even with an increasing rate of automation and technological innovation, manufacturing facilities will depend on humans for many years to come. As such, the system must be designed with an understanding of human capabilities and motivations as well as existing technology. The system has not been invented that a human cannot incapacitate if so inclined. In fact, it is difficult with complex systems to devise a system design that even the well-meaning employee can operate reliably and effectively. Remember, *no system is foolproof; fools are too ingenious*. Lighting, heating, noise, and ventilation can all be important determinants of human performance.[1] Simply keeping a clean and tidy workplace can improve morale and performance. The clean workplace is naturally better organized, and a sense of pride in the shop flourishes.

The facility layout problem is easy to state but difficult to solve. The problem consists of assigning each department to a specific location in the facility. Several measures of problem difficulty have been proposed. Let M be the number of facilities that must be located. Clearly, the larger the value of M, the more difficult the problem. Of equal importance is the variability of interaction relationships between departments. If one dominant flow pattern exists, such as in an assembly process where all parts go to department 1, then 2, and so on up to department M, then the solution is obvious. Arrange the departments in this same order. At the opposite extreme, if the flow from every department to every other department is identical, that is, there is no dominant flow pattern, then the problem is also simple.[2] Any arrangement will do; unfortunately, they are all equally ineffective. This situation offers a good opportunity for redefinition of departments, product design, and process routings to establish a more rational flow pattern.

Most problems lie in between the extremes of dominant and equal flows. Lewis and Block [1980] proposed a measure of problem difficulty based on flow dominance. Define the cost parameters w_{ij} to be the weights for material flow between departments i and j. Normally,

$$w_{ij} = f_{ij}h_{ij} \tag{7.1}$$

where f_{ij} is the flow volume measured in (trips/time) and h_{ij} is the material handling system cost factor with units *cost/unit distance*. The h_{ij} include apportioned

---

[1] Readers interested in such human performance issues should consult Konz [1985].

[2] The terms "from-to" and "between" have distinct meanings. From-to implies direction. Between flows can be going in either direction and generally are measured by the sum of the directed flows.

fixed and variable cost of material handling equipment, labor cost, and inventory cost for time in transit. Alternatively, these factors can be used as subjective weights including safety, customer importance, and other factors as well as standard accounting costs. If all movement is of equal concern, all h_{ij} may be set to 1. (In most cases $w_{ii} = 0$.) A simple flow dominance measure is f, the coefficient of variation of the w_{ij} for M departments:

$$f = \frac{\left[\dfrac{\sum\limits_{i=1}^{M} \sum\limits_{j=1}^{M} w_{ij}^2 - M^2 \overline{w}^2}{M^2 - 1} \right]^{1/2}}{\overline{w}} \tag{7.2}$$

where $\overline{w} = \sum_{i=1}^{M} \sum_{j=1}^{M} w_{ij}/M^2$ is the average flow cost parameter. Experience has shown that problems with $f > 2$ have a few dominant flows and are therefore easy to solve. Small values of f ($f \approx 0$) indicate many nearly equal flows, problems for which many solutions are near optimal. Problems with $f \approx 1$ are difficult and generally require computer and interactive graphics assistance to find good solutions.

The measure f does not consider problem size. As M increases, the facility layout problem becomes more difficult. The measure f of equation 7.2 is maximized when a single w_{ij} dominates, that is, only one weight is nonzero. f is minimized when all weights are equal. Application of these conditions to expression 7.2 results in an upper and lower bound on f of

$$f_u = M \sqrt{\frac{M^2 - M + 1}{(M-1)(M^2-1)}} \qquad f_l = M \sqrt{\frac{1}{(M-1)(M^2-1)}} \tag{7.3}$$

respectively. f_u is strictly increasing in M while f_l is strictly decreasing in M. Table 7.1 summarizes these trends.

	Table 7.1	Bounds on Flow Dominance			
M	2	5	10	20	30	50
f_u	2.00	2.34	3.20	4.46	5.50	7.07
f_l	1.15	0.51	0.34	0.23	0.19	0.14

To include the effect of problem size in measuring difficulty, we can use the modified measure

$$f' = \frac{f_u - f}{f_u - f_l} \tag{7.4}$$

Problems with f' near 0 or 1 are easy.

EXAMPLE 7.1

Our plant produces industrial gas dryers. Several models are made, adding up to a production of about 100 units per day spread over two shifts. The company began as designers and manufacturers of small electronic controls. Several years ago, after years of making the controls for the gas dryers, the company purchased the gas dryer manufacturing plant being studied. A simplified model of the plant consists of six departments: Shipping and Receiving (SR), Pipe Construction (PC), Painting Stations (PS), Instrument Construction (IC), External Frame Construction (XT), and Assembly and Test (AT). Raw materials including purchased motors are received at SR. Parts are inspected and accepted here and then transferred by forklift to their point of use. A finished dryer essentially consists of piping subassemblies made in PC, instruments made in IC, a sheet metal frame made in XT, and a motor. Frames are first sent to PS and then transferred to final assembly. Frames are moved by overhead crane. Other subassemblies are taken to final assembly by forklift. Final assembly assembles the motor and subassemblies into a finished dryer. Dryers are then tested. Any necessary repairs are performed at that time. Dryers are assembled on carts that can be rolled to SR, where they are packed and shipped. It takes about three days on average for parts to be shipped as a dryer after they pass receiving inspection. Testers make several trips per day on foot to IC, XT, and PC to discuss failed parts and to obtain replacements. On an average day, two frames will be returned to PS for touchup.

Table 7.2 shows daily flows between departments. An estimated five trips per day are made by testers for replacement parts to PC, IC, and XT. Let us measure flow dominance.

Table 7.2   Material Loads Moved per Day

From-To	SR	PC	PS	IC	XT	AT
SR	—	40	10	30	10	50
PC		—				100
PS			—			102
IC				—		100
XT			100		—	
AT	100	5	2	5	5	—

Solution

Table 7.2 contains 14 nonzero flow values and 22 zero values (including the diagonal elements). Summing across rows, we first obtain

$$\overline{w} = \frac{40 + 10 + \cdots + 5}{36} = \frac{659}{36} = 18.3056$$

The $\sum w_{ij}^2 = 55,683$, and using equation 7.2,

$$f = \frac{\left[\dfrac{55,683 - 36(18.3056)^2}{35}\right]^{1/2}}{18.3056} = 1.9285$$

An f of 1.9 indicates the presence of some strong flows but not a simple dominant relationship. If we contemplate this procedure, we will see one weakness of this flow dominance measure. The location of the larger flows in the table did not affect the result. In practice, serial flows of 100, where product flows from SR to PC to IC to XT to AT, would be easy to plan. However, we obtain the same measure if flows are 100 from SR to PC, 100 from SR to IC, 100 from SR to XT, and 100 from SR to AT, a much different situation for the layout planner. The moral: LOOK AT THE FLOWS! No single summary measure will fully classify the problem for you.

Since we have come this far, we should compute f'. Using equation 7.4, we obtain

$$f' = \frac{f_u - f}{f_u - f_l} = \frac{2.525 - 1.9285}{2.525 - 0.4536} = 0.2879$$

An f' of 0.3 again indicates that some large flows exist but not complete dominance of these flows. ∎

Before proceeding to models for layout, we need to understand the variety of situations that occur. Layout problems assume that the facilities to be located occupy space. This sounds self-evident, but the assumption distinguishes layout problems from location problems. Location problems include the related activities of placing a small machine into an operating plant or adding a new production facility into a production and distribution chain. In locating either a small machine in a large plant or a new plant on a map of a large geographical region, the facility can be assumed to reside at a point. A 10-machine group, however, must have sides and area (square feet) when located next to another machine group. Space must be considered in the layout decision model to measure interdepartmental distances and to determine adjacencies. We concentrate on layout problems, although the last section of the chapter contains a discussion of in-plant location problems.

Another classification factor is the distance metric. Distance is often rectilinear (movement along horizontal and vertical directions) or euclidean (movement in a straight line between initial and final points). Rectilinear distance generally applies to aisle travel. Euclidean distance is sometimes appropriate for overhead conveyors. If we let department i be located at coordinates (x_i, y_i), we may compute rectilinear and euclidean distances between facilities i and j by

$$d_{ij}^R = |x_i - x_j| + |y_i - y_j| \tag{7.5}$$

and

$$d_{ij}^E = \left[(x_i - x_j)^2 + (y_i - y_j)^2 \right]^{1/2} \tag{7.6}$$

respectively. Both these measures are special cases of the general convex l_p norm

$$d_{ij}^p = \left[|x_i - x_j|^p + |y_i - y_j|^p \right]^{1/p} \qquad p > 0 \tag{7.7}$$

Rectilinear and euclidean distance correspond to p of 1 and 2, respectively. (As $p \to \infty$ our interest is in the maximum move along the x or y axis. Such a measure would be appropriate for an AS/RS machine that moves in both directions simultaneously.) The department location (x_i, y_i) is often assumed to be its centroid. If a specific I/O point is known for the department, the distance between I/O points should be used. If separate input and output points are used for a department, then distance between departments need not be symmetric. In a plant, product may have to move along a fixed set of aisles. In this case rectilinear measures may underestimate actual distance. Material heading in one direction may be forced to backtrack in the opposite direction at some point to stay on the existing aisles. More sophisticated distance measures are required in this instance. As an alternative distance measure, we may use the adjacency indicator δ_{ij}. Facilities i and j are either touching, $\delta_{ij} = 1$, or distant from one another, $\delta_{ij} = 0$. We illustrate several distance metrics in Figure 7.3.

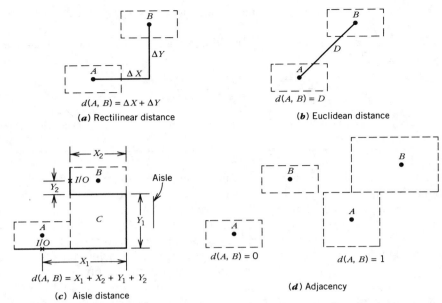

Figure 7.3   The distance measures from Department A to Department B.

Three approaches to layout are discussed in the next three sections. First, systematic layout planning is covered, because of its historical importance and to provide a systematic foundation for how other partial methods fit together. Second, we look at quantitative distance measure methods based on the quadratic assignment problem model. Third, we look at graph theoretic approaches that rely on adjacency measures. The quadratic assignment problem and graph theoretic approaches are technical methodologies that could be integrated into systematic layout planning. We will embellish the next section with several simple quantitative methods, but these more technical methodologies deserve their own section. We then turn to developing net layouts, which incorporate aisles and I/O points, and finally move on to related location problems.

7.2
SYSTEMATIC LAYOUT PLANNING

Muther [1973] moved the science of facility layout forward with a Systematic Layout Planning (SLP) approach to manual design of facilities. The overall philosophy is related in Figure 7.4. Succeeding work has concentrated on formalizing the details of the layout process and developing analytical models for certain steps of the SLP procedure. We describe the steps of SLP, pointing out the relevant data, decisions, and methodologies at each step. In practice, a hierarchical layout approach is often used. The first pass through SLP may select the basic layout (block) plan. A detailed pass through SLP is then made to perform the detailed layout for each department, indicating locations for individual machines, staging areas, workbenches, and other entities.

STEP 0.   DATA COLLECTION. The process of collecting the relevant data in an accurate form is critical. The political nature of large organizations often makes this a difficult process. The key here is collection of data on Product (what

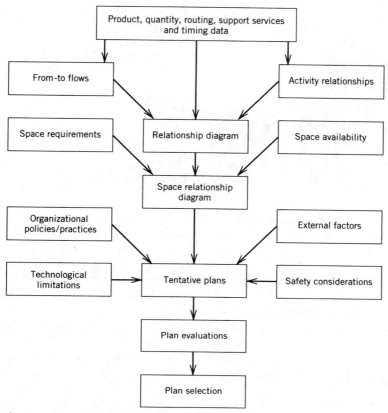

Figure 7.4 Overview of Systematic Layout Planning.

is to be produced), Quantity (volume to be produced), Routing (how it is to be produced), Support services (with what will we produce), and Timing/Transport (when will we produce and how will we move parts in and out). The quantity and variety to be produced often dictate the proper layout type (product, process, or group). Individual or families of products can be shown on a Product–Quantity chart, as shown in Figure 7.5. This Pareto analysis of product importance can be used to determine which products justify their own lines or which families justify their own cells. It is not uncommon to find 20 percent of items constituting 80 percent of demand. These *vital few* deserve special attention. The remaining products may warrant only lumping into a general job shop area.

By concentrating on the relatively quantifiable PQRST items, the modeler can hopefully avoid being an unwitting accomplice in the building of fiefdoms and space misallocation. Data should be collected at the machine level for subsequent aggregation into departments. For instance, data from part-route sheets and long-term production plans can be used to determine utilizations of each machine type (the number of machines and workers needed) and flows between machines. Route sheets are produced during process planning where product dimensions, tolerances, and features are used to select the sequence of machines or workcenters visited by the part. Ultimately, square footage must be allocated to each department. The modeler can expect "generous" estimates of required space from

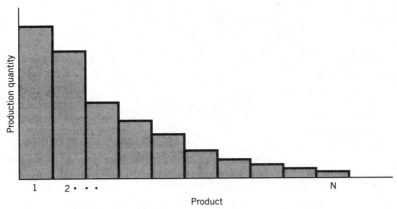

Figure 7.5   A sample Product-Quantity chart.

departmental staff. However, if the modeler understands the essential activities of each department, then industrial norms of space, labor, and capital requirements for these activities can be combined with appropriate local adjustments and quantitative resource usage estimates to obtain objective estimates of space requirements.

STEP 1. FLOW ANALYSIS. Prior to tabulating and coalescing data into information, the analyst must specify the physical workcenters that are to be spatially arranged. This involves a decision of layout type. Department definitions can be based around products, processes, or cells of similar parts. Flow volumes and patterns must be established. Operation process charts are helpful in determining the move patterns for each product. The chart shows all major operations, inspections, moves, and storages of the product. If a tentative layout exists, the flow pattern can be illustrated pictorially. Figure 7.6 gives an example of an operation chart. Figure 7.7 shows a flow process chart. Quantitative flow data can be summarized with From-To charts. Such a chart was shown in Table 7.2. The chart shows the number of material handling loads transported *from* workcenter i *to* workcenter j per period. Total flow volume between workcenters can be obtained by checking the route sheets for each interoperation move of each part and, in each instance, adding to the appropriate location in the From-To chart the ratio of period demand to unit handling load. Total cost is the sum over all workcenter pairs of the product of unit load volumes in both directions, handling cost rate, and movement distance. Defining the distance between departments i and j as d_{ij}, we obtain the cost of material movement from workcenter i to j as

$$c_{ij} = w_{ij} d_{ij} \qquad (7.8)$$

and total cost[3] as

$$C = \sum_{i=1}^{M} \sum_{j=1}^{M} c_{ij}$$

---

[3]Note that $w_{ij} = f_{ij} b_{ij}$, terms that were defined earlier. The reader should be careful to note that whereas From-To charts can be considered arrays, the multiplicative operation defined in equation 7.8 indicates multiplication element by element and is not equivalent to matrix multiplication.

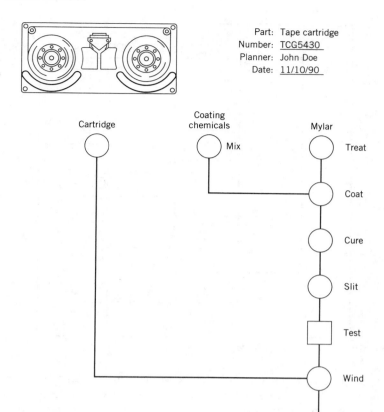

Figure 7.6   An operation chart for tape cartridges.

There are organizational and modeling reasons for maintaining three separate From-To charts, that is, charts for flow volumes, movement cost, and distance between workcenters. From expressions 7.1 and 7.8, c_{ij} is in units of *dollars per period*. Flow volumes are determined by department definition and production plans. Unit distance handling costs are set by choice of the material handling equipment and methods. Distances are fixed by the layout decision. Thus, we can experiment with different production plans or dynamic changes and need be concerned only with f_{ij}. Likewise, sensitivity analysis to material handling system specifications need be concerned only with tracing out the effects of changes in b_{ij}, and alternative layout proposals can be evaluated by the proper modifications to d_{ij}. This organizationally meaningful decomposition of costs simplifies data maintenance and modeling analysis for large facilities.

A number of flow patterns have been used for movement within and between departments. Figure 7.8 shows several possible patterns. A popular

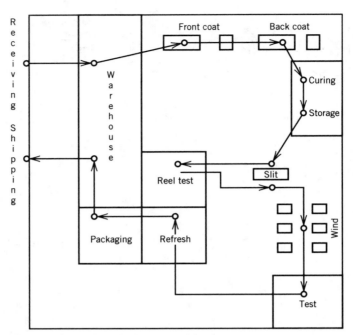

Figure 7.7   A flow process chart for tape cartridges.

approach is to use a spine-oriented straight-line pattern between departments with a complementary U-shaped pattern within departments. However, this approach places shipping and receiving on opposite ends of the building. This approach is demonstrated in Figure 7.9. If this is undesirable for the application, departments can be combined into a U-shaped pattern for the entire facility. In the general case, many parts fabrication departments feed into a final assembly area. Departmental flows can either be oriented parallel to one another and converge into assembly or be oriented perpendicular to assembly, entering where required (Figure 7.10). Remember that even single-floor facilities are three-dimensional. Overhead conveyors can be used to feed parts and subassemblies; however, long conveyors can become sinks for in-process inventory if they are not carefully designed. The key is to *design* a rational flow pattern that avoids confusion and interference.

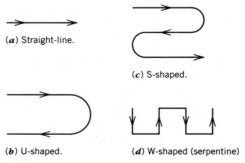

Figure 7.8   Basic flow patterns.

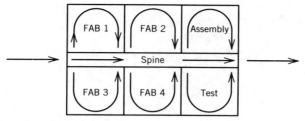

Figure 7.9   A plant flow pattern.

STEP 2.   QUALITATIVE CONSIDERATIONS. It is common to have significant information that is difficult to quantify on the desirability of locating two workcenters together. For instance, shipping and receiving may share common facilities, such as loading docks or common management. It may be desirable to have engineering and purchasing together to ensure that quality materials are obtained if, in the same proximity, the two groups could be expected to communicate freely or, at least, more effectively and frequently than if located on opposite ends of the building. Environmental factors may make it desirable to separate production facilities such as delicate testing from processes that induce vibration or heat in the environment. Historically, such information has been summarized in a relationship or REL chart. An example of a REL chart is shown in Figure 7.11. The chart was developed from the flows in Table 7.2 with the extra condition that the presence of flammable materials in Paint Services prevented that department from being located near Parts Construction.

   An upper triangular matrix that is often displayed as a triangle, the REL chart contains a unique diamond for each pair of workcenters. Each diamond is usually marked with a value of $A, E, I, O, U,$ or X. The chart

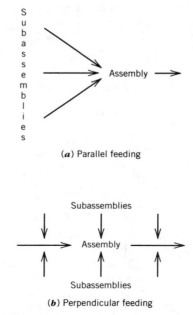

Figure 7.10   Assembly flow patterns.

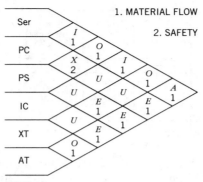

Figure 7.11   An integrated REL chart.

symbols indicate the degree of desirability of locating the two associated departments adjacent to each other in the production facility. As indicated in the figure, the six categories range from "Absolutely necessary" to "Undesirable." It should be remembered that these ratings are qualitative, and although they provide ordinal rankings of workcenter relationships, they cannot validly be used in the standard algebraic operations of addition and summation. One cannot assume, for instance, that an A is worth two E's or even that two A's are equivalent. Nonetheless, REL charts are widely used and will be an integral part of Section 7.4. A coded explanation of the relationship value may also be included in the diamond. A value of 1 may, for instance, indicate environmental incompatability.

STEP 3. RELATIONSHIP DIAGRAM. The relationship diagram combines quantitative and qualitative relationship data to initiate the determination of the relative location of facilities. A square template is created for each department. Templates are arranged in a logical order. Templates are then connected by lines that communicate the relationship (A, E, I, O, U, or X) between department pairs. A sample relationship diagram is shown in Figure 7.12. An iterative process is then invoked to visually rearrange the templates to better adhere to the relationships, that is, departments that share an A

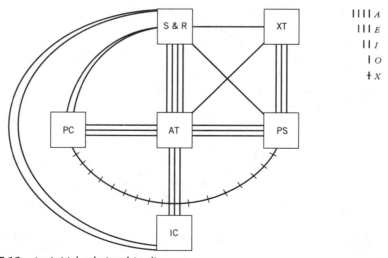

Figure 7.12   An initial relationship diagram.

relationship should be very close and preferably adjacent. As with any design problem, many candidate arrangements should be attempted. Several of the best layouts should be saved for subsequent refinement.

Development of the relationship diagram reflects the two basic steps of many heuristic design procedures: construction followed by improvement. Construction refers to the initial arrangement of the templates. Let r_{ij} be the relationship between departments i and j. We can assign some arbitrary value to achieving an adjacency of importance r_{ij}. We let $V(r_{ij})$ be this function. Of course, the $V(r_{ij})$ have only ordinal meaning, and we should be careful not to assign too much importance to the value. A hypothetical assignment function might be $V(A) = 81, V(E) = 27, V(I) = 9, V(O) = 3, V(U) = 1, V(X) = -243$. Departments can be initially ordered by their **total closeness rating**, defined as

$$TCR_i = \sum_{j=1, j \neq i}^{M} V(r_{ij}) \tag{7.9}$$

Large total closeness ratings indicate that the department has significant interactions with a number of other departments and, hence, should probably be near the center of the facility so that it may be close to many other departments. To construct an initial solution, place departments in nonincreasing order of TCR_i. Iteratively add the next department to the periphery of the layout, always placing the newly added department to the location that either maximizes the sum of the closeness ratings with adjacent departments or minimizes the sum of the closeness rating times the distance between departments. To maximize the total closeness rating for adjacent departments our objective is

$$\text{maximize } V = \sum_{i=1}^{M-1} \sum_{j=i+1}^{M} \delta_{ij} V(r_{ij}) \tag{7.10}$$

where δ_{ij} is 1 if departments i and j are touching and 0 otherwise. To minimize distance, the objective is

$$\text{minimize } Z = \sum_{i=1}^{M-1} \sum_{j=i+1}^{M} V(r_{ij}) d_{ij}$$

EXAMPLE 7.2

Construct a relationship diagram for the gas dryer plant.

Solution

In computing TCR_i values we choose to ignore the X relationships. The negative $V(X)$ would tend to indicate a lowering in importance for this department, which is clearly not true when selecting a location.[4] From equation 7.9 and Figure 7.11

$$TCR_{SR} = V(I) + V(O) + V(I) + V(O) + V(A) = 9 + 3 + 9 + 3 + 81 = 105$$

---

[4]The reader who objects should remember that we are just looking for an aid to start the process. Including X's will lower priority for the associated departments and consequently add them late in the construction process. This restricts them to the periphery of the facility. If this seems desirable for your problem, go ahead and include X relationships.

indicating the relationships of *SR* with *PC* through *AT*. In the same fashion, we find that $TCR_{PC} = 38, TCR_{PS} = 58, TCR_{IC} = 39, TCR_{XT} = 35,$ and $TCR_{AT} = 165$. AT is thus ranked first and placed in the center of the layout. SR is next. SR could be located on top, right, bottom, or left of AT. All positions increase the adjacency score by $V(r_{SR,AT})$. We arbitrarily place SR on top of AT. Third is PS. We could envision six possible locations, above SR, right of SR, right of AT, below AT, left of AT, or left of SR. The best location is next to AT, because of the *E* relationship. We select the position right of AT. Fourth is IC. Seven positions could be selected (right of AT is lost but right of PS and below PS are now possible). Figure 7.13*a* shows the adjacency score gain for each possible placement. For a gain of 27, we place IC below AT. (Note that if diagonal adjacencies were allowed, a better score would be obtained by placing IC right of SR and above PS. The reader who does not mind the extra calculation is free to use this extended measure in the future.) In the same manner, we then decide to add PC left of AT, and finally XT above PS. The final result is shown in Figure 7.13*b*.   ∎

Improvement involves looking for better arrangements than the constructed layout. A typical improvement strategy is to try to improve the layout score by switching department locations. Improvement can be performed with either the adjacency or distance measure. A solution is termed *k-opt* if no switch of *k* or fewer variables (in our case variables are department assignments) exists that will improve the solution. Many heuristic procedures are concerned with finding 2-opt solutions. Such solutions are local optimums but, as we shall demonstrate shortly, not necessarily globally optimal.

CRAFT (Armour and Buffa [1963]) is probably the most popular computer-based improvement procedure. Simply described, CRAFT searches for switches that lead to 2-opt or 3-opt solutions. A number of effective heuristics exist for finding improved solutions (see Francis and White [1974] for several). One straightforward approach is steepest descent pairwise interchange. Steepest descent considers every possible pair of departments and switches the pair that yields the largest improvement in the objective function. The procedure terminates when no improving switch can be found.

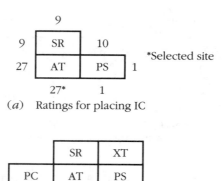

(*a*)   Ratings for placing IC

(*b*)   Final construction

Figure 7.13   The layout construction for Example 7.2.

To simplify calculations we derive an expression for updating the value of the objective function after switching departments i and j. Recall that w_{ij} is the flow cost per unit distance of separation between departments i and j. We now assume that w_{ij} includes flow in both directions; thus, $w_{ij} = w_{ji}$. Prospective department locations constitute a grid, for the REL diagram the grid consists of equal-sized squares usually forming a rectangle representing approximate building length and width. Having defined a grid, it is more natural to talk about the distance between specific grid squares. This distance, for grids r and s will be denoted by $d(r, s)$. Grid squares are numbered from 1 to M. (If fewer departments than grid squares exist owing to extra space, simply include dummy departments with all flow weights equal to 0). A feasible solution to the layout problem is any vector $\boldsymbol{a} = (a_1, a_2, \ldots, a_M)$, where a_i is the grid square to which department i is assigned. In this context we have

$$\text{total flow cost} \equiv C(\boldsymbol{a}) = \sum_{i=1}^{M-1} \sum_{j>i} w_{ij} d(a_i, a_j) \tag{7.11}$$

Suppose we exchange departments u and v. If $\boldsymbol{a}$ is the original assignment and $\boldsymbol{a}'$ is the modified layout, then we can compute the cost savings from the exchange by

$$\Delta C_{uv}(\boldsymbol{a}) = C(\boldsymbol{a}) - C(\boldsymbol{a}')$$

$$= \sum_{i=1}^{M-1} \sum_{j>i} w_{ij} d(a_i, a_j) - \sum_{i=1}^{M-1} \sum_{j>i} w_{ij} d(a_i', a_j') \tag{7.12}$$

The only terms of equation 7.12 that do not cancel out in the double sums are those involving either department u or v. We assume that distances are symmetric so even the $w_{uv}d(a_u, a_v)$ and $-w_{uv}d(a_u', a_v')$ terms cancel, since $a_u' = a_v$ and $a_v' = a_u$. Including only the relevant terms, we obtain

$$\Delta C_{uv}(\boldsymbol{a}) = \left[\sum_{i=1}^{M} w_{iu}d(a_i, a_u) + \sum_{i=1}^{M} w_{iv}d(a_i, a_v) - w_{uv}d(a_u, a_v) \right]$$

$$- \left[\sum_{i=1}^{M} w_{iu}d(a_i', a_u') + \sum_{i=1}^{M} w_{iv}d(a_i', a_v') - w_{uv}d(a_u', a_v') \right]$$

Again noting that $a_i = a_i'$ for $i \neq u, v$ and $a_u' = a_v, a_v' = a_u$, then

$$\Delta C_{uv}(\boldsymbol{a}) = \sum_{i=1}^{M} w_{iu}d(a_i, a_u) + \sum_{i=1}^{M} w_{iv}d(a_i, a_v) - w_{uv}d(a_u, a_v)$$

$$- \left[\sum_{i \neq v} w_{iu}d(a_i, a_v) + \sum_{i \neq u} w_{iv}d(a_i, a_u) + w_{uv}d(a_u, a_v) \right]$$

Noting that the factors $d(a_u, a_u)$ and $d(a_v, a_v)$ in the first two sums are 0 and grouping terms, we obtain

$$\Delta C_{uv}(\boldsymbol{a}) = \sum_{i=1}^{M} (w_{iu} - w_{iv})[d(a_i, a_u) - d(a_i, a_v)] - 2w_{uv}d(a_u, a_v) \tag{7.13}$$

Equation 7.13 provides a comparatively efficient computational method for evaluating departmental exchanges. The equation could be equivalently expressed with the sum running from $i \neq u, v$ and the last term removed. The last term neutralizes the $i = u$ and $i = v$ terms in the summation. Second, an analogous expression 7.14 can be derived for the adjacency measure in equation 7.10:

$$\Delta V_{uv}(\boldsymbol{a}) = \sum_i [V(r_{ui}) - V(r_{vi})][\delta_{ui}(\boldsymbol{a}) - \delta_{vi}(\boldsymbol{a})] - 2\delta_{uv}(\boldsymbol{a})V(r_{uv}) \qquad (7.14)$$

Last, note that a necessary condition for a candidate assignment $\boldsymbol{a}$ to be optimal is that $\Delta C_{uv}(\boldsymbol{a}) \leq 0$ for all pairs (u, v).

EXAMPLE 7.3

Find a good block layout for the layout constructed in Example 7.2.

Solution

We start by transforming the relationship diagram of Figure 7.13 into a rectangular layout, as shown in Figure 7.14. The locations that compose the facility are numbered 1 to 6. Distance will be rectilinear between centroids with all departments being one unit square. Thus, $d(1, 2) = 1$, and $d(1, 6) = 3$. Suppose the suggestion is made to switch AT and XT.

Consider first the distance measure 7.13. Weights are the flows from Table 7.2. $u = $ AT and $v = $ XT; thus, $a_u = 5$ and $a_v = 3$. Inserting these values into equation 7.13 we obtain

$$\Delta C_{\text{AT,XT}}(\boldsymbol{a}) = \sum_{i=\text{SR}}^{\text{AT}} (w_{i,\text{AT}} - w_{i,\text{XT}})[d(a_i, 5) - d(a_i, 3)] - 2w_{\text{AT,XT}}d(5, 3)$$

Following through the computation we find

$$= (150 - 10)[1 - 1] + (105)[2 - 2] + (4)[1 - 1]$$
$$+ (105)[1 - 3] + (5)[2] + (-5)[-2] - 2(5)2 = -210$$

The negative value for the switch indicates that a worse solution would be obtained, and so the switch should not be made by itself.

Suppose we were interested in adjacency. Using equation 7.14, we calculate

$$\Delta V_{\text{AT,XT}}(\boldsymbol{a}) = (1 - 1)[81 - 3] + (0 - 0)[27 - 1](1 - 1)[27 - 27]$$
$$+ (1 - 0)[27 - 1] + (0 - 1)[3 - 0] + (1 - 0)[0 - 3]$$
$$- 2(0)3 = 26$$

A positive ΔV implies a decrease in adjacency score from the switch. Since we want to maximize the adjacency score, once again we conclude not to switch AT and XT.

To execute steepest descent we would need to compute ΔC_{uv} for all 15 pairs of departments. The pair with the largest positive value would be switched. The procedure would then start over, evaluating all 15 pairs again. ∎

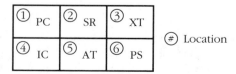

Figure 7.14   One rectangular realization of the relationship diagram.

STEP 4. SPACE REQUIREMENTS. Departmental space requirements can be estimated by several methods. Industrial standards on space requirements per unit of resource can be applied to the number of workers or machines in the department. This approach requires prior listing of the elements, such as machines and storage racks, that will be occupying space. This includes listing machine types and the number required. If this list is available, a second option involves making rough sketches of workplaces for each element and developing local standards. If a similar layout exists and is to be scaled up or down, current space *needs* (as opposed to assignment) can be used as a basis to be appropriately scaled. A fourth option allows *x* square feet per unit produced. When using this method, the analyst must be careful to modify standards for local conditions and dynamic technology. The ratio may be on a per-worker or per-value-added basis instead of per unit.

STEP 5. SPACE AVAILABILITY. If an existing facility is to be utilized, space may be limited. If a new facility is to be acquired, the facility should be designed to accommodate the flow of resources: raw materials, energy, humans, and items consumed in manufacturing. However, even in this case financial limitations may restrict the amount of space to be made available. The layout planner must determine the minimum space required and be prepared to defend this estimate with an explanation of consequences for the manufacturing system if such space if not provided.

STEP 6. SPACE RELATIONSHIP DIAGRAM. The relationship diagram simplifies the layout problem by assuming all departments are of equal size. This assumption facilitates switching of any pair of departments in solution improvement procedures. Of course, in the real problem the tool crib may be significantly smaller than the testing department, which is significantly smaller than the assembly area. The space relationship diagram replaces the equal-sized templates of the relationship diagram with templates proportional in size to departmental space requirements. The templates can then be rearranged to find an improved solution to this more realistic problem model.

Executing departmental switches is not as simple for space relationship diagrams. Individual departments may be represented on a grid by a number of grid squares proportional to departmental space needs. Figure 7.15 shows a 20 × 30 grid and an initial space relationship diagram derived directly from the previous example and integrated with the space needs summarized in Table 7.3.

In making departmental switches, all grid squares corresponding to the departments involved must be swapped to maintain the shape integrity of the departments. CRAFT and other programs solve the question of how to execute departmental switches on the grid by permitting only departments that are adjacent or of the same size to be switched. If departments are of the same size, simply swap grid squares. If departments are adjacent and of different size, we select the grid squares of the larger department that are farthest from the centroid of the smaller department. We select enough of these grids to house the smaller department and move it to those grids. A check is made to ensure that neither of the resulting departments becomes disjoint. Figure 7.16 illustrates such a switch. The "*" in the figure indicates the first grid location selected to become part of PS. This location was

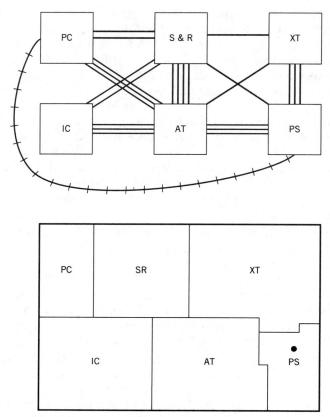

Figure 7.15   A space relationship diagram and grid.

selected since it is farthest from the centroid of the original PS. A strict rule of selecting the farthest 45 grid locations would produce the departmental boundary indicated by the solid stepped line. Although an integral department is formed, the shape is clearly undesirable. Instead, starting from the "*" location a growing spiral could be used. Moving clockwise, we include grids encountered by the spiral that are part of the original departments until 45 locations are found. In this case the spiral consumes the locations indicated by the "...." contour. In this process, squares of increasing size

**Table 7.3   Space Requirements
for Example Problem**

Department	Square Footage
SR	10,000
PC	6,000
PS	4,500
IC	12,000
XT	15,000
AT	12,500
Total	60,000

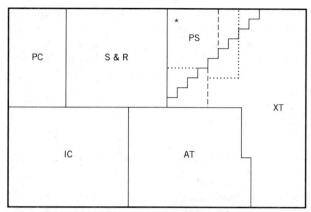

Figure 7.16   The grid after the switch of XI and PS.

are gobbled up. After filling the 6×6 square, 9 more locations are filled as the spiral traces the outside of the 6×6 square to fill 45 locations. The final alternative marked by "- - -" indicates a desire to maintain equal-length rectangles for easy insertion of aisles. The analyst is free to select his or her procedure of choice, the first two being easier to code, the last being perhaps more realistic and easily done manually. One problem with computerized procedures is that after several switches, departmental shapes can become so irregular as to make calculations meaningless.

With the relationship diagram, departmental centroids were exchanged when departments were exchanged. This is no longer necessarily true. Certainly the alternative locations of the new PS in Figure 7.16 have different centroids, all of which differ from that of XT in the original layout. To avoid the need to compute the precise centroid locations for every possible switch, it is possible to estimate new centroids by switching centroids in the initial layout. If a switch looks promising, the actual centroids can be calculated and the actual distance savings computed.

STEPS 7 AND 8. MODIFYING CONSIDERATIONS AND LIMITATIONS. To this point, layout planning has been performed without regard to the details of system implementation and operation. Site-specific and operations-specific conditions may require adjustment to the layout. For instance, the terrain or location of external transportation systems such as rail, road, or river access at the facility site may restrict the location of shipping and receiving. Floors in an existing building may have variations in load-bearing capacity. Existing limitations on access to utilities in certain areas may require expensive renovation if ignored. Utilities here includes heating, ventilation, and air conditioning (HVAC), lighting, embedded waste recycling networks, compressed air, electricity, and other power sources. Ceiling heights and support columns may come into play.

The final block plan must also consider the design of a rational flow system. Aisles should be straight and close to the point where move requests are generated without obstructing manufacturing activities.

STEP 9. EVALUATION. At this point several reasonable alternatives exist. A pictorial display of each with superimposed flows should be drawn. A list of the advantages and disadvantages of each layout can be made. Where possible, costs should be attributed to the layout. Flow volumes times distances should fit into this category, along with an estimate of construction or relayout costs. A rating (perhaps even A, E, I, O, U, or X) should be assigned for each qualitative factor for each layout alternative. These factors include flexibility, maintainability, expandability (modularity), safety, and operational ease.

7.3
QUADRATIC ASSIGNMENT PROBLEM APPROACH

The general facility layout problem of finding a relationship diagram or block plan can be modeled as a **Quadratic Assignment Problem** (QAP). The quadratic assignment problem is to find the minimum cost assignment of M departments to M locations where the cost to assign department i to location k and department j to location l is c_{ijkl}. The c_{ijkl} represent the cost of movement and other interactions between departments i and j. This cost is allowed to depend on the locations of the departments as well as their function. Often we can factor c_{ijkl} as the product of w_{ij} and $d(a_i, a_j)$ where by definition $k = a_i$ and $l = a_j$. The model is stated as

$$\text{minimize} \sum_{i=1}^{M} \sum_{k=1}^{M} \sum_{j=1}^{M} \sum_{l=1}^{M} c_{ijkl} x_{ik} x_{jl} \qquad (7.15)$$

subject to

$$\sum_{i=1}^{M} x_{ik} = 1 \qquad \text{for all locations } k \qquad (7.16)$$

$$\sum_{k=1}^{M} x_{ik} = 1 \qquad \text{for all departments } i \qquad (7.17)$$

$$x_{ik} \ 0 \text{ or } 1$$

The decision variables are the x_{ik}. x_{ik} is 1 if department i is assigned to location k and 0 otherwise. Objective 7.15 measures cost of material flow for any desired distance measure. Constraints 7.16 ensure that each location is assigned an activity. Dummy activities are used if extra space exists. Constraints 7.17 ensure that each department is assigned to a unique physical location. As described, our model appears to be solving for a relationship diagram, since locations are considered indistinguishable in the constraints. However, if we let actual departments be represented in the model by a set of pseudo-subdepartments relative in number to the square footage of the actual department, then this expanded model resembles a space relationship or block plan problem. Unfortunately, the quadratic assignment problem is difficult to solve and, generally, is not solved to optimality for more than 15 to 20 departments (locations). In view of the complexity induced by the quadratic objective, we will present an empirically proven heuristic as well as a branch and bound approach for finding optimal solutions.

Equations 7.13 and 7.14 provide simple mechanisms for evaluating department switches. To avoid duplicating the discussion, we will hereafter assume that total flow is the measure of importance. It might seem that a simple way to solve the QAP would be to start with a solution a and proceed to switch pairs of departments that

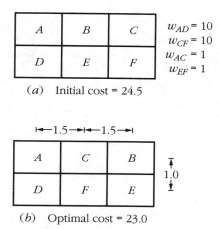

$w_{AD} = 10$
$w_{CF} = 10$
$w_{AC} = 1$
$w_{EF} = 1$

(*a*)   Initial cost = 24.5

(*b*)   Optimal cost = 23.0

Figure 7.17   The failure of pairwise switches.

improve the total flow cost until no improving switches exist. Indeed, if at each step we make the best possible switch, we have a steepest descent approach. Unfortunately, for a general quadratic function this method does not guarantee finding an optimal solution. An example of a local optimum is given in Figure 7.17. Let vertical moves have a unit distance of 1.0 and horizontal moves a distance of 1.5. Weights are 10 between AD and CF and 1 between AC and EF. The assignment in Figure 7.17*a* has cost 24.5. Assignment $\boldsymbol{a} = (1, 3, 2, 4, 6, 5)$ in Figure 7.17*b* has cost 23.0. However, as shown in Table 7.4, every single pair switch from the initial assignment has a cost greater than 24.5; thus, no improving single pair switch exists.

Despite the lack of guaranteed optimality, pairwise exchanges provide a useful mechanism for improving layouts. Still, it can become tedious to calculate the savings of every possible switch at each step. We will first present a more efficient heuristic, which has performed well in practice, and then will discuss a branch and bound strategy for finding optimal solutions.

Table 7.4   Failure of Pairwise Exchanges

Potential Switch	Resultant Cost
A-B	38.0
A-C	84.5
A-D	25.5
A-E	31.5
A-F	63.5
B-C	38.0
B-D	29.5
B-E	25.5
B-F	29.0
C-D	62.5
C-E	28.5
C-F	26.5
D-E	41.0
D-F	84.5
E-F	39.5

7.3.1 VNZ Heuristic

A popular heuristic approach was first introduced by Vollmann et al. [1968]. The heuristic makes use of a familiar idea. Departmental flow costs can be used to rank order the importance of departments. We previously used the total closeness rating concept (equation 7.9) to rank departments. In VNZ we assume that an initial solution exists. The total flow cost for department i is

$$TFC_i = \sum_{j=1}^{M} c_{i,a_i,j,a_j} \qquad (7.18)$$

or, if the factoring assumption holds,

$$TFC_i = \sum_{j \neq i} w_{ij} d(a_i, a_j)$$

Departments are placed in nonincreasing order of TFC_i, that is, $TFC_{[1]} \geq TFC_{[2]} \geq \cdots \geq TFC_{[M]}$.

The VNZ procedure uses two phases. In phase one, the two most important departments, measured by total between-department flow cost (equation 7.18), are selected and all possible exchanges involving these departments are sequentially considered. In phase two, we make two passes through the pairs of departments making any switch that shows improvement when encountered. In steepest descent, $M(M-1)/2$ exchanges must be evaluated at each step. In VNZ we examine $2M$ exchanges in phase one and a total of at most $M(M-1)$ exchanges in phase two. We can describe the concept of the procedure as follows:

Phase 1

1. Set $M_1 = [1]$ and $M_2 = [2]$, the two facilities with largest TFC_i. Set $m = M_1$.
2. Determine the list of departments i with $\Delta C_{im}(\mathbf{a}) > 0$. Order this list by nonincreasing $\Delta C_{im}(\mathbf{a})$. Proceed through this list making each switch in turn provided that $\Delta C_{is}(\mathbf{a}') > 0$ where $\mathbf{a}'$ is the updated assignment vector as switches are made.
3. Set $m = M_2$ and go to 2 if Step 3 not previously executed.

Phase 2

4. Starting with department pair 1 and 2, then pair 1 and 3, and so on up to pair $M-1$ and M, evaluate $\Delta C_{ij}(\mathbf{a}')$ where i and j constitute the pair, and exhange i and j if cost is reduced. Continue until every pair has been examined without making an exchange, or until each pair has been examined twice.

EXAMPLE 7.4

Demonstrate the use of VNZ on the six-department problem considered in the previous examples.

Solution

It would be wise to use the initial assignment $\mathbf{a} = (2, 1, 6, 4, 3, 5)$ of Figure 7.14. However, if we started from this assignment, the VNZ procedure would terminate quickly, failing to find any improvements. Let's therefore pretend that this solution is unknown and start at $\mathbf{a} = (1, 2, 3, 4, 5, 6)$. Computing the TFC_i with the weights of Table 7.2 and unit square locations, we obtain

i	TFC_i
SR	560
PC	250
PS	324
IC	240
XT	225
AT	979

As an example

$$TFC_{SR} = (1) \cdot 40 + (2) \cdot 10 + (1) \cdot 30 + (2) \cdot 10 + (3) \cdot 150 = 560$$

reflecting the interactions with PC, PS, IC, XT, and AT, respectively. We see that AT is the most important department with SR a distant second. We now check to see which departments can offer improvement by switching with AT or SR. Corresponding flow cost differences are

Switch	$\Delta C_{uv}(1, 2, 3, 4, 5, 6)$
SR-AT	51
PC-AT	220
PS-AT	235
IC-AT	32
XT-AT	346
SR-PC	35
SR-PS	32
SR-IC	5
SR-XT	290
SR-AT	51

Noting that all switches yield positive improvement, we realize how poorly our initial assignment scored. The first switch is AT and SR. This yields assignment (6, 2, 3, 4, 5, 1) with cost 1238. We then attempt to switch AT and PC. Starting from our updated $\mathbf{a}$, the switch still yields a lower score, and we proceed to $\mathbf{a} = (6, 1, 3, 4, 5, 2)$ with cost 1124. Now we attempt to switch AT and PS. The new assignment $\mathbf{a} = (6, 1, 2, 4, 5, 3)$ has cost 1099 and is accepted. With this layout, an AT and IC switch no longer is desired. The AT-XT switch is accepted though, lowering cost to 889. Next, we try to switch SR and PC. This would raise cost to 904 and is rejected. A switch of SR and PS lowers cost to 809 with the assignment (2, 1, 6, 4, 3, 5), precisely the layout of Figure 7.14. The three remaining phase one switches all fail to be accepted. We switch to phase two. Fifteen (all possible combinations of two departments) switches are attempted and rejected at which point we stop with assignment (2, 1, 6, 4, 3, 5). Is this assignment optimal? Unfortunately, we still do not know. Let's now consider an approach for guaranteeing optimality.   ■

7.3.2* Branch and Bound

Several branching and node bounding schemes have been proposed for solving quadratic assignment problems optimally. We will discuss one such algorithm described in Francis and White [1974]. The motivation for this approach is that linear assignment problems (LAPs) of the form

$$\text{minimize} \sum_{i=1}^{M} \sum_{j=1}^{M} c_{ij} x_{ij}$$

subject to

$$\sum_{i=1}^{M} x_{ij} = 1 \qquad \text{for all } j$$

$$\sum_{j=1}^{M} x_{ij} = 1 \qquad \text{for all } i$$

$$x_{ij} \; 0 \text{ or } 1$$

are relatively easy to solve. The linear assignment problem is a special case of the classic Transportation problem. Special purpose algorithms such as the Hungarian algorithm also exist for the LAP. The special structure of the constraint matrix ensures that the LAP can be solved with nonnegativity restrictions on the x_{ij} replacing the integer restrictions in the last set of constraints (Bazaraa and Jarvis [1977]). We will hereafter assume that the reader can solve linear assignment problems (if not, read ahead to the first section of the next chapter).

In solving the QAP we will build up gradually more complete, partial assignments. Initially all departments are unassigned. It is helpful to think of the progression of the algorithm as a tree. Branches in our tree will be created by assigning a department to a specific location. Departments will be assigned in order so that if we are at the rth level in the tree, then departments $1, 2, \ldots, r$ have assigned locations. Each branch ends in a node that represents a partial solution. We refer to nodes with no branches emanating out from them as terminal nodes. As branches spread out into more complete partial solutions, new terminal nodes replace old ones.

If we can obtain a good lower bound on the cost of any completion of the partial solution, then this may serve as a bound on the node. When we reach the Mth level in the tree, all departments are assigned, implying a complete solution. Once a complete solution is found with cost no greater than that of every currently terminal node in the tree, we may rest in the shade assured that we have the optimal solution for the model. No new branches can spring forth and drop more fruitful assignments.

Before seeing how to obtain bounds on partial solutions, we will generalize the problem slightly by permitting fixed assignment costs c_{ik} of assigning department i to location k. This will allow us to include some of the considerations mentioned above such as the need to add access to utilities or strengthen floors. As before, w_{ij} are between flow weights ($w_{ij} = w_{ji}$). Total cost of assignment **a** is

$$TC(\mathbf{a}) = \sum_{i=1}^{M} c_{i a_i} + \frac{1}{2} \sum_{i=1}^{M} \sum_{j \neq i} w_{ij} d(a_i, a_j) \qquad (7.19)$$

The 1/2 factor appears, since each weight term will appear twice in the double sum of equation 7.19.

We will start by finding a lower bound on the problem without any partial assignment, that is, no decisions have yet been made. Recall that the constraint sets are the same for QAP and LAP. If we can find a lower bound on the quadratic cost of assigning department i to location k, then the solution to the linear assignment problem with these costs is a lower bound on the QAP. One such lower bound is given by

$$g_{ik} = c_{ik} + \frac{1}{2}\mathbf{w}_i\mathbf{d}'_k \tag{7.20}$$

where $\mathbf{w}_i$ is the row vector of w_{ij} values in nonincreasing order with w_{ii} excluded, and $\mathbf{d}'_k$ is the column vector of $d(k, h)$ values in nondecreasing order with $d(k, k)$ excluded. The $w_{i1}, \ldots, w_{iM}$ contain the flows into and out of department i. The $d(k, 1), \ldots, d(k, M)$ contain the distances these flows must travel if department i is assigned to location k. The idea here is that we must match up the flow weights in $\mathbf{w}_i$ with the distances in $\mathbf{d}_k$. The cheapest way to do this is to use the shortest distances for the highest flow volumes—hence, the reverse ordering of the vectors. Solving the linear assignment problem with cost matrix $G = [g_{ik}]$ yields a lower bound on QAP and is denoted $V(0)$. $V(0)$ is only a lower bound, since we have implicitly allowed departments other than i to be located differently when computing the separate g_{ik}. For each cost coefficient, we myopically allowed each interacting department to be in its best location, that is, we may have placed department 3 in location 3 when computing g_{11} and then may have let it be in location 4 when computing g_{21}, or even when computing g_{12}. Thus, our costs are unfairly optimistic.

This same idea can be extended for bounding partial solutions. Of course, we better be sure to obtain tighter bounds as the solution grows. Let partial assignment $\mathbf{a}_q = (a_{q1}, a_{q2}, \ldots, a_{qq})$ denote the locations of departments $1, 2, \ldots, q$ where $q \le M$. Total flow cost can be broken up into the three parts:

$$TC(\mathbf{a}_q) = \sum_{i=1}^{q} c_{ia_i} + \sum_{i=1}^{q-1} um_{i \le j \le q} w_{ij} d(a_i, a_j) \tag{7.21}$$

$$+ \sum_{j=q+1}^{M} \left[c_{ja_j} + \sum_{i=1}^{q} w_{ij} d(a_i, a_j) \right] \tag{7.22}$$

$$+ \sum_{i=q+1}^{M-1} \sum_{j>i} w_{ij} d(a_i, a_j) \tag{7.23}$$

The right-hand side of equation 7.21 accounts for costs involving fixed departments with other fixed departments. Equation 7.22 accounts for the cost of the yet-to-be-assigned departments with the fixed departments. Equation 7.23 contains the interaction cost of unassigned departments. Given $\mathbf{a}_q$, equation 7.21 is known exactly. For a hypothesized a_j value, $j > q$, equation 7.22 is also known exactly. We can order the remaining departments and locations for each ij combination in equation 7.23 and obtain a bound as we did for $V(0)$. That is, let $\mathbf{w}_i(\mathbf{a}_q)$ be the row vector formed by taking elements $q + 1$ to M in the ith row of the weight matrix $W = [w_{ij}]$ and placing them in nonincreasing order. Let $\mathbf{d}_k(\mathbf{a}_q)$ be the row vector formed by taking the values $d(k, h)$ for unused locations h and placing them in nondecreasing order. Now define the cost coefficients

$$g_{i,k}(\mathbf{a}_q) = c_{ik} + \sum_{j=1}^{q} w_{ij} d(a_{qj}, k) + \frac{1}{2}\mathbf{w}_i(\mathbf{a}_q)\mathbf{d}'_k(\mathbf{a}_q) \tag{7.24}$$

The first two terms on the right-hand side of equation 7.24 correspond to equation 7.22, and the last term corresponds to equation 7.23. To find the lower bound on any node $\mathbf{a}_q$, we solve the $(M - q) \times (M - q)$ linear assigment problem for the unassigned departments and locations. Use the cost coefficients of equation 7.24. Let the LAP solution have value $G^*(\mathbf{a}_q)$. Then the node bound is

$$V(\mathbf{a}_q) = G^*(\mathbf{a}_q) + F^*(\mathbf{a}_q)$$

where $F^*(\mathbf{a}_q)$ is the cost of the fixed decisions given by equation 7.21.

Let us summarize the procedure for finding optimal assignments.

STEP 1. We begin by computing $V(0)$. To compute $V(0)$, we solve a LAP by using the cost coefficients defined by equation 7.20.

STEP 2. Create and bound the M branches $\mathbf{a}_1 = (r), r = 1, \ldots, M$. Each bound is found by solving a LAP with departments $2, \ldots, M$ and excluding location a_1. Set $q = 1$.

STEP 3. Select the open node with the smallest $V(\mathbf{a}_q)$. Set q to the length of this node's partial assignment. If, for this node $q = M - 3$, go to 4. If $q < M - 3$, create the next $M - q$ partial assignment nodes that can be formed by assigning department $q + 1$. Solve a LAP to bound each of these nodes. Go to 3.

STEP 4. Set $q = M$. Create the three new nodes possible by assigning department $M - 2$. Only two possible assignment completions exist for each node, since only departments $M - 1$ and M are unassigned. Complete both possible assignments; $V(\mathbf{a}_M)$ is the cost of the best assignment. If any $V(\mathbf{a}_M)$ is lower than the previous upper bound, set upper bound to $V(\mathbf{a}_M)$. If no open node has a bound less than the upper bound, stop; otherwise go to 3.

EXAMPLE 7.5

Find an optimal layout for the six-department example problem.

Solution

For our example problem it is convenient to summarize the weights and distances as follows:

| | Weights | | | | | | | Distances | | | | | |
From-To	1	2	3	4	5	6	From-To	1	2	3	4	5	6
1	—	40	10	30	10	150	1	—	1	2	1	2	3
2	40	—	0	0	0	105	2	1	—	1	2	1	2
3	10	0	—	0	100	104	3	2	1	—	3	2	1
4	30	0	0	—	0	105	4	1	2	3	—	1	2
5	10	0	100	0	—	5	5	2	1	2	1	—	1
6	150	105	104	105	5	—	6	3	2	1	2	1	—

The details of even a six-department problem would be extensive. To illustrate the procedure while conserving time, we will assume that the partial assignment $\mathbf{a}_2 = (2, 1)$ has already been established, perhaps by constraints on utilities and the presence of shipping docks. To bound node $(2, 1)$, we must first find the cost coefficients for the associated LAP. Extracting the appropriate terms from the tables we have

$$\mathbf{w}_3(21) = (104, 100, 0) \quad \mathbf{d}_3(21) = (1, 2, 3)$$
$$\mathbf{w}_4(21) = (105, 0, 0) \quad \mathbf{d}_4(21) = (1, 2, 3)$$
$$\mathbf{w}_5(21) = (100, 5, 10) \quad \mathbf{d}_5(21) = (1, 1, 2)$$
$$\mathbf{w}_6(21) = (105, 104, 5) \quad \mathbf{d}_6(21) = (1, 1, 2)$$

$\mathbf{w}_3(21)$, for instance, is formed by taking the flow weights w_{34}, w_{35}, and w_{36} and placing these in nonincreasing order. We use the weights from department 3 to 4, 5, and 6 because these are the three departments still unassigned. In a similar vein, $\mathbf{d}_5(21)$ contains distances from location 5 to the unused locations 3, 4, and 6. These distances are placed in nondecreasing order. The cost coefficients for the LAP are

$$G = \begin{bmatrix} 162.0 & 172.0 & 112.0 & 122.0 \\ 82.5 & 112.5 & 82.5 & 112.5 \\ 65.0 & 75.0 & 62.5 & 72.5 \\ 524.0 & 569.0 & 469.5 & 724.5 \end{bmatrix} \tag{7.25}$$

The costs in G include terms from equations 7.22 and 7.23. The rows and columns correspond to departments and locations 3, 4, 5, and 6. For instance, the cost of assigning department 3 to location 4 is, at least,

$$g_{34} = w_{13} \times d(a_1, a_3) + w_{23} \times d(a_2, a_3) + \tfrac{1}{2}\mathbf{w}_3(21)\mathbf{d}_4'(21)$$

Knowing $a_1 = 2, a_2 = 1$, and setting $a_3 = 4$ gives

$$g_{34} = 10(2) + 0(1) + \tfrac{1}{2}(304) = 172$$

as shown in the previous G matrix. Solving the LAP equation 7.25, we find $G^*(\mathbf{a}_2) = 749$. The predetermined cost (equation 7.21) is $F^*(\mathbf{a}_2) = w_{12} \times d(2, 1) = 40$, yielding a lower bound of 789 for (2, 1). We now branch from node (2, 1) and create the four possible extensions from assigning department 3. The four new nodes must be bounded. Consider node (2, 1, 3). We have

$$\mathbf{w}_4(213) = (105, 0) \quad \mathbf{d}_4(213) = (1, 2)$$
$$\mathbf{w}_5(213) = (5, 0) \quad \mathbf{d}_5(213) = (1, 1)$$
$$\mathbf{w}_6(213) = (105, 5) \quad \mathbf{d}_6(213) = (1, 2)$$

$\mathbf{w}_4(213)$ contains the ordered weights for department 4 with the yet unassigned departments 5 and 6. Likewise, $\mathbf{d}_4(213)$ contains the (oppositely) ordered distances from location 4 to the unused sites 5 and 6. The new cost coefficients are

$$G_{213} = \begin{bmatrix} 112.5 & 82.5 & 112.5 \\ 322.5 & 212.5 & 122.5 \\ 774.5 & 623.0 & 776.5 \end{bmatrix}$$

Solving this LAP yields a cost of $G^*(2, 1, 3) = 858$. Fixed costs also exist for interaction between departments 1 and 2 $[w_{12}d(2, 1)]$ and 1 and 3 $[w_{13}d(2, 3)]$ of 50, bringing the bound of the node to 908. A similar analysis is performed for the other three nodes created. Results are shown in Figure 7.18. Partial

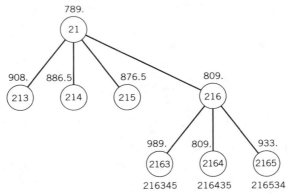

Figure 7.18 Nodes and bounds for the example problem.

assignment (2, 1, 6) has the lowest bound, and we branch from this point. Three new nodes are created. Since each node has only two possible solution completions, we cost these solutions and select the best in each case. To bound node (2, 1, 6, 3), for instance, we evaluate the cost of assignments (2, 1, 6, 3, 4, 5) and (2, 1, 6, 3, 5, 4). This best solution along with its cost is shown in each case. Since these solutions are complete, they represent upper bounds. Node (2, 1, 6, 4) produces an upper bound of 809. As no terminal node exists with lower bound less than 809, we stop the algorithm with optimal assignment (2, 1, 6, 4, 3, 5). ■

For this relatively small problem, both our construction method and VNZ from a random starting assignment also found this solution. Although this is encouraging to know, we still cannot be sure this will happen in general. In practice, we will not know whether those approaches happened to find an optimal solution unless bounds are computed. As with all heuristic procedures, it is important to estimate how good the solution is, and how much better another alternative might be. With that information, a rational decision can be made as to whether it is worthwhile to continue looking for a better solution. With branch and bound we might, for instance, decide to stop prematurely if our upper bound (best current feasible solution) is within a managerially acceptable x percent of the lower bounds on all the terminal nodes.

7.4
GRAPH THEORETIC APPROACH

In describing SLP we pointed out the need to incorporate both quantitative flow data and qualitative factors in the layout process. Although we presented the adjacency score objective in equation 7.10, up to this point we have concentrated our analysis on minimization of total cost-weighted, flow distance. In this section we will adopt the optimization criterion of maximizing the adjacency score. Two departments will be considered adjacent if they share a common border of positive length (this definition excludes departments that are diagonally separated and meet at a single point). The desirability of any adjacency is assumed to be given in the form of a REL rating: *A, E, I, O, U,* or *X.* We will see that it may not be possible to satisfy all desired department adjacencies, and so our goal is to include the most important in the layout. Our approach to maximizing the adjacency score will be to use some basic results from graph theory. We begin by defining what is meant by a graph and showing how it relates to our block plan layouts.

A physical map of departments can be represented as a graph, $G(N, A)$. The node set (N) corresponds to departments; hence, *card* $(N) = M$. The arcs (A) connect departments that are adjacent. Arcs are undirected. Figure 7.19 shows an example of the physical layout of departments and the corresponding graph model. The exterior of the building is referred to as department EX. The reader should select a department from Figure 7.19*a* and see that it is connected in 7.19*b* to the same set of departments with which it shares a border.

The set of possible physical layouts corresponds to a special type of graph, namely, **planar graphs**. A planar graph is one that can be drawn in two dimensions with no arcs crossing. The key word in the definition is "can." Figure 7.20 gives an alternate graph model of our example; however, the graph is not drawn in a planar format. Nevertheless, this is a planar graph, since the identical graph in terms of nodes and arcs can be represented by its isomorphism, shown in Figure 7.19*b*. We make this point here because it is important to realize that finding a planar representation is not always easy. Yet, it is precisely the planar representation in which the layout planner is interested. We must keep this in mind when developing our procedures later in this section.

Planar graphs have the nice feature that there exists a **dual** graph to each planar graph that converts nodes to bounded regions and arcs to common boundaries. Thus, for a planar graph there exists a dual graph that gives positive area to each department and maintains the adjacencies indicated by the arcs in the original graph. The graphs in Figure 7.19 are duals of each other. Bounded regions are called **faces** and can be

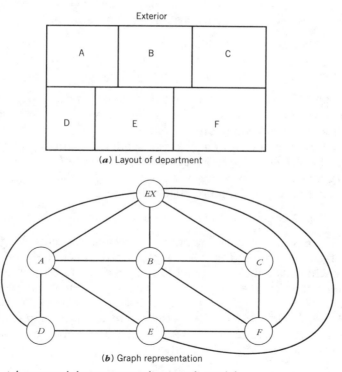

(*a*) Layout of department

(*b*) Graph representation

Figure 7.19   A layout and the corresponding graph model.

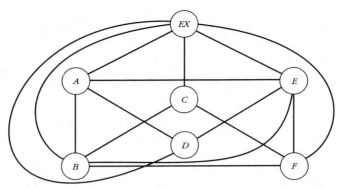

Figure 7.20   A nonplanar representation of the layout (arcs cross).

denoted by the nodes, which imply the boundary. For instance, the boundary of face
ABE in Figure 7.19*b* would be formed by arcs (A, B), (B, E), and (E, A). The exterior
of the graph is also a face.

Every block plan can be represented as a planar graph. To simply illustrate the
validity of this statement take an arbitrary plan. Locate a node on the interior of
each department. For any adjacent departments, connect the corresponding nodes
by drawing a line between the nodes crossing only the common boundary from the
original plan. Such a path must exist for contiguous departments. If we change the
word *department* to "closed region," we have a general procedure for finding the dual
of a planar graph. To illustrate the procedure once again consider Figure 7.19 and
attempt to obtain Figure 7.19*b* by taking the dual of Figure 7.19*a*. Each department
in Figure 7.19*a* becomes a node. Department *A* is adjacent to departments *B, D, E,*
and the *EXterior*. Thus, in Figure 7.19*b*, arcs are added from *A* to *B, D, E,* and *EX*.

Several other results for planar graphs can help in our design of the layout.

Property 1. The dual of a planar graph is planar. For G^D = dual of G, we have
$e_G = e_{G^D}$, $V_G = f_{G^D}$, and $f_G = V_{G^D}$ where e_G, V_G, and f_G are the number of
edges, vertices, and faces in graph G, respectively.

This result follows from our description of how to construct a planar graph, that
is, by connecting nodes of adjacent departments we will cross (complement) each
existing arc exactly once. The importance of this result lies in the resultant ability to
form planar adjacency graphs with nodes representing departments and to guarantee
the existence of a planar dual to represent a relationship diagram. Moreover, the dual
of the dual of a planar graph is the original graph. This allows us to go back and
forth freely between the model representations.

Property 2. The maximum number of arcs in a planar graph is $3M - 6$ for $M \geq 3$.

This property has monumental implications. Since arcs represent adjacencies, we
can satisfy at most $3M - 6$ requests for department adjacency. If $M = 10$, this means
24 pairs of departments at most can be adjacent. A 10-department problem has 55 possible departmental adjacencies, and thus, at least, 31 adjacencies will go unachieved.
If $M = 20$, at most 54 of 210 possible adjacencies can be included in *any* final block
plan. A planar graph with $3M - 6$ edges is called a **maximally planar graph** (MPG).
One caution: just because a graph has $3M - 6$ edges does not mean it is planar, that
is, we are not free to pick any $3M - 6$ edges.

Property 3. A maximally planar graph has $2M - 4$ faces and each face is triangular.

All faces require at least three bordering edges, and edges are used in two faces. Thus with $3M - 6$ edges, we can have at most $2(3M - 6)/3 = (2M - 4)$ faces. It is then easy to see why this result must hold true. Suppose a face had 4 or more edges. We could add an edge connecting any two diagonal vertices. This would add an extra edge without violating planarity in contradiction to Property 2. The importance of this property is that we need only consider triangular faces when trying to construct maximally planar graphs.

We need one final definition before proceeding. A **maximally planar weighted graph** (MPWG) is a MPG whose sum of arc weights is, at least, as large as the sum for all other MPGs. If we assign importance weights to departmental adjacencies, our objective of maximizing the adjacency score is thus equivalent to finding a MPWG.

7.4.1 General Graph Theoretic Approach

The general graph theoretic layout procedure can now be summarized as:

General Procedure

1. Find a MPWG based on REL chart weights. Add a pseudo-department vertex to form the building exterior if desired.
2. Find the dual of the MPWG.
3. Convert the dual into a block plan.

We will examine several approaches for finding a near MPWG. The procedure for finding the dual of a graph was described above. Unfortunately, the third step requires manual intervention and is best performed by the human analyst.

7.4.2 Finding a Maximally Planar Weighted Graph

The ultimate objective is to find a MPWG where nodes correspond to departments and edge weights are measures of the desirability of locating the two departments next to each other. For the simplicity of this graph model, we sacrifice the ability to indicate interdepartmental distances resulting from space and aisle travel considerations. Of course, at this stage we simply want basic configuration ideas so the loss is not a problem. Later when specifying exact dimensions, we can include these details. For now, let's just consider basic conceptual layouts. We will look at a quick strategy for constructing a MPWG and then two improvement strategies. Unfortunately, it is hard to find a MPWG. It may seem like we could just select the most important $3M - 6$ relationships as implied by Property 2. However, it is a hard problem just to determine if a planar graph representation of these relationships exists. Even if one exists, we still have to find it. We will take an easier path.

Construction: Column-Sum Insertion

To be effective, our construction heuristic must include high weight edges yet avoid the need to test for planarity. To avoid planarity testing we employ a strategy that maintains planar, triangular faces. Our approach will be to iteratively add departments to the layout. A single pass of M steps completes construction. First, as before, order the departments by nonincreasing TCR_i. Then form a tetrahedron from the first four departments. These departments will always be adjacent. Next, iteratively insert nodes (departments) into the best available face. The new department is connected to the three departments that determine the selected face. Inserting the new department adds three new edges and two new faces. Planarity is thereby ensured and, if we are lucky, optimality may also be achieved.

We now describe what has been termed the Column-Sum Insertion heuristic.

1. Rank departments in nonincreasing order of TCR_i (equation 7.9), $i = 1, \ldots, M$.
2. Form a tetrahedron from departments 1 through 4.
3. For $i = 5, \ldots, M$: Insert department i into the face with maximum sum of weights of i with the three nodes of the face.

EXAMPLE 7.6

Reconsider our six-department example using graph theoretic decision aids.

Solution

Total closeness ratings were computed in Example 7.2. The four largest were associated with departments SR, AT, PS, and IC. We therefore start with the tetrahedron of Figure 7.21a. This produces an adjacency score of

$$
\begin{aligned}
\text{score} &= V(r_{\text{SR,AT}}) + V(r_{\text{SR,PS}}) + V(r_{\text{SR,IC}}) \\
&\quad + V(r_{\text{PS,AT}}) + V(r_{\text{PS,IC}}) + V(r_{\text{AT,IC}}) \\
&= 81 + 3 + 9 + 27 + 1 + 27 = 148
\end{aligned}
$$

The maximum score obtainable is given by the $3(6) - 6 = 12$ largest relationships. These include 1-A, 4-E, 2-I, 3-O, and 2-U for an upper bound of 218.

Department PC enters next. PC can be inserted into face SR-AT-PS, SR-PS-IC, AT-PS-IC, or SR-AT-IC, For each face, determine the addition to the adjacency

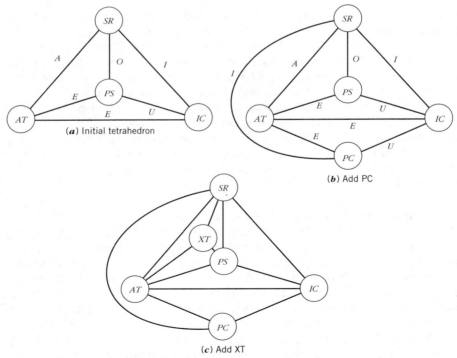

(**a**) Initial tetrahedron

(**b**) Add PC

(**c**) Add XT

Figure 7.21   An example of column sum insertion.

score for selection of that face. For instance, adding to face SR-AT-PS augments the score by $V(r_{AT,PC}) + V(r_{PS,PC}) = 9 + 27 - 243 = -207$, not a favorable decision. The best choice is face SR-AT-IC for a gain of 37. This step is shown in Figure 7.21b. The new score is 185. Last, we add XT. The increment in the score for each face is

Face	Score Increase
SR-AT-PS	33
SR-PS-IC	31
SR-PC-IC	5
SR-AT-PC	7
AT-PS-IC	31
AT-PC-IC	5

Thus, we select face SR-AT-PS as shown in Figure 7.21c. The new score is $185 + 33 = 218$, and we have found an optimal adjacency diagram.

The next step involves finding the dual graph. Figure 7.22 demonstrates this procedure. We have not been concerned with the building exterior. Department PC forms the exterior of the graph in Figure 7.22b. After some reshaping, we obtained the block plan of Figure 7.22c. Departments AT and PC are irregularly shaped, and space is not yet incorporated. The reader is challenged to find a better plan. Clearly, this last step is the most challenging. Often, the planner will be forced to sacrifice certain relationships to achieve a sensible layout. By keeping the REL chart close by for consultation, the planner should be able to obtain a good plan in reasonable time. The dual graph provides a starting point for developing a block plan with desired relationships.   ■

In general, the constructed layout may not be optimal in that a maximally planar graph of greater weight may exist. If the adjacency score is less than the upper bound, improvements can be attempted. In attempting to improve the score of the planar weighted graph, we again must maintain planarity. Two possible approaches are **edge replacement** and **vertex relocation**.

Edge Replacement

Consider an edge AB that forms part of two triangular faces, ABC and ABD as in Figure 7.23. Edge CD may or may not be in the complete adjacency graph. If absent from the graph, it may not be possible to add CD (as in the figure) without losing planarity. However, if edge AB is removed, CD is now permissible. This change is preferable provided $V(r_{CD}) > V(r_{AB})$. The associated gain is $V(r_{CD}) - V(r_{AB})$. The updated portion of the graph is shown in Figure 7.23b. Thus, for each edge such as AB that forms a quadrilateral such as $ABCD$, we consider replacing edge AB with edge CD if CD is absent.

Vertex Relocation

The solution was constructed by inserting departments into triangular faces. For any vertex of degree 3 (the degree of a vertex is the number of edges that are incident

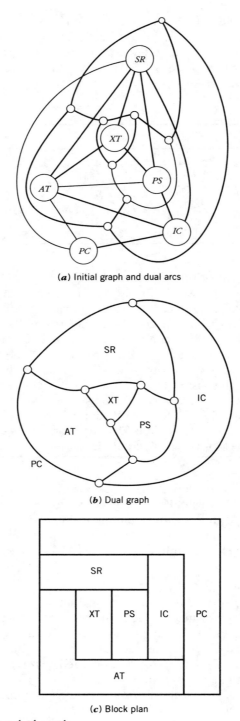

(*a*) Initial graph and dual arcs

(*b*) Dual graph

(*c*) Block plan

Figure 7.22   Finding a dual graph.

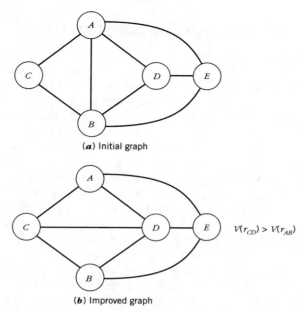

(**a**) Initial graph

(**b**) Improved graph $V(r_{CD}) > V(r_{AB})$

Figure 7.23   An example of edge replacement.

at the vertex), we might attempt moving the vertex into any other triangular face. For instance, suppose vertex D is located in what would otherwise be face ABC as in Figure 7.24. Consider face FGH elsewhere in the graph. Moving vertex D to face FGH yields a net gain of $\Delta = V(r_{DF}) + V(r_{DG}) + V(r_{DH}) - [(V(r_{AD}) + V(r_{BD}) + V(r_{CD})]$. If $\Delta > 0$, vertex D should be moved to face FGH.

Thus, given a tentative solution, we attempt to make edge replacements or vertex relocations or both until no improvements can be found or the upper bound on adjacency weight is achieved.

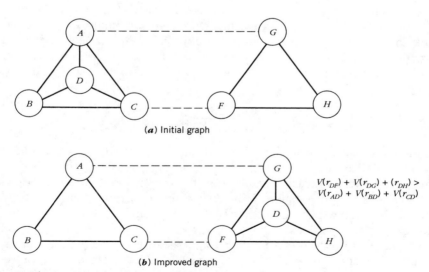

(**a**) Initial graph

(**b**) Improved graph $V(r_{DF}) + V(r_{DG}) + (r_{DH}) >$
$V(r_{AD}) + V(r_{BD}) + V(r_{CD})$

Figure 7.24   An example of vertex relocation.

7.5
DECOMPOSITION OF LARGE FACILITIES

In analyzing large systems it is often helpful to decompose the system into nearly independent entities. In addition to the lower computational burden, it is easier to grasp the essential trade-offs in smaller problems. Model solutions are also less sensitive to changes in the minor relationships that are excluded. Solving quadratic assignment problems and finding maximally planar weighted graphs fall into this category. The question is how to decompose the problem. This question is similar to that of finding machine groups in Group Technology. Given the large set of M departments, we wish to divide these departments into subproblems where departments share high interactions with other departments in their subproblem and are nearly independent of departments in other subproblems. One approach is to use a graph model where each department represents a node. Arc weights are the degree of interaction between department pairs, either flow volumes or activity relationships. We then partition this graph into clusters by finding the cheapest cuts of the graph. The hierarchical clustering method of Section 6.4.4 is a natural approach for creating subproblems of departments by eliminating the weakest (lowest relationship or volume) interdepartmental links. Starting with all arcs, iteratively remove the next weakest arc until no connected subgraph exceeds a maximum allowable subproblem size. The graph partitioning construction and improvement techniques of Section 6.4.5 could also be used. As we are not overly concerned with the exact size of the resulting subproblems, dummy nodes can once again be used to let partition sizes vary as we peel off department groups.

Once the graph is separated into subgraphs, each subgraph is treated as a complete layout problem and solved with the techniques already discussed in this chapter. The last step involves reassembling the subproblems. Each subproblem can be treated as a single department. All flows out of any of the departments that constitute a subproblem emanate from the centroid of the solved subproblem. Thus the overall problem becomes one of laying out the fixed subproblem solutions relative to each other.

This decompostion procedure offers a reasonable approach to very large problems, many of which will have natural subproblems of related departments. When space considerations are investigated in the final step of recombining subproblems, it is possible to consider various rotations of the subproblem layouts. Flows are now from department centroid to department centroid and, hence, minor improvements can be obtained by optimal orientation of the subproblem layouts relative to each other.

7.6
NET AISLE AND DEPARTMENT LAYOUT

Once the basic block plan is formulated, detail flow patterns must be established. Aisles must be determined and departmental I/O locations must be specified. We assume all interdepartmental travel is from the output location of the originating department to the input location of the terminating department and is constrained to stay on aisles. Flow between departments will naturally follow the shortest path that satisfies these conditions.

Montreuil [1988] proposed a modeling approach for producing the **net layout** consisting of detailed shapes for departments, aisles, and flow paths. Input to the model consists of relative locations of departments including adjacencies, aisle corridors, and bounds on department dimensions. Model output gives aisle widths, coordinates of department boundaries, and I/O locations. By concentrating on department perimeter instead of area for modeling size, a linear programming model can be used to perform net layout. We will assume that departments are rectangular in shape, but more general shapes can be modeled. We use the following notation:

l_{zi}, u_{zi}	lower and upper bounds on z-axis length of department i
P_{li}, P_{ui}	lower and upper limits of perimeter of department i
X_i, Y_i	I/O location for department i (decision variable)
X_{li}, Y_{li}	lower left corner of department i (decision variable)
X_{ui}, Y_{ui}	upper right corner of department i (decision variable)
x_{la}, y_{la}	lower left corner for aisle segment a
x_{ua}, y_{ua}	upper right corner for aisle segment a
w_a	required width for aisle segment a

The linear program can then be stated as

$$\text{minimize} \sum_{i=1}^{M-1} \sum_{j>i} w_{ij}(X_{ij}^+ + X_{ij}^- + Y_{ij}^+ + Y_{ij}^-) \tag{7.26}$$

subject to

$$X_i - X_j = X_{ij}^+ - X_{ij}^- \qquad \text{if } w_{ij} > 0 \tag{7.27}$$

$$Y_i - Y_j = Y_{ij}^+ - Y_{ij}^- \qquad \text{if } w_{ij} > 0 \tag{7.28}$$

$$l_{xi} \le X_{ui} - X_{li} \le u_{xi} \qquad \text{for all } i \tag{7.29}$$

$$l_{yi} \le Y_{ui} - Y_{li} \le u_{yi} \qquad \text{for all } i \tag{7.30}$$

$$P_{li} \le 2[(X_{ui} - X_{li}) + (Y_{ui} - Y_{li})] \le P_{ui} \qquad \text{for all } i \tag{7.31}$$

$$X_{li} \le X_i \le X_{ui} \qquad \text{for all } i \tag{7.32}$$

$$Y_{li} \le Y_i \le Y_{ui} \qquad \text{for all } i \tag{7.33}$$

$$x_{ua} - x_{la} \ge 0 \qquad \text{for all horizontal aisles } a \tag{7.34}$$

$$y_{ua} - y_{la} = w_a \qquad \text{for all horizontal aisles } a \tag{7.35}$$

$$y_{ua} - y_{la} \ge 0 \qquad \text{for all vertical aisles } a \tag{7.36}$$

$$x_{ua} - x_{la} = w_a \qquad \text{for all vertical aisles } a \tag{7.37}$$

$$Y_{ui} \le Y_{lj} \qquad \text{for all horizontal } ij \text{ adjacencies} \tag{7.38}$$

$$X_{ui} \le X_{lj} \qquad \text{for all vertical } ij \text{ adjacencies} \tag{7.39}$$

$$X_{li}, X_{ui}, X_i, Y_{li}, Y_{ui}, Y_i, x_{la}, x_{ua}, y_{ua}, y_{la} \ge 0$$

The objective function 7.26 minimizes the sum over all between flows of the product of flow weight and rectilinear distance. Rectilinear distance is determined by the constraints 7.27 and 7.28. These expressions make use of the standard trick for handling absolute values in the objective function of a linear program, that is, replace $|s|$ by the difference of two positive variables, s^+ and s^-. The assumption here is that no backtracking is required. If this assumption is not valid, the objective must be expanded to count flow along every aisle segment. By using the length of aisle segments instead of absolute difference in coordinates, any desired flow path can be modeled for each move.

Constraints 7.29 and 7.30 restrict the horizontal and vertical lengths of each department, respectively. The center of each two-sided inequality is the difference between the upper and lower borders of the department, that is, its length in the associated direction. Enclosed area is restricted by perimeter constraints 7.31. I/O locations must be within the department; this is represented in constraints 7.32 and 7.33. Constraints 7.34 and 7.35 force each planned horizontal aisle segment to have nonnegative length and requisite width. These widths must be known and should be based on the type and volume of flow along the aisle segment. For horizontal aisle segments, the range of the x coordinates is forced to be nonnegative, and the range of the y coordinates is constrained to be the aisle width (bounds could alternatively be used if an acceptable range exists). Vertical aisle segments are similarly defined in constraints 7.36 and 7.37. The last set of constraints is perhaps the most tedious. For every department-to-department or department-to-aisle or aisle-adjacency indicated by the block plan, a constraint is included that forces the adjacent edges to have a common coordinate value. Expressions 7.38 and 7.39 suggest the form of these constraints for department-to-department adjacencies. Analogous expressions can be used for aisle-to-department and aisle-to-aisle adjacencies.

The general net layout model is reasonably flexible. As indicated above, more general departmental shapes can be modeled by inclusion of the appropriate variables for department corners. Existing buildings can be modeled by constraining department corners to fit within the existing enclosure. Alternatives for aisle placement and department shape can be quantitatively evaluated by adjusting the constraint and variable sets.

EXAMPLE 7.7

Consider the plan in Figure 7.14. Only one aisle will be used, a main aisle down the center of the plant. Restrictions on department sizes are given in Table 7.5. These restrictions are derived from the space requirements of Table 7.3 and the layout planners' knowledge of the activities performed within each area. Lengths in the y direction are effectively constrained by x and perimeter constraints. The aisle will be 12 feet wide. Find a net layout.

Table 7.5   Department Size Restrictions

Department	X Length (Min, Max)	Perimeter (Min, Max)
SR	(90, 110)	(380, 420)
PC	(70, 80)	(280, 320)
PS	(60, 80)	(240, 280)
IC	(100, 120)	(400, 450)
XT	(100, 150)	(480, 500)
AT	(100, 120)	(420, 460)

Solution

To simplify the model several unnecessary variables were removed. For instance, the placement of SR to the right of PC ensures $X_{SR} \geq X_{PC}$. Thus the absolute value signs in the objective function term for movement between SR and PC were eliminated along with the need for the $\pm$ tandem of extra variables. Only variables for SR-AT, PC-IC, and XT-PS were included owing to their vertical alignment in the input layout. Fifty-six constraints were required. Three constraints were needed to define the $\pm$ variables created for the vertically aligned departments. The SR-AT constraint was

$$X_{SR} - X_{AT} - X_{SRAT}^{+} + X_{SRAT}^{-} = 0$$

Twelve constraints placed the two-sided bounds on department length. The upper bound on SR is

$$X_{uSR} - X_{lSR} \leq 110$$

Twelve additional constraints were used to bound department perimeters. Another twelve constraints of the form

$$X_{SR} - X_{uSR} \leq 0$$

ensured that I/O locations were within the horizontal length of the department. Since only one aisle exists and all departments border this aisle, the optimal Y_i must all be on the aisle. Thus, the model was reduced with knowledge that $Y_i = y_l$ for the lower departments IC, AT, and PS and $Y_i = y_u$ for the upper departments. Six constraints were required to ensure vertical consistency of the department definitions, that is, $Y_{lSR} - Y_{uSR} \leq 0$. Horizontal adjacencies required four constraints. As an example, relative placement of PC and SR was achieved through

$$X_{uPC} - X_{lSR} \leq 0$$

One constraint was required for each department to place it above or below the aisle as in $y_u - Y_{lSR} \leq 0$. The last constraint set aisle width by $y_u - y_l = 12$.

The objective function simply aggregated the flow weights times the inter I/O distances. For instance, one contributor was $40(X_{SR} - X_{PC})$. The complete function is

$$\text{minimize } 50X_{SR} - 145X_{PC} + 114X_{PS} - 135X_{IC} + 150X_{SRAT}^{+} + 150X_{SRAT}^{-}$$
$$+ 10X_{XT} + 106X_{AT} + 100X_{PSXT}^{+} + 100X_{PXST}^{-} + 5X_{XTAP}^{+} + 5X_{XTAP}^{-}$$
$$+ 145Y_{lSR} - 155Y_{uAT} + 105Y_{lXT} - 130Y_{uPS} - 30Y_{uIC} + 100Y_{lPC}$$

The solution is shown algebraically in Table 7.6 and graphically in Figure 7.25 on page 245. An objective value of 20,950 was obtained. In addition to the desired shape for departments, the optimal locations of I/O points are placed. The E relationships of AT with PC and IC pulled the AT I/O point, and hence the ST I/O point, to the left.

To investigate the effect of an existing building, suppose the facility is constrained to a 200×350 ft structure. The model was rerun with upper bounds of 350 on X_{uXT} and X_{uPS} and 200 on Y_{uPC}, Y_{uSR}, and Y_{uXT}. A reasonable block layout is produced as shown in Figure 7.25b. Extra space (for expansion) is shown. The I/O location for XT is placed above PS because of their dominant E relationship. ∎

Table 7.6 Solution to Net Layout

Variable	Value	Variable	Value
X_{SR}	100	Y_{lPS}	50
X_{PC}	100	Y_{uPS}	110
X_{PS}	200	X_{lPS}	200
X_{IC}	100	X_{uPS}	260
X_{XT}	200	Y_{lIC}	10
X_{AT}	100	Y_{uIC}	110
X_{SRAT}^1	0	X_{lIC}	0
X_{SRAT}^-	0	X_{uIC}	100
X_{PSXT}^1	0	Y_{lXT}	122
X_{PSXT}^-	0	Y_{uXT}	212
X_{XTAT}^1	100	X_{lXT}	200
X_{XTAT}^-	0	X_{uXT}	350
Y_{lSR}	122	Y_{lAT}	0
Y_{uSR}	212	Y_{uAT}	110
X_{lSR}	100	X_{lAT}	100
X_{uSR}	200	X_{uAT}	200
Y_{lPC}	122	y_l	110
Y_{uPC}	182	y_u	122
X_{lPC}	20		
X_{uPC}	100		

7.7
LOCATING NEW FACILITIES

We have been concentrating on techniques for locating space-consuming entities relative to each other. A related problem involves adding one or more new entities into a system where they will interact with the existing facilities. If the new entities are either new machines to be added to a large facility or new plants/warehouses to be added to a large distribution network, then it may be reasonable to assume that the area to be occupied by the new entities is not important. Instead, our interest centers on finding the ideal point location for each new entity. Using the motivation of adding a new machine to a large production facility, we will use the term "machine" to indicate an entity. The key differences between the current model and those discussed earlier in the chapter are the presence of fixed locations with which the new machine will interact and the assumption that sufficient space exists in the previous system to insert the new machine without worrying about its space requirements. We first discuss the case of a single new machine. We then look at the addition of several new facilities. In both cases we will restrict our attention to rectilinear movement. Many other distance metrics have been examined. The interested reader should ·consult the references at the end of the chapter, in particular, Francis and White [1974] and Love et al. [1988].

7.7.1 Single Facility Location

Let us assume that we know the locations $P_1, \ldots, P_M; P_i = (x_i, y_i)$ of the M existing facilities with which the new machine will interact. The decision variables are the coordinates $X = (x, y)$ of the new machine. (x^*, y^*) will denote the optimal loca-

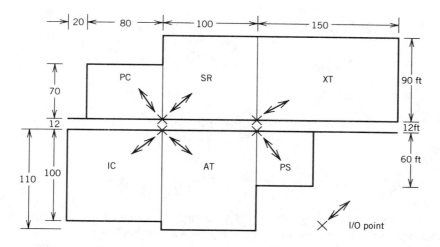

(*a*) Unconstrained layout

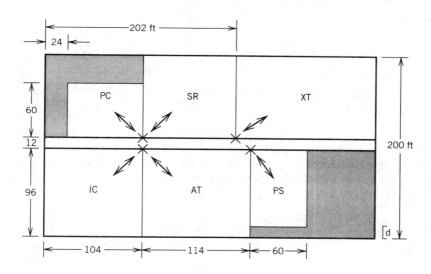

(*b*) 200 × 350-foot constrained layout

Figure 7.25   A net layout example.

tion. The new machine will have an interaction cost with each existing machine that is proportional to its distance $d(X, P_i)$ to P_i. The proportionality constant is w_i, the product of (trips/period) and (cost/unit distance moved) for movement between the new and existing machine i. Although we are not concerned with the size of the new machine directly, there may be locations in the plant that are too crowded to support X. Thus, we may want to restrict X to some feasible set of regions Ω. A simple statement of the problem is then

$$\underset{(x,y) \in \Omega}{\text{Minimize}} \; f(x, y) = \sum_{i=1}^{M} w_i d(x, p_i) \qquad (7.40)$$

We will again assume rectilinear distance; hence, equation 7.40 becomes

$$\underset{(x,y) \in \Omega}{\text{Minimize}} \ f(x, y) = \sum_{i=1}^{M} w_i \left(|x - x_i| + |y - y_i| \right) \qquad (7.41)$$

The objective function in equation 7.41 can be rewritten as

$$f(x, y) = f_1(x) + f_2(y)$$

where

$$f_1(x) = \sum_{i=1}^{M} w_i |x - x_i|; \ f_2(y) = \sum_{i=1}^{M} w_i |y - y_i|$$

indicating the separability of $f(x, y)$. The implication is that we can solve for x^* and y^* independently. Thus, hereafter we worry only about finding x^*.

Let's first assume that $\Omega = R^2$, that is, there are no constraints on the solution. Define a **median location** to be any point such that no more than one half of total flow is to the left or right, above or below the location. (We visualize the problem as locating on a two-dimensional plane). A median location (x, y) then satisfies

$$\sum_{x_i < x} w_i \leq \sum_{i=1}^{M} \frac{w_i}{2} \quad \text{and} \quad \sum_{x_i > x} w_i \leq \sum_{i=1}^{M} \frac{w_i}{2}$$

FACT. Every median location solves the single facility rectilinear location problem.

JUSTIFICATION.

Assume that there exists an optimal location x^* that is not a median location. We will show that this statement leads to a contradiction with the definition of a median location. Divide the existing locations into the three sets $S_1 = \{i : x_i < x^*\}$, $S_2 = \{i : x_i > x^*\}$, and $S_3 = \{i : x_i = x^*\}$. Since x^* is not a median location, either

$$\sum_{S_1} w_i > \sum_{i=1}^{M} \frac{w_i}{2}$$

or

$$\sum_{S_2} w_i > \sum_{i=1}^{M} \frac{w_i}{2}$$

Assume

$$\sum_{S_1} w_i > \sum_{i=1}^{M} \frac{w_i}{2}$$

(the alternative case follows similarly). In the neighborhood of x^*, the objective can be written

$$f_1(x) = \sum_{S_1} w_i (x - x_i) + \sum_{S_2} w_i (x_i - x) + \sum_{S_3} w_i |x_i - x|$$

Now consider a small positive step to the left such that $x = x^* - \varepsilon$ where $\varepsilon > 0$.

$$f_1(x) - f_1(x^*) = \left[-\sum_{S_1} w_i + \sum_{S_2} w_i + \sum_{S_3} w_i \right] \varepsilon$$

Using the fact that

$$\sum_{S_1} w_i > \sum_{i=1}^{M} \frac{w_i}{2}$$

we must have

$$-\sum_{S_1} w_i + \sum_{S_2} w_i + \sum_{S_3} w_i < 0$$

Thus, $f_1(x) < f_1(x^*)$, a contradiction to the assumed optimality of x^*.

Our result says that if we were making more than one half of our trips in any direction (right or left), then it would be better to locate farther in that direction. Moving ε would save this distance on more trips than it increases, thus reducing total movement. An immediate consequence of this result is that an optimal x^* must coincide with some x_i. Otherwise, we could move in the direction of extra weight, at least until the next existing facility is encountered, and save total distance. If weight is equal on both sides, movement in either direction until a facility is encountered leaves total distance unchanged. Thus, alternate optimum in the form of a line segment between the coordinates of the two median locations may exist. In two dimensions, this can expand to become a rectangle.

EXAMPLE 7.8

A test station is to be added to an existing plant. The station will receive 4 loads per day from location (10, 8), 2 from (1, 4), 3 from (4, 6), and 2 from (4, 9). Where should the station be located to minimize total distance traveled?

Solution

The $\sum_i w_i = 4 + 2 + 3 + 2 = 11$. Thus the median location must have no more than 11/2 trips in any direction. First, order the x coordinates.

Facility	x	Weight	Cumulative Weight
2	1	2	2
3	4	3	5
4	4	2	7 > 5.5
1	10	4	11

Since cumulative weight first exceeds 11/2 at $x = 4$, $x^* = 4$. Similarly, ordering the y locations results in $y^* = 8$. Thus, locate at $(x^*, y^*) = (4, 8)$.

Figure 7.26 shows the effect of the median location property. As we move from x^* in either direction, $f_1(x)$ increases with slope equal to the difference in weights between those facilities farther to the right and those to the left. ∎

The optimal location found may not be feasible. Other machines may be there, access to the location may be insufficient, or organizational constraints may dominate. Accordingly, we would like an efficient method to explore nearby locations. Contour lines provide a simple graphical tool for performing sensitivity analysis. A contour line is a line of constant value for the objective function. The objective function is strictly increasing as we move farther away from the optimum location(s). A complete contour line of value greater than $f(X^*)$ will enclose an area with all points inside the line having lower objective value and all points outside the region having higher value. We know that the slope in the x (and, therefore, y because of separability of the objective) direction is constant between coordinates of existing facilities. Unless we pass the vertical line of some x_i or the horizontal line through some y_i, nothing changes with regard to derivatives of the objective function. This implies that contour lines will have constant slope within rectangles determined by the coordinates of the existing facilities. Let's see what these contours look like.

Let p be the number of unique x_i locations and q the number of y_i locations. We can simplify the objective by redefining the existing locations to occur at x coordinates $c_1, \dots, c_j, \dots, c_p$ and y coordinates $r_1, \dots, r_j, \dots, r_q$. The weights C_j and R_j are associated with the coordinates and are the sum of the weights of all facilities with the corresponding x or y coordinate. Our objective is then

$$\underset{(x,y)}{\text{minimize}} \; = \; \sum_{j=1}^{p} C_j |x - c_j| + \sum_{j=1}^{q} R_j |y - r_j|$$

For any point (x, y) in what we will refer to as rectangle $[s, t]^5$ defined by $c_t \le x \le c_{t+1}$ and $r_s \le y \le r_{s+1}$, we have

$$
\begin{aligned}
f(x, y) &= \sum_{j=1}^{t} C_j(x - c_j) + \sum_{j=t+1}^{p} C_j(c_j - x) \\
&\quad + \sum_{j=1}^{s} R_j(y - r_j) + \sum_{j=s+1}^{q} R_j(r_j - y) \\
&= \left(\sum_{j=1}^{t} C_j - \sum_{j=t+1}^{p} C_j \right) x - \left(\sum_{j=1}^{t} C_j c_j - \sum_{j=t+1}^{p} C_j c_j \right) \\
&\quad + \left(\sum_{j=1}^{s} R_j - \sum_{j=s+1}^{q} R_j \right) y - \left(\sum_{j=1}^{s} R_j r_j - \sum_{j=s+1}^{q} R_j r_j \right)
\end{aligned}
$$

(7.42)

The first and third terms in parentheses are differences of column and row weights, respectively, and will be abbreviated by Θ_t and Φ_s. The second and fourth terms in parentheses are constants and can be replaced by K_{st}. Thus, we can write equation 7.42 as

$$f(x, y) = \Theta_t x + \Phi_s y + K_{st}$$

---

[5] "Rectangles" on the periphery are unbounded as we assume $c_0 = r_0 = -\infty$ and $c_{p+1} = r_{q+1} = \infty$.

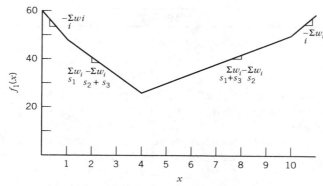

Figure 7.26   The sensitivity of objectives to solution.

To form an isocost (contour) line we define a constant k and find $(x, y) \in [s, t]$ such that

$$k = \Theta_t x + \Phi_s y + K_{st}$$

Rearranging terms,

$$y = -\frac{\Theta_t}{\Phi_s}x + \frac{k - K_{st}}{\Phi_s}$$

Differentiating we find that within the block $[s, t]$,

$$\frac{dy}{dx} = -\frac{\Theta_t}{\Phi_s} \qquad (7.43)$$

Expression 7.43 shows the trade-off required to offset the gain (loss) in objective function as we move along x with a loss (gain) in y. The appropriate trade-off is simply the ratio of the difference in the weights for the box. Perhaps an easier way to look at expression 7.43 is the equivalent form

$$\left(\sum_{j=1}^{t} C_j - \sum_{j=t+1}^{p} C_j\right) dy - \left(\sum_{j=1}^{s} R_j - \sum_{j=s+1}^{q} R_j\right) dx = 0$$

Expression 7.43 leads to a straightforward algorithm for constructing contour lines. The technique is as follows:

STEP 1. Draw p vertical lines to intersect all x_i and q horizontal lines to intersect all y_i.

STEP 2. Label vertical lines by C_j and horizontal lines by R_j, the sum of the weights of facilities intersected by the lines.

STEP 3. Set $\Theta_0 = \Phi_0 = -\sum_{i=1}^{M} w_i$. Label columns $j = 1, \ldots, p$ by $\Theta_j = \Theta_{j-1} + 2C_j$ and rows $j = 1, \ldots, q$ by $\Phi_j = \Phi_{j-1} + 2R_j$

STEP 4. For each rectangular segment $[r, s]$ compute the slope of isocost lines by $S_{rs} = -\Theta_s/\Phi_r$.

STEP 5. Select any point (x, y) and draw the contour that starts and ends at (x, y) using slope s_{rs} in each segment. Step 5 can be repeated as many times as desired to produce the contour map.

EXAMPLE 7.9

Construct contour lines for Example 7.8.

Solution

We start by drawing three vertical and four horizontal lines through existing facilities and labeling each by the sum of the weights for facilities on the line. The outcome is shown in Figure 7.27a. Starting at $-\sum_i w_i = -11$ for the lower left corner, rows and columns are marked as in step 3. Each region is then assigned the slope of the negative of its column divided by its row value. Figure 7.27b shows the result. Next, pick an arbitrary point. We start at (4, 5). Proceed through rectangles, always moving at the slope for the rectangle. Eventually we terminate at the starting point with the contour shown in Figure 7.27c. The optimal solution has cost $\sum_i w_i d(X^*, P_i) = 2(7) + 3(2) + 2(1) + 4(6) = 46$. All values on the contour have cost equivalent to that at (4, 5), namely, $2(4) + 3(1) + 2(4) + 4(9) = 55$.   ∎

7.7.2 Multifacility Location

We will briefly look at the case of several new machines. Our intent here is to show how this problem can be modeled as a linear program. We allow for N new facilities to be located. In addition to the w_{ji} weights for interaction between new machine j and old machine i, we have weights v_{jk} for interactions between new machines j and k. These weights would typically be trips per period between the machines. New facility locations will be denoted $\mathbf{X}_j^n = (x_j^n, y_j^n)$. Our new objective function is

$$\text{minimize } f(\mathbf{X}_1^n, \ldots, \mathbf{X}_j^n) = \sum_{1 \le j < k \le N} v_{jk} \left(|x_j^n - x_j^n| + |y_j^n - y_k^n| \right)$$

$$+ \sum_{j=1}^{N} \sum_{i=1}^{M} w_{ji} \left(|x_j^n - x_i| + |y_j^n - y_i| \right) \tag{7.44}$$

Once again, the objective 7.44 is separable in x_j^n and y_j^n, leading to two problems. The problem for x_j^n becomes

$$\text{minimize } f_1(x_1^n, \ldots, x_N^n) = \sum_{1 \le j < k \le N} v_{jk} \left(|x_j^n - x_k^n| \right) + \sum_{j=1}^{N} \sum_{i=1}^{M} w_{ji} \left(|x_j^n - x_i| \right) \tag{7.45}$$

We again employ the standard linear programming trick for absolute values. By using the pairs p_{jk}^+, p_{jk}^- for the v_{jk} terms and q_{ji}^+, q_{ji}^- for the w_{ji} terms, we obtain the model:

$$\text{minimize } \sum_{1 \le j < k < N} v_{jk} \left(p_{jk}^+ + p_{jk}^- \right) + \sum_{j=1}^{N} \sum_{i=1}^{M} w_{ji} (q_{ji}^+ + q_{ji}^-) \tag{7.46}$$

subject to

$$x_j^n - x_k^n = p_{jk}^+ - p_{jk}^- \qquad 1 \le j < k \le N$$

$$x_j^n - x_i = q_{jk}^+ - q_{jk}^-; \qquad i = 1, \ldots, M \qquad j = 1, \ldots, N$$

$$p_{jk}^+, p_{jk}^-, q_{ji}^+, q_{ji}^- \ge 0$$

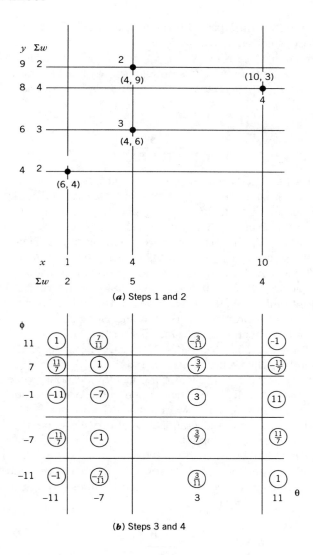

(a) Steps 1 and 2

(b) Steps 3 and 4

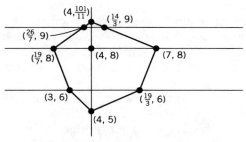

(c) Step 5 from point (4, 5)

Figure 7.27   An example of contours for a single facility (rectilinear location).

The linear dependence of the paired variables prevents their ever appearing simultaneously in a basic solution.

The current formulation has $[N(N-1)/2] + NM$ constraints and $N^2 + 2NM$ variables. For large problems more efficient solution approaches are available for the dual of problem 7.46. The interested reader should consult Francis and White [1974].

7.8
FURTHER READING

Numerous articles have been written on facility layout and location. These articles can be found in industrial engineering, operations research, and architectural publications. In the last decade or so, authors have applied mathematical programming techniques (Bazaara and Elshafei [1979], Bazaara and Sherali [1980]), simulated annealing (Wilhelm and Ward [1987]), heuristic improvement techniques (Liggett [1981], Picone and Wilhelm [1984]), and other approaches to the Quadratic Assignment Problem. Evans et al. [1987] attacked the layout decision problem by using fuzzy sets. Graph theory methods (Levin [1964], Roth et al. [1982, 1985], Foulds et al. [1985], Foulds and Giffin [1985], Montreuil et al. [1987]) have evolved rapidly in recent years. A review of these developments is provided in Hassan and Hogg [1987]. Computer-aided approaches to layout have been pursued since the early CRAFT and CORELAP programs (Moore [1974], Tompkins and Moore [1977], Nozari and Enscore [1981], Khator and Moodie [1984], and Ketcham and Malstrom [1984]).

We have only scratched the surface of the facility location problem. Papers such as Hakimi [1964] and Minieka [1981] have opened the door for quickly finding optimal locations on a network. Love, Morris, and Wesolowsky [1988], Brandeau and Chiu [1989], and Mirchandani and Francis [1990] provide general references on facility location. A number of important and interesting problems have been studied. Efroymson and Ray [1966] consider adding extra facilities to reduce material movement. Others such as Van Roy and Erlenkotter [1982] have considered the dynamic expansion and contraction of the set of available facilities through time.

7.9
SUMMARY

Material flow is a major concern in the design of manufacturing facilities. Simplified flow patterns facilitate the entire manufacturing planning and control function. Unstructured flow patterns tend to hide the process and complicate attempts to improve efficiency. Flow patterns should be simple and keep backtracking to a minimum. One approach is to use a central spine as a conduit of materials and production resources.

Systematic layout planning offers a strategy for obtaining first a block plan and subsequently a detailed layout for each "block" of a facility. Quantitative and qualitative data are integrated with space and other considerations to obtain a good, feasible layout. Graph theoretic methods and quadratic assignment problem procedures are available for assisting in the determination of good relative layouts, but optimal solutions are hard to come by for large problems. As part of layout planning, linear programming can be used to aid in the specification of exact department dimensions including aisles and I/O locations.

In some instances we may want to add a new entity to an existing layout. With rectilinear flow patterns, the new object should be located near the median of the existing objects with which it interacts. When certain locations are prohibited or other considerations exist, contour lines can aid in selecting the final location. Linear programming can be used for adding multiple facilities.

REFERENCES

Armour, G. C., and E. S. Buffa (1963), "A Heuristic Algorithm and Simulation Approach to Relative Location of Facilities," *Management Science,* 9(1), 294–309.

Bazaara, M. S., and A. N. Elshafei (1979), "An Algorithm for the Quadratic Assignment Problem," *Management Science,* 26(1), 109–120.

Bazaara, M. S., and J. J. Jarvis (1977), *Linear Programming and Network Flows,* John Wiley & Sons, Inc., New York.

Bazaara, M. S., and H. Sherali (1980), "Bender's Partitioning Scheme Applied to a New Formula of the Quadratic Assignment Problem," *Management Science,* 27(1), 29–41.

Brandeau, M. L., and S. S. Chiu (1989), "An Overview of Representative Problems in Location Research," *Management Science,* 35(6), 645–674.

Efroymson, M. A., and T. L. Ray (1966), "A Branch and Bound Algorithm for Plant Location," *Operations Research,* 14(3), 361–368.

Evans, G. W., M. R. Wilhelm, and W. Karwowski (1987), "A Layout Design Heuristic Employing the Theory of Fuzzy Sets," *International Journal of Production Research,* 25(10), 1431–1450.

Foulds, L. R., P. B. Gibbons, and J. W. Giffin (1985), "Facilities Layout Adjacency Determination: An Experimental Comparison of Three Graph Theoretic Heuristics," *Operations Research,* 33, 1091–1106.

Foulds, L. R., and J. W. Giffin (1985), "A Graph-Theoretic Heuristic for Minimizing Total Transport Cost in Facilities Layout," *International Journal of Production Research,* 23(6), 1247–1257.

Francis, Richard L., and John A. White (1974), *Facility Layout and Location,* Prentice-Hall, Inc., Englewood Cliffs, NJ.

Hakimi, S. L. (1964), "Optimum Locations of Switching Centers and the Absolute Centers and Medians of a Graph," *Operations Research,* 12, 450–459.

Harmon, Roy L., and Leroy D. Peterson (1990), *Reinventing the Factory,* The Free Press, New York.

Hassan, Mohsen, and Gary L. Hogg (1987), "A Review of Graph Theory Application to the Facilities Layout Problem," *OMEGA, International Journal of Management Science,* 15(4), 291–300.

Hillier, Frederick S., and Michael M. Connors (1966), "Quadratic Assignment Problem Algorithms and the Location of Indivisible Facilities," *Management Science,* 13(1), 42–57.

Khator, Suresh, and Colin Moodie (1984), "Computer Assisted Plant Layout Using a Graphics Editor," *Computers & Industrial Engineering,* 8(3), 171–179.

Ketcham, Ronald L., and Eric M. Malstrom (1984), "A Computer Assisted Facilities Layout Algorithm Using Graphics," *Proceedings of the IIE Fall Conference.*

Khator, Suresh, and Colin Moodie, (1984), "Computer Assisted Plant Layout Using a Graphics Editor," *Computers & Industrial Engineering,* 8(3), 171–179.

Konz, Stephan (1985), *Facility Design,* John Wiley & Sons, New York.

Levin, P. H. (1964), "Use of Graphs to Decide Optimum Layout of Buildings," *Architects Journal,* 140, 809–815.

Lewis, W. P., and T. E. Block (1980), "On the Application of Computer Aids to Plant Layout," *International Journal of Production Research,* 18(1), 11–20.

Liggett, Robin S. (1981), "The Quadratic Assignment Problem: An Experimental Evaluation of Solution Strategies," *Managment Science,* 27(4), 442–458.

Love, Robert F., James G. Morris, and George O. Weslolwsky (1988), *Facilities Location: Models & Methods,* North-Holland, New York.

Minieka, Edward (1981), "A Polynomial Time Algorithm for Finding the Absolute Center of a Network," *Networks,* 11, 351–355.

Mirchandani, Pitu B., and Richard L. Francis (eds). (1990), *Discrete Location Theory,* John Wiley & Sons, New York.

Montreuil, Benoit (1988), "From Gross to Net Layouts," Working Paper, School of Industrial Engineering, Purdue University, West Lafayette, IN.

Montreuil, Benoit, H. Donald Ratliff, and Marc Goetschalckx (1987), "Matching Based Interactive Facility Layout," *IIE Transactions*, 19(3), 271–279.

Moore, James M. (1974), "Computer Aided Facilities Design: An International Survey," *International Journal of Production Research*, 12(1), 21–44.

Muther, Richard (1973), *Systematic Layout Planning*, Cahners Books, Boston.

Nozari, Ardavan, and E. Emory Enscore, Jr. (1981), "Computerized Facility Layout with Graph Theory," *Computers & Industrial Engineering*, 5(3), 183–193.

O'Brien, C., and S. E. Z. Abdel Barr (1980), "An Interactive Approach to Computer Aided Facility Layout," *International Journal of Production Research*, 18(2), 201–211.

Picone, C. J., and W. E. Wilhelm (1984), "A Perturbation Scheme to Improve Hillier's Solution to the Facilities Location Problem," *Management Science*, 30(10), 1238–1249.

Roth, J., R. Hashimshony, and A. Wachman (1982), "Turning a Graph into a Rectangular Floor Plan," *Building and Environment*, 17(3), 163–173.

Roth, J., R. Hashimshony, and A. Wachman (1985), "Generating Layouts with Non-Convex Envelopes," *Building and Environment*, 20(4), 211–219.

Tompkins, James A., and James M. Moore (1977), *Computer Aided Layout: A User's Guide*, Facilities Planning and Design Division, Institute of Industrial Engineers, Atlanta, GA.

Van Roy, T. J., and D. Erlenkotter (1982), "A Dual-Based Procedure for Dynamic Facility Location," *Management Science*, 28(10), 1091–1105.

Vollmann, T. E., C. E. Nugent, and R. L. Zartler (1968), "A Computerized Model for Office Layout," *Journal of Industrial Engineering*, 19(7), 321–329.

Wilhelm, Mickey R., and Thomas L. Ward (1987), "Solving Quadratic Assignment Problems by Simulated Annealing," *IIE Transactions*, 19(1), 107–119.

PROBLEMS

7.1. Describe the steps of the systematic layout planning procedure.

7.2. List the objectives that you believe should be considered in determining the layout of a manufacturing plant.

7.3. Is material handling movement cost a reasonable objective for determining the best facility layout? Why or why not?

7.4. A new facility consisting of 10 departments is to be designed. Data have been collected as indicated in Figure 7.28.

    a. Construct an initial layout ignoring space considerations.

    b. Improve your layout using steepest ascent pairwise exchanges.

    c. Integrate space relationships to obtain a block layout.

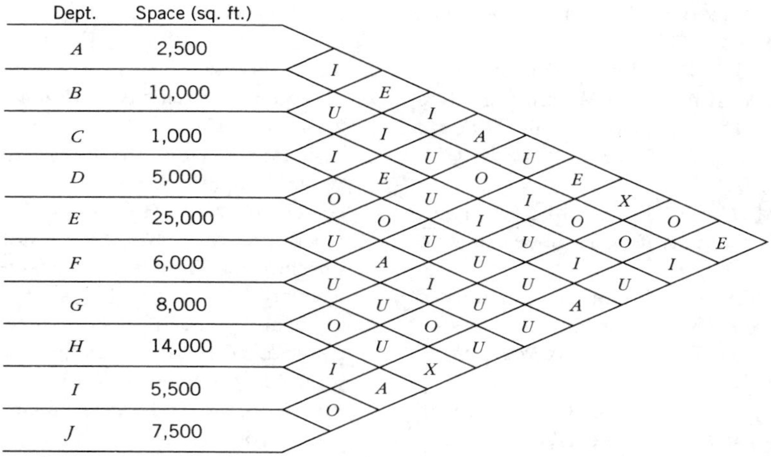

Figure 7.28   REL chart data for Problem 7.4.

7.5. Using the flow data in Table 7.7 select a random starting solution and apply the VNZ heuristic. Assume that all departments have equal size and will be constructed as squares in a 2 × 3 grid.

Table 7.7   Example Flow Data

From-To	A	B	C	D	E	F
A	—	10	12	0	40	0
B	34	—	59	0	0	12
C	0	10	—	25	0	0
D	0	23	12	—	21	0
E	9	14	0	0	—	32
F	0	27	14	0	10	—

7.6. Measure the flow dominance for the data in Table 7.7. Is this an easy or hard problem to solve?

7.7. List several environments in which rectilinear distance would be appropriate. Describe a situation for which euclidean distance would be appropriate.

7.8. The problems in this chapter minimize total or average distance. Describe a situation for which it might be more appropriate to minimize maximum distance between any two departments.

7.9. Using the flow data in Table 7.7, find the optimal layout via branch and bound.

7.10. Using the flow data in Table 7.7 and the added information that departments 3 and 5 must be separated, create a REL chart. Solve the problem all the way to a block layout using the graph theoretic methods of Section 7.4.

7.11. Suppose fixed costs exist to locate a department in a specific location and these costs vary with location. For instance, extra ventilation may be required, which makes location on an exterior wall cheaper, or certain locations may already have

required utilities available and, hence, will save the need to run pipe. Derive the required modifications to equations 7.11 and 7.13 for total cost and change in total cost on exchanging facilities in this environment. Let f_{ik} be the fixed cost of assigning department i to location k.

7.12. Suppose linear costs c_{ik} for assigning department i to location k exist in addition to quadratic flow costs in the quadratic assignment problem formulation of facility layout. Show how the linear cost can be absorbed directly into the quadratic costs by proper redefinition of c_{ijkl}. Show that any feasible solution yields the correct sum of linear and quadratic costs.

7.13. Consider the flow data in Table 7.8. All departments are similar in size. By using a 4 × 2 grid propose an initial layout. Apply the VNZ heuristic to find a better layout.

7.14. The four major areas in a plant are Receiving, Fabrication, Assembly, and Test. Fabrication and Assembly are twice as large as the other areas. The facility is 300 ft long and 200 ft wide. Using the flow data in Table 7.9, solve for the optimal block layout.

Table 7.9   Interdepartmental Flows for Problem 7.14

From-To	Rec	Fab	Assembly	Test
Rec	—	120	240	10
Fab	0	—	5	20
Assembly	0	250	—	30
Test	5	20	15	20

7.15. Use the REL chart of Figure 7.29 (see page 256) and find the Total Closeness Ratings for all departments. Let the values for 81, 27, 9, 3, 1, and −81 for A, E, I, O, U, and X ratings.

Table 7.8   Flow Data for Problem 7.13

From-To	A	B	C	D	E	F	G	H
A	—	0	10	35	0	50	12	0
B	0	—	24	12	0	0	8	10
C	0	0	—	45	10	0	0	45
D	0	21	0	—	0	0	32	24
E	12	0	0	24	—	0	12	34
F	0	16	46	0	0	—	12	0
G	40	21	0	0	0	21	—	0
H	21	45	19	0	0	0	32	—

7.16. Using the REL chart in Figure 7.29, construct a planar graph layout representation using column sum insertion. Find the dual of this graph and form it into a block layout.

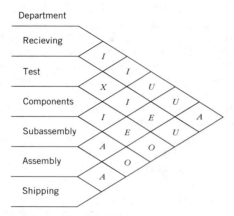

Figure 7.29   An REL chart for Problem 7.15.

7.17. Resolve Problem 7.16, first trying to improve your constructed graph by edge replacement and vertex relocation.

7.18. A new product line is to be manufactured in an existing 200 × 300 ft facility. One main aisle will run the length of the facility. The aisle will be 12 ft wide with 94 ft of production area on either side. Departments A, B, and C will occupy the northern portion of the plant with B in the center. Departments D and E will occupy the southern portion. All interdepartmental flow will be along the main aisle. Using the constraints in Table 7.10, first formulate a linear program for net space allocation. Second, solve the linear program.

Table 7.10   Constraints for Problem 7.18

Department	X Length (Min, Max)	Perimeter (Min, Max)
A	(100, 150)	(320, 450)
B	(50, 80)	(280, 320)
C	(80, 100)	(340, 360)
D	(140, 180)	(450, 500)
E	(120, 150)	(400, 500)

7.19. Describe the importance of flexibility and modularity in facilities layout design.

7.20. The data for a four-department layout problem is given in Table 7.11. The table contains fixed location costs, flow volumes, and distances between locations.

    a. Setup and solve the assignment problem to find an initial lower bound on total location and flow cost.

    b. Suppose we are given the partial assignment $\mathbf{a} = (3, 1)$. Find the lower bound for this node in the branch and bound tree.

Table 7.11a   Fixed Location Costs for Problem 7.20

	Location			
Dept.	1	2	3	4
1	10	5	5	10
2	20	20	20	20
3	15	10	10	10
4	10	10	5	15

Table 7.11b   Interdepartmental Flow Volumes

From-To	1	2	3	4
1	0	10	20	5
2	10	0	15	20
3	20	15	0	10
4	5	10	20	0

Table 7.11c   Interlocation Distances

From-To	1	2	3	4
1	—	1	2	1
2	1	—	1	2
3	2	1	—	1
4	1	2	1	—

7.21. Consider the REL chart of Figure 7.30.

    a. Find an upper bound on REL score for any maximally planar weighted graph solution.

    b. Find a complete initial solution using column-sum insertion.

c. Attempt to improve the solution using vertex relocation.

d. For the solution found in part c, take the dual and construct a block layout.

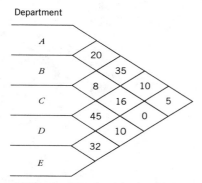

Department

Figure 7.30   An REL chart for Problem 7.21.

7.22. A new machine is to be added to a manufacturing system. Ten loads of parts per day will be received at the new machine from location (10,5). Five loads per day will arrive from (1,5). On completion at the new machine, each load of parts will be sent to a workcenter located at (3,6). Where should the new machine be placed if flow is along horizontal and vertical aisles?

7.23. Machines are located at (3,8), (6,5), (2,9), and (12,10). A conveyor is to be built from (0,4) to $(x,4)$. The conveyor will cost $1000 + 10$ per unit length per period for depreciation and operation. Material handlers will carry five loads per day at a cost of one dollar per unit distance to conveyor per trip from each machine to the nearest point on the conveyor. How long should the conveyor be; that is, find the optimal x.

7.24. Existing machines one through five are at coordinates (100, 40), (50, 50), (20, 65), (40, 12), and (16, 43), respectively. Two new machines will be added. The first new machine will receive one load of material per day from the first three machines. Six loads per day will then be sent from the first new machine to the second. The second machine will then send one load per day to existing machines one, two, and three, and five loads per day to existing machines four and five.

a. Write a linear programming formulation for this multimachine location problem.

b. Solve the linear programming problem.

c. Suppose demand and, hence, all flow volumes double. How would this affect the solution?

d. Suppose only the number of loads from existing machines one, two, and three to new machine one double. How would this affect the optimal solution?

7.25. Suppose several new machines are to be added to an existing layout. It is assumed that enough free space exists to add the machines wherever desired. Show that the overall objective function can be separated into independent, single-machine problems if the new machines do not interact, that is, if $v_{jk} = 0$.

7.26. It is known from Section 7.6.1 that any newly placed facility must be at a median location with respect to other facilities with which it interacts. Use this fact to develop a heuristic for sequentially placing new facilities in an existing facility. Does your procedure assure an optimal solution? (*Hint:* Be careful: Must all solutions that place new machines at a median with respect to other machines be optimal? Might we find a better solution by moving several at the same time?)

7.27. Two new facilities are to be located with respect to three existing facilities. Existing locations are $P_1 = (8, 15)$, $P_2 = (10, 20)$, and $P_3 = (30, 10)$. Interaction cost parameters are $v_{12} = 8$ and

$$w = \begin{pmatrix} 6 & 3 & 5 \\ 0 & 7 & 2 \end{pmatrix}$$

Assuming rectilinear distance, write out the linear program for finding the optimal Y coordinates.

PART III

SUPPORTING COMPONENTS

In addition to the overall flow of materials, the manufacturing system must plan for the productive activities within workstations, precise transportation methods between workstations, and the storage of component parts and finished products. Each of these has its own peculiar problems and solution methodologies.

Chapter 8 provides an introduction to modeling techniques that are useful for sequencing activities within a production cell. Classic operations research models, such as the assignment problem and traveling salesman problem, have direct applications in many production environments. These standard problems may also be combined and enhanced to accurately model more complex workcenter sequencing problems. A selection of problems from mechanical and electrical industries are considered to indicate the almost universal applicability of these techniques.

Chapter 9 addresses material handling problems. Conservation requires that all of the entering material eventually leaves the system. Although planning for all the possible contingencies may be tedious, it is randomness that most complicates the planning process. Products and transport requests may arrive in bulk and at random intervals. The planning of an Automated Guided Vehicle transport system is used to illustrate the process of breaking a large problem into its constituent parts and the modeling of each of these separately in a manner that facilitates aggregation.

Although value is not added by having product sit idle, warehousing is still a necessary activity in many situations to guard against breakdowns, demand variability, and transport costs and delays. Chapter 10 describes the various alternatives for storing items within a facility and presents models for optimal use of space and storage/retrieval machines.

CHAPTER 8

MACHINE SETUP AND OPERATION SEQUENCING

Excellence is never granted to man
but as the reward for labour.
—Sir Joshua Reynolds

8.1
INTRODUCTION

The three previous chapters dealt with major system design strategies. We now narrow our scope and discuss the activities that occur within an individual machine cell or workstation. We are interested in finding the most efficient operating mode for the cell. In particular, we address the problem of tooling the machine and then sequencing production activities. We assume that a set of jobs is waiting to be produced. A job is a batch of one or more similar parts. Our objective entails maximizing productivity by minimizing the time required to produce the job set. Several opportunities may exist. It may be possible to sequence batches to minimize tooling changeovers. We may be able to sequence activities within the cell to minimize idle time of material handling devices and machines. Alternatively, productivity may be improved by optimal layout of the workcell, placing part feeders and tooling to minimize assembly time. In Chapter 4 we discussed scheduling and sequencing of jobs assuming processing times were known. In this chapter we examine methods for minimizing the processing time required by jobs. Thus, this chapter bridges the gap between manufacturing system modeling and the traditional methods analysis techniques of industrial engineering.

We could not possibly describe all of the sequencing problems encountered in manufacturing cell planning. Instead, we include several typical problems along with solution approaches that have been found to be useful for many sequencing problems in manufacturing. We will learn that many of these problems fit into the form of classic optimization problems. Recognizing these common mathematical forms can lead to the quick development of solution methodologies for newly encountered problems.

The specific details of the manufacturing sequence of an individual item may seem less important than the overall system design. Adopting such an attitude can have serious consequences in a competitive environment. Despite the benefits derived from a proper system design, it is difficult to stay in business if variable production cost exceeds the market price. A well-designed system provides the opportunity for reduced production cost. Efficient operation of the system results in the realization of this opportunity. Note that when a salesman increases sales, the net return to the organization is several cents on the dollar. The remainder goes to material and production costs. In manufacturing, a penny saved is truly a penny earned.

It seems appropriate at this point to reiterate the difference between design and operation problems. When selecting a system design, we can afford to use large models that require considerable computational resources for solution. Having to wait overnight for the computer to crunch through an algorithm is not a problem; in fact, we would still be allowed to perform reasonable sensitivity analysis. Actual implementation of the solution may be months away. On the other hand, sequencing and material handling control decisions are often required on a real-time basis. Many such decisions are made each day. This chapter falls between these extremes. We are interested in short-term planning problems. Normally, a problem must be solved for each part type or daily product mix. In the former case, solutions may be saved for use in later product runs. Implications of any one model output are relatively minor, not justifying extensive modeler time for adjustment and sensitivity analysis. However, in the long run the impact of short-term plans adds up, justifying the implementation of simple but effective automated decision models.

8.2
TASK ASSIGNMENT

The basic worker–task assignment problem is one of the fundamental problems of operations research, fitting the framework of the Linear Assignment Problem (LAP). Strictly speaking, this problem does not fall into the category of machine setup and operation sequencing; however, the LAP is an essential building block of the procedures to be described later in this chapter. Moreover, the LAP has applications of its own in manufacturing planning.

Suppose we have N tasks to be performed and each task requires a worker. Worker capabilities differ as do task requirements. The value of assigning a particular worker to a particular task depends on the worker and task combination. Problem terminology is flexible. For instance, tasks may correspond to machines, workcenters, tools, or jobs. Workers may instead be machines or spatial locations. The key to our model is that we must mate objects from two sets. Objects cannot be used twice, and costs depend solely on the matched pairs selected.

We will assume that the number of tasks (N) and workers (M) are equal. Otherwise, dummy workers are added with their assigned tasks being left undone or dummy tasks are added with their assigned workers being idle. Accordingly, we let N represent the number of tasks and workers. Define c_{ij} as the cost (or time) if worker i is assigned to task j. The c_{ij} reflect the matchup between worker skill levels (or machine capability) and task requirements. We can summarize the assignment costs in a matrix C, with one row for each worker and one column for each task. c_{ij} is a lateness penalty for task j if i is a dummy worker. c_{ij} is idle worker cost if j is a dummy task. The basic LAP can be formulated as

$$\text{LAP}: \quad \text{minimize} \sum_{i=1}^{N} \sum_{j=1}^{N} c_{ij} x_{ij} \quad (8.1)$$

subject to

$$\sum_{i=1}^{N} x_{ij} = 1 \quad \text{for all } j \quad (8.2)$$

$$\sum_{j=1}^{N} x_{ij} = 1 \qquad \text{for all } i \tag{8.3}$$

$$x_{ij} = 0 \quad \text{or} \quad 1$$

Objective 8.1 sums the cost to perform each task. Each task has a constraint of the form of constraint 8.2. The constraint assigns exactly one worker to the task. Constraints 8.3 ensure that each worker receives exactly one task assignment.

LAP is one of the "nice" integer programs. It is a special case of the more general minimum cost network flow problem. Even if we relax (remove) the integer restrictions and solve the resulting problem, the solution turns out to be an integer anyway. However, there are even easier approaches to solving the LAP and general transportation problems. We briefly review the Hungarian algorithm presented in many operations research texts. Ahuja et al. [1989] review several other solution approaches.

The algorithm makes use of two basic facts.

Fact 1. If we add any constant to every element in a row or column of C, the optimal solution does not change and the value of the optimal solution changes by an amount equal to the added constant.

Fact 2. If all $c_{ij} \geq 0$, then any feasible solution of cost 0 must be optimal.

To understand why these facts are true, think of the cells of C as corresponding to the decision variables x_{ij}. Constraint sets 8.2 and 8.3 state that all feasible solutions must select exactly one cell from each row and column. Thus, adding a constant to a row or column must change any feasible solution by the same amount. The second fact follows from the objective 8.1 and the nonnegativity of the x_{ij}.

The algorithm works by adding and subtracting constants from rows and columns so as to maintain a nonnegative cost matrix. As soon as we can find a feasible solution using only 0 cost cells, we must be optimal.

Hungarian Algorithm

STEP 1. Cost reduction. Subtract the minimum element in each row from all row entries. Subtract the minimum column element in each column from all column entries. The resulting matrix contains the reduced costs.

STEP 2. Attempt to find a feasible solution using only 0 entries in the reduced cost matrix. If this succeeds, stop; this solution is optimal. If all 0 elements can be covered with less than N horizontal and vertical lines, go to 3.

STEP 3. Further reduction. Find the minimum uncovered element. Subtract this reduced cost from each uncovered element and add it to each twice-covered element. Go to 2.

Although step 3 seems more complex than step 1, in actuality the procedure is simply adding constants to certain rows and columns without letting any elements become negative.

EXAMPLE 8.1

A production foreman has five workers and five machines. Workers differ in experience and machines differ in ease and consistency of operation. Potential assignments of workers to machines are illustrated in Figure 8.1. The foreman has estimated the profits per day, shown in Table 8.1, from the assignment of workers to machines. These values might be derived from estimates of jobs completed per day.

<table>
<thead>
<tr><th></th><th colspan="5">Machine</th></tr>
<tr><th>Worker</th><th>A</th><th>B</th><th>C</th><th>D</th><th>E</th></tr>
</thead>
<tbody>
<tr><td>1</td><td>20</td><td>15</td><td>40</td><td>8</td><td>0</td></tr>
<tr><td>2</td><td>20</td><td>15</td><td>35</td><td>10</td><td>−5</td></tr>
<tr><td>3</td><td>10</td><td>10</td><td>30</td><td>5</td><td>5</td></tr>
<tr><td>4</td><td>10</td><td>10</td><td>25</td><td>3</td><td>0</td></tr>
<tr><td>5</td><td>5</td><td>10</td><td>25</td><td>3</td><td>−2</td></tr>
</tbody>
</table>

Table 8.1   Profits per Day

How should workers be assigned?[1]

Solution

The first step is to convert profits into a cost matrix for minimization. This is easily done by taking the negative of profit. An outline of the formulation is

$$\text{minimize} \; -20x_{1A} - 15x_{1B} + \cdots + 2x_{5E}$$

subject to

$$\begin{bmatrix} 1 & 0 & 0 & 0 & 0 \\ 0 & 1 & 0 & 0 & 0 \\ 0 & 0 & 1 & 0 & 0 \\ 0 & 0 & 0 & 1 & 0 \\ 0 & 0 & 0 & 0 & 1 \\ I & I & I & I & I \end{bmatrix} \begin{bmatrix} x_{1A} \\ . \\ . \\ . \\ . \\ x_{5E} \end{bmatrix} = \begin{bmatrix} 1 \\ . \\ . \\ . \\ . \\ 1 \end{bmatrix} \tag{8.4}$$

$\underline{1}$ denotes row vector $(1, 1, 1, 1, 1)$ and I is the 5×5 identity matrix. The special structure shown in the constraint matrix of equation 8.4 makes the assignment problem easy. The matrix is **unimodular:** every square submatrix has a determinant of $1, 0,$ or -1. This property holds for affiliated problems such as transportation and network flow and is useful for developing solution algorithms. Unimodularity ensures that basic feasible solutions will be integer. Thus, we can apply procedures like the simplex algorithm without worrying about the integer restrictions.

To convert the problem into minimization format, multiply all profit coefficients by -1. This produces the matrix

$$\begin{bmatrix} -20 & -15 & -40 & -8 & 0 \\ -20 & -15 & -35 & -10 & 5 \\ -10 & -10 & -30 & -5 & -5 \\ -10 & -10 & -25 & -3 & 0 \\ -5 & -10 & -25 & -3 & 2 \end{bmatrix} \begin{matrix} -40 \\ -35 \\ -30 \\ -25 \\ -25 \end{matrix}$$

---

[1]We will ignore the advantages of cross-training and seek to maximize profit for this period. After all, the poor foreman is judged on his department's short-term performance. Cross-training is desirable, but someone farther up the chain of command would have to take responsibility for such a policy and implement an appropriate evaluation scheme.

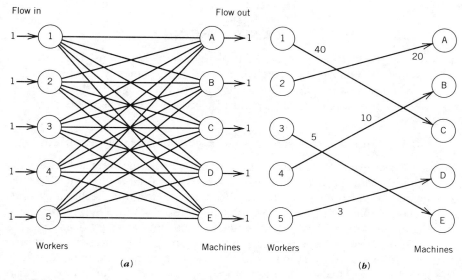

Figure 8.1 A network representation of the assignment problem.

The minimum value is shown to the right of each row. These values are subtracted from each element in the row, producing

$$
\begin{bmatrix}
20 & 25 & 0 & 32 & 40 \\
15 & 20 & 0 & 25 & 40 \\
20 & 20 & 0 & 25 & 25 \\
15 & 15 & 0 & 22 & 25 \\
20 & 15 & 0 & 22 & 27
\end{bmatrix}
$$

The new matrix has minimum column values of $(15, 15, 0, 22, 25)$. These values are subtracted from each column. The resulting matrix is

$$
\begin{bmatrix}
5 & 10 & 0 & 10 & 15 \\
0 & 5 & 0 & 3 & 15 \\
5 & 5 & 0 & 3 & 0 \\
0 & 0 & 0 & 0 & 0 \\
5 & 0 & 0 & 0 & 2
\end{bmatrix}
$$

Using the 0 elements, we can form the feasible solution that assigns workers $(1, 2, 3, 4, 5)$ to machines (C, A, E, B, D), respectively. This solution has a daily profit of \$78. Note that summing the column and row reduction coefficients produces the cost $C = -40 - 35 + \cdots + 25 = -78$ in agreement with the computed profit. The optimal solution is depicted in Figure 8.1b. ∎

Now, is it reasonable to expect foremen to be solving assignment problems? Probably not by hand. However, it would not be unreasonable to have the foreman interactively input suitable parameter estimates to a user-friendly PC package that could provide scheduling advice. Even without such software support, if this discussion makes the reader recognize the interaction between scheduling decisions for resources with overlapping capabilities, it has served its primary purpose. Moreover, the LAP provides a good introduction to the problems encountered in the remainder of this chapter.

8.3
TASK SEQUENCING

In this section we look at two types of task-sequencing problems. The first class of problems analyzes changeover costs when costs are completely determined by the current job setup on a process and the next job to be loaded. In the second set of problems the entire sequence of jobs must be known to evaluate costs.

8.3.1 Complete Changeovers—Traveling Salesman Problem

Suppose we are given N jobs to be performed on one machine. Total processing time is fixed by unit processing time and batch size. Total setup time depends on the sequence in which jobs will be run. For instance, in group technology, parts in the same family may require little or no changeover. Parts in different families may require complete breakdown and rebuilding of machine tooling. Changing sizes in a packaging line may be relatively easy, but changing package contents may require extensive cleaning and switching of machinery. Common to both these examples is the fact that individual changeover times depend only on the current and next product. For bottleneck facilities, maximizing the production of salable product is the prime objective. This situation corresponds to the Traveling Salesman Problem (TSP). In the TSP, a salesman must visit each city in his territory and then return home. In our case, the worker must perform each job and then return to the starting condition. The problem can be visualized on a graph. Each city (job) becomes a node. Arc lengths correspond to the distance between the attached cities (job changeover times). The salesman wants to find the shortest tour of the graph. A tour is a complete cycle. Starting at a home city, each city must be visited exactly one time before returning home. Each leg of the tour travels on an arc between two cities. The length of the tour is the sum of the lengths of the arcs selected. Figure 8.2 illustrates a five-city TSP. Trip lengths are shown on the arcs in Figure 8.2a, the distance from city i to j is denoted c_{ij}. We have assumed in the figure that all roads (arcs) are bidirectional. If arc lengths differ depending on the direction of the arc, the TSP is said to be asymmetric. A possible tour is shown in Figure 8.2b. The cost of this tour is $c_{12} + c_{24} + c_{43} + c_{35} + c_{51}$.

Several mathematical formulations exist for the TSP. One approach is to let x_{ij} be 1 if city j is visited immediately after i and 0 otherwise. A formal statement is then

$$\text{minimize} \sum_{i=1}^{N} \sum_{j=1}^{N} c_{ij} x_{ij} \tag{8.5}$$

subject to

$$\sum_{j=1}^{N} x_{ij} = 1 \qquad \text{for all } i \tag{8.6}$$

$$\sum_{i=1}^{N} x_{ij} = 1 \qquad \text{for all } j \tag{8.7}$$

$$\textit{No subtours} \tag{8.8}$$

$$x_{ij} = 0 \quad \text{or} \quad 1$$

The objective function accumulates time as we go from city i to j. Constraints 8.6 ensure that we leave each city. Constraints 8.7 ensure that we visit (enter) each

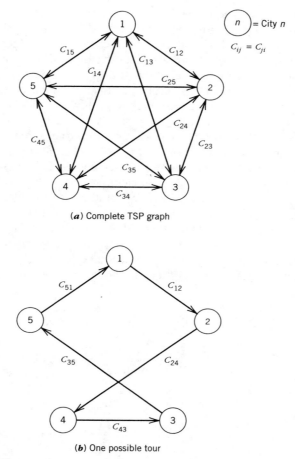

(*a*) Complete TSP graph

(*b*) One possible tour

Figure 8.2   TSP illustrated on a graph.

city. A subtour occurs when we return to a city prior to visiting all other cities. Without restriction 8.8, our formulation looks like an LAP. Unfortunately, the subtour elimination constraints significantly complicate model solution.

The obvious question now is how do you solve a TSP. The TSP belongs to a class of hard problems that become increasingly difficult to solve as the problem size N grows. For large problems, optimal solutions are difficult to obtain. Heuristics are commonly used. The assignment problem formed by relaxing (temporarily eliminating) constraints 8.8 can be solved to find a lower bound on the TSP, and thus provide a weak measure of how close we are to optimality. Small to medium and even large (1000-city) problems with special structures can be solved optimally by using branch and bound or other optimization techniques. Some of the TSP applications in this chapter are unfortunately messy, however. Therefore, we consider heuristic procedures. Heuristics exist that can usually find "good" solutions with a number of computations proportional to N^2 [i.e., $O(N^2)$]. Good here means within about 5 percent of optimal. With a little more effort, solutions can usually be found within about 1 percent of optimality (see Lin and Kernighan [1973] and Lawler et al. [1985]).

One reasonable construction procedure is the **closest insertion algorithm**. The closest insertion algorithm starts by selecting any city. We then proceed through $N-1$ stages, adding a new city to the sequence at each stage. Thus a partial sequence is always maintained, and the sequence grows by one city each stage. At each stage we select the city from those currently unassigned that is closest to any city in the partial sequence. We add the city to the location that causes the smallest increase in the tour length. The closest insertion algorithm can be shown to produce a solution with a cost no worse than twice the optimum when the cost matrix is symmetric and satisfies the triangle inequality (see Reingold et al. [1977], p. 144). Symmetric implies $c_{ij} = c_{ji}$ where c_{ij} is the cost to go from city i directly to city j. Unfortunately, symmetry need not exist in our changeover problem. Normally, the triangular inequality ($c_{ij} \leq c_{ik} + c_{kj}$) will be satisfied, but this alone does not suffice to ensure the construction of a good solution, so be careful. We may also try repeated application of the algorithm choosing a different starting city each time and then choose the best sequence found. Of course, this increases our workload by a factor of N.

We now state the algorithm formally. Let S_a be the set of available (unassigned) cities at any stage. S_p will be the partial sequence in existence at any stage and is denoted $S_p = \{s_1, s_2, \ldots, s_n\}$, implying that city s_2 immediately follows s_1. For each unassigned city j, we use $c(j)$ to keep track of the city in the partial sequence that is closest to j. [We store $c(j)$ only to avoid repeating calculations at each stage.] Last, bracketed subscripts $[i]$ refer to the ith city in the current partial sequence.

Closest Insertion Algorithm

STEP 0. Initialize.

$$n = 1.\ S_p = \{1\}.\ S_a = \{2, \ldots, N\}.\ \text{For } j = 2, \ldots, N\, c(j) = 1.$$

STEP 1. Select new city. Find $j^* = \underset{j \in S_a}{\text{argmin}}\{c_{j,c(j)}, \text{ or } c_{c(j),j}\}.$ Set $n = n + 1$.

STEP 2. Insert j^*, update $c(j)$. $S_a = S_a - j^*$. Find city $i^* \in S_p$ such that $i^* = \text{argmin}_{[i] \in S_p}\{c_{[i]j^*} + c_{j^*,[i+1]} - c_{[i][i+1]}\}$. Update $S_p = \{s_1, \ldots, i^*, j^*, i^*+1, \ldots, s_n\}$. For all $j \in S_a$, if $\min\{c_{j,j^*}, c_{j^*,j}\} < c_{j,c(j)}$ then $c(j) = j^*$. If $n < N$, go to 2.

EXAMPLE 8.2

A machine is finishing batch 1098A. Four other batches must be completed during this period. Table 8.2 contains changeover times. Use the closest insertion routing to find a job sequence.

Table 8.2   Changeover Times

From/To	1098A	1102A	321B	310B	316B
1098A	—	1.1	2.8	2.4	2.6
1102A	0.7	—	2.9	2.5	2.7
321B	2.6	3.1	—	0.4	0.6
310B	2.6	3.1	0.8	—	0.6
316B	2.7	3.2	2.9	0.5	—

Solution

This problem actually calls for a TSP path rather than a tour. We need not return to the setup condition for 1098A after completing the five jobs. In fact, the changeover times *to* 1098A are irrelevant, since we are already there. To handle this aspect of the problem, we may simply replace the costs in column one of Table 8.2 with 0. The modeled cost is 0 because in practice we do not return. When picking a city to enter [evaluating $c(j)$], we will ignore the 0 costs of returning to city 1. Otherwise, all cities would be equally desirable for entry to the schedule at each stage, since each could be added last and be associated with this 0 arc cost.

For simplicity, renumber jobs 1098A through 316B as jobs 1 through 5, respectively. Now, let's do closest insertion.

STEP 0.   Start with job 1. $n = 1$. $S_p = \{1\}$. $S_a = \{2, 3, 4, 5\}$.
$c(2) = c(3) = c(4) = c(5) = 1$, each city is closest to city 1.

STEP 1.   Select new city.
For city 2, $c(2) = 1$; the associated cost is $c_{12} = 1.1$. (Remember to ignore c_{21} for the path problem.)
For city 3, $c(3) = 1$; the associated cost is $c_{13} = 2.8$.
For city 4, $c(4) = 1$; the associated cost is $c_{14} = 2.4$.
For city 5, $c(5) = 1$; the associated cost is $c_{15} = 2.6$.
City 2 has the minimum associated cost and we set $j^* = 2$. Set $n = 2$.

STEP 2.   Insert city 2, and update $c(3)$, $c(4)$, $c(5)$.
$S_a = \{3, 4, 5\}$. We place city 2 after city 1. The incremental cost is 2 after 1: $c_{12} + c_{21} - c_{11} = 1.1 + 0 - 0 = 1.1$. The new partial tour is shown in Figure 8.3a.

Now we check to see if any member of S_a would prefer to be linked to our new entry, job 2. However, in each case $c_{1,j} < c_{j,2}$ and $c_{1,j} < c_{2,j}$. Thus, no changes are made.

STEP 1.   Select new city.
The $\min_{j \in S_a}\{c_{j,c(j)}, c_{c(j),j}\} = \min\{c_{1,3}, c_{1,4}, c_{1,5}\}$ occurs at $c_{1,4} = 2.4$. (Recall that arcs returning to job 1 are excluded in this path problem.) Thus, $j^* = 4$. Set $n = 3$.

STEP 2.   Insert job 4.
Updating, $S_a = \{3, 5\}$. We can enter job 4 after job 1 or 2. Incremental costs are
4 after 1: $c_{14} + c_{42} - c_{12} = 2.4 + 3.1 - 1.1 = 4.4$
4 after 2: $c_{24} + c_{41} - c_{21} = 2.5 + 0 - 0 = 2.5$
The preferable choice is 4 after 2, as shown in Figure 8.3b. Updating $c(j)$ for remaining unassigned jobs 3 and 5 we find
Job 3: $\min\{c_{13}, c_{34}, c_{43}\} = c_{34} = 0.4$, thus $c(3) = 4$.
Job 5: $\min\{c_{15}, c_{45}, c_{54}\} = c_{54} = 0.5$, thus $c(5) = 4$.

STEP 1.   Select new job.
Since $c_{3,4} < c_{5,4}$ job 3 wins out and $j^* = 3$. Set $n = 4$.

STEP 2.   Insert job 3.
Updating, $S_a = \{5\}$. Our insertion choices and incremental costs are
3 after 1: $c_{13} + c_{32} - c_{12} = 2.8 + 3.1 - 1.1 = 4.8$;
3 after 2: $c_{23} + c_{34} - c_{24} = 2.9 + 0.4 - 2.5 = 0.8$;
3 after 4: $c_{43} + c_{31} - c_{41} = 0.8 + 0 - 0 = 0.8$.

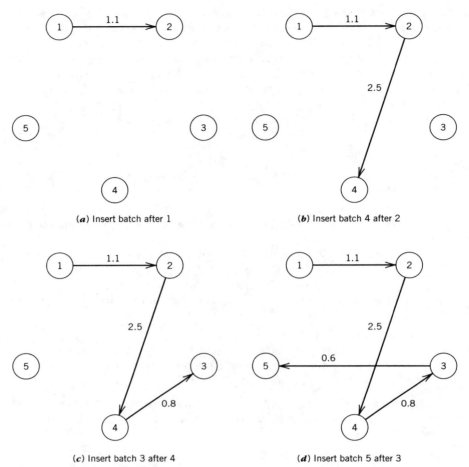

(*a*) Insert batch after 1   (*b*) Insert batch 4 after 2

(*c*) Insert batch 3 after 4   (*d*) Insert batch 5 after 3

Figure 8.3   Tour construction.

We will break the tie arbitrarily by placing 3 after 4. Updating $c(5)$, we have

Job 5: $\min\{c_{54}, c_{35}, c_{53}\} = c_{54} = 0.5$ and $c(5)$ is unchanged.

STEP 1.   Select new job.
Only job 5 remains, $j^* = 5$.

STEP 2.   Insert job 5.
Updating, $S_a = \{\}$. Insertion choices are
5 after 1: $c_{15} + c_{52} - c_{12} = 4.7$;
5 after 2: $c_{25} + c_{54} - c_{24} = 0.7$;
5 after 4: $c_{45} + c_{53} - c_{43} = 0.7$; and
5 after 3: $c_{35} + c_{51} - c_{31} = 0.6$.
Insert job 5 after 3. The final sequence is $\{1, 2, 4, 3, 5\}$ with cost
$C = c_{12} + c_{24} + c_{43} + c_{35} = 5.0$.

STEP 1. S_a is empty. We set $n = 5$ and stop.   ■

Entire books have been written on the TSP. Indeed it is one of the most studied operations research problems. Problems with hundreds, even a thousand, cities are now routinely solved. We have just touched on this subject with our Closest Insertion

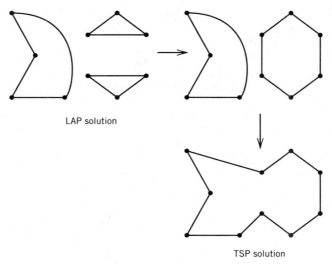

Figure 8.4   Subtour integration.

construction heuristic. A number of other construction heuristics have been proposed that use the concept of a **minimum spanning tree.** A minimum spanning tree is any set of $N - 1$ arcs that touch each node and have the smallest sum of costs for any such set. Minimum spanning trees are easy to find. If this set of arcs can be modified into a tour such that each city is touched by exactly two arcs in a connected tree, then we have a TSP solution. We could also start with the solution of the assignment problem and try to connect two subtours at a time by switching arcs as shown in Figure 8.4. Once all LAP subtours are combined, we have a TSP solution. Improvement strategies start with a constructed tour and seek to make improvements. The most common approach is to improve by removing k edges from the tour and replacing them with k edges that maintain a tour but have lower cumulative cost. Lin and Kernighan [1973] describe an $O(N^{2.2})$ improvement heuristic for the symmetric TSP that performs very well, operating along the lines of the graph-partitioning heuristic described in Chapter 6. Modifications are possible for the asymmetric case (Kanellakis and Papadimitriou [1980]). Approaches such as branch and bound can be used to find optimal solutions. The interested reader should consult the end-of-chapter references. Our primary purpose here is to present the model.

8.3.2 Partial Changeovers

We now look at sequencing problems with more complex changeover cost structures. As an example, consider an NC machine with a tool magazine capable of holding M tools. N jobs are to be processed on this machine. Job j requires a specific set of tools A_j. Tools will normally be required by several jobs but the total number of tools required exceeds M. Hence, tool changes must occur. We assume that no job requires more than M tools. Figure 8.5 graphically depicts the situation. Jobs will be sequenced onto the machine. At each job completion, any tools required by the next job and not currently on the machine must be added. Certain unnecessary tools will be removed. Our objective is to order the jobs and tool changeovers to minimize the total number of tools changed on the machine. A tool change involves removing one tool and adding another. This problem differs from the previous section in that the entire sequence of changes is relevant. Instead of completely breaking down the

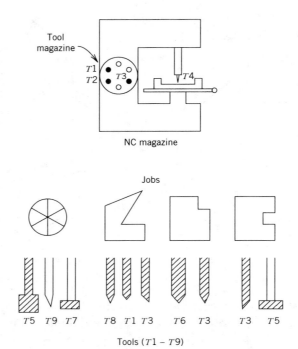

Figure 8.5   The NC tool machine problem.

setup each time, unneeded tool slots are left occupied. Knowing the current job on the machine only partially specifies the tooling setup.

Several results help simplify our problem. First, assuming that tools are not needed elsewhere, there is no advantage in ever having less than M tools on the machine. Second, assume an ordering for jobs on the machine is given. Tang and Denardo [1988] prove that for changing the tool magazine, the common-sense rule Keep Tool Needed Soonest (KTNS) is optimal. In KTNS we only remove as many tools as necessary to make way for the next job. The tools removed are those that will not be needed again until the longest time (most jobs) in the future. Intuitively, this avoids taking off tools that must soon be added back. Ties for future usage can be broken arbitrarily. Hence, given the job sequence we load the tools required for the first job. Remaining tool slots are used for additional tools needed for job 2, and so on until all slots are full. Jobs are manufactured and, when m new tools are needed, the m "longest until next use" tools are removed.

We still have the problem of how to order the jobs. We can simplify the problem slightly. Suppose job r uses only a subset of the tools of job s. Placing job r immediately after s cannot increase changeovers and must be optimal. Hence, initially we screen the set of tools required for each job. For all cases where A_r is a subset of A_s, we place r immediately after s. Each occurrence of the dominance relationship reduces the size of the problem by one job.

The problem of sequencing jobs here is similar to that of finding groups in GT. In fact, we can form a tool–job matrix with a 1 in row i and column j indicating that tool i is used for job j. By ordering the rows and columns according to their binary string, we obtain a job sequence that puts jobs with common tool requirements together. Alternatively, we can use the TSP procedure. The number of tools required for job k that are not required for job i gives a pessimistic estimate of the number of tools to be changed if k follows i. We say these estimates are pessimistic because, in reality, we may be fortunate and find some of these tools left over on the machine. In any event, we can use these estimates as costs for the TSP.

Our solution procedure involves three steps. In step 1, jobs are combined to reduce problem size. In step 2, jobs are ordered. Either the TSP or binary clustering approach could be used. Finally, step 3 performs tool setup planning by applying the KTNS rule. KTNS tells us which tool will be needed soonest if the proposed job order is followed. It is possible, however, that jobs exist lower in the ordered list that can be performed with the current tool setup. When planning the sequence of tool setups, we should keep in mind the dominance relation that there is nothing to be gained by postponing execution of a job if all its required tools are loaded. Thus, before removing any tools, we check the list of remaining jobs. Any job that can be performed with the tools currently on the machine is moved up in the order and run.

We demonstrate the general method with an example. We will first use binary clustering and then the TSP approach just to see if there is a difference in this case. Both approaches give intuitively appealing job orders, but neither is necessarily optimal when imbedded in the overall sequencing and tooling problem.

EXAMPLE 8.3

Consider the set of jobs and tool requirements shown in Table 8.3. The machine can hold $M = 4$ tools. The tool magazine is currently empty.

Table 8.3   NC Machine Job Set for Example

Tool	1	2	3	4	5	6	7	8	9	Row Value
a				1			1			36
b		1						1	1	131
c			1			1				72
d		1			1	1			1	153
e		1	1			1	1			216
f	1			1						288
g	1			1			1	1	1	295
h			1			1				72
Col. Value	256	128	64	32	16	8	4	2	1	

Solution

Binary Clustering Approach

STEP 1. Reduction. Visually, dominated jobs are easier to find looking at an ordered matrix. As such, we will skip this step until after clustering. The reader should, however, glance at Table 8.3 and note that job 4 dominates job 1 and 6 dominates 5. As we will learn, other dominance relations exist as well.

STEP 2. Job Ordering. Binary clustering (Section 6.4.2) is used to order jobs. Values for the initial row (tool) ordering are shown in Table 8.3. The new tool order is g, f, e, d, b, c, h, a. After two complete iterations of row and column ordering, the procedure converges to the ordering of Table 8.4. At this point row order has stabilized.

Table 8.4   Final Tool-Job Ordering for Example

Tool	Job									
	4	1	7	9	8	2	6	5	3	
g	1	1	1	1	1					496
f	1	1								384
a	1		1							320
b				1	1	1				54
d				1		1	1	1		46
e						1	1	1	1	15
c							1		1	5
h							1		1	5
	256	128	64	32	16	8	4	2	1	

In accord with our heuristic, job order is 4, 1, 7, 9, 8, 2, 6, 5, 3. Additional dominated sets are now obvious. Jobs 4, 1, and 7 are combined using tools g, f, a. Jobs 9 and 8 are combined and require tools g, b, d. Last, jobs 6, 5, and 3 are combined. (The careful reader will note that we could have placed job 5 after 2 instead.)

TSP Approach

STEP 2. Job Ordering. After reducing the problem we have the four-city TSP shown in Table 8.5. S/E refers to the start and ending position. We will not require tools to be removed on completion of the last job; thus, costs to return to the S/E position are 0.

Table 8.5   TSP Costs for Example 8.3

From Job(s)	To Job(s)				
	S/E	417	98	2	653
S/E	—	3	3	3	4
417	0	—	2	3	4
98	0	2	—	1	3
2	0	3	1	—	2
653	0	3	2	1	—

Costs from S/E to a job set are the number of tools used by that job set. Job set 417 uses tools g, f, and a, for instance. To see how other costs were determined consider $c_{417,98}$. Switching from jobs 417 to 98 requires potentially adding two tools, b and d. Tool g must already be loaded. Hence, $c_{417,98} = 2$. Other costs are computed in the same fashion. We will solve this TSP by using the closest insertion algorithm.

Initially, the tour consists of S/E. $S_a = \{417, 98, 2, 653\}$. Job set 417 is as close as any set to the start so that we will select it to be inserted. Possible tours are $S/E \rightarrow 417 \rightarrow S/E$ and $417 \rightarrow S/E \rightarrow 417$. Both have cost 3. We will select the first, since it agrees with our visualization of the tour. Updating, $S_a = \{98, 2, 653\}$.

Since 98 is closest ($c_{98,417} < \min\{c_{2,417}, c_{653,417}, c_{S/E,98}, c_{S/E,2}, c_{S/E,653}\}$), we insert job set 98. We must place this job set either before or after set 417. In either case the increase in tour cost is 2. Arbitrarily place 98 after 417.

The smallest cost coefficient from outside the partial tour to a job set in the tour (or vice versa) is $c_{2,98} = 1$. Thus, we insert job 2.

Job 2 can go after S/E, 417, or 98. Inserting 2 after S/E increases cost by $c_{S/E,2} + c_{2,417} - c_{S/E,417} = 3$. Inserting 2 after 417 yields incremental cost $c_{417,2} + c_{2,98} - c_{417,98} = 2$. Inserting 2 after 98 increases costs $c_{98,2} + c_{2,417} - c_{98,417} = 1 + 3 - 2 = 2$. We have two positions to choose from. We choose the partial sequence $\{S/E, 417, 98, 2\}$.

Only job set 653 remains. The smallest incremental cost is 2 to add 653 after 2. Placing 653 after S/E would increase cost by 4, placing 653 after 417 would have increased cost by 4, and placing it after 98 increases cost by 3. Our solution is thus $\{S/E, 417, 98, 2, 653, S/E\}$ with cost

$$C = c_{S/E,417} + c_{417,98} + c_{98,2} + c_{2,653} + c_{653,S/E} = 8$$

Our pessimistic cost estimate calls for eight tool changes. Note that the sequence is the same as that found by clustering. This will not happen in all problems.

STEP 3. Tool Setup Planning. The first four (M) tools used are loaded onto the machine. These are tools g, f, a, and b. Jobs 4, 1, and 7 are run with this setup. Job 9 requires tool d. Either tool f or a could be removed as neither will be used again. We arbitrarily select f. We can now make jobs 9 and 8. The next job requires tool e, which is added to replace g. After job 2, tools a and b are removed to make way for c and h. The sequence of setups can be summarized as follows:

Tools Added	Tools Removed	Tools on Machine	Jobs Run
g, f, a, b	—	g, f, a, b	4, 1, 7
d	f	g, a, b, d	9, 8
e	g	a, b, d, e	2
c, h	a, b	d, e, c, h	6, 5, 3

Since each tool must be loaded at least once, the loading of eight tools in total is a lower bound on setup for this problem. Our solution is clearly optimal, therefore, as no tool is ever reloaded. ∎

A problem very similar to our NC machine problem occurs in the assembly of printed circuit boards. Component insertion machines place electrical components into designated locations on boards. Insertion machines hold a limited number of component feeders. A feeder is basically a roll of components of the same type. The production sequence of boards must be determined and components must be allocated to feeders to minimize the number of changeovers of feeders. Baker [1988] and Maimon and Shtub [1991] consider such problems.

The problem becomes more complex when tools (component feeders) are loaded into groups or heads for common loading onto a machine, and machines hold a limited number of heads. Using surface mount technology, machines insert components onto circuit boards. The machine may hold roughly 6 to 10 part bins, with each part bin capable of holding several component feeders. Each day the machine is assigned a set of circuit cards to assemble. Our problem becomes the determination of the order in which to run circuit cards, the assignment of feeders to bins, and the order in which bins are set up onto the machine. Except for the new aspect of assigning feeders to bins, the problem structure is the same. If bins only hold one feeder, the problem reduces to the partial changeover problem just studied.

8.4
INTEGRATED ASSIGNMENT AND SEQUENCING

In the previous section we touched on the relationship between cell setup and job order. Cell setup (tooling) was explicitly determined by the job sequence. In this section we look at problems where these decisions are related, but neither automatically dictates the other. This relationship is examined through two specific problems. The problems addressed in this section are characterized by the importance of both making a sequencing decision for the order in which a set of jobs (or operations) will be performed *and* a decision on how to set up the tool magazine for the machine. Tool setup may include both the selection of tools and the location of tools in the magazine.

The first problem occurs when planning the setup and the operation of assembly cells. The second problem considered is that of the setup and the sequencing of an NC punch press. In both instances, the objective is to minimize the cycle time. The problems differ in the level of interaction between tool location decisions.

Both problems can be formulated as 0–1 mathematical programming problems. However, actual problem sizes are rather large and the need for frequent resolution for new part types makes it difficult to find optimal solutions. We will look at using heuristic approaches that have proved useful for these and similar problems. The heuristics discussed will explicitly consider both aspects of the problem. Generally, these heuristics provide very good solutions, significantly better than strategies that naively ignore either the sequencing or setup aspect.

8.4.1 Assembly Cell Layout and Sequencing

A typical final step in the production of a product involves the assembly of parts onto a "frame." Several assembly environments are common. These include (1) a single product being mass produced; (2) multiple products produced in alternating lots with possible changeover required; and (3) a mixture of part types being assembled in the cell simultaneously without changeover. Chapter 2 discussed the assignment of work to individual workstations. In this section we are concerned with how to perform the set of tasks that have been assigned to the workstation. We will discuss first the case of a single product and then show how the procedure can be generalized to include a part mixture. The multiple product case is handled by repeated solution of the single product problem. This section is based on the work of Drezner and Nof [1984] for setup and sequencing on a single machine. Crama et al. [1990] describe a comprehesive, hierarchical decision model for planning a multiple machine flow line with sequence-dependent task setup times.

Single Part Type

We are presented with a product type to be assembled. The product frame has N locations where parts are to be added. The workstation has N locations (bins) where part feeders can be placed. Each location can hold one part type. Our problem is to assign the part feeder for each of the N part types to a unique bin location. Figure 8.6 shows a schematic of the workstation. In addition to assigning feeders to bins, we must determine the order in which to add parts to the frame. We will refer to these two decisions as cell layout and sequencing, respectively.

This problem can be modeled as a $2N$-city Traveling Salesman Problem. The TSP analogy to our problem is as follows. The assembler (robot or worker) begins at some "home" position. The assembler must travel to a bin location for a part, then to the appropriate frame location, then back for another part. This continues until all parts are added and the arm returns home to wait for the next product frame. Cities are thus the N bin locations and N frame locations. (We hereafter assume that the home position can be ignored, since time to travel to the home position and back to the first feeder is exceeded by the time to remove the finished frame and fix a new frame into place. If not, treat the home position as an extra city.) The salesman is normally allowed to move between any pair of cities in the TSP. Our assembler must alternate between bin location cities and frame location cities. To avoid inappropriate moves, insert a large cost c_{ij} to go between two bin locations, or between two frame locations. The solution to the TSP is the $2N$-element sequence of location visits. As bin locations will precede each frame location in the solution, and each frame location requires a specific part type, the location of each part feeder is implied. The feeder is placed in the bin location visited immediately before the frame location that requires that part.

This model is not very flexible. The required assumptions are restrictive. Suppose some part types are used more than once, that is, the number of frame locations exceeds the number of required bins. This implies the need to visit certain "cities" or part locations more than once in a complete production cycle. This is not permitted in the standard TSP. Precedence restrictions on assembly order are also difficult to manage with this formulation. We can overcome these problems by decomposing

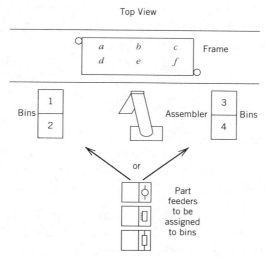

Figure 8.6   A workstation schematic.

the problem into two stages. First, we will assign part types to bin locations. Second, the sequence in which parts are to be added will be decided.

Bin Assignment

A reasonable objective in assigning parts to bins is to minimize the loaded travel time (distance) for the assembler. The total loaded time is simply the sum of the travel times to the frame locations from the feeder holding the corresponding part. If we know where parts are located, we can easily compute the total loaded travel time of the assembler for this part type. If a part type is used more than once, simply sum its trips to obtain the cost of assigning part type i to bin location j. Allocating parts to bins is thus a linear assignment problem.

EXAMPLE 8.4

A robotic assembly cell has four possible locations for loading part magazines (see Figure 8.6). Only three part types are used, however. The product base will be fixtured in the center of the cell, directly in front of the robot. A total of six parts, two of part type A, one part type B, and three part type C, must be added to the product. A's are required in product locations a and c. B is required in product location e, and C in locations b, d, f. A precedence restriction exists such that location e must be filled before f. Because of the size of the product, it is difficult (time consuming to avoid obstacles) to move from certain magazine locations to certain product locations. Travel times for the robot arm to pick a part from the indicated bin, travel, and place it into a product location are given in Table 8.6. Assign part magazines to bin locations.

Table 8.6   Travel Times from Bins to Frame Locations

	\multicolumn{6}{c}{Frame Location}
Bin	a	b	c	d	e	f
1	10	12	15	11	20	21
2	10	10	13	15	19	21
3	14	10	10	21	16	16
4	21	12	10	26	13	13

Solution

The first step involves computing the cost coefficients, the loaded travel time resulting from assigning each part type to each bin. As an example, consider assigning part A to bin 1. Let t_{kl} be the travel time given in the table above to move from bin k to frame location l. Total loaded travel time of assigning part type A to bin 1, c_{A1}, is then $c_{A1} = t_{1a} + t_{1c} = 25$. c_{A1} reflects time to travel from location 1 to both locations a and c, which require part type A. The remaining values in Table 8.7 are found in a similar fashion. Part type D represents a dummy magazine that will not be used in practice but serves to balance the number of locations and part types in the model.

**Table 8.7   Total Travel Times
for Part Type Assignments**

Part	Bin Location			
	1	2	3	4
A	25	23	24	31
B	20	19	16	13
C	44	46	47	51
D	0	0	0	0

Table 8.7 represents a 4×4 assignment problem. An optimal solution with a cost of 81 time units results from assigning part type magazines (A, B, C) to locations $(2, 4, 1)$, respectively. Location 3, which would appear in Figure 8.6 to be a good location, is left empty. This is because of the need to travel around the part to insert components $d, e,$ and f.   ∎

Insertion Sequencing

We now consider the sequencing of assembly tasks. Bin assignment minimized loaded travel time, but the unloaded travel time from frame insertion points back to a bin location to retrieve the next part was ignored. Now that feeder locations are fixed, we can concentrate on sequencing part insertions to minimize unloaded travel time. Once again, we are faced with a TSP. Frame insertion locations represent cities. Costs are unloaded travel times to move from a frame location to obtain the next part from its feeder magazine. (Total loaded travel time back to the frame is now fixed by bin assignment and can be excluded.) The TSP may be further complicated by the existence of precedence constraints, reflecting that certain parts must be added before others. Assembly precedence constraints now have a direct interpretation, since assembly activities are seen to follow one another. Our TSP tour must have a starting location (possibly the home location for the machine). When developing a sequence, we must satisfy precedence constraints.

The reader may be wondering why we bothered with the bin assignment step, since we are once again left with a TSP. There are two reasons. First, we now only have N product frame locations to order instead of $2N$ bin plus frame locations. This makes the problem less than one half as hard to solve. Second, as we have discussed, special considerations, such as multiple uses of a part, can be handled.

The closest insertion algorithm can be modified to accommodate precedence constraints. Remember that a city corresponds to a location on the product frame, and visiting a city corresponds to inserting the proper part in that frame location. When selecting tasks, we choose only from those with all predecessors already in the partial schedule. Start with any city without predecessors. At each stage, when a city is selected for addition to the partial sequence, we set incremental costs to ∞ for all positions in front of any predecessor. Thus, a task can never be placed before a predecessor.

EXAMPLE 8.5

Sequence the insertions from Example 8.4. Recall that location e must be filled before f.

Solution

We add a dummy city to indicate the home location of the assembler at the start of assembly. Costs are empty travel times from the product frame to feeder bins as shown in Table 8.8. To help us understand how costs were generated, the part type and fixed bin location for each part type are included. The cost to follow insertion a with b is the time to travel from location a back to the feeder holding the part for insertion at b. Insertion b uses a part type C. In the previous example we decided to place part type C in bin 1. From Table 8.6, the travel time from location a to bin 1 is 10 units. This is the value that appears in Table 8.8 for traveling from a to b.

Let us assume for safety reasons that the assembler must be in its fixed home position while the product frame is loaded and unloaded at the work station. Hence, from-to costs for dummy task 0 are nonzero. The bottom row of Table 8.8 gives travel times from the home point to the indicated bins. The last column contains the travel times from the product frame to the home point.

Table 8.8 Empty Travel Times for Example 8.5

| Part: | A | C | A | C | B | C | |
| Bin: | 2 | 1 | 2 | 1 | 4 | 1 | |
Insertion:	a	b	c	d	e	f	0
a	—	10	10	10	21	10	12
b	10	—	10	12	12	12	15
c	13	15	—	15	10	15	9
d	15	11	15	—	26	11	8
e	19	20	19	20	—	20	16
f	21	21	21	21	13	—	13
0	5	6	5	6	8	6	—

Once again we utilize the closest insertion algorithm. The only difference is that city f is not eligible to enter until city e has entered. When city f is entered, it must be placed after city e. The order of entry is $0, a, c, b, d, e, f$. The sequence of partial tours is $0 \rightarrow a, 0 \rightarrow a \rightarrow c, 0 \rightarrow a \rightarrow b \rightarrow c, 0 \rightarrow a \rightarrow d \rightarrow b \rightarrow c, 0 \rightarrow a \rightarrow d \rightarrow b \rightarrow c \rightarrow e$, and $0 \rightarrow a \rightarrow d \rightarrow b \rightarrow c \rightarrow e \rightarrow f$. This constructed sequence has cost (time) of 79.

It is informative to study this solution. For insertions a, b, f, and 0, the exiting arc is that with the minimum cost. We incur cost of 10 to go from c to e, which exceeds c's lowest exiting arc (c to 0) by 1; a cost of 11 from d to b, which is 3 more than b's minimum; and a cost of 20 from e to f, 4 more than e's minimum. Thus, the most we could reduce cost by optimizing would be $1 + 3 + 4 = 8$. However, this would require going from e to 0, which is impossible; f must be between e and 0. Noting this added restriction, the minimum cost out of e is 19 and the worst we can be is 5 from optimal. We can continue to play this game. For instance, f must come last. Otherwise, cost out of f would be 21, necessarily raising cost by 8, more than the 5 we could possibly gain over our current solution. If f must be last, suddenly d and c

exiting arcs are now at their lower bound. Thus, we can be at most 1 unit from optimal! Perhaps there is a tour with cost 78; nonetheless, closest insertion has done its job. We have quickly found an at least near-optimal solution.   ■

Mixed Product Extensions

A typical environment has several product types being produced in the same cell. Frames enter the cell in a seemingly random order but maintain a long-term relative demand rate. The demand proportions are determined by the bill of materials for the end products and their demand. The cell layout must accommodate all products. To analyze this problem we must know whether the line is paced or unpaced.

Consider first the case of an unpaced line. We assume that the relative demand rates for the products are known. Our goal will be to minimize the average time for a product to be assembled. This approach is justified in the case where there is always a supply of products waiting to enter the cell. Without this assumed input, however, our approach still minimizes average flow time and maximizes cell capacity.

Suppose M frame types are to be made and p_m is the proportion of type m frames. Insertion sequencing can be solved separately for each part type, but bins must be kept in the same location for all part types. A table of travel times as in Table 8.7 can be compiled for each part type. These tables are then combined by the weighted average

$$c_{ij} = \sum_{m=1}^{M} p_m c_{ijm} \tag{8.9}$$

where c_{ijm} is the total travel time per type m frame if part type i is assigned to bin j. An assignment problem is then solved with costs c_{ij} to locate parts.

EXAMPLE 8.6

Suppose a second frame type must be made in addition to that of Table 8.7. Approximately 20 percent of demand will be of the new type. The new frame uses all four part types. Total travel times are given in Table 8.9 for the new frame.

Table 8.9   Total Travel Times for the New Frame

Part	Bin Location			
	1	2	3	4
A	15	19	25	41
B	20	18	31	16
C	17	19	10	21
D	28	16	12	25

Solution

A new travel time matrix for the bin assignment problem is generated by taking 80 percent of the value in Table 8.7 and 20 percent of the Table 8.9 value. For instance, average travel time per frame if part type A is assigned to bin 1 is

$$c_{A1} = 0.8(25) + 0.2(15) = 23$$

Mixed frame times are shown in Table 8.10.

Table 8.10   Average Times for Part Type Assignments

Part	Bin Location			
	1	2	3	4
A	23.0	22.2	24.2	33.0
B	20.0	18.8	19.0	13.6
C	38.6	40.6	39.6	45.0
D	5.6	3.2	2.4	5.0

Solving the LAP, the optimal solution places parts (A, B, C, D) in locations (2, 4, 1, 3), respectively, with average loaded travel time of 76.8 per frame.[2]   ■

If the line is paced, the objective should be minimization of the longest assembly time. A common cycle time must be allowed for all parts and this cycle time must permit completion of the slowest frame. This problem is more difficult to solve. Alternative bin assignments can be evaluated and the best selected. For each assignment, the insertion sequence must be found for each frame type. The bin assignment with the smallest assembly time for any frame is best.

8.4.2 Cell Layout and Sequencing with Interdependent Tools

Consider the setup of an NC punch press such as that shown in Figure 8.7. The press uses various tools (punches) to impart features (holes) to sheet metal parts. The tool turret must be loaded with all punches needed to make a specific part (the turret on the machine shown in Figure 8.7 holds up to 36 tools). A part may require 200 "hits" or punches, each in a specific location and using a specific tool. The hit sequence may have to satisfy precedence constraints. To operate the machine, a part is loaded along with its required tools. Starting from the home position, the machine bed moves to place the location of the next scheduled hit under the tool turret. At the same time, the turret rotates, if necessary, to place the proper tool into the hit position. A hit is then made. The process continues until all hits are completed and the bed returns to the home position. The objective is to load tools onto the turret and sequence hits to minimize the production cycle. The time to make a hit once the bed and turret are positioned is fixed. Thus, cycle time varies only as a result of the time to reposition between hits. For each hit, this time is the maximum of the time to reposition the bed holding the part or the turret holding the tools. Generally, repositioning the turret takes longer if it must be rotated.

Walas and Askin [1984] give a mathematical formulation of this problem. The model involves two hard problems that we have considered before. If we know the sequence in which tools are to be used, then we can determine their best placement on the

---

[2]For this particular cost matrix, an optimal solution is easily found. Each part type can be assigned to its lowest cost bin. Since each part must be assigned, this assignment is optimal.

Figure 8.7   An NC punch press.

turret by solving a quadratic assignment problem. The cost to place tool k in position i and l in position j is the time to rotate from position i to j multiplied by the number of times we switch from tool k to l or vice versa. Note that once the hit sequence is set, then the sequence of tool uses is known, since each hit requires a specific tool. Likewise, suppose we know the tool assignments on the turret. We then need to sequence hits. The time to go between hits is a traveling salesman problem. Each hit (and the home location) is a city. The cost to travel between cities is the maximum of the turret and bed rotation times. Treating the problem as a combination of QAP and TSP and developing an integrated heuristic, Walas and Askin found improvements up to 25 percent in cycle time as compared with simplistic TSP-based heuristics previously used.

The basic structure of the procedure used by Walas and Askin involves iteration between the TSP and QAP subproblems. Initially, a TSP cost matrix is formed for hits. In computing costs, it is assumed that all tool switches will be between the closest feasible locations on the turret if a change of tools is required. Given the hit sequence, the tools are placed by solving the QAP. Actual tool locations are then used to update the TSP cost matrix, and that problem is resolved. As is typical of large problems, several other steps can be used to speed processing. For instance, the hit sequencing TSP is difficult to solve optimally with 200 hits. Thus, a quick heuristic was used. However, the solution from the heuristic was divided into "tool-use subsets." A tool-use subset is a set of consecutive hits using the same tool. This subset was combined with the immediate preceding and succeeding hits to form a series of smaller TSP problems. An optimal or, at least, more effective heuristic can then be applied to each of these smaller subsets to reorder hits. Subsets can also be treated as individual cities, and a TSP can be resolved to sequence the use of these tools. It is possible that tool use subsets can be rearranged or even that two subsets using the same tool can be combined. We will illustrate the basic approach with a small example.

EXAMPLE 8.7

Consider the part shown in Figure 8.8. Lowercase letters indicate the tool required for each hit. Although some machine beds can move in horizontal and vertical directions simultaneously, ours cannot. Travel time between hit

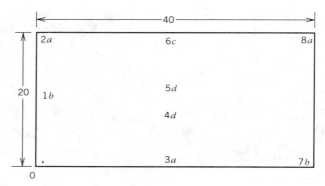

Figure 8.8 The part for Example 9.7.

locations are as given in Table 8.11. Zero is the home position. The turret can hold five tools in any order. The turret can rotate in either direction, taking 35 time units to move to an adjacent location, 55 time units to move two locations, and 75 to rotate three positions. Assign tools and sequence hits. Since our part requires only four tools, one turret location will be left empty.

Table 8.11   Interhit Travel Times for Example 8.7

					Hit				
Hit	0	1	2	3	4	5	6	7	8
0	—	10	20	20	28	32	40	50	60
1	10	—	10	30	20	20	30	40	50
2	20	10	—	42	32	28	20	50	40
3	20	30	42	—	8	12	20	30	42
4	28	20	32	8	—	5	20	22	32
5	32	20	28	12	5	—	20	22	28
6	40	30	20	20	20	20	—	30	20
7	50	40	50	30	22	22	30	—	20
8	60	50	40	42	32	28	20	20	—

Solution

Before applying our integrated TSP–QAP approach, let us look at how we might otherwise find a "quick and dirty" solution. A reasonable approach might start at the home location 0 and go to the closest hit, 1. This hit requires tool b. We will make all the b hits, always going to the closest hit location still to be visited. Thus, we proceed $0 \rightarrow 1 \rightarrow 7$. No other hits require tool b. The closest hit is 8 so we will proceed there and rotate to tool a. After completing all of a's hits (8, 2, and 3), we go to the nearest remaining hit location, 4 and load tool d. Continuing in this fashion we obtain the hit sequence $0 \rightarrow 1 \rightarrow 7 \rightarrow 8 \rightarrow 2 \rightarrow 3 \rightarrow 4 \rightarrow 5 \rightarrow 6 \rightarrow 0$ using tools $b \rightarrow a \rightarrow d \rightarrow c$ and consuming $10 + 40 + 35 + 40 + 42 + 35 + 5 + 35 + 40 = 282$ time units. Transition times assume no tool change time associated with the home location. The tool can be rotated back to position b while the part is being unloaded and the next part loaded. This seems like a reasonable procedure. Can a more thoughtful approach do better?

Let us consider an integrated approach.

STEP 0. Form the TSP cost matrix assuming that all tool changes take 35 time units. The 35 time units represent a lower bound and are valid if all tooling changeovers are to adjacent turret locations. The appropriate cost matrix is given in Table 8.12. Consider the travel time for hits 0 to 8. The 60 represents the sum of the 20 and 40 units moved along the two axes as seen in Figure 8.8. Sixty is used since it exceeds the 35 time units to change the tool. On the other hand, the travel time for changing from hit 4 to 1 is 35 because we change tools but part bed travel time is only about 25 units.

Table 8.12   Interhit Travel Times for Example 8.7

					Hit				
Hit	0	1	2	3	4	5	6	7	8
0	—	10	20	20	28	32	40	50	60
1	10	—	35	35	35	35	35	40	50
2	20	35	—	42	35	35	35	50	40
3	20	35	42	—	35	35	35	35	42
4	28	35	35	35	—	5	35	35	35
5	32	35	35	35	5	—	35	35	35
6	40	35	35	35	35	35	—	35	35
7	50	40	50	35	35	35	35	—	35
8	60	50	40	42	35	35	35	35	—

STEP 1. Solve the TSP. One heuristic solution to this TSP is the tour $0 \rightarrow 2 \rightarrow 5 \rightarrow 4 \rightarrow 3 \rightarrow 8 \rightarrow 7 \rightarrow 6 \rightarrow 1 \rightarrow 0$ with cost 252.

STEP 2. Assign tools. Although tool changes did not seem desirable, the tool sequence here is $a \rightarrow d \rightarrow a \rightarrow b \rightarrow c \rightarrow b$. We need to solve a QAP at this point. We switch between a and d twice, a and b once, and b and c twice. The cost to assign a to slot A and d to slot D is thus $2 \cdot 35 = 70$, twice the turret rotation time from A to D. Other cost coefficients are similarly computed. Locating the tools $a, d, empty, c, b$ on the turret gives a solution that satisfies the adjacency assumption. Thus, we need not look for a better QAP assignment. This solution is feasible and has cost 252. ∎

8.5
SUMMARY

A competitive system must use its resources in an efficient as well as effective manner. Machine operation problems can often be modeled as binary integer programming problems. Although these problems are generally difficult to solve, many of the problems fall into well researched classes such as assignment and traveling salesman problems. With current technology, very good if not optimal solutions can be found even for real-world instances of these problems.

Linear assignment problems occur when a set of task activities must be assigned to a set of resources. Activities could be jobs, and resources could be workers/machines, but the model is just as valid if tasks are tools or part feeders and the resources are physical locations. Linear assignment problems are relatively easy to solve.

Task sequencing generally takes the form of a traveling salesman problem. The TSP is difficult to solve optimally, but good heuristic methods exist even for problems with hundreds of "cities." When changeover costs require more information than just the pairs of adjacent tasks in the sequence, then special-purpose algorithms must be used instead. The binary clustering algorithm used in GT is another method for sequencing activities.

Modern technology has produced automated machines with large tool storage capability. Such technology requires joint consideration of tool planning and task sequencing. Procedures exist to combine the individual solution approaches for finding better, integrated problem solutions. These procedures often involve use of the assignment and sequencing problems as subproblems.

REFERENCES

Ackoff, Russell L., and Maurice W. Sasieni (1968), *Fundamentals of Operations Research,* John Wiley & Sons, Inc., New York.

Ahuja, R. K., T. L. Magnanti, and J. B. Orlin (1989), "Network Flows," in *Optimization,* G. L. Nemhauser, A. H. G. Rinnooy Kan, and M. J. Todd, eds., North-Holland Publishing Co., New York.

Baker, K. R. (1988), "Scheduling the Production of Components at a Common Facility," *IIE Transactions,* 20, 32–35.

Crama, Y., A. W. J. Kolen, and A. G. Oerlemans (1990), "Throughput Rate Optimization in the Automated Assembly of Printed Circuit Boards," *Annals of O. R.,* 26, 455–480.

Drezner, Zvi, and Shimon Y. Nof (1984), "On Optimizing Bin Packing and Insertion Plans for Assembly Robots," *IIE Transactions,* 16(3), 262–270.

Kanellakis, P. C., and C. H. Papadimitriou (1980), "Local Search for the Asymmetric Travelling Salesman Problem," *Operations Research,* 28, 1086–1099.

Lawler, E. L., J. K. Lenstra, A. H. G. Rinnoy Kan, and D. B. Shmoys (eds.) (1985), *The Traveling Sales-*

man Problem: A Guided Tour of Combinatorial Optimization, John Wiley & Sons, New York.

Lin, S., and B. W. Kernighan (1973), "An Effective Heuristic Algorithm for the Traveling Salesman Problem," *Operations Research,* 21(2), 498–516.

Maimon, Oded, and Avraham Shtub (1991), "Grouping Methods for PCB Assembly," *International Journal of Production Research,* 29(7), 1379–1390.

Reingold, Edward M., Jurg Nievergelt, and Narsingh Deo (1977), *Combinatorial Algorithms: Theory and Practice,* Prentice-Hall, Inc., Englewood Cliffs, NJ.

Tang, Christopher S., and Eric V. Denardo (1988), "Models Arising from a Flexible Manufacturing Machine, Part I: Minimization of the Number of Tool Switches," *Operations Research,* 36(5), 767–777.

Taha, Hamdy A. (1971), *Operations Research: An Introduction,* The Macmillan Co., New York.

Walas, Robert A., and Ronald G. Askin (1984), "An Algorithm for NC Turret Punch Press Tool Location and Hit Sequencing," *IIE Transactions,* 16(3), 280–287.

PROBLEMS

8.1. Discuss why it is important to find good yet quick solutions to short-term planning problems. Do you think it is easier to automate solution procedures for short- or long-term planning? Why?

8.2. A certain firm makes long-range plans annually. Optimal decisions may save more than

$200,000 annually with respect to judgmental decisions made without the use of sophisticated models. On a typical workday, the firm makes about 20 cell-design and sequencing decisions. Each decision can easily make a difference of at least $50. Which type of decision is more important?

8.3. An R&D manager must assign four teams to four projects. The manager has assessed the expected long-term profit from each project for each group that might be assigned. Using the profit data in Table 8.13, find the optimal assignment of teams to projects.

Table 8.13   Expected Long-Term Profits (Millions) for Problem 8.3

	Project			
Team	1	2	3	4
A	25.0	12.5	5.0	1.5
B	15.0	8.0	7.0	1.5
C	5.0	3.0	1.5	0.0
D	15.0	5.0	5.0	2.0

8.4. A production manager has five workers and five tasks that need to be completed. She is interested in minimizing total worker time. Based on skill levels and task requirements, she has estimated the times for each worker to perform each task. Find the best assignment by using the times in Table 8.14.

Table 8.14   Task Times by Worker for Problem 8.4

	Task				
Team	1	2	3	4	5
A	10	8	20	15	3
B	13	12	25	2	10
C	4	15	16	5	8
D	10	12	14	2	12
E	8	11	28	3	9

8.5. A work cell produces three families of parts denoted by A, B, and C. Individual items can share family setup but also require some individual setup and tear down time. Using the data below, first construct the matrix of changeover times between all jobs. Second, find a good sequence. The machine is currently set up for type C jobs. The C setup can be taken apart in 0.5 hours. Only type A and B jobs remain to be produced in this period.

Family setup for type A jobs takes 4.0 hours. The setup can be taken apart in 1.0 hour. Family B takes 6.0 hours to setup and 1.0 hour to take down. The remaining jobs are given in Table 8.15.

Table 8.15   Job Data for Problem 8.5

Job	Individual Setup Hours	Individual Take Down Hours
A12	1.0	0.5
A31	0.5	0.2
B13	0.6	0.1
B21	1.0	0.0
B14	0.4	0.1
A64	0.5	0.1

Can you think of any general sequencing rules to give the machinist?

8.6.

a. Find a lower bound on the optimal solution of Problem 8.5 by solving the associated linear assignment problem.

b. Compare this bound to the solution found in Problem 8.5.

8.7. Using the changeover times in Table 8.16, find a good sequence of jobs to minimize makespan.

8.8. Six holes are to be punched in a part. The time to rotate the part into position is proportional to the rectilinear distance between punch locations. Using the coordinates in Table 8.17, sequence punches. The home position must be the inital and ending location.

Table 8.16   Job Changeover Times for Problem 8.7

From/To	Start/End	1	2	3	4	5
Start/End	0	10	3	4	20	12
1	10	0	8	15	4	7
2	3	8	0	25	15	9
3	4	15	25	0	17	6
4	20	4	15	17	0	9
5	12	7	9	6	9	0

**Table 8.17   Punch Locations
for Problem 8.8**

Punch	X Coordinate	Y Coordinate
Home	0.0	0.0
1	2.4	4.2
2	1.7	8.3
3	4.5	9.1
4	0.4	6.2
5	7.2	3.5
6	4.5	7.1

8.9. A turret lathe can hold six cutting tools. Twelve jobs are waiting in queue at the lathe. The tools required by each job are given below. Currently the lathe holds tools (b, c, d, f, g, m). Sequence the jobs. Are you satisfied with your solution? Using Table 8.18, find a lower bound on the number of tool changes required.

**Table 8.18   Tools by Job
for Problem 8.9**

Job	Tools	Job	Tools
1	a, b, d, e, f	7	f, m, p
2	m, p	8	a, e, f
3	a, d, p, m	9	c, m, p
4	a, d, e, f, g	10	c, e, i, h
5	c, e, i, j	11	m, p
6	a, e, p	12	e, i, j, p

8.10. Eight different circuit boards are assembled in a shop by an automatic insertion machine. The machine can hold only six component feeders. For the component requirements given in Table 8.19, find a good assembly sequence for the eight boards. Indicate when component feeders should be changed. Feeder locations are currently empty.

**Table 8.19   Components by Board Type
for Problem 8.10**

Board	Components	Board	Components
1	a, c, d, e	5	c, d, e, f, g, h
2	d, f, g, h	6	a, c, f, h
3	a, d, e, f	7	a, c, g
4	a, d, f, g	8	a, c, e, g, h

8.11. Resolve Example 8.2 assuming that we must end up with a setup for batch 1098A, that is, solve for a TSP tour. Explain how this could be used provided that we had a required end of period setup, whether that setup coincided with the initial conditions of 1098A or not.

8.12. Consider the History-Dependent (Partial) Changeover problem of Section 8.3.2. An alternative policy to that used in the text would be to set up M tools and then perform all operations on all jobs that require those tools. M new tools are then loaded and all job operations using these tools are run. In this fashion tools are never loaded more than once, but jobs may need to be loaded onto the machine and run several times. Discuss the relative merits of both approaches.

8.13. Consider the two policies used in Problem 8.12. Develop a model to decide which policy should be used. Define any notation used such as costs of tool setups, job loading, and work in process.

8.14. Binary clustering and a pessimistic TSP cost method were suggested in Section 8.3.2 for ordering jobs. Describe a procedure for using a TSP model for ordering jobs based on similarity coefficients. (*Hint:* The key is to define the distance between jobs. An alternative distance measure would be $1 - s_{ij}$, where s_{ij} is a measure of similarity between jobs i and j.)

8.15. Solve Problem 8.10 using the method developed in Problem 8.14.

8.16. Describe a 2-opt procedure for improving a solution sequence for a TSP. Define a sequence as $\boldsymbol{a} = (x_1, \ldots, x_n)$ where x_i is the ith city visited. Explain the list of neighborhoods that can be reached in one step from any sequence $\boldsymbol{a}$. Give an efficient equation for computing the new cost of a potential two-city switch, given only the cost of $\boldsymbol{a}$ and the c_{ij} coefficients.

8.17. Read the paper by Lin and Kernighan listed in the references. Write a computer program to solve TSPs using the improvement procedure described.

8.18. A workcell is being set up to produce two types of circuit boards. The same raw board is used for both circuits; only the inserted parts differ. The workcell will need four part feeders (a, b, c, d) to make the two board types. Five bin locations are available in the workcell. Each bin can hold one part feeder, and each feeder can hold one part type. The workcell has a robot that grasps parts at the part feeders and then inserts the part into the board. Only one part can be grasped and inserted at a time. The demand rate for the first board is twice that of the second board. Other relevant data is provided in Table 8.20.

a. Construct the assignment cost matrix for locating part feeders in bin locations so as to minimize total loaded travel distance of the robotic inserter.

b. Assuming part feeders have been assigned to bin locations, explain how you would determine the insertion sequence for board types 1 and 2. You need not solve this problem, but tell what type of problem this is, state an appro-

Table 8.20b   Distances from Bin Locations to Board Locations

	Board Location			
Bin	1	2	3	4
1	10	15	12	17
2	12	15	15	21
3	15	16	16	21
4	12	14	8	11
5	9	10	11	13

Table 8.20a   Required Part Types in Board Locations in Problem 8.18

Board Type	Board Location			
	1	2	3	4
1	a	b	c	d
2	b	a	c	b

priate algorithm for solving it, and set up the necessary input data for board type 1. The input data can be in the form of a labeled matrix with the proper cost coefficients. Assume that the decision has been made to assign part feeders (a, b, c, d) to bins (1, 5, 4, 2), respectively.

8.19. Using the data in Table 8.20, assign part feeders and sequence insertions to make board type 1 only.

CHAPTER 9

MATERIAL HANDLING
SYSTEMS

He had forty-two boxes, all carefully packed,
With his name painted clearly on each:
But, since he omitted to mention the fact,
They were all left behind on the beach.
—Lewis Carroll, from *The Hunting of the Snark*

9.1
INTRODUCTION

The material handling system implements the flow path planned during facility layout. The system must control the flow of parts, mobile resources (e.g., tools), and wastes (e.g., chips removed during machining). Flows must be controlled both between departments and within departments. Departments may appear as "islands," each with its own internal material handling system. Departmental systems may differ in degrees of automation and integration into the overall manufacturing system. Department A in Figure 9.1 uses workstations integrated by a loop conveyor. Deliveries from and to the department can be made by many alternative methods. Within the department, kits are dispensed along the conveyor from the central storage area to appropriate workstations. Workstations return their completed kits to the conveyor for delivery to another workstation or back to the I/O and storage area. A similar type of arrangement is shown in Figure 9.2. Department B in Figure 9.1 uses manual trucks with staging areas at each workstation.

Historically, material handling simply meant the handling of material. It was easy to agree that when it came to the frequency, length, and equipment cost, the maxim *the less the better* ruled. After all, manufacturing is the business of adding value to materials, and value is not imparted by simply moving parts around the plant. Today, material handling is properly viewed as an integral part of the total manufacturing system. As with sales or maintenance functions, it would be foolish to minimize cost without regard to the broader implications.

The material handling system acts as the circulatory system of the plant, distributing vital material to all of the plant's cells. The system may use data acquisition and communication components that allow it to also serve as the nerves, carrying important information to the system controller (brain). Our objective should not be to find the minimal cost material handling system but, rather, the system that satisfies all of our requirements to be effective and efficient manufacturers. The human body could

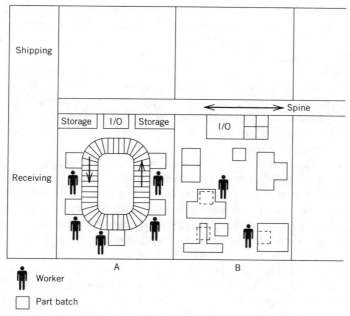

Figure 9.1   Within and between department flow of material.

survive with thinner arteries, fewer blood cells, and less frequent signals between the brain and muscles, but could it run a marathon in 2 hours or slam-dunk a basketball?

With traditional manual methods it may have sufficed to issue the shop order to "move that batch to Department X." In today's environment, location is only one of the relevant characteristics for the material handling system. Part orientation is also important for automated systems. It can be difficult to orient parts properly for feeding to automated machines. Once the parts are oriented, there is value, in terms of avoiding future work, of maintaining knowledge of the part's position. Orientation is only one of the keys, however; the material handling system must also ensure the right product, location, condition, quantity, and timing of deliveries.

As with any design problem we must first ask the question *why*. Why will the system exist and what do we expect it to accomplish? This leads us to the description of *what material* is involved and *where and when moves* will occur. The description of the

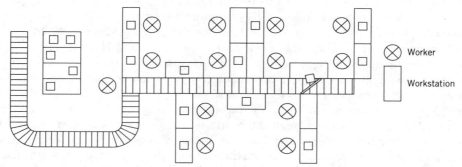

Figure 9.2   A department layout with central storage and control and a bilevel or bidirectional conveyor.

move includes the frequency and path from origin to destination as well as the keys mentioned in the previous paragraph. This input is used to determine the *method,* namely, *who* moves the material and *how.* The additional system requirements dictate the *monitoring* level desired.

If you venture into the older portions of some U.S. cities, you may still see the remnants of old warehouses that were built straddling either side of the railroad tracks. This technique minimized the distance parts had to be moved when unloaded. More recently, ideas such as **point of use storage** have been adopted to reduce the number of material moves. In point of use storage, parts are moved directly from workcenter to workcenter instead of returning to a storage area between production operations. Not only is the number of moves cut in half, but inventory levels suddenly become apparent, along with wasteful scheduling and dispatching practices.

Another concept is the use of a **unit load**. The unit load of a part type is the quantities of that part which are combined and transported as a single item. Instead of moving each item, an entire unit load is moved. The unit load concept offers significant savings for items requiring considerable material handling effort. A typical unit load would be a full pallet or container of items. In addition to reducing the number of moves, the aggregation of items can reduce the risk of damage during movement and permit the use of standard handling equipment. The disadvantage is the need to return empty containers/pallets to their point of use and the cost of containers/pallets. When selecting unit load size, it should be kept in mind that the unit load cannot be larger than the batch size for parts in-process. In some instances it is desirable to split batches into sublots so that as soon as the first sublot is completed on a machine, these units can be moved and started at the next workstation. In this situation, the sublot limits the unit load. Choice of unit load depends on the layout of the facility. When parts are transferred from machine to adjacent machine, it may be economical to transfer one part at a time. When part moves are long, it is often necessary to use a larger unit load such as a pallet. In general, the shorter the move, the smaller the proper load size. Last, weight and size limits exist on unit loads to ensure safe handling with the available equipment and aisles.

Before progressing to analytical design models in subsequent sections, we will briefly describe the basic types of material handling equipment and list the historically accepted Principles of Material Handling.

9.1.1 Basic Equipment Types

A large industry exists dedicated to the design and implementation of material handling systems. Technology levels range from high-tech controls to traditional push-carts and pallets. In this section we list some of the common types of equipment. More complete descriptions are available in texts such as Tompkins and White [1984].

Conveyors can be used for moving materials of relatively uniform size and weight with moderate to high frequency between a specified set of locations over a fixed path. In addition to material movement, conveyors can be used as a positioning fixture for workers along a production line. Although the fixed path is normally considered a prerequisite to conveyor use, we need only watch our luggage being unloaded from an airplane to realize that short portable conveyors exist as well. As with the other types of equipment to follow, many types of conveyors are available. One defining characteristic is the surface that supports the product. Common surfaces for parts to rest on include belts, chutes, wheels, or rollers. Belts circulate, keeping the part in the same position relative to the belt. Chutes, wheels, and rollers stay in place (or rotate about their longitudinal axis), while parts move along the conveyor.

Trolley conveyors, often located overhead, allow parts to be transported at equal space increments. Parts are suspended on hooks or placed on carriers connected to the conveyor. Figure 9.3 shows a roller and trolley conveyor. Power and free is a modification of the trolley conveyor. Two tracks allow carriers to be disengaged from the power chain for processing or storage. The carrier is then reengaged to the chain for transport. Tow line conveyors allow variable path trucks to be temporarily connected to a powered conveyor for automatic movement over a fixed path. The trucks can then be disconnected and moved to their final destination. Conveyors can be nonpowered, gravity powered, or powered electronically, pneumatically, or by a

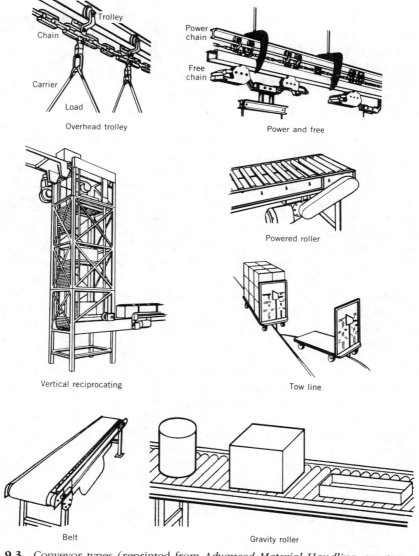

Figure 9.3 Conveyor types (reprinted from *Advanced Material Handling,* courtesy of the Material Handling Institute).

chain or belt drive. In short, conveyors can be in the floor, raised with supports from the floor, or overhead. They can carry bulk materials, pallet loads, small boxes, or individual parts. They can be connected to bowl feeders, and diverters can be added to channel and orient parts. They can move continuously, intermittently, or asynchronously with parts moving on and off.

Cranes and **hoists** are overhead lifting devices used for intermittent moves of varying size and weight within a fixed space. Hoists lift material vertically, generally while it is suspended from a hook. Cranes move horizontally, the product normally being suspended from a hoist while the crane traverses. The space covered is determined by the guide rails. A bridge crane consists of a hoist on a horizontal beam. The hoist is free to move along the beam. The beam is attached at each end to runners that allow the beam to move perpendicular to its length. Freedom to move along the runners and the beam provides access to a rectangular area. (The hoist's vertical movement makes access three-dimensional.) A gantry crane is similar but with ground level runways and a vertical support beam that connects to the boom. Monorail cranes have a lifting device that can run along a single beam. Stacker cranes have a vertical beam fitted with a support platform and move along aisles for storing and retrieving items from racks.

Automated storage/retrieval systems (AS/RS) combine storage, picking equipment, and controls with various levels of automation for fast and accurate storage and retrieval of products/materials. Systems can be designed for large pallet loads, but miniload systems for smaller items, such as boxes, and even microload systems that can deliver small containers of parts to individual workstations are also popular. Microload systems, for instance, are often found in assembly of small products such as circuit cards. Many systems have high, deep racks with narrow aisle vehicles. Narrow aisle vehicles move horizontally and vertically through the aisle to access specific rack locations. The fork moves perpendicular to the vehicle's movement, avoiding the need for 90° vehicle rotation when grasping or depositing a load. A load pickup and dropoff point exists at the end of the aisle. Automated **storage carousels** provide an alternative design where the storage bins revolve, providing end-of-aisle depositing and picking. Storage and retrieval orders are sequenced based on location to minimize the rotation time between orders. Figure 9.4 illustrates a storage carousel.

Industrial trucks constitute another equipment class. This class of equipment includes fork lifts, hand carts, and tractor-trailer rigs like those used for transporting luggage in airports. Trucks are useful for intermittent moves over varying paths but require adequate aisles. The high-tech end of this class is occupied by **automated guided vehicles (AGVs)** (see Figure 5.4). We discuss AGVs in Section 9.3.

Containers such as **pallets** and **tote pans** are useful for aggregating product into a unit load for handling. In addition to increasing the amount of product transferred in a single move, the use of totes for unit loads keeps parts from becoming lost or separated and allows use of standard methods for the range of products that can fit into a single container size. A modicum of protection is provided as well.

Last, numerous miscellaneous auxiliary equipment types are available. A wide variety of attachments are available for lift vehicles and other equipment, thus allowing for controlled securing and movement of various product forms. Stations for palletizing and/or wrapping products are widely used in distribution. **Robots** have been widely used for pick and place operations. Automatic identification equipment qualifies as auxiliary equipment. Bar code and RF (radio frequency) systems play an important role in material control. We briefly discuss both of these. Other automatic identification technologies include machine vision and imaging, voice recognition, optical character recognition, and magnetic stripe scanning.

Figure 9.4 A storage carousel system for order picking. (Photo courtesy of The Buschman Company.)

Bar codes provide a quick, economical, nearly error-free method for automatically identifying a part type or the contents of a container. Bar codes can be identified with an error rate of less than one character per million compared with an error rate in the thousands per million for a keypunch operation. The bar code system consists of the code labels, bar code printer for constructing the codes, and an optical code reader and signal receiver for identifying the code and transmitting the information to a central computer. Code readers can be hand-held wands such as light pens. Light pens require direct contact between the code and the pen when being read. Alternatively, laser scanners can automatically read a code when it passes by in a range of up to several feet. Laser scanners are **fixed beam**, requiring the code to be passed in front of the beam, or **moving beam**, which surround a surface and search out a code. Bar codes use a series of dark, nonreflective bars and light, reflective spaces of varying widths to represent alphanumeric symbols. Codes consist of an initial blank space, a start digit, data characters, an end digit, and a trailing blank space. Several codes are in existence. The reader has undoubtedly seen the UPC (Uniform Product Code) used by food manufacturers and retailers for inventory and checkout purposes. Each code contains 12 digits, 1 for product type, 5 for the manufacturer, 5 for the specific item, and a parity check digit. The same code with a 2 to 5 digit addendum is used in the publishing industry. The manufacturing industry tends to use Code 39 (also called 3 of 9) for product distribution. This is also the standard code for the Department of Defense. The code draws its name from the fact that each character has exactly 3 wide elements (bars or spaces) and 6 narrow elements. Forty-three symbols are used: A, . . . ,Z, 0, . . . ,9, , - , . ,$, /,+, and %. The code can be of variable length and is scanned in either direction.

RF systems allow remote transmission of data to control computers. RF tags with thousands of bits of data can be attached to each product. The tags are also programmable. Information can be added and read from the tags as the product proceeds through its set of production operations. Basically, a product can carry its own process plan for guiding each machine visited and the output on each operation performed can be added to the encoded data to automatically construct a process and quality audit for the individual product unit.

9.1.2 Principles of Material Handling

The College-Industry Council on Material Handling Education (CICMHE) has adopted 20 Principles of Material Handling. The principles are meant as a guide or checklist for use when designing or modifying material handling systems. Although not absolute, they provide a benchmark for comparison of alternatives. Taken from the Material Handling Institute (the sponsor of CICMHE) with a paraphrased explanation, the principles are as follows:

1. **Orientation Principle.** Study the system relationships prior to specification to determine problems, constraints, and goals.
2. **Planning Principle.** Plan to meet requirements efficiently while maintaining flexibility for contingencies.
3. **Systems Principle.** Coordinate and integrate the receiving, inspection, storage, production, assembly, packaging, warehousing, and distribution systems.
4. **Unit Load Principle.** Use a large but practical unit load.
5. **Space Utilization Principle.** Effectively use all cubic space.
6. **Standardization Principle.** Use standard equipment and methods whenever possible.
7. **Ergonomic Principle.** Recognize human limitations and human–mechanical interactions when designing the system.
8. **Energy Principle.** Consider energy consumption in economic comparisons.
9. **Ecology Principle.** Minimize adverse environmental effects.
10. **Mechanization Principle.** Mechanize where feasible to increase efficiency.
11. **Flexibility Principle.** Use methods and equipment that can perform a variety of activities under a variety of conditions.
12. **Simplification Principle.** Simplify and eliminate handling steps where possible.
13. **Gravity Principle.** Gravity is free; use it.
14. **Safety Principle.** Provide safe methods and equipment that adhere to codes and acquired experiential knowledge.
15. **Computerization Principle.** Consider computerization and on-line data acquisition for handling and storage systems to improve control.
16. **System Flow Principle.** Integrate data and material flows.
17. **Layout Principle.** Analyze multiple viable sequencing and layout solutions, and select the most efficient and effective.
18. **Cost Principle.** Compare alternatives on the basis of cost per unit handled (delivered).
19. **Maintenance Principle.** Use preventive maintenance on handling equipment.
20. **Obsolescence Principle.** Prepare an economic plan for equipment replacement based on life-cycle costs.

The Simplification Principle forms perhaps the most basic rule of material handling procedure design and improvement. Moving material does not add value to a product, but it does cost money, risk damage, and take time. Thus, whenever possible, moves should be eliminated. Every move should be questioned as to why it is necessary. The point of use storage concept is the recognition of this fact.

9.2
EQUIPMENT SELECTION

In designing the material handling system, an equipment type must be specified for each material movement. This decision is partly based on economics but, as mentioned above, the best system is that which supports the overall mission of the production facility. However, once alternative methods have been identified that are acceptable with respect to satisfying the system needs, an economic model can be useful for selecting, or at least comparing, the alternatives. We will assume that M equipment types are available. N product moves are to be planned. The unit load size is assumed to be known for each part type and a tentative layout exists. Thus, the frequency and distance required for each move j, $j = 1, \ldots, N$ is known. Our decision variables are

$$X_{ij} = \begin{cases} 1 & \text{if equipment type } i \text{ is used for move } j \\ 0 & \text{otherwise} \end{cases}$$

We must also determine Y_i, the number of units of equipment type i acquired. Adding the cost parameters and technological coefficients:

c_{ij} = total variable operating cost per period for equipment type i to perform move j

C_i = fixed cost per unit-period for equipment type i

t_{ij} = time per move for i to perform move j

T_i = available time per unit-period for equipment type i

If equipment type i cannot perform move j, then set either c_{ij} or t_{ij} to a large number. The c_{ij} can normally be obtained by the product of the three multiplicative factors: (trips/period) × (distance/trip) × (cost/unit-distance).

EXAMPLE 9.1

A powered lift truck is available at a cost of $25,000. The engineer estimates a 7-year life with a $3000 salvage value. Operating costs of driver, and maintenance are estimated to be $2000 per month.[1] Plant experience suggests trucks are available about 75 percent of the time. Estimate the fixed and variable costs for this truck. Company policy sets a 10 percent internal rate of return for such calculations.

Solution

Using an interest rate of 10 percent per year, the present worth factor for a payment in 7 years is $(1 + 0.10)^{-7} = 0.5132$. The present worth of the $3000 salvage value is thus $(0.5132) 3000 = 1540$. This translates to an effective truck cost of $25,000 - 1540 = 23,460$. C_{truck} is found by converting this cost to a monthly equivalent. For simplicity, we now assume monthly compounding of interest. Dividing the interest by the 12 months per year and extending the time horizon accordingly, the 7 years at 10 percent translates to 84 periods at 10/12 percent. The capital recovery factor for finding the monthly equivalent with an interest rate of r for t years is

$$\left[\frac{r(1 + r)^t}{(1 + r)^t - 1} \right]$$

---

[1] The driver may be assigned other tasks when the truck is not needed and is therefore considered a variable cost.

With $r = 0.008333$ and $t = 84$, the factor computes to 0.0166. Thus, $C_{truck} = 0.0166(23,460) = \$389.50$ per month.

Since the exact moves are not yet known, the best we can do is to find variable cost in dollars per move time. Allocating the maintenance cost to moves, variable cost is $\$2000/0.75 = \$2666.67$ per month. Monthly costs $c_{truck,j}$ can be found by multiplying $\$2666.67$ times the estimated move time for j.   ∎

Given the ability to estimate costs, we can formulate the decision model

$$\text{minimize cost/period} = \sum_{i=1}^{M}\sum_{j=1}^{N} c_{ij}X_{ij} + \sum_{i=1}^{N} C_i Y_i \qquad (9.1a)$$

subject to

$$\sum_{i=1}^{M} X_{ij} = 1 \qquad \text{for all } j \qquad (9.1b)$$

$$\sum_{j=1}^{N} t_{ij}X_{ij} \le T_i Y_i \qquad \text{for all } i \qquad (9.1c)$$

$$X_{ij}\; 0 \quad \text{or} \quad 1 \quad Y_i \text{ integer} \qquad (9.1d)$$

The objective function 9.1a accumulates both variable-move and fixed-equipment costs per period. A variety of material handling types can be accommodated by this cost model. In general, the depreciation cost of trucks and containers would be contained in C_i. Costs based on the time equipment is used, such as walking by a material handler, or labor and battery recharging for a forklift would be placed in c_{ij}. Constraints 9.1b assure that all moves are assigned. The formulation selects a unique method for each move. Constraints 9.1c guarantee that sufficient material handling units are purchased for each type of equipment.

EXAMPLE 9.2

Three alternatives (pushcart, powered truck, and conveyor) are being considered for six interdepartmental moves. Problem data are given in Table 9.1. Costs are on a monthly basis. We have placed all conveyor costs in the variable component c_{3j} and set the $t_{ij} = 1$ to indicate that conveyors are fixed and cannot be used for more than one point-to-point move. Because of interference with other plant operations, a conveyor is impractical for move 6. Accordingly, we use a large cost. Move 3 is rather short and not an efficient use of a powered truck. Conveyor cost is determined by depreciation and load/unload operations. Utilizations t_{ij} are given as proportions of an equipment unit taking availability into account. $T_i = 1$ in this case. Model the decision problem and solve for the optimal material handling system.

Solution

With this data, formulation 9.1 becomes

$$\text{minimize } 1286.40X_{11} + 2680.00X_{12} + \cdots + 1050.00X_{35}$$
$$+ 9999.00X_{36} + 75.00Y_1 + 389.50Y_2$$

Table 9.1   Problem Data for Example 9.2

	Equipment Type					
	1		2		3	
Move	c_{1j}	t_{1j}	c_{2j}	t_{2j}	c_{3j}	t_{3j}
1	1286.40	0.72	640.00	0.24	890.00	1.0
2	2680.00	1.50	1333.33	0.50	3600.00	1.0
3	268.00	0.15	320.00	0.12	325.00	1.0
4	375.20	0.21	186.67	0.07	1200.00	1.0
5	268.00	0.15	133.33	0.05	1050.00	1.0
6	643.20	0.36	320.00	0.12	9999.00	1.0
C_i	75.00	389.50	0.00			

subject to:

$$X_{11} + X_{21} + X_{31} = 1$$

$$\ldots$$

$$X_{16} + X_{26} + X_{36} = 1$$

$$0.72X_{11} + 1.50X_{12} + 0.15X_{13} + 0.21X_{14} + 0.15X_{15} + 0.36X_{16} \leq Y_1$$

$$0.24X_{21} + 0.50X_{22} + 0.12X_{23} + 0.07X_{24} + 0.05X_{25} + 0.12X_{26} \leq Y_2$$

$$X_{ij} \ 0 \quad \text{or} \quad 1 \quad Y_1, Y_2 \quad \text{integer}$$

Note that the Y_3 variable for conveyors was not needed in the model, since its cost coefficient is 0. Solving the integer program we find that the solution $x_{21} = x_{22} = x_{24} = x_{25} = x_{26} = x_{33} = 1$, $Y_2 = 1$ and all other variables are 0. This solution implies purchasing a conveyor for move 3 and a lift truck for all other moves. Solution cost is \$3327.83 per month. The lift truck is used 98 percent of the time. Suppose it was then decided that planned truck usage could not exceed 90 percent due to maintenance, breakdowns, demand variability, and occasional additional work assignments. The model is then resolved with a coefficient of 0.9 added to the Y_2 variable in the last constraint. The modifed model is optimized by acquiring a conveyor for move 1 and using the lift truck for the other 5 moves. Monthly cost is \$3572.83, a \$245 increase, but the lift truck is now used only 86 percent of the time on average.   ∎

For large problems, the mathematical program 9.1 can be difficult to solve because of the integer Y_i variables (equation 9.1d). Suppose we relaxed the integrality constraints on the Y_i. The implication is that we could purchase fractional pieces of equipment. Assuming $C_i \geq 0$, we will purchase only the amount of equipment type i needed to make the capacity constraints 9.1c feasible. From constraints 9.1c, we find that

$$Y_i = \frac{\sum_{j=1}^{N} t_{ij} X_{ij}}{T_i} \tag{9.2}$$

Thus, we can remove variables Y_i from the model. If we substitute the relation 9.2 into relation 9.1a, we obtain the objective

$$\text{minimize} \sum_{i=1}^{M} \sum_{j=1}^{N} c'_{ij} X_{ij}$$

where $c'_{ij} = c_{ij} + C_i(t_{ij}/T_i)$. The remaining mathematical program has a very simple structure. A method can be selected for each move independently, since only constraints 9.1b and 9.1d remain. The optimal solution occurs when each move is assigned to the equipment type with $\min_i c'_{ij}$. (Ties are broken arbitrarily.) This trivial 0–1 program provides two types of information. First, we have a lower bound on the optimal solution to the actual integer problem. Second, we know the best equipment type for each move if, by chance, each piece of equipment were to be fully utilized. Suppose we made the same X_{ij} assignments indicated by the relaxed problem. We can then compute the actual number of pieces of each equipment type needed. Rounding up this quantity to the next integer, we obtain a feasible solution to the equipment selection problem. This offers an upper bound against which any solution can be compared.

EXAMPLE 9.3

Find a lower and upper bound on the solution to the problem of Example 9.2 by first solving the relaxed problem and then rounding up all Y_i values.

Solution

Adjusted variable costs found by adding a t_{ij} portion of c_i to each c_{ij} are shown in Table 9.2.

Table 9.2   Adjusted Material Handling Costs for Example 9.3

Move	c_{1j}	c_{2j}	c_{3j}
1	1340.40	733.48	890.00
2	2792.50	1528.08	3600.00
3	279.25	366.74	325.00
4	390.95	213.94	1200.00
5	279.25	152.81	1050.00
6	670.20	366.74	9999.00

A lower bound is found by selecting the least expensive alternative in each row. This produces the solution of a fork truck for moves 1, 2, 4, 5, and 6 and a manual cart for move 3. Summing the adjusted costs yields a lower bound of $3274.30. The implied integer solution is $x_{13} = x_{21} = x_{22} = x_{24} = x_{25} = x_{26} = Y_1 = Y_2 = 1$, and all other variables are 0. The cost of this feasible solution gives the upper bound of $3345.83. This is $18.00 per month more than the optimal solution found earlier.   ■

Rounding the Y_i values to the next larger integer yields a heuristic solution. It is desirable to find out if improved solutions can be found. Assuming that we wish to keep the same set of equipment types, an improved solution can be found only by transferring moves to eliminate one or more pieces of equipment. If we specify the number of pieces of each equipment type (the Y_i), we obtain the 0–1 integer program 9.1, where the Y_i are now constants. Consider what happens if we ignore the 0–1 restrictions on the X_{ij} and treat the new model as a linear program. The X_{ij} now become the proportion of move j that is performed by material handling equipment type i. If we solve this linear program, our basic, feasible solution can have at most $N + M$ positive X_{ij}, one per constraint. Since at least one method is required for each move, this leaves at most M instances in which multiple methods are used for a move. Typically, $M \ll N$. Since the LP model can be easily resolved for changes in any Y_i, we can play "what-if" games. After finding the upper bound solution, we can search for better solutions by selecting an equipment type i, reducing Y_i by one unit, and solving the linear program 9.1. Note the second term in the objective $(\sum_i C_i Y_i)$ is now a constant. In addition to removing a handler of type i, we can try to replace a unit of i with an additional unit of any other handler i' where $C_i > C'_i$. This alternative allows some addition to capacity while still reducing fixed equipment cost.

EXAMPLE 9.4

The analyst solving this problem had obtained the heuristic solution of Example 9.3 but questioned the need to buy a manual cart and have a manual material handler who would be used only 15 percent of the time. What happens if these options are restricted?

Solution

Out of curiosity the analyst set $Y_1 = 0$, removing the first handling equipment type. At the same time, recognizing that a second lift truck would be even less utilized, the analyst set $Y_2 = 1$. The analyst proceeded to solve the linear program using the 18 variables x_{11} to x_{36}. The solution was to purchase 0.833 of the conveyor for move 3 and let the lift truck handle all other movement. The cost of this solution would be \$3327.00. Of course, we cannot acquire only 0.833 of a conveyor. To find a feasible solution we must add the other $(1 - 0.833)C_3 = \$54.17$. However, if we purchase this conveyor, we can save the variable portion of the lift truck cost used for move 3, since all of move 3 will be by conveyor. This gives a savings of $0.167 \cdot 320.00 = \$53.33$. The net increase is only 0.83. This is the optimal solution found in Example 9.2. Of course, the analyst did not know this solution was optimal, only that it was within $[100(3327.83 - 3274.29)]/3274.29 = 1.6$ percent of the lower bound. Concluding this to be close enough, the analyst worked on the final details of preparing this recommendation.   ∎

A secondary objective in the equipment selection problem is to minimize the number of material handling equipment types that are used. This simplifies system operation. Our model allows a consideration of these options by completely removing any undesired equipment type. We saw this when the manual material handler was eliminated in the previous example.

Before proceeding, it is important to note that the equipment selection model described in this section made a key assumption. Given information included move volumes. How many plant engineers know precisely the number of interdepartmental moves for each product over the anticipated life of new handling equipment? Many expensive handling systems have been installed only to sit idle as demand shifted between products or withered away altogether. In a dynamic world, equipment flexibility and modularity are important. **Flexibility** refers to the ability to handle different product sizes, shapes, weights, paths, and volumes with the same equipment. **Modularity** concerns the ability to change flow path and throughput capacity by replicating equipment. We may need to buy an extra truck or parallel conveyor, or to add another section to the AGV track or to link another department to the conveyor. The safest decision is to acquire equipment that can handle a variety of product types and can be reconfigured as the layout and move requirements vary. Likewise, purchasing modular equipment allows handling capability to closely track requirements. Most equipment types are amenable to modular usage; the layout engineer must, however, be careful in designing the system to allow for future changes in scale without major disruption of operations.

9.3
RECEIVING BULK LOADS

A common scenario is for a large load of material to be received at the same time for processing. This may occur at the loading dock, a warehouse, or a workstation. Results are known (Tompkins and White [1984]) for the case of when orders arrive according to a Poisson process (exponentially distributed interarrival time) and each unit requires an exponential processing time. Let b be the number of loads per arrival, a random variable. We assume that b is independent of the time between arrivals and service time for the loads in the arrival. λ is the rate at which bulk arrivals occur, and μ is the rate at which individual loads are serviced. Let L be the average number of loads waiting to be serviced and W be the average time duration between when a load arrives and its service is completed. Subscript q indicates that the measure relates only to time waiting in the queue for service to begin. With a single server, our situation corresponds to an $M^b/M/1/\infty$ queueing system for which

$$L + \frac{\lambda[V(b) + E^2(b) + E(b)]}{2[\mu - \lambda E(b)]} \tag{9.3a}$$

$$L_q + L - \frac{\lambda E(b)}{\mu} \tag{9.3b}$$

$$W = \frac{V(b) + E^2(b) + E(b)}{2E(b)[\mu - \lambda E(b)]} \tag{9.3c}$$

$$W_q = W - \frac{1}{\mu} \tag{9.3d}$$

EXAMPLE 9.5

Batches of parts arrive at a warehouse for storage. The number of pallet loads in a batch has mean 5 and variance 4. Batches arrive at the rate of 10 per day. A single stacker is available for storing loads. Average time to store a load is exponentially distributed with mean of 0.01 days. Find the expected number of loads waiting to be stored.

Solution

We are given $\lambda = 10$, $\mu = 100$, $E(b) = 5$, and $V(b) = 4$. From equation 9.3a,

$$L = \frac{10[4 + 25 + 5]}{2[100 - 10(5)]} = 3.4 \text{ loads}$$

Unfortunately, instead of the mean number of loads, we are probably more interested in a reasonable upper bound on the number of loads, as we must supply accumulation space for the maximum number of loads. A more complex system state model is necessary to find the probabilities of various numbers of loads waiting. Approaches for constructing such models are shown in Chapter 11.   ■

9.4
CONVEYOR ANALYSIS

Conveyors are loaded with material. The material travels along the conveyor and is eventually unloaded. During that trip, material may visit workstations. While at a workstation the material may remain on the conveyor for processing or be temporarily removed for processing. The conveyor must be designed to provide the desired levels of performance in the intended environment. In a recirculating (closed-loop) conveyor, for instance, an overloaded unload station may cause material to recirculate, passing through the loading and processing stations one or more times before finally being unloaded.

Standard decision variables for conveyor design include speed, spacing of carriers, conveyor length, carrier capacity, and number of load and unload stations. Researchers have applied deterministic and probabilistic modeling techniques to analyzing conveyor performance in many environments. Muth and White [1979] review many of these models. We will discuss representative models for some of the more widely studied situations. Models vary based on the assumptions regarding the number of loading and unloading stations, interarrival times at loading, and service times at loading and unloading. Interarrival times may be deterministic or stochastic. Even if deterministic, interarrival times may be constant or cyclical.

9.4.1 Closed Loop Conveyors

Closed loop conveyors, which revolve at a constant speed along a fixed path, have found a place in both industrial use and analytical modeling. Part carriers are equally spaced along the length of the conveyor. Consider the conveyor shown in Figure 9.5. Parts are loaded onto carriers at M_l loading stations. Parts are unloaded from carriers at M_u unload stations. In between, parts may have operations performed at any or all of M_w workstations. The conveyor is assumed to travel at speed v ft/min, and there are N carriers or storage locations equally spaced along the conveyor. Each carrier can hold c parts.

Load/Unload Capacity

As an example of a closed loop conveyor model, we will examine the situation studied by Bastani [1988]. Units arrive at a single loading station every λ^{-1} minutes

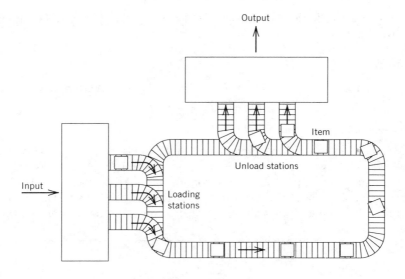

Figure 9.5   A closed-loop conveyor.

(deterministic interarrivals). Each carrier can hold one unit, and carriers are spaced a distance d apart on the conveyor. If the next carrier passing the loading station at the time the unit arrives is full or if the loading station is busy loading a previous arrival, the arriving unit is set aside to be handled by an alternate method. The time to unload a unit is μ^{-1}, also deterministic. We will allow $M_u \geq 1$ unload stations. The first time a unit passes an idle unload station, it is unloaded. If all M_u stations are busy, the unit recycles past the load station.

    If units arrive faster than they can be loaded, blocking will occur on a regular basis. This condition occurs if $\lambda > v/d$, that is, arrivals per time exceed the number of carriers passing the load station per time. In fact, we can only load one of every k consecutive arrivals, where k is the smallest integer greater than or equal to $\lambda \cdot d/v$. If $k > 1$, then each time a load operation begins, the next $k - 1$ arrivals are set aside because of the unavailability of a carrier. We will assume that loading time is less than interarrival time. Otherwise, a similar blocking argument owing to the load station being busy is necessary.

EXAMPLE 9.6

A 100-foot conveyor has carriers spaced every 10 feet. Parts arrive at the rate of two per minute to be loaded. Actual loading takes very little time. Find the minimum conveyor speed to prevent blocking of incoming arrivals due to unavailability of a carrier.

Solution

The required condition to prevent blocking is $v \geq d \cdot \lambda$. Thus, $v \geq d \cdot \lambda = 10 \cdot 2 = 20$ feet per minute.   ■

It may seem that it would be foolish to design a conveyor with $k > 1$. If the system were truly constant through time, $k < 1$ would certainly be a sensible criterion.

However, suppose the conveyor is fed by a production process with periodic runs due to setup. Periods of arrivals with rate λ may be offset by periods without arrivals. During arrivals, some parts must be stored at conveyor loading. These parts are then added when arrivals cease. In this case, conveyor feasibility may be based on the long-term arrival rate. Planning should provide for the accumulation of $(k-1)/k$ proportion of product at the input station during the production run.

Unloading stations must be designed to handle conveyor traffic. This concern is satisfied provided that $M_u\mu \geq \lambda/k$, that is, the unload stations can keep pace with the load stations and prevent recycling. Even when capacity is sufficient, some banking of arrivals may be necessary at unloading. With discrete carriers, arrivals need not be evenly spaced. If $\lambda \cdot d/v = \frac{2}{3}$, for instance, then a pattern will develop of two full followed by one empty carrier. The time for the conveyor to advance from one carrier to the next is d/v. A single unload station that takes between d/v and $1.5d/v$ to unload a part will have sufficient capacity but will not be ready to unload the second consecutive unit in each pair when it first arrives. Thus, one half the units will recycle back to the loading station, causing congestion, while the unload station operates at less than full capacity. Eventually a steady-state situation will develop, but not necessarily without some blocking of loading due to recycled units.

Setting Carrier Capacity

The previous model concentrated on feasibility of the load/unload stations. Alternatively, the major issue may be one of determining the sufficiency of conveyor (carrier) capacity. The carrier may be a formal holding device or simply a length of the conveyor capable of being accessed at the same time by a workstation. We will consider a deterministic model. By deterministic we mean that the volume and timing of load/unload requests are known.

A conveyor is shown in Figure 9.6. As before, let N be the number of carriers. M is the number of load/unload stations. The amount of material loaded onto the jth carrier to pass station i is given by $f_i(j)$. Negative values of $f_i(j)$ indicate attempts to unload the carrier. We assume a repeating period of length p for load/unload activities such that

$$f_i(j) = f_i(j + p)$$

As an example, if a load station adds a standard carton to every second carrier passing the station, we would have $p = 2$ and $f_i(j) = \{1, 0\}$.

Since conveyor stability requires loading to equal unloading over the cycle p, we must have

$$\sum_{i=1}^{M} \sum_{j=1}^{p} f_i(j) = 0$$

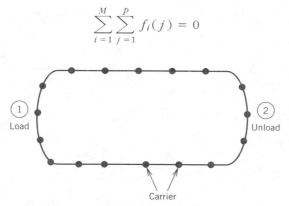

Figure 9.6   An example of a deterministic loop conveyor.

In addition to each carrier being one of N, carriers are also identified by their current position j in the load/unload cycle. We assume that the $f_i(j)$ are defined such that they are coordinated between workstations. A carrier corresponding to $j = 1$ at the first workstation would also correspond to $j = 1$ at each subsequent station visited on the same trip around the conveyor. On any trip around the conveyor, a carrier in position j will change its load by $\sum_{i=1}^{M} f_i(j)$, the sum of the loads (unloads) at each station.

Consider an arbitrary carrier n just leaving station i. The N carriers to pass station i before carrier n returns represent $\lfloor N/p \rfloor$ complete cycles of length p plus r additional carriers. Each time around the conveyor, a carrier's position increases by $r = N \bmod p$. Suppose we have seven carriers and a load/unload cycle of five; then if on the previous visit a carrier received $f_i(1)$ loads, it will now receive $f_i(1 + 7)$ loads. Station i sees the carrier as the eighth to arrive. With a five period cycle, $f_i(8)$ is the same as $f_i(3)$, that is, $r = 2$ and $r +$ old position $= 2 + 1 = 3$.

We will determine the required carrier capacity by taking a ride on the carrier and recording its load levels. Start with an empty carrier in position 1 arriving at station 1. We progress through all workstations maintaining position 1. On the next cycle around the conveyor, the carrier is in position $1 + r$. We continue the trip until we return empty to position 1 at station 1. This is a complete rotation of states for the carrier. If we continued, we would just be repeating the past. Minimum required carrier capacity is then the difference between the maximum and minimum levels of loads observed.

EXAMPLE 9.7

Consider again the two-station conveyor of Figure 9.6. The conveyor has 17 equally spaced carriers. The load/unload cycle has length 6 given by $f_1(j) = \{1, 2, 0, 3, 0, 0\}$, $f_2(j) = \{-2, 0, -2, 0, -2, 0\}$. Consider an empty carrier just arriving empty at load station 1 at the start of a cycle. Track the contents on the conveyor.

Solution

We must find $r = N \bmod p$. For this example, $r = 17 \bmod 6 = 5$. Let b be the load on the carrier. Starting with an empty carrier ($b = 0$) in position 1, we visit station 1. Since $f_1(1) = 1$, we increase b to 1. Our next stop is the unload station. As $f_2(1) = -2$, our load is reduced and $b = -1$. When we return to station 1, it is 17 carriers since our last visit. Essentially we are the eighteenth carrier seen by the station. This puts us sixth in position. Equivalently, $(1 + r) = 6$. In position 6, we receive $f_1(6) = 0$ and b remains at -1. Continuing, we encounter the states shown in the table.

	Cumulative Loads (b)	
Position	Station 1	Station 2
1	1	-1
6	-1	-1
5	-1	-3
4	0	0
3	0	-2
2	0	0

The maximum change in b is from 1 to -3, a change of 4. Thus, carriers need a capacity of at least 4. Actual load levels will be 3 above the values shown in the table. This makes all values nonnegative and still less than or equal to 4, the carrier capacity. ■

9.5
AGV SYSTEMS

Conveyors are used for fixed-path, point-to-point material transfer. Manned trucks provide full path flexibility. In between these two extremes lies the AGVS. The system can transport material between a finite number of locations. The locations must be designated and programmed into the system controllers in advance, but locations can be changed with some effort. AGVs are important in that they support asynchronous assembly (unpaced lines) and can be integrated into unmanned manufacturing systems. Using data acquistion and transmittal capablilites, AGV systems provide material control as well as transport. A central computer assigns transport tasks to vehicles. Monitoring of vehicle positions and traffic control can be performed by the central computer or through local controllers, each of which controls a section of the path. The local controllers can store complete status and path information on their domain and pass off vehicles to one another at intersection points. Controllers on each vehicle execute instructions and monitor their immediate area to ensure safety. Most existing systems navigate by an **inductive guidepath:** A wire embedded in the floor carries an alternating current to induce a magnetic field, which is detected by antennae mounted on the bottom of vehicles. Open loop dead reckoning, optical/chemical guidepaths, beacon systems with vehicle-mounted receivers and stationary (ceiling mounted) beam transmitters, and inertial guidance systems (position determined by double integration of rotational and translational acceleration) have also been used for navigation. Vehicles may also normally travel on a tow line embedded in the floor and then occasionally be disconnected to roam to off-line locations.

AGVs can be used either to "pick up and drop off" loads or as mobile part fixtures. In the first instance, loads are picked up at a location and transported to another location. Locations correspond to workstations or storage locations. After completing service at the workstation, the load, consisting of one or more parts, waits at the pickup point for an AGV. The pickup point is the workstation's output buffer. Dropoff points are workstation input buffers. In the second instance, the part stays on the AGV during processing. The AGV forms the base for holding the part. Gould [1990] describes the use of AGVs in two automobile assembly plants. Each AGV carries an automobile through the assembly process. More than one thousand vehicles serve over a hundred workstations. The vehicles interface with monorails, robots, tow lines, and power and free conveyors throughout the production of the automobile. The vehicles receive RF signals at fixed communication points to direct the in-process automobile to the proper parallel workstation at each stage of the process. Line balancing can thus be accomplished in real time.

We will now discuss specifying the AGVS for a pickup and dropoff environment. In specifying the AGV system, we must determine the location of pickup points P, drop off points D, the path that connects the P and D locations, the number of vehicles to place in the system, and the routes that these vehicles will take. The path layout and the location of the P and D points are design issues; the number of vehicles and routing rules are operational issues. We begin with a few comments on design and then proceed to operation.

9.5.1 AGV System Design

The general location of P and D points is determined by the layout of the I/O and processing stations served by the AGV system. We may only have pickups at the system's input point and dropoffs at the output point, but most workcenters will have both pickups and dropoffs. In terms of hardware, the P and D locations may or may not be the same physical location. (For instance, a machining center may have an indexing rotary table handling all I/O, or it may have separate input and output buffers.) The specific location of these points and the connections between them have a fundamental effect on the cost and performance of the system.

The guide path normally follows the existing aisles in the facility. The aisle intersections and pickup/delivery points can be considered as nodes on a graph that describes the AGV flow path. Directed arcs between nodes indicate the direction of vehicle flow. The decision problem concerns the choice of direction on each arc. It is assumed that vehicles will follow the shortest directed path (or a suitable uncongested path) once the path is defined. Gaskins and Tanchoco [1987] and Kaspi and Tanchoco [1990] present a 0–1 programming model for assisting in the choice of travel direction for each arc on the graph. The objective is to select a set of arcs connecting P and D points to minimize the total loaded travel. Major constraints include those ensuring that (1) if an arc enters a node then another arc must leave that node; and (2) arc choices allow travel from any node to any other node.

In practice, several designs should be considered and evaluated for congestion. Figure 9.7 gives a schematic layout of a facility and two possible AGV paths. Both alternatives adopt the traditional approach of a continuous path wherein vehicles can traverse the entire system. Daily flows between departments are provided in Table 9.3.

The path with the shortest total length is not necessarily best. Spurs, such as arc a,b in Figure 9.7b, may be useful for shortening trips and reducing vehicle interference. Several principles of good design can be helpful in path layout. As the rules are examined, one should refer to the alternatives to observe how the design rules were implemented.

Path Design Rules

1. Travel should be unidirectional unless traffic is very light. (This minimizes blocking.)
2. Pickup stations should be downstream of dropoff stations. (The vehicle should be able to drop off its load and then pick up a new load.)

Table 9.3   Interdepartmental AGV Flows

From-To	1	2	3	4	5	6	7	8	9	Sum
1	—	40	25	30	10	10	20	5	10	150
2		—	40		30		10	10		90
3			—				50		10	60
4		5	10	—		10				25
5				100	—					100
6				60		—				60
7						40	—		40	80
8				10		5		—		15
9					60				—	60
Sum	0	45	75	200	100	65	80	15	60	640

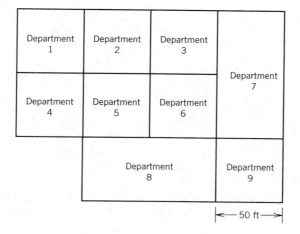

(*a*) General facility layout

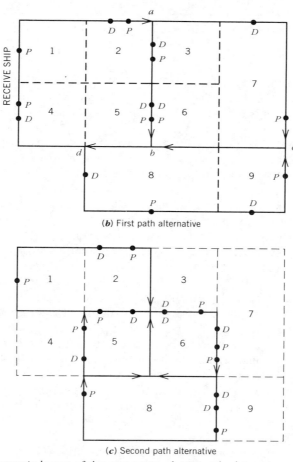

(*b*) First path alternative

(*c*) Second path alternative

Figure 9.7   A schematic layout of departments and AGV path alternatives.

3. For each pickup point along a segment, total dropoffs from the start of the segment to this pickup point should be at least as large as total pickups to this point in the segment.[2] (The goal is dual command operation for each segment transversal.)
4. Place P and D points on low usage segments. (This avoids blocking of vehicles attempting to bypass a P or D point.)
5. During operation, if empty vehicles enter and stop on a segment to pick up, then no vehicles should leave the segment empty after dropping off a load in the segment. The two trips through the segment should be combined. Similarly, if full vehicles enter to drop off a load and then leave empty, then no vehicles should enter empty and leave full. (If possible, use dual command visits to a segment to reduce empty travel time.)
6. Spurs (bypasses) and shortcuts may be considered for reducing trip distances and vehicle blocking; however, the price of this additional flexibility and performance is increased complexity of vehicle control.

EXAMPLE 9.8

Consider the data in Table 9.3. Generate two possible material handling paths.

Solution

Two possible paths were shown in Figures 9.7*b* and 9.7*c*. In developing the flow paths it was helpful to have knowledge of the prevailing flow pattern in the facility. Material is received at department 1 and distributed throughout the facility. Although not a flow shop, material tends to proceed along the length of the facility and then reverse direction, eventually ending at department 4. In each case, undirected segments were first constructed to ensure at least one face in common with each department. The common face(s) would have to include feasible P and D locations based on the layout within each department. Once a proposed path was set, segments were given direction in accord with prevailing material flow patterns and directional compatibility (we cannot have all of the arcs beginning or ending at an intersection point).

We could have first specified P and D points and then applied the above-referenced 0–1 program to select directed arcs. Instead, we utilized potential flexibility in the P and D locations during system design and first constructed a directed path between departments. Note that all D points are upstream of P points. An attempt was made to minimize the number of stops on segments that are expected to be heavily utilized. Consider the second alternative. The vertical segments between departments 2 and 3 and between 5 and 6 do not have any P or D points for this reason. Virtually all incoming flows will take the vertical segment between 2 and 3, and hence excessive stopping could cause blocking. A similar problem could arise for department 2 pickups and deliveries. However, moving these points to the segment separating departments 2 and 5 could be equally problematic. At this point, quantitative flow data are needed to evaluate the proposed path.

The first path has a total length of 950 feet compared to 800 feet for the second path. Of greater importance, however, is the number of vehicles needed

---

[2]A segment of the material handling path is any portion of the path from one intersection point to another. Once a vehicle enters a segment, it must traverse the entire segment. An example is the link from node *a* to *b* in Figure 9.7*b*.

and the ability of the layout to adapt to new requirements. The number of vehicles needed will be determined later in the chapter. Congestion on segments
and distances between departments, particularly those with high flow volumes,
are major determinants of vehicle requirements.   ■

Considerable theory exists for aiding the location of points along networks (Handler and Mirchandani [1979]). Our problem of locating P and D points along the
guidepath falls into this category. Typical objectives are minimization of average or
maximum trip length. Unfortunately, congestion is usually not easily included in these
models, and blocking is a primary concern in AGV path design.

Bozer and Srinivasan [1989, 1991] propose an alternative layout approach, as shown
in Figure 9.8. The tandem system approach is composed of a set of interconnected
loops, each loop having one vehicle. To transfer a load from one loop to another,
either a buffer is needed at the intersection of the loops or both vehicles must meet
simultaneously to effect the transfer. The advantage of this approach is the elimination
of vehicle blocking and the ease of traffic control for the system. Disadvantages
include the inability to increase throughput by adding vehicles and the need to
transfer loads. While additional vehicles could be added to a loop, blocking would
become a problem.

9.5.2 Estimating Vehicle Requirements

Given a guidepath, we next determine the number of vehicles needed to support the
material handling requirements. Although the actual problem is stochastic owing to
breakdowns, random transport requests, and vehicle blocking, we will use a simpler
deterministic model to estimate requirements. Total vehicle utilization time can be
divided into five components: loaded travel time, unloaded travel time, blocked
time, load time, and unload time. We assume that the number of loads to be picked
up at each P and deposited at each D is known. These would be determined by
production and process plans. As the tentative layout path of the system has been
specified, it is relatively straightforward to find the shortest path between P and D

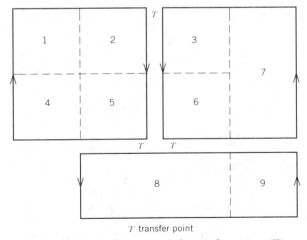

T transfer point

Figure 9.8   Three-vehicle tandem AGV system with transfer points (T).

locations.[3] Combining the shortest paths with specifications on AGV travel speeds and acceleration, loaded travel time can be computed. Likewise, specifications on load and unload time can be multiplied by the number of loads to find total load and unload time for vehicles.

EXAMPLE 9.9

For the data in Example 9.8, assume vehicles require 30 seconds per load or unload operation and travel at 5 feet per second. The 30 seconds includes time lost in acceleration and deceleration. Find vehicle time per day for loading, unloading, and loaded travel.

Solution

Using alternative 1 in Figure 9.7b, we can construct a matrix of travel distances from P to D points. The distances in Table 9.4 were computed using the directed path segments and a grid superimposed on Figure 9.7b, which produced the coordinate locations

Department	D	P
1	(0,120)	(0,130)
2	(70,150)	(80,150)
3	(100,130)	(100,120)
4	(0,70)	(0,80)
5	(100,80)	(100,70)
6	(100,80)	(100,70)
7	(170,150)	(200,70)
8	(50,25)	(100,0)
9	(175,0)	(200,30)

Table 9.4   Loaded Travel Distances (Feet) for Alternative 1

From P to D	1	2	3	4	5	6	7	8	9
1	—	90	140	340	190	190	190	295	445
2	290	—	40	240	90	90	90	195	345
3	240	340	—	180	40	40	440	145	295
4	40	140	190	—	240	240	240	345	495
5	190	290	340	140	—	390	390	95	245
6	190	290	340	140	390	—	390	95	245
7	290	390	440	240	490	490	—	195	345
8	420	520	570	370	620	620	620	—	75
9	290	390	440	240	490	490	490	195	—

---

[3] If shortest paths are not obvious from inspection of the guidepath, then model the path as a network with P, D, and intersection points being nodes and path segments being arcs. Arc lengths are segment traversal times. A matrix of shortest from-to distances between nodes can be initialized with arc distances for connected nodes and ∞ for other node pairs. A dynamic programming recursion is used to construct shortest distances between all node pairs (i, j). At iteration $k, k = 1, \ldots, N$ where N is the number of nodes, replace d_{ij}, the distance from i to j, by the smaller of the old d_{ij} and the alternative $d_{ik} + d_{kj}$. After N iterations, all shortest paths are known.

As an example of distances, consider the trip from department 9 to 6. The load is picked up at (200, 30). The vehicle travels 20 feet and then turns left to travel along the aisle separating departments 7 and 9. The vehicle proceeds 200 feet before turning right. The vehicle then proceeds 100 feet along the border of the facility before turing right again at (0, 150). The vehicle travels another 100 feet before turning right just as it reaches department 3. The vehicle must then travel 70 feet to location (100, 80), where it deposits its load at department 6. Total distance = 20 + 200 + 100 + 100 + 70 = 490 feet.

Total loaded travel distance can be found by multiplying the value in each cell of Table 9.3 by the corresponding value in Table 9.4. Summing over all cells gives a total loaded travel distance of

$$\text{total loaded travel distance} = 40(90) + 25(140) + \cdots + 60(490)$$
$$= 159,925 \text{ feet}$$

At 5 feet per second, 159,925 feet requires 8.88 vehicle hours per day.

From Table 9.3, vehicles carry 640 loads per day. At 30 seconds delay per load or unload, this yields 320 minutes or 5.33 hours per day spent loading and 5.33 hours per day unloading. ∎

Next, we consider the amount of empty travel time experienced by the AGVs. In a highly utilized system, empty travel consists mostly of travel from the dropoff point to the next pickup location. Our approach will be to find the minimum amount of empty travel time possible to satisfy load movement requirements. Decision variables are the amount of empty trips between locations in the system. We will assume that all pickups and deliveries are available at the start of the period, or at least that we can order our trips in accord with production so wasted movement can be avoided.

Maxwell and Muckstadt [1982] use a transportation model to find a lower bound on empty travel. Once again view the guidepath as a network. Although we could treat each P and D point as a node of the network, we will make a simplifying assumption that allows us to treat each department as a single node. When a load is delivered, the now empty vehicle will first attempt to pick up a load in the same department. Only if no loads are waiting will the vehicle be sent elsewhere to pick up a load. Provided that P stations are close but downstream of D stations, this is a reasonable policy. Other locations of interest, such as recharge stations or path intersections, can also be included as nodes. Let t_{ij} be the shortest travel time from node i to j and v_{ij} be the number of loads moved from node i to j per period (see Table 9.3). Let I_i and E_i be the desired initial and ending number of vehicles at node i. Our decision variables are x_{ij}, the number of empty moves from node i to j. Vehicle conservation constraints then yield the relations

$$E_i = I_i + \sum_j v_{ji} - \sum_j v_{ij} + \sum_j x_{ji} - \sum_j x_{ij} \qquad (9.4)$$

Equation 9.4 states that the ending number of vehicles at node i is equal to the starting number of vehicles plus those that arrive full or empty minus those that leave full or empty. Rearranging terms to isolate the decision variables, we obtain

$$\sum_j x_{ij} - \sum_j x_{ji} = I_i - E_i + \sum_j v_{ji} - \sum_j v_{ij} \qquad (9.5)$$

Our objective is to

$$\text{minimize} \sum_i \sum_j t_{ij} x_{ij} \qquad (9.6)$$

Since all $t_{ij} \geq 0$ and $x_{ij} \geq 0$ and t_{ij} represent shortest paths, it is clear that empty carts should not enter and leave any node, that is, at most one of the sums $\sum_j x_{ij}$ and $\sum_j x_{ji}$ is positive for each i. Otherwise we have some i, j, k such that $x_{ij} > 0$ and $x_{ki} > 0$. Suppose that, instead, we increase x_{kj} by minimum $\{x_{ij}, x_{ki}\}$ and reduce both x_{ij} and x_{ki} by this same amount. The LHS of equations 9.5 are unchanged numerically, and the objective function is reduced by $(t_{ki} + t_{ij})x_{kj} - t_{kj}x_{kj}$. If the shortest path from k to j passes through i, then the change does not affect the solution; otherwise, $t_{ki} + t_{ij} - t_{kj} > 0$ and the solution is improved. Setting $g_i = I_i - E_i + \sum_j v_{ji} - \sum_j v_{ij}$, (the RHS of equations 9.5), we are left with the problem

$$\text{minimize } \sum_i \sum_j t_{ij} x_{ij} \tag{9.7a}$$

subject to

$$\sum_j x_{ij} = g_i \qquad \text{for all } i \text{ such that } g_i > 0 \tag{9.7c}$$

$$\sum_j x_{ji} - g_i \qquad \text{for all } i \text{ such that } g_i < 0 \tag{9.7d}$$

Equations 9.7 constitute a classical transportation problem. Constraints 9.7b represent sources with excess vehicles to be shipped to the destinations of constraints 9.7c. Existing algorithms can easily solve problems of the size needed for the typical AGV system.

EXAMPLE 9.10

For the alternative shown in Figure 9.7b, find a lower bound on the empty travel time. There is no desired accumulation or depletion of vehicles for any station.

Solution

Since no change in vehicle location is desired, $I_i = E_i = 0$. Using the $\sum_j v_{ij}$ row and $\sum_j v_{ji}$ column in Table 9.3, we can compute the g_i as follows:

$$g_1 = I_1 - E_1 + \sum_j v_{j1} - \sum_j v_{1j} = 0 - 0 + 0 - 150 = -150$$

The $\sum_j v_{ij}$ are the row sums and the $\sum_j v_{ji}$ are the column sums of Table 9.3. The remaining g_i are found in a similar manner yielding:

Department	g_i
1	-150
2	-45
3	15
4	175
5	0
6	5
7	0
8	0
9	0

Based on the g_i, we must ship 150 empty vehicles to department 1 and 45 to department 2. These vehicles come from departments 3 (15 vehicles), 4 (175 vehicles), and 6 (5 vehicles). Using the distances of Table 9.4 corrected for the fact that we are now transporting from D to P points, we obtain the transportation cost tableau

From-To	1	2	Capacity
3	260	360	15
4	60	160	175
6	210	310	5
Reqts.	150	45	

The fixed 100-foot distance from 1 to 2 makes all feasible solutions optimal for this specific problem. One solution is to ship 15 empty vehicles from 3 to 1, 135 empty vehicles from 4 to 1, 40 empty vehicles from 4 to 2, and 5 empty vehicles from 6 to 2. Total distance is 19,950 feet.

Ideally, all other moves involve dropping off a load at a department's D point and then proceeding to its P location to pick up a load. We have not yet accounted for the within department movement from D to P locations. Each such move is 10 feet for departments 1 through 6, 110 feet for department 7, 75 feet for department 8, and 55 feet for department 9. The number of such moves for departments 1 through 9, respectively, is 0, 45, 60, 25, 100, 60, 80, 15, and 60, the minimum of $\sum_j v_{ji}$ and $\sum_j v_{ij}$ for each department i. This adds an extra 16,125 feet, making an empty travel total of 36,075 feet. At 5 feet per second this takes 2.00 hours. ■

The last component of vehicle utilization is blocking. The typical traffic control system divides the total path into zones (see Malmborg [1990, 1991]). The traffic controller prevents two vehicles from ever being in the same zone. A vehicle is halted and said to be blocked if it attempts to enter a zone that already contains another vehicle. From our loaded and unloaded travel computations, we have an estimate of the usage of each zone, since we know the number of times each route will be used. Let N be the number of vehicles in the system, M the number of zones, and t_z the average time to traverse zone $z, z = 1, \ldots, Z$. For each route let Y_{ijz} be 1 if zone z is used to travel from i to j and 0 otherwise. θ_z and ϕ_z are the sets of pickup and dropoff locations in zone z, respectively. Apart from blocking, total time spent in zone z is then

$$T_z = t_z \sum_i \sum_j (x_{ij} + v_{ij}) Y_{ijz} + p_i \sum_{i \in \theta_z} \sum_j v_{ij} + d_j \sum_{j \in \phi_z} \sum_i v_{ij} \qquad (9.8)$$

where p_i and d_j are load pickup and dropoff times, respectively, and the last two terms in equation 9.8 account for pickup and dropoff time in zone z. Ignoring blocking, total utilization is $T = \sum_{z=1}^{Z} T_z$. Each vehicle spends approximately T_z/N time active in zone z each period. Blocking time is difficult to estimate accurately. As a rough approximation, we will assume vehicle locations are independent. Thus for a period of length S, the probability that an arbitrary vehicle is already in zone z when a vehicle arrives can be approximated by T_z/NS. The probability zone z is empty is $[1 - (T_z/NS)]^{N-1}$. If the zone is blocked, we will further simplify the analysis by assuming that the vehicle will remain in the zone for one half its average traversal time. Average traversal time will be denoted t'_z. The total number of visits to zone z in the period is given by

$$V_z = \sum_i \sum_j (v_{ij} + x_{ij})Y_{ijz}$$

Thus, excluding the time spent blocked,

$$t'_z = \frac{T_z}{V_z}$$

Total blocking time due to vehicles attempting to enter zone z is

$$B_z = V_z \cdot \left[1 - \left(1 - \frac{T_z}{NS}\right)^{N-1}\right] \cdot \frac{t'_z}{2} \qquad (9.9)$$

The three RHS terms in equation 9.9 represent the number of visits to zone z, the probability of being blocked on arrival, and the expected time blocked given blocking occurs. Total blocking time for the system, B, is then approximated by summing over all zones.[4]

$$B = \sum_{z=1}^{Z} B_z \qquad (9.10)$$

If more exact estimates are desired, a simulation model can be built.

We now have all of the ingredients of vehicle utilization. Total time required per period is approximated from equations 9.8 and 9.10 by

$$U = \sum_{z=1}^{Z} T_z + B$$

EXAMPLE 9.11

For the AGV system of Figure 9.7b, estimate the number of vehicles needed without considering blocking. Vehicles are estimated to be available 12 hours per period (periods are 16 hours, and several hours per period are spent recharging and in maintenance). The control system divides the path into zones of 25 feet in length. Next estimate blocking time for vehicles. Does this change the number of vehicles required?

Solution

From previous examples, we know that loading, unloading, loaded travel, and empty travel time totals $5.33 + 5.33 + 8.88 + 2.00 = 21.54$ hours per period. At 12 hours per vehicle, this suggests the need for two vehicles.

---

[4]This blocking model is meant as only a rough approximation. An alternative approach would be to use an (M/M/1):(GD/N/N) queuing model. The single server is the zone itself. N represents the number of vehicles. Waiting time in such a system corresponds to blocked vehicle time. In the example that follows, the (M/M/1) model results in an estimate approximately twice that obtained by equation 9.9. The difference stems from the forgetfulness property of the exponential. Under the exponential assumption, the time for the vehicle ahead of you to vacate the zone has mean t'_z. The model in equation 9.9 takes the approach that if you arrive and see the zone occupied, it is more reasonable to assume that the vehicle has completed one half its deterministic transit time in the zone. An alternate queuing model is given by Malmborg [1990]. If more accurate estimates are needed, simulation modeling (see Chapter 12) may be necessary.

The control system has 38 zones (1950/25). Since the number of trips between each pair of P and D points is known, we could compute the time spent in each specific zone. Instead, we will save time and space by approximating all T_z by their average value of

$$\overline{T_z} = \frac{\sum_z T_z}{Z} = 21.54 \frac{\text{hours}}{38} = 0.57 \text{ hours}$$

Using this approximation for all zones,

$$B = \sum_z \cdot V_z \cdot \left[1 - \left(1 - \frac{\overline{T_z}}{NS} \right)^{N-1} \right] \cdot \frac{t_z'}{2} \qquad (9.11)$$

$$= 38 \cdot \frac{7840}{38} \cdot \left[1 - \left(1 - \frac{0.57}{2(16)} \right) \right] \cdot \frac{0.002747}{2} = 0.19 \text{ hours}$$

The values in equation 9.11 need some explanation. We have approximated V_z in equation 9.10 by the average value 7840/38. The 7840 zone visits are derived from the total 159,925 feet loaded plus 36,075 feet unloaded travel divided by 25 feet per zone. This is further divided by the number of zones to estimate the number of visits to an arbitrary zone z. Because we are using average values as an approximation, the summation over zones was replaced by the factor 38, the number of zones.

Adding 0.19 hours to the loaded, unloaded, loading, and unloading times, we find a total utilization of 21.73 hours per day. Thus, two vehicles should still suffice. Notice that blocking was not a major problem for this system because only two vehicles were needed. ∎

Before leaving the topic of AGV system design, we must note that the future may be radically different. Truly flexible, low-inventory production will require fast, small, free-roaming vehicles. Systems of this kind are already under development (see Kim and Tanchoco [1991]). One approach has vehicles being guided by dead reckoning with refined positional updates as fixed beacons are passed. Alternatively, vehicles can keep in radio contact with strategically placed transmitters that permit a continuous calculation of position. Such vehicles, coupled with sophisticated system controllers and suitable facility layout, may be free to visit any location. As vehicle prices drop, smaller production and transfer batch sizes become feasible. A review of current results for AGVS design and operation is given in King and Wilson [1991]. Goetschalckx and McGinnis [1989] and Bohlander et al. [1991] describe an engineering workstation developed to assist in the design of AGVSs.

9.5.3 AGVS Operation

During system operation, pickup and delivery demands must be met at each P and D point each shift. We will consider two cases. The first case assumes relatively constant demand rate for pickup and delivery requests. This would be true of a continuous flow assembly/manufacturing process with constant demand. The second case examined is a job shop where shop status is more dynamic in nature.

Static Flow

We assume that the time between pickup and delivery requests is nearly constant at each P and D location. In this environment it is appropriate to develop a set of routes for vehicles that can be continuously repeated at a time interval that satisfies the service requests of the departments. Define a **cycle** as a path that starts at a P or D location and alternates from P to D locations, eventually returning to its starting point. Our objective is to determine a set of cyclical routes and a frequency for each route so as to satisfy all of the loaded and unloaded moves determined in the previous section. Routes are easy to construct. Simply start at any pickup point. From the set of loaded moves originating at this point, select a destination. The move with largest volume is a good choice. From the corresponding dropoff point select a subsequent pickup point that receives empty vehicles from this dropoff location. Continue this process until the original pickup point is revisited. This is a route. The number of trips to assign to this route is the minimum number of trips to be planned for any move along this route. Assign this number of trips to this route. To update move requirements, each move along this route must then have its number of required moves reduced by the number of trips just planned. The process continues until all moves are accounted for.

Once the routes are formed and the required number of trips per period along each route is known, trips can be assigned to vehicles. Trips of each route should be evenly spread across the planning period. Also, to avoid extra empty travel time, it is preferable to assign trips with the same starting point to the same vehicle. Once a prospective period plan is finalized, it can be checked for blocking by simulation (see Chapter 12).

EXAMPLE 9.12

Plan the set of routes for the AGV path studied in the last section.

Solution

Combining loaded trips from Table 9.3 with the planned empty moves from Example 9.10, we find the total moves between departments as shown in Table 9.5a. Note that each department has a balanced number of flows in and out.

Table 9.5a Total Interdepartmental AGV Flows

From-To	1	2	3	4	5	6	7	8	9	Sum
1	—	40	25	30	10	10	20	5	10	150
2		—	40		30		10	10		90
3	15		—				50		10	75
4	135	45	10	—		10				200
5				100	—					100
6		5		60		—				65
7						40	—		40	80
8				10		5		—		15
9					60				—	60
Sum	150	90	75	200	100	65	80	15	60	835

To create routes, we iteratively start at the largest table value and cycle. For each cycle, we reduce trips on that cycle by the maximum number of moves. For instance, we start with the 135 trips from department 4 to 1. From 1, the best we can do is take 40 loads to department 2. From 2, we can take 40 loads to department 3. From 3, we take 40 loads to department 7. We cannot take all 50 loads because this route has already been restricted to at most 40 loads by the desired moves from 1 to 2. From 7, we take 40 loads to department 6. From 6, we can take 40 loads back to department 4. Thus, we have 40 trips on the route $4 \rightarrow 1 \rightarrow 2 \rightarrow 3 \rightarrow 7 \rightarrow 6 \rightarrow 4$. The remaining flows to be planned are shown in Table 9.5b.

Table 9.5b   Reduced Interdepartmental AGV Flows

From-To	1	2	3	4	5	6	7	8	9	Sum
1	—	0	25	30	10	10	20	5	10	110
2		—	0		30		10	10		50
3	15		—				10		10	35
4	95	45	10	—		10				160
5				100	—					100
6		5		20		—				25
7						0	—		40	40
8				10		5		—		15
9					60				—	60
Sum	110	50	35	160	100	25	40	15	60	835

The next route will start at department 5, since its flow of 100 loads to department 4 dominates the table. From 4, the largest flow is to department 1. Since only 95 more loads are planned from 4 to 1, the maximum trips on this route is reduced to 95. From 1, we return to 4, since the 30 trips from 1 to 4 is largest for department 1. We are now restricted to, at most, 30 loads on this route. The process continues until we return to department 5. Only 15 trips on this route are scheduled, since a move from department 3 to department 1 is part of the route. The complete route and all subsequent routes are summarized in Table 9.6.

Table 9.6   Planned AGVS Routes

Departmental Path	Number of Trips
$4 \rightarrow 1 \rightarrow 2 \rightarrow 3 \rightarrow 7 \rightarrow 6 \rightarrow 4$	40
$5 \rightarrow 4 \rightarrow 1 \rightarrow 4 \rightarrow 1 \rightarrow 3 \rightarrow 1 \rightarrow 7 \rightarrow 9 \rightarrow 5$	15
$4 \rightarrow 1 \rightarrow 4 \rightarrow 1 \rightarrow 3 \rightarrow 7 \rightarrow 9 \rightarrow 5 \rightarrow 4$	10
$5 \rightarrow 4 \rightarrow 2 \rightarrow 5$	30
$5 \rightarrow 4 \rightarrow 1 \rightarrow 5$	10
$4 \rightarrow 1 \rightarrow 6 \rightarrow 4$	10
$9 \rightarrow 5 \rightarrow 4 \rightarrow 1 \rightarrow 9$	10
$9 \rightarrow 5 \rightarrow 4 \rightarrow 1 \rightarrow 4 \rightarrow 2 \rightarrow 7 \rightarrow 9$	5
$9 \rightarrow 5 \rightarrow 4 \rightarrow 3 \rightarrow 9$	10
$2 \rightarrow 8 \rightarrow 4 \rightarrow 1 \rightarrow 7 \rightarrow 9 \rightarrow 5 \rightarrow 4 \rightarrow 2$	5
$6 \rightarrow 4 \rightarrow 6$	10
$5 \rightarrow 4 \rightarrow 1 \rightarrow 8 \rightarrow 4 \rightarrow 2 \rightarrow 7 \rightarrow 9 \rightarrow 5$	5
$2 \rightarrow 8 \rightarrow 6 \rightarrow 2$	5

> In planning the routes, ties, in the sense that two destinations allowed the same number of trips for the route, were broken arbitrarily. For each route, we try to distribute its trips evenly over the period without adding significant levels of extra empty travel to tie routes together.   ∎

Dynamic Flow

The uncertain and ever-changing nature of a job shop makes it virtually impossible to plan moves ahead of time. Instead, dynamic dispatching rules are needed. Requests for service can be of two types. A move request occurs when a part finishes at a workstation. If more than one vehicle is idle, the control system must select the vehicle to service this request. (If one vehicle is idle, the decision is easy. If no vehicles are available, the request goes into queue.) The second situation occurs when a vehicle becomes available but multiple move requests are queued. This is the more typical situation. The control system must decide which move has highest priority given system status, particularly the status of workstation input and output queues, the location of the available vehicle, and the location of the loads waiting to be picked up. Useful rules include prioritizing pickups based on a FCFS request by workstations or on the number of remaining spaces available in the output queue (Egbelu and Tanchoco [1987]). The latter rule is designed to avoid workstation blocking. In fact, once workstations become blocked, input queues can become full and the entire system begins to "lock," preventing workstations from operating and vehicles from moving. To avoid locking wherein neither vehicles nor workstations can operate, a central storage area can be used to hold extra loads that cannot be delivered.

A cyclic routing policy could also be used in the job shop environment. Empty vehicles travel from pickup point to pickup point making a complete cycle of the system until a load waiting for delivery is found. (This is, of course, a Traveling Salesman Problem for determining the shortest cycle through the network of pickup points.) The first load encountered is delivered. The vehicle then moves to the next pickup point in search of a deliverable load. Bartholdi and Platzman [1989] show that this "greedy" cycling policy works well for simple loop systems. A simple loop system is one in which vehicles repeatedly travel a fixed tour that covers all arcs in the guidepath.

Several rules have been proposed for just-in-time based AGV control (Egbelu [1987], Occena and Yokota [1991]). The objective is to employ "demand-driven" move priorities. This translates into advancing parts to a machine only when its input and output queues are nearly empty. The justification follows from the recognition that if the output queue is sufficiently occupied, we do not want the machine to be processing, since this would only produce unneeded inventory. If the input queue has waiting jobs, the machine cannot use any more raw material.

9.6
PALLET SIZING AND LOADING

At the beginning of this chapter we discussed unit load considerations, but generally we have assumed that the load and any corresponding pallet size were known. In actuality, choosing the proper pallet size and loading the pallet are important problems, which, unfortunately, are difficult to solve optimally. In many manufacturing facilities, products are packed into boxes as they are completed. These boxes are then stacked onto pallets to create the unit handling load. In some cases, boxes are placed into containers before stacking. The *Manufacturer's* Pallet Packing Problem involves selecting the optimal sizes of the box, container, and pallet for high volume (large batch size) product. (Matson and Naik [1991] discuss choosing container sizes for small batch production environments.) Steudel [1979] proposed using dynamic programming for loading identical containers onto pallets. The real objective is to maximize the number of product units that can fit on a pallet. Constraints include the pallet dimensions, stability of the load (Carpenter and Dowsland [1985]), and weight-bearing capacity of containers.

In the *Distributor's* Pallet Packing Problem, the objective is to load a fixed set of containers of varying dimensions onto as few pallets as possible. The set of containers corresponds to a customer order. Dynamic programming approaches to the two-dimensional (single layer at a time) and three-dimensional pallet loading problems are described in Hodgson [1982] and Tsai et al. [1991], respectively. Conceptually, the pallet-loading problem is similar to the widely studied bin packing and cutting stock problems (Golden [1976], Garey and Johnson [1981], and Dyckhoff [1990]).

The interaction between unit load size and other system aspects, such as batch sizes, storage space, and handling costs, have also been considered. The interested reader might begin by investigating Tanchoco and Agee [1981], Tanchoco et al. [1983], and Trevino and Daboub [1990].

9.7
SUMMARY

Material handling systems exist to support the overall manufacturing process. Conveyors, trucks, cranes, storage systems, containers, and auxiliary equipment should be specified to move the right product to the right location in the right quantity, orientation, and condition at the right time *and* to provide important data acquisition and transmittal. The material handling system is responsible for movement within each department and for connecting departmental "islands." However, part movement does not add value, and moves should be eliminated wherever possible.

When alternative methods exist for satisfying system needs, an integer programming model can be helpful for selecting equipment for each individual move. Although utilization levels of individual units of equipment are important in this decision, good solutions can often be found by relaxing the integer restrictions on the number of units required.

Conveyors can be used for both in-process buffer storage and movement over fixed paths. Conveyors should be designed to have sufficient capacity to avoid blocking of workstations. This requires that the unloading rate is at least as large as loading. As a secondary objective, conveyors are sometimes used as in-process storage buffers. Although some buffer inventory is often necessary, conveyors can be improperly used to store large stocks of inventory and to hide process problems. Before installing long conveyors with large in-process inventories, attempts should be made to reduce move distances, balance workstation cycle times, and improve workstation availabilities.

AGV systems can aid in the design of flexible, automated facilities. Guidepaths are limited in locations that can be visited, but the path can be periodically modified. The choice of guidepath is important for determining the total travel and congestion in the system. The future will see faster, free-roaming vehicles that increase system flexibility. Guidepaths should be designed to minimize vehicle blocking while serving all relevant locations. Vehicles must be routed in real time to avoid congestion on the path while also keeping input buffers of bottleneck machines stocked and output buffers clear. Some systems have AGVs repeatedly following a fixed tour of the guide path, whereas others allow dynamic, variable movement in response to move requests.

REFERENCES

Apple, J. M. [1977], *Plant Layout and Material Handling*, John Wiley & Sons, Inc., New York.

Bartholdi, John J., and Loren K. Platzman [1989], "Decentralized Control of AGVs on a Simple Loop," *IIE Transactions*, 21(1), 76–81.

Basics of Material Handling [1973], The Material Handling Institute, Charlotte, NC.

Bastani, A. S. [1988], "Analytical Solution of Closed-Loop Conveyor Systems," *European Journal of Operational Research*, 35, 187–192.

Bohlander, Ronald A., Wiley D. Holcombe, and James W. Larsen [1991], "An Advanced AGVS Control System: An Example of Integrated Design and Control," *Progress in Material Handling and Logistics*, J. A. White and I. Pence, eds., Springer-Verlag, Heidelberg, Germany, 471–487.

Bozer, Yavuz A., and Mandyam M. Srinivasan [1989], "Tandem Configurations for AGV Systems Offer Simplicity and Flexibility," *Industrial Engineering*, Feb., 23–27.

Bozer, Yavuz A., and Mandyam M. Srinivasan [1991], "Tandem Configurations for AGVS and the Analysis of Single Vehicle Loops," *IIE Transactions*, 23(1) 72–82.

Carpenter, H., and W. B. Dowsland [1985], "Practical Considerations of the Pallet Loading Problem," *Journal of the Operational Research Society*, 36(6), 489–497.

Dyckhoff, H. [1990], "A Typology of Cutting and Packing Problems," *European Journal of Operational Research*, 44, 145–159.

Egbelu, P. J. [1987], "Pull vs. Push Strategy for AGV Load Movement in a Batch Manufacturing System," *Journal of Manufacturing Systems*, 6, 209–220.

Egbelu, P. J., and J. M. A. Tanchoco [1987], "Characteristics of AGV Dispatching Rules," in *Automated Guided Vehicle Systems*, R. H. Hollier, ed., IFS Publications Ltd., New York, 125–142.

Garey, M. R., and D. S. Johnson [1981], "Approximation Algorithms for Bin Packing Problems: A Survey," in *Analysis and Design of Algorithms for Bin Packing in Combinatorial Optimization*, G. Ausieloo and M. Lucertini, eds., Springer-Verlag, Heidelberg, Germany, 147–172.

Gaskins, R. J., and J. M. A. Tanchoco [1987], "Flow Path Design for Automated Guided Vehicle Systems," *International Journal of Production Research*, 25(5), 667–676.

Glover, Fred, Darwin D. Klingman, Nancy V. Phillips, and Robert F. Schneider [1985], "New Polynomial Shortest Path Algorithms and Their Computational Attributes," *Management Science*, 31(9), 1106–1128.

Goetschalckx, Marc, and L. F. McGinnis [1989], "Engineering Work Station Is a Design Tool for Computer-Aided Engineering of Material Flow Systems," *Industrial Engineering*, 21(6), 34–38.

Golden, B. L. [1976], "Approaches to the Cutting-Stock Problem," *AIIE Transactions*, 8(2), 265–272.

Gould, Les [1990], "Partners in Productivity: AGV Systems and Simple Controls," *Modern Materials Handling*, 45(1), 55–57.

Handler, G. Y., and P. Mirchandani [1979], *Location on Networks*, MIT Press, Cambridge, MA.

Hassan, M. M. D., G. L. Hogg, and D. R. Smith [1985], "A Construction Algorithm for the Selection and Assignment of Materials Handling Equipment," *International Journal of Production Research*, 23(2), 381–392.

Hodgson, Thom J. [1982], "A Combined Approach to the Pallet Loading Problem," *IIE Transactions*, 14(3), 175–182.

Hodgson, Thom J., R. E. King, S. K. Monteith, and S. R. Schultz [1987], "Developing Control Rules for AGVS Using Markov Decision Processes," *Material Flow*, 4, 85–96.

Hollier, R. H., ed. [1987], *Automated Guided Vehicle Systems*, Springer-Verlag, New York.

Kaspi, M., and J. M. A. Tanchoco [1990], "Optimal Flow Path Design of Unidirectional AGV Systems," *International Journal of Production Research*, 28(6), 1023–1030.

Kim, Chang Wan, and J. M. A. Tanchoco [1991], "Prototyping the Integration Requirements of a Free-Path AGV System," *Progress in Material Handling and Logistics*, J. A. White and I. Pence, eds., Springer-Verlag, Heidelberg, Germany, 545–557.

King, R. E., and C. Wilson [1991], "A Review of Automated Guided-Vehicle Systems Design and

Scheduling," *Production Planning and Control,* 2(1), 44–51.

Malmborg, Charles J. [1991],"A PC Based Implementation of the Control Zone Model for AGVS Design," *Progress in Material Handling and Logistics,* J. A. White and I. Pence, eds., Springer-Verlag, Heidelberg, Germany.

Malmborg, Charles J. [1990], "A Model for the Design of Zone Control AGVSs," *International Journal of Production Research,* 28(10), 1741–1758.

Matson, Jessica O., and Girish N. Naik [1991], "A Group Technology Approach for Container Size Selection," *Progress in Material Handling and Logistics,* J. A. White and I. Pence, eds., Springer-Verlag, Heidelberg, Germany, 57–66.

Maxwell, W. L., and J. A. Muckstadt [1982], "Design of Automated Guided Vehicle Systems," *IIE Transactions,* 14(2), 114–124.

Maxwell, W. L., and R. C. Wilson [1981], "Dynamic Network Flow Modelling of Fixed Path Material Handling Systems," *IIE Transactions,* 13(1), 12–21.

Muth, E. J. [1975], "Modelling and System Analysis of Multistation Closed Loop Conveyors", *International Journal of Production Research,* 13(6), 559–566.

Muth, E. J., and J. A. White [1979], "Conveyor Theory: A Survey," *IIE Transactions,* 11(4), 270–277.

Occena, L. G., and T. Yokota [1991], "Modelling of an AGVS in a JIT Environment," *International Journal of Production Research,* 29(3), 495–511.

Sharp, G. P., and F. F. Liu [1990], "An Analytical Method for Configuring Fixed-Path Closed-Loop Material Handling Systems," *International Journal of Production Research,* 28(4), 757–783.

Soltis, D. J. [1985], "Automatic Identification Systems: Strengths, Weaknesses and Future Trends," *Industrial Engineering,* 55–59.

Steudel, Harold J. [1979], "Generating Pallet Loading Patterns: A Special Case of the Two Dimensional Cutting Stock Problem," *Management Science,* 25(10), 997–1004.

Tanchoco, J. M. A., and Marvin H. Agee [1981], "Plan Unit Loads to Interact with All Components of a Warehouse System," *Industrial Engineering,* 13(6), 36–48.

Tanchoco, J. M. A., R. P. Davis, P. J. Egbelu, and R. A. Wysk [1983], "Economic Unit Loads for Multi-Product Inventory Systems with Limited Storage Space," *Material Flow,* 1, 141–148.

Tompkins, James A. and John A. White [1984], *Facilities Planning,* John Wiley & Sons, New York.

Torok, Douglas B. [1990], "Scanning the Options: Hand-Held, Non-Contact Laser Scanners," *Material Handling Engineering,* 45, 75–79.

Trevino, Jaime, and Juan J. Daboub [1990], "Computer Aided Design of Unit Loads: A Design and Selection Procedure," *Progress in Material Handling and Logistics,* Proceedings of the 1990 Material Handling Research Colloquium, Hebron, KY, 81–109.

Tsai, Russell D., Eric M. Malstrom, and Way Kuo [1991], "A Three Dimensional Dynamic Palletizing Heuristic," in *Progress in Material Handling and Logistics,* J. A. White and I. Pence, eds., Springer-Verlag, Heidelberg, Germany, 181–204.

PROBLEMS

9.1. When delivering an item for use, what characteristics must be ensured by the material handling system?

9.2. If you were assigned the task of updating the Principles of Material Handling, what changes would you suggest?

9.3. Explain why the cheapest material handling system capable of moving material between locations may not be the best system.

9.4. List the basic types of material handling equipment. For each type, describe the environment where its usage is appropriate.

9.5. Consider the four products described in Table 9.7. If the plant has six departments (A through F), which manufacture only those four products, find the From-To flow volumes for the plant.

Table 9.7   Product Data for Problem 9.5

Product	Weekly Demand	Size of Unit Load	Department Visited 1	2	3	4
1	100	50	A	C	E	D
2	200	40	A	B	C	D
3	25	1	B	C	D	F
4	75	5	A	E	F	—

9.6. What are the important factors to consider in selecting unit load size?

9.7. A forklift incurs a depreciation cost of $2500 per year. The driver earns $19,000 per year. Annual maintenance and miscellaneous expenses are $1200. A work sampling study has shown that the truck spends 30 percent of its time traveling loaded, 25 percent traveling empty, 15 percent picking up or dropping off loads, 20 percent idle, and 10 percent in maintenance. When traveling, the truck averages 3 miles per hour. Estimate the cost of the truck and driver per foot of material transport.

9.8. Three equipment alternatives are available to perform seven product moves. Cost and resource data are shown in Table 9.8. Fixed costs for equipment types 1, 2, and 3 are $125, $348, and $21.50 per unit per period, respectively. Formulate the decision problem of product selection as a mathematical program.

9.9. Solve Problem 9.8.

9.10. Suppose in Problem 9.8 that the same method must be used for all moves in accord with the standardization principle. Which method would you select?

9.11. Trucks arrive at a receiving dock at the rate of 12 per day. A truck can have anywhere between 1 and 20 unit loads to deliver. A single crew unloads and stores product. Loads can be unloaded at the rate of 150 loads per day. Find the average number of loads waiting to be unloaded and the average time a load spends waiting to be unloaded.

9.12. Orders arrive randomly at the rate of 11 per week. Each order requests a specified number of units of various part numbers. The number of part numbers requested has a mean of 8 and variance of 16. The time to make the required number of units for a part on a single order is 0.01 weeks. If the entire plant is scheduled as a single entity, that is, the plant works on one part for one order at a time, estimate the mean number of parts waiting to be manufactured.

9.13. A conveyor is 500 feet long with carriers placed every 2 feet. The arrival rate of incoming parts to the loading station is 3 parts per minute. Loading time is 15 seconds. How fast must the conveyor move to avoid stockpiling of parts at the load station?

9.14. A closed-loop conveyor rotates at the rate of 0.3 meters per second. A front end loading station can load one item in 12 seconds. An unloading station at the opposite end of the plant can unload an item in 12 seconds. Each item will take up at least 5 feet of linear space on the conveyor. The load and unload stations will operate two shifts. During the eight hours of shift 1, loads will arrive at the rate of 6 per minute. In shift 2, 3 loads per minute will arrive.

    a. At what speed would you run this conveyor during each shift?
    b. How much material will accumulate at the load station during the first shift?
    c. Can this conveyor design handle the day's workload?

9.15. A conveyor has 26 equally spaced carriers. The three load/unload stations have cyclical carrier loading functions of $f_{1(j)} = \{0, 1, 2, 1, 2\}$, $f_2(j) = \{0, -1, -1, 0, 0\}$, $f_3(j) = \{-1, -1, 0, -1, -1\}$. Find the necessary carrier capacity.

Table 9.8   Data for Problem 9.8

Move	Equipment Type					
	1		2		3	
	c_{1j}	t_{1j}	c_{2j}	t_{2j}	c_{3j}	t_{3j}
1	126.00	0.32	23.50	0.12	213.50	0.95
2	43.50	0.12	12.45	0.10	98.00	0.83
3	12.40	0.13	4.56	0.05	31.45	0.20
4	31.40	0.20	32.50	0.50	49.56	0.30
5	11.23	0.14	7.56	0.12	32.40	0.34
6	148.50	0.90	29.45	0.56	345.26	0.95
7	45.50	0.35	19.50	0.20	108.00	0.60

9.16. Batches of material arrive at a loading station according to a Poisson arrival process at the rate of three arrivals per hour. Each batch is composed of approximately 10 pallets. A worker individually places each pallet onto a conveyor and enters pallet contents into an information system. The time to service a pallet depends on the number of items it contains, but service time is roughly exponential with mean of 1.5 minutes. Analyze the station.

9.17. Batches of material arrive at a loading station every 10 minutes for the first 4 hours of an 8-hour shift. Nothing arrives in the next 4 hours. Each batch contains 10 pallets of material, and it takes 1.5 minutes to enter a pallet into the storage and information system. Construct a graph showing the number of pallets at the loading station (not yet entered) throughout the 8-hour shift.

9.18. Generate a third alternative AGV path for the facility defined in Figure 9.7 and Table 9.3. Discuss the strengths and weaknesses of your path and the *P* and *D* locations.

9.19. Consider alternative 2 in Figure 9.7*c*. Compute the components of AGV utilization for this alternative. How many AGVs would be required?

9.20. A forklift is used to move pallet loads between departments. Table 9.9 contains interdepartmental flow volumes per shift. Move distances are given in Table 9.10. Production is for a line of high-volume parts, and thus interdepartmental flows rates are fairly constant through time.

  a. Determine a set of cyclical routes for the forklift to minimize wasted travel.

  b. Schedule arrivals during the shift in accord with demand.

9.21. Describe the difference in just-in-time and traditional rules for control of AGVs.

9.22. AGV vehicles circulate on a simple loop. The operating policy is as follows. Vehicles travel around the loop. When empty they stop and pick up the first load they pass that is waiting for transport. The load is dropped off at its destination and the vehicle continues. Does this seem

Table 9.9   Interdepartmental Flow Volumes for Problem 9.20

From-To	Department						
	1	2	3	4	5	6	7
1	—	10	34	0	0	19	24
2	10	—	0	0	34	12	0
3	28	12	—	54	0	0	0
4	0	24	0	—	21	4	21
5	34	12	0	21	—	34	18
6	0	12	27	27	0	—	6
7	12	18	12	0	24	24	—

Table 9.10   Travel Distances (Feet) for Problem 9.20

From-To	Department						
	1	2	3	4	5	6	7
1	—	100	225	125	250	250	200
2	250	—	125	25	200	137	100
3	75	175	—	200	325	312	275
4	225	325	100	—	125	112	75
5	100	200	325	225	—	337	300
6	112	212	37	212	362	—	312
7	150	250	75	175	50	37	—

and the vehicle continues. Does this seem like a good policy? Can you prove how close to optimal it must be? (*Hint:* Assume at the start of a period that a fixed set of pickups and dropoffs are given. What is the minimum number of complete loops the vehicle must make to accomplish all moves? What is the maximum number required by this policy?)

9.23. Re-solve Example 9.9 assuming that only the first six departments in Tables 9.3 and 9.4 exist.

9.24. Re-solve Example 9.10 assuming that only the first six departments in Tables 9.3 and 9.4 exist.

9.25. Re-solve Example 9.11 assuming that only the first six departments in Tables 9.3 and 9.4 exist.

9.26. An AGV path is being designed. A tentative layout is shown in Figure 9.9. Interdepartmental flows per day are shown in the Table 9.11. AGVs travel 1 meter per second and take 1 minute to

load or unload. Each department has one location where loads are picked up and delivered. The pickup/delivery point for department 1 is fixed; others could be relocated.

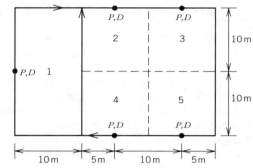

Figure 9.9   The AGV layout for Problem 9.26.

Table 9.11   Interdepartmental Flows per Day for Problem 9.26

From-To	1	2	3	4	5
1	—	10	25	0	0
2	0	—	10	0	25
3	15	0	—	10	0
4	0	40	0	—	20
5	24	10	0	50	—

a. Discuss the advantages and disadvantages of the proposed layout.

b. Compute the travel distance from department 2 to 5. How much time is spent per day by loaded vehicles moving from department 2 to 5?

c. For each department find the minimum number of empty vehicles that will be entering and leaving the department daily.

d. Either accept this design or propose a better one of your own.

WAREHOUSING: STORAGE AND RETRIEVAL SYSTEMS

*New opinions are always suspected, and usually opposed,
without any other reason but because they are not already
common.*

*It is one thing to show a man that he is in error,
and another to put him in possession of truth.*
—**John Locke**, *An Essay Concerning Human Understanding*

10.1
INTRODUCTION

Manufacturers have spent much time and energy in recent years reducing work-in-process inventory. Flexible automation, setup time reduction programs, and point of use storage have all contributed heavily. Nevertheless, we still find it necessary to build warehouses, particularly for distribution and spare parts provisioning. Product batches must be stored and assembled into customer orders for delivery. Critical spare parts may be stocked to avoid long, costly shutdowns. Regional distribution warehouses permit high-volume, focused factories to serve a geographically dispersed market with quick delivery lead time. In this chapter we consider the problems of warehouse design and operation.

We often think of a warehouse as a place to store goods. In actuality, as shown in Table 10.1, a variety of facilities and activities is associated with warehousing. These activities require considerable data acquisiton and transmission as well as the physical material handling activities that are listed. In addition to the tie that all manufacturing operations have to labor reporting and inventory systems, warehousing activities and their outcomes have impact on purchasing, billing, and accounts payable.

Warehouses vary in the level of automation. The warehouse may be a narrow-aisle, fully automatic storage and retrieval system (AS/RS), a manually controlled, forklift-based operation, or somewhere in between. A top view of these two extreme cases is diagrammed in Figure 10.1. In all cases, operating problems include selection of storage locations for loads, design of product storage (stacking) patterns, and sequencing of storage and retrieval requests.

10.1.1 Warehouse Components

A warehouse is composed of the **building shell, storage medium, storage/retrieval transport mechanism,** and its **controls/policies.** We briefly describe each of these before moving on to warehouse design and operation problems.

Table 10.1   Warehousing Systems

Facility	Activity
Storage racks/slots	Storage
Quality control	Receiving inspection
Receiving docks	Unloading, depackaging, identifying, sorting
Store/retrieve	Location selection, transport, S/R request sequencing
Manufacturing	Parts preparation (for shop floor release)
Shipping docks	Order assembly, packing, labeling, loading

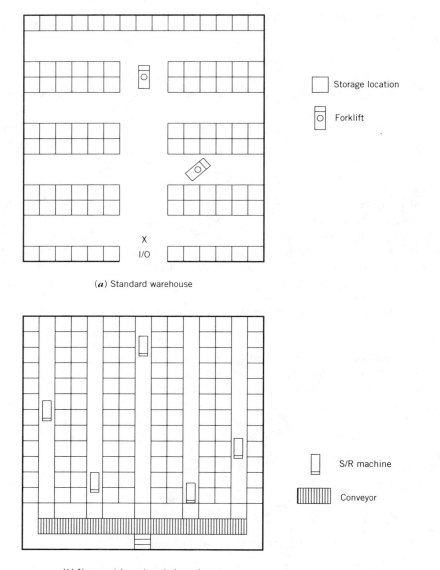

(**a**) Standard warehouse

(**b**) Narrow-aisle, automated warehouse

Figure 10.1   Top views of standard manual and narrow-aisle automated warehouses.

Building Shell

The building shell supplies space and a controlled environment for product, equipment, and personnel. Wares may be housed in a portion of the manufacturing plant or in a separate building. Modern automated storage and retrieval systems (AS/RS) often consist of a lightweight shell supported by the storage rack structure, thus emphasizing the fact that the building exists only to protect the internal operations. The racks may be 80 feet or higher and as little as 4 feet of building height may be lost to clearance between the ceiling and the top unit storage level.

Storage Medium

Storage racks are commonly used to hold loads. **Pallet racks** (Figure 10.2a) are formed from vertical support beams that are connected by longitudinal cross beams. The racks are designed with storage locations along their height and length. Each location can hold a loaded pallet. The vertical spacing between support shelves limits load height. Pallets can vary in size, but standard sizes range from 24×32 inches to 48×48 inches (Tompkins and White [1984]).

Placing pallet racks back to back allows storage depths of more than one pallet. Individual locations can be easily identified by their row (opening along the main

Figure 10.2a   Pallet storage racks (photo courtesy of Altman-Hall Associates).

aisle), bay (column), and vertical level. Gravity conveyor racks can be used to minimize aisle space and to ensure FIFO operation. Loads are inserted at one end of the rack depth and removed from the other end. While being stored, loads work their way to the front of the queue as the loads in front are removed. Cantilever racks (Figure 10.2*b*) are used to store long materials such as bar stock. Tall, stacker crane racks (Figure 10.2*c*) are used for high-rise AS/RS. Bins (Figure 9.4) are commonly used for small items. Rotating carousels can be used to facilitate end-of-aisle storage and retrieval.

Transport Mechanisms

A mechanism, either automated, semiautomated, or manual, must be used to transport loads from the I/O location to storage locations and to later retrieve loads. We will refer to this as the **S/R machine.**

Forklifts are often used to store and retrieve pallets. Such warehouses may have 50 percent or more of their floor space dedicated to aisles. Pallet racks can be placed back to back, but the two racks facing each other must be sufficiently separated to allow a forklift to maneuver.

Narrow-aisle vehicles are often used, particularly in high-rise AS/RS systems, to reduce floor space requirements. Traveling along rails in the floor or suspended overhead, these vehicles travel up to 500 feet per minute along aisles. Aisle widths may be as little as 6 inches wider than the depth of unit loads. (Loads are transported with their depth being perpendicular to the aisle direction, thus foregoing the need for load rotation when storing or retrieving.) Once the targeted location is reached, the vehicle's forks move perpendicular to the base to store or retrieve a load.

Figure 10.2*b*   Cantilever storage racks (photo courtesy of Altman-Hall Associates).

Figure 10.2c Stacker crane and rack system (courtesy of Abell-Howe Company).

The S/R machine must be compatible with the material handling equipment used to move material into and out of the warehouse area. A staging area must be provided to deposit incoming loads for storage and to pick up loads being retrieved from storage.

Controls/Policies

Warehouses vary from fully automated to manual. Stacker cranes can be automated or have a *man-on-board* for vehicle movement control or part movement. The system may automatically select storage locations for loads, including content recognition through bar code or RF identification, or the system may require manual input of load contents. Increased levels of automation require greater capital investment and are less flexible but generally more accurate. Later in this chapter we discuss the choice of control policies, such as how storage and retrieval location decisions should be made.

The basic policy decision is whether to have **dedicated** or **open** storage. In dedicated storage, each product has its own permanent storage area. When a load is to be retrieved, the location is known based on the product type and zonal storage allocation. Open storage, often referred to as randomized storage, allows loads to be placed in any location. An information system must keep track of each load in storage and its location as well as the currently available locations.

The type of warehouse selected depends on the product to be stored, the volume of the product to be stored, and the rate of storage and retrieval requests.

Automated storage and retrieval systems (AS/RSs) offer significant space and material control advantages. Deep-lane, unit-load, miniload, and carousel systems are available. Deep-lane and unit load systems store large unit loads, differing only in that deep-lane systems have greater load depth. Rack-supported, high-rise AS/RSs offer an economical means for adding new warehousing space. Miniload and carousel systems are used for small-parts order-picking environments. Order picking involves retrieval of either a small load or portion of a load to fill a customer order. The order may be a kit of parts for a production line or a shipment. A miniload storage location may contain a compartmentalized bin, for instance. Each compartment holds the stock of a different part number. The miniload and

Table 10.2   Normal Limits for Automated Storage and Retrieval Systems

System	S/R Horizontal Speed	S/R Vertical Speed	Max Load Wt.	Max Height	Depth
Deep-lane	400 fpm	90 fpm	4000 lb	100 ft	>1 load
Unit-load	400 fpm	90 fpm	4000 lb (or more)	100 ft	1 load
Miniload	350 fpm	80 fpm	500 lb	10–40 ft	1 load
Carousel	60 fpm rotation		5000 lb	6–10 ft	1–2 ft

carousel systems are generally placed in existing facilities, in close proximity to the manufacturing lines they support. Table 10.2 summarizes normal load limits for these automated systems.

10.2
WAREHOUSE DESIGN

White and Kinney [1982] note two principles for warehouses. The first is called the **85 percent rule.** The designer should plan on a maximum of 85 percent occupation of slots and 85 percent fill of the storage cube in the occupied slots. The S/R machine should not be utilized more than 85 percent of the time. This sounds like a nice practitioner rule of thumb; however, the second rule is that there are no valid rules of thumb for warehouses! If you are interested in finding optimal solutions, each situation must be analyzed as a unique problem.

The standard warehouse is a rectangular facility with I/O operations at one end. Storage racks are oriented along the longitudinal axis, or maybe perpendicular to this axis. Travel is along rectilinear (two-dimensional perpendicular) aisles. If we want to move from coordinate (x_1, y_1) to (x_2, y_2) on the floor, the travel distance is

$$D = |x_1 - x_2| + |y_1 - y_2| \tag{10.1}$$

Suppose we were to build a warehouse with I/O location at point (0,0) on a grid. Our objective might be to design the warehouse to minimize average storage or retrieval time. Let us look at a typical design, as shown in Figure 10.3. Storage locations are assumed to be square. Warehouse length is αa and width βb, where a gives the number of single location storage rows and b is the number of bays in a row. In terms of effective storage locations, the warehouse is $a \times b$. α and β are relative adjustment factors to incorporate aisle space based on the internal layout scheme of the warehouse. For instance, in Figure 10.3 we could approximately identify $\alpha = 2$ and $\beta = 1$. One half the longitudinal movement is across aisles. Along the width, we overlook the main aisle, considered to be a minor portion of total width. We could set β to a best guess value slightly greater than one to incorporate aisle width, but for illustration purposes $\beta = 1$ will suffice.

The warehouse has ab storage locations per level. We will assume that the warehouse is n levels high. If a total of K locations is needed, then we must have $abn \geq K$. If all locations are used equally, the average trip has a one-way distance of $\alpha a/2 + \beta b/4$ (the longest trip is $\alpha a + \beta b/2$). Warehouse shape can be found from the model

$$\text{minimize } \frac{\alpha a}{2} + \frac{\beta b}{4} \tag{10.2a}$$

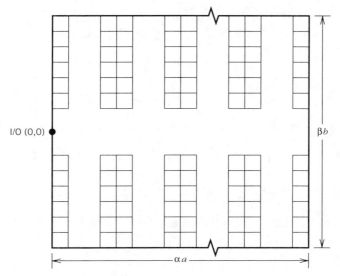

Figure 10.3   A standard warehouse design.

subject to:

$$a \cdot b \geq \frac{K}{n} \tag{10.2b}$$

$$a, b \text{ integer} \tag{10.2c}$$

We will solve problem 10.2 by assuming that a and b are sufficiently large that little is lost by letting them be continuous variables. In this case, expression 10.2b can be written $a = K/bn$. Substituting this relation into the objective 10.2a and differentiating with respect to b, we find the necessary condition:

$$\frac{-\alpha K}{2nb^2} + \frac{\beta}{4} = 0 \tag{10.3}$$

Solving equation 10.3

$$b^* = \left(\frac{2\alpha K}{\beta n}\right)^{1/2} \quad a^* = \left(\frac{\beta K}{2\alpha n}\right)^{1/2} \tag{10.4}$$

This gives the interesting design rule:

For $\alpha = 2\beta$, we obtain $b^* = 4a^*$ and warehouse length (αa) equals one-half warehouse width (βb). Interestingly, if $\alpha = \beta$, warehouse length should still equal one-half warehouse width. However, in the first case there are four times as many storage locations along the width as along the length. In the second case, $b^* = 2a^*$.

EXAMPLE 10.1

A rectanglar warehouse is being designed. Sixteen hundred storage locations are needed. Rack heights are such that four vertical levels will be accommodated. The I/O point will be located in the middle of the west wall. All loads are on 36 × 36-inch pallets. Bays will be 42 × 42 inches to allow adequate clearance. Find both warehouse dimensions and average one-way travel distance for a store or retrieve. As in Figure 10.3, assume $\alpha = 2$ and $\beta = 1$.

Solution

We first solve for warehouse dimensions. Using equations 10.4

$$a^* = \left(\frac{1600}{2 \cdot 2 \cdot 4}\right)^{1/2} = 10$$

Likewise, $b^* = 40$. Ten rows require 42 inches $\cdot 10 = 420$ inches or 35 feet. Adding aisle space we obtain a warehouse length of 70 feet. Warehouse width is 42 inches $\cdot$ 40 or 140 feet.

Next, find one-way travel distance. From expression 10.2a

$$\text{average one-way distance} = \frac{\alpha a}{2} + \frac{\beta b}{4} = \frac{2 \cdot 10}{2} + \frac{40}{4} = 20 \text{ locations}$$

Since storage locations are 3.5 feet long, this translates to a trip length of 70 feet. ∎

In problem 10.2 it was assumed that n was known. But how high should the warehouse be built? The answer may depend on building costs or storage equipment limitations. Building costs tend to increase with building height, and forklifts and S/R machines have limited reach. Expressions 10.4 can be solved with several values of n and total construction, equipment, and operating costs estimated for each option.

Many narrow-aisle S/R machines can move vertically and horizontally simultaneously. For such machines, the time to access a storage location along a row is given by the Chebyshev measure Travel time $= \max \{z/v_z, x/v_x\}$ where z and x are the vertical and horizontal distances to the location and v_z, v_x are the associated travel speeds. For Chebyshev travel, the best rack design is square-in-time such that the rack length/rack height ratio is equal to the *horizontal speed/vertical speed* ratio. With this design the S/R machine reaches the top and end of the rack at the same time, and any interior location can be reached more quickly.

The rectangular warehouse shape examined thus far is commonplace. It is optimal for Chebyshev travel or when the I/O point is at the end of each aisle. This would be true, for instance, if an AGV can drop off the incoming load at any aisle, as shown in Figure 10.4a

In general, contours of constant travel time determine optimal warehouse shape. If travel is rectilinear from a single I/O point, these contours are triangular, suggesting a warehouse shape as shown in Figures 10.4b and 10.4c. All locations along the exterior (apart from the side with the I/O point) have equal distance as measured by equation 10.1. Of course, building costs and conventional practice may override minimum travel distance considerations.

Rack Orientation

We have not yet discussed the proper orientation of racks within the warehouse. Several studies have looked into this issue. Let c_p be the cost of the building per time per unit length of warehouse perimeter. Using a detailed cost model, Bassan et al. [1980] showed that when the product of *throughput loads/time · handling cost/ unit distance moved* $> 2\,c_p$, racks should be oriented perpendicular to the longitudinal axis of the warehouse, as shown in Figure 10.3. If perimeter cost is dominant, however, the longitudinal orientation shown in Figure 10.5 is preferable.

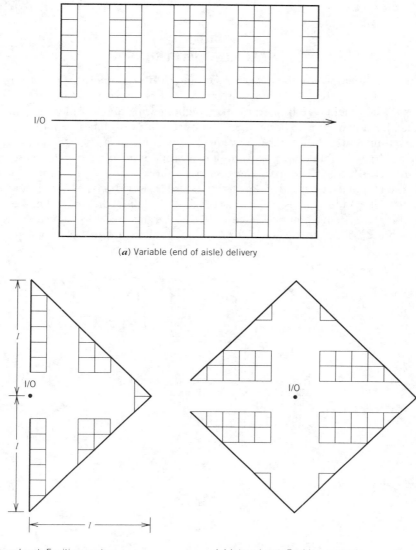

(a) Variable (end of aisle) delivery

(b) External port: Equitime contour
for rectilinear travel

(c) Internal port: Equitime contour
for rectilinear travel

Figure 10.4 Warehouse patterns.

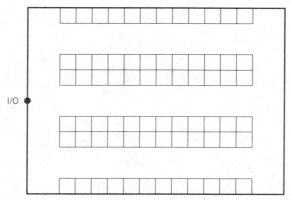

Figure 10.5 Longitudinally oriented racks.

10.3
STACKING PATTERNS

It is not necessary to employ racks for storage. In **block stacking** pallets are placed on top of one another. Investment is minimized and space remains flexible. However, stacking heights are limited by the weight-bearing and stability capability of the product. Each row contains only one type of product. Product is removed from one row at a time. Once the row is empty, the row becomes available for storing another product. The unusable empty space in a row while it is being emptied is referred to as **honeycomb loss.** The loss is necessary to allow access to the oldest loads.

Rows are generally filled from a single order and then are slowly emptied as loads are demanded. Thus, on the average one half of the row will be unavailable while it is being emptied. Figure 10.6 illustrates this loss. Fourteen loads are to be stored. Loads can be stacked three high. Pallets are to be retrieved one at a

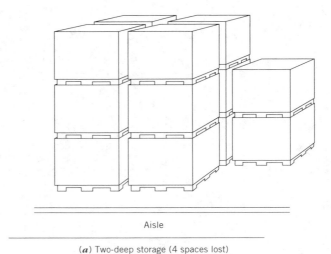

Aisle

(*a*) Two-deep storage (4 spaces lost)

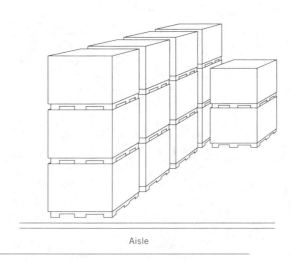

Aisle

(*b*) Four-deep storage (10 spaces lost)

Figure 10.6   Block stacking and honeycomb loss.

time. Space requirements are shown for two-deep and four-deep storage. Row depth constitutes the major decision for block storage. Each product will have an average of one half a row empty at any time. This will be the row from which retrieval requests are currently being filled. When product lot sizes are large in terms of loads, storage depths may be large, on the order of 10 to 20 loads. Ratio of lot size to stack height, aisle widths, storage/retrieval times, row clearances, and pallet size all affect storage depth. Models for determining floor space requirements and optimal stacking patterns for block and other rack arrangements are given in DeMars et al. [1981].

10.4
LOCATION IN WAREHOUSES

In this section we consider the problem of assigning incoming loads to storage locations. The recurring theme will be that the quicker the expected turnaround of the load, the closer it should be located to the I/O point. Two situations are considered. First we assume that each product has its own dedicated storage area. Next, we assume a sophisticated controller with the ability to assign parts to any open location and to keep track of currently stored products.

10.4.1 Dedicated Storage

In this section we assume that N products must be allocated to storage locations in a warehouse. A contiguous region of the warehouse is set aside for each product. Thus, whenever a product is needed, we know where it can be found. Dedicated storage simplifies warehouse control. Inventory status checking is also simplified. Product orders can be initiated by visual inspection of inventory levels. Storage locations can also be designed for specific product sizes or environmental requirements. On the other hand, dedicated storage requires that enough space be allocated to each product to store its maximum inventory level. Since maximum inventories may be twice the average level, much of the space will have low occupation levels.

EXAMPLE 10.2

Seven products must be stored for distribution. Table 10.3 shows the production batch size and safety stock for each product. (Safety stock is the amount of inventory planned to be on hand when a recently produced batch becomes avail-

**Table 10.3   Product Storage Volumes
for Example 10.2**

Product	Batch Size	Safety Stock
1	1,000	200
2	2,500	400
3	11,600	1,250
4	850	300
5	1,750	500
6	4,500	1,000
7	2,600	950

able for distribution.) Values are cubic feet of storage space required. Find the storage volume required for a warehouse with dedicated space for each product.

Solution

For each product, we must allow for safety stock plus a full batch. Thus, required space is

$$\text{space}_{\text{dedicated}} = (1000 + 200) + (2500 + 400) + \cdots + (2600 + 950)$$
$$= 29,400$$

Using principle one, we should set this as the 85 percent level of capacity to allow for growth and dynamic fluctuations. ∎

We begin the process of allocating products to space by dividing the warehouse into square grids. A grid contains one or more storage locations, but all grids have the same storage capacity. Product $i, i = 1, \ldots, N$ requires a maximum of A_i grid squares for storage. The total number of grids is M and we assume $\sum_{i=1}^{N} A_i = M$. (Otherwise add a dummy product that requires the extra warehouse space.) The warehouse will be allowed to have P shipping and receiving ports. All storage and retrieval requests occur at a port. In fact, we will assume that we know the number of loads per time that must pass through each port for each product. We let w_{ip} be proportional to the cost per period for sending product i through port p per unit distance traveled per storage or retrieval request. Normally, w_{ip} will be *trips/period*. If load transport costs vary for the products, then the w_{ip} should include a factor for *cost of product i/unit distance*. Thus, the product of w_{ip} with *distance/trip* will give the total period cost of moving i through port p. Distances will not be known until the assignment of products to grid squares is finalized, but the allocation of storage space into grids allows us to define the parameters d_{pj} as the distance from the center of grid j to port p. Our goal then is to find the set of A_i grids to assign to each product i. We will denote this set as S_i. Note that if item i is assigned to grid j ($j \in S_i$), the corresponding travel cost per period due to storage of i in j is c_{ij} where

$$c_{ij} = \frac{1}{A_i} \sum_{p=1}^{P} w_{ip} d_{pj} \tag{10.5}$$

Expression 10.5 indicates that $1/A_i$ of product i's flow is to grid j. This expression implies the assumption that all grids for item i use all ports in the same proportion. This may not be true; each port may be served by the closest grid with product i. However, assuming equal port use across a product's grids, we can model the grid assignment problem as a 0–1 program. Let x_{ij} be 1 if product i is assigned to grid j and 0 otherwise. Then,

$$\text{minimize} \sum_{i=1}^{N} \sum_{j=1}^{M} c_{ij} x_{ij} \tag{10.6a}$$

subject to:

$$\sum_{j=1}^{M} x_{ij} = A_i \qquad \text{for all } i \tag{10.6b}$$

$$\sum_{i=1}^{N} x_{ij} = 1 \qquad \text{for all } j \tag{10.6c}$$

$$x_{ij} \in \{0, 1\}$$

The objective 10.6a accumulates costs as we assign products to grids. Constraints 10.6b guarantee that i is assigned to A_i grids. Constraints 10.6c ensure that each grid is used. Otherwise, the model might naively place several products in the same grid. The generalized assignment problem 10.6 is a special case of the Transportation Problem and can be solved relatively easily. The transportation analogy is that each of the N products is a source. Source i must ship A_i units. Each of the M grid destinations is required to receive one unit.

EXAMPLE 10.3

The small warehouse of Figure 10.7 has 20 storage locations. Each side of the aisle has five bays and the bays have two levels. The storage aisle can be accessed from either end. Four product types will be stored. Flow volumes are shown in Table 10.4a and travel times are shown in Table 10.4b. Travel times are proportional to distance. Both sides of the aisle can be accessed freely, but the warehouseman must climb a step stool to reach the second level. Climbing time is reflected in the table values. Right-side grids are even numbered and left-side grids are odd numbered. Since right-and left-side grids are equivalent in terms of cost, only odd grids are listed in the tables. Products must be allocated to locations.

Solution

The transportation tableau for the problem is shown in Table 10.5.

As an example of the computation of the c_{ij} cost coefficients, consider source 1 and grid 1.

$$c_{11} = \frac{1}{A_1}(w_{11} \cdot d_{11} + w_{12} \cdot d_{21}) = \tfrac{1}{6}(120 \cdot 1 + 240 \cdot 5) = 220$$

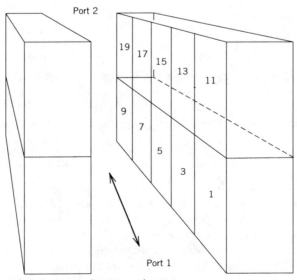

Figure 10.7   A small rack system for Example 10.3.

Table 10.4a   Product Space Requirements for Example 10.3

Product	No. of Grids Required	Product Trips per Period	
		Port 1	Port 2
1	6	120	240
2	7	70	105
3	2	100	100
4	5	20	40

Table 10.4b   Port to Grid-Travel Times for Example 10.3

Port	Grid									
	1	3	5	7	9	11	13	15	17	19
1	1	2	3	4	5	2	3	4	5	6
2	5	4	3	2	1	6	5	4	3	2

Table 10.5   Transportation Costs for Example 10.3

Product Source	Grid Destination										Supply
	1	3	5	7	9	11	13	15	17	19	
1	220	200	180	160	140	280	260	240	220	200	6
2	85	80	75	70	65	110	105	100	95	90	7
3	300	300	300	300	300	400	400	400	400	400	2
4	44	40	36	32	28	56	52	48	44	40	5
Capacity	1	1	1	1	1	1	1	1	1	1	20

To assign product 3 to grid 17,

$$c_{3,17} = \frac{1}{A_3}(w_{31}d_{1,17} + w_{32}d_{2,17}) = \tfrac{1}{2}(100 \cdot 5 + 100 \cdot 3) = 400$$

Solving this transportation problem we obtain the assignments $S_1 = \{5, 6, 7, 8, 9, 10\}$, $S_2 = \{3, 4, 16, 17, 18, 19, 20\}$, $S_3 = \{1, 2\}$ and $S_4 = \{11, 12, 13, 14, 15\}$ with a total travel time of 2454. Product 1 is located on the bottom level near port 2. Product 2 wants to be next closest to port 2, which tends to place it above product 1. Product 3 is located close to port 1 and product 4 goes above it. ∎

Although the current model is computationally tractable, Francis and White [1974] point out that one special case makes solution almost trivial. Assume that all products use all ports in the same proportion. Such would be the case, for instance, if all loads enter through one port and leave through another or, of course, if there were only one port. This important **factoring assumption** can be stated $w_{ip} = c_i \cdot w_p$. Ordinarily, c_i will be the total volume of product i moving in and out of storage per time, possibly weighted by a cost per unit distance moved for product i. Then, w_p is the proportion of loads that use port p. Note that w_p must be independent of both product and grid selected. In this special case we have

$$c_{ij} = \frac{1}{A_i} \sum_{p=1}^{P} w_{ip} d_{pj} = \frac{c_i}{A_i} \sum_{p=1}^{P} w_p \cdot d_{pj}$$

Letting

$$f_j = \sum_{p=1}^{P} w_p d_{pj} \qquad (10.7)$$

we see that c_{ij} factors into the product of two terms, one based on the product only and one based on the grid only. The total objective becomes $\sum_{i=1}^{N} \sum_{j \in S_i} c_i f_j / A_i$. Each f_j is matched with a c_i / A_i. Now suppose that you are given two sets of numbers and you desire to order the numbers in each set such that the vector product is minimized. The vector product is minimized by matching small values in the first set with large values in the second set, and vice versa. This provides a simple solution algorithm when the factoring assumption holds. The algorithm puts the products with the highest throughput per grid into the lowest cost grids.

Solution Under Factoring Assumption

STEP 1. ORDER GRIDS. Compute $f_j, j = 1, \ldots, M$ using equation 10.7. Place the grids in nondecreasing order of f_j, that is, $f_{[1]} \le f_{[2]} \le \cdots \le f_{[M]}$.

STEP 2. ORDER PRODUCTS. Put products in nonincreasing order, that is,

$$\frac{c_{[1]}}{A_{[1]}} \ge \frac{c_{[2]}}{A_{[2]}} \ge \cdots \ge \frac{c_{[N]}}{A_{[N]}}$$

STEP 3. ASSIGN PRODUCTS. For $i = 1, \ldots, N$ assign product $[i]$ to the first $A_{[i]}$ grid squares still available.

Ordering products based on throughput per storage volume dates back at least to Heskett [1963] and has proved to be an effective procedure.

EXAMPLE 10.4

Consider a five-row, five-column warehouse with a major aisle and I/O locations along the south wall. Mapped to a two-dimensional grid, loads are received at (0,0). Product is shipped from (5,0). Three products are stored in the warehouse. All products make four times as many trips to the shipping point as from the receiving point (each input batch serves an average of four customer orders). Storage and throughput requirements are given in Table 10.6. Assign products to storage locations.

Table 10.6   Product Data for Example 10.4

Product	No. of Grids Required	Total Loads Moved per Day
1	10	100
2	5	150
3	8	160

8.2	7.6	7.0	6.4	5.8
7.2	6.6	6.0	5.4	4.8
6.2	5.6	5.0	4.4	3.8
5.2	4.6	4.0	3.4	2.8
4.2	3.6	3.0	2.4	1.8

(0.0)   (5.0)
Input   (a) f_j values   Output

Vacant	Vacant	1	1	1
1	1	1	1	3
1	1	3	3	3
1	3	3	2	2
3	3	2	2	2

(0.0)   (5.0)
Input   Output
(b) Optimal assignments

Figure 10.8   Warehouse assignments for Example 10.4.

Solution

Since $w_1 = \frac{1}{5}$ and $w_2 = \frac{4}{5}$ for all three products, the factoring assumption holds true.

STEP 1. Compute f_j from $f_j = 0.2d_{1j} + 0.8d_{2j}$. For instance, numbering grids starting at the lower left corner and moving across rows one at a time, we have $f_1 = 0.2(1) + 0.8(5) = 4.2$. Likewise, $f_2 = 0.2(2) + 0.8(4) = 3.6$. Figure 10.8a shows all f_j values.

STEP 2. Ordering products we have c_i/A_i values of (10, 30, 20). Thus, product 2 is most important, followed by product 3 and finally 1.

STEP 3. Product 2 is assigned the best $A_2 = 5$ locations. Product 3 then gets the next best 8 locations. The final optimal layout is shown in Figure 10.8b.   ∎

Using dedicated space can result in relatively low utilization of prime (close to I/O) locations. As we have learned, if safety stocks are small relative to batch sizes, locations will be empty almost one half of the time. As an alternative, we could employ a "supermarket" storage policy. Each product has a small storage area of good locations. The overflow stock is kept in the less desirable locations. Retrieval requests are serviced from the good, dedicated spaces. When time permits, the overflow stock is moved to occupy the dedicated slots that have been vacated. This policy makes more effective use of the S/R machine capability and allows quick retrievals to prevent backlogging of storage and retrieval requests.

10.4.2 Random (Open) Storage

In computer controlled warehouses, unit loads can be stored independently. The computer keeps track of the contents of each location. Status information can also be accessed by product so that the set of locations containing a specific product type is available. When a load arrives, the computer selects an open storage location. (If loads are small, loads of several part types may be aggregated into a storage pallet. We will assume, however, that each pallet to be stored contains a single product type.) When a retrieval is requested, the computer retrieves a pallet of the proper product type. If several pallets of the same product are currently in storage, the decision can be based on pallet attributes like required retrieval time or length of time in storage. Open warehouses need fewer storage locations than dedicated warehouses. The warehouse can be sized to hold the maximum level of total inventory, a value significantly less than the sum over all products of their maximum inventory levels, as was required with dedicated storage.

EXAMPLE 10.5

Suppose that the warehouse of Example 10.2 was upgraded such that individual loads could be stored in any location. Assume that products are produced on a rotational basis. Determine space requirements.

Solution

At any point in time the most recently produced product will have safety stock plus a nearly full batch stored. The product currently in production will have barely more than its safety stock in inventory. On average, each product requires space for its safety stock plus one half a batch. Required space is

$$\text{space}_{\text{open}} = (500 + 200) + (1250 + 400) + \cdots + (1300 + 950)$$
$$= 17,000$$

Note the substantial reduction in space requirements as compared to the 29,700 value for the dedicated warehouse.   ∎

Let us consider an AS/RS with random storage. Random storage indicates that arriving loads can be placed in any available location. If rack utilization is high, then loads are evenly distributed throughout the racks. In designing storage systems, throughput is an important measure. Throughput is the number of storage and retrieval requests that can be handled by the S/R machine per time. We first look at single commands. A single command is a simple storage *or* retrieval activity. Single command time is travel time to the rack location, pickup time, return time, and deposit time. Travel time to the rack location and return time are equivalent. Consider the rack structure shown in Figure 10.9.[1] Let X and Z be the horizontal and vertical lengths of the storage racks; let v_x and v_z be the horizontal and vertical speeds of the S/R vehicle; and let t_{pd} be the fixed time per pickup or deposit. To estimate time per leg of the trip, we will integrate over the entire rack structure. Expected one-way travel time is

$$(XZ)^{-1} \int_0^Z \int_0^X \max\{\frac{x}{v_x}, \frac{z}{v_z}\} \, dx \, dz \tag{10.8}$$

Equation 10.8 uses the assumptions that (1) loads are randomly stored; (2) the S/R machine moves in horizontal and vertical directions simultaneously (hence, the max travel time measure); and (3) the S/R machine is located at a corner of the rack

---

[1]The actual structure may have storage locations on both sides of the aisle; however, this would not affect average travel time provided both sides are equally accessible.

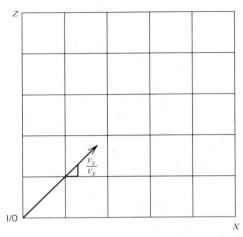

Figure 10.9 A rack structure for random storage travel time calculation.

structure, normally the lower level at the beginning of the aisle. Assume further that $X \geq Z(v_x/v_z)$. This last assumption will simplify our integration by assuring that maximum horizontal travel time is at least as large as maximum vertical travel time. (If this condition is not met, we simply reverse the x and z coordinates, i.e., switch X and Z and v_x and v_z.) With this convention, expression 10.8 becomes

$$(XZ)^{-1} \int_0^Z \int_0^X \max \left\{ \frac{x}{v_x}, \frac{z}{v_z} \right\} dx\, dz$$

$$= (XZ)^{-1} \left[\int_0^Z \int_0^{z\frac{v_x}{v_z}} \frac{z}{v_z} dx\, dz + \int_0^Z \int_{z\frac{v_x}{v_z}}^X \frac{x}{v_x} dx\, dz \right] \quad (10.9)$$

$$= (XZ)^{-1} \left[\int_0^Z \frac{z^2 v_x}{v_z^2} dz + \int_0^Z \left(\frac{X^2}{2v_x} - \frac{z^2 v_x}{2v_z^2} \right) dz \right]$$

and integrating again yields

$$E(\text{one-way-travel-time}) = \frac{v_x Z^2}{6 v_z^2 X} + \frac{X}{2 v_x}$$

Since a single command takes two one-way trips plus the pickup and deliver activities, total single command trip time is

$$T_{sc} = \frac{v_x Z^2}{3 v_z^2 X} + \frac{X}{v_x} + 2 t_{pd} \quad (10.10)$$

EXAMPLE 10.6

A storage aisle is 10 locations high and 20 locations long. Vertical travel is 10 locations per minute, and horizontal travel is 25 locations per minute. Thirty seconds is required per pickup or dropoff. The S/R machine may travel in both directions simultaneously. Find S/R cycle time for executing a single command.

Solution

Our model parameters are $X = 20, Z = 10, v_x = 25, v_z = 10$, and $t_{pd} = 0.5$ where distance units are locations and time units are minutes.

First, check the condition $X \geq Z\,(v_x/v_z)$. For our problem the condition becomes $20 \geq 10 \cdot (25/10) = 25$. Clearly, the condition is violated. Thus, we will switch the coordinate axis definition such that $X = 10, Z = 20, v_x = 10$, and $v_z = 25$. Now, from equation 10.10

$$T_{sc} = \frac{10(20^2)}{3(25^2)10} + \frac{10}{10} + 2(0.5) = 2.21 \text{ minutes}$$ ∎

When both storage and retrieval requests are waiting, it is more efficient to perform dual commands. The S/R machine takes a load to be deposited in the rack structure, but before returning to the I/O location, the S/R machine moves to a retrieval location to pick up a load. Assuming retrievals are FCFS, retrievals are also randomly distributed in the rack structure and locations are independent of the storage move. The trip now contains three legs: a move to store, an interleave to retrieve, and a return to I/O. White and Kinney [1982] provide the result that for a dual command, total travel time is

$$T_{dc} = \frac{X}{30v_x}\left(40 + \frac{15v_x^2 Z^2}{v_z^2 X_2} - \frac{v_x^3 Z^3}{v_z^3 X^3}\right) + 4t_{pd} \tag{10.11}$$

EXAMPLE 10.7

Find the average dual-command cycle time for the storage system defined in Example 10.6.

Solution

From equation 10.11,

$$T_{dc} = \frac{20}{30(25)}\left(40 + \frac{15(25^2)10^2}{10^2(20^2)} - \frac{25^3(10^3)}{10^3(20^3)}\right) + 4(0.5) = 3.64 \text{ minutes}$$

Using dual commands, a complete storage and retrieve cycle takes 3.64 minutes on average. In comparison, the same activity would require 4.42 minutes using two single commands. In this instance, dual commands yield an 18 percent reduction in travel time. ∎

Dual commands can yield even more significant increases in throughput when requests need not be served on a first-come basis. Han et al. [1987] note that whereas storages may need to be performed FCFS because of the method in which loads are presented to the system, retrievals can normally be sequenced. It was found that by selecting the retrieval nearest to the storage location, throughput increases of up to 22 percent are possible.

10.4.3 Class-Based Storage

In addition to lowering space requirements, open warehouses allow us to more effectively practice the principle of throughput-based storage. In our discussion on dedicated storage, the entire batch of an item was located in the same region. It is unlikely, however, that the entire batch will be withdrawn at the same time. More likely, the loads will be retrieved smoothly over the cycle time of the product. Thus, although the first load may be retrieved soon, the last load may remain in storage for quite some time. In this case, it would be more effective to locate the first load close to the I/O and the last load near the rear of the warehouse.

Class-based storage provides increased throughput capacity with respect to random storage. In class-based storage, we need know only the expected length of stay for each pallet. Storage locations are split into classes based on their distance from the I/O port. Incoming pallets are likewise assigned to a class based on expected storage time. Three classes, reflecting the standard A, B, C inventory breakdown, are often used. After deciding the break points between A and B, and B and C lengths of stay, the number of locations needed for the storage of each class of items is determined. Locations are assigned as in Section 10.4.1. The A items receive the best locations, and then class B locations are picked. The remainder form class C. Within each class, products are randomly assigned a location. If an A item arrives and no open A spaces are available, the system should store the item in an available B location. In a series of papers in the mid-1970s, Graves, Hausman, and Schwarz [1977] studied the use of class sizes based on the length-of-stay distribution. **Interleaving**, or using dual-command trips, was also found to be helpful. Use of three class divisions can reduce round-trip time an additional 20 to 40 percent when item throughputs vary. The 40 percent reductions correspond to the case where 20 percent of items generate 90 percent of requests. The 20 percent reductions are for the case of those top 20 percent of items generating 60 percent of requests. Optimal sizes of the partitions also vary with the throughput distribution, but as a general rule, class A should account for about 1 percent of locations and class B should account for about 25 percent. The general problem of modeling the performance of S/R machines has continued to receive attention, with results being extended by Elsayed and Stern [1983], Bozer and White [1984], Elsayed and Unal [1989], and Foley and Frazelle [1991].

10.4.4 Storing Complementary Items

The two previous sections assumed individual loads were stored and retrieved. In other situations, pickers are presented with a list of items to be picked from a warehouse. The list contains all of the items from one or more customer orders. The picker moves through the warehouse, visiting each location on the pick list. From each storage bin only a small volume of parts will be retrieved. The picker will visit many locations before returning to the system I/O point. This scenario is common for small items stored in bulk. If items tend to be ordered together, this information should be utilized in making storage location decisions (see Frazelle and Sharp [1989] for advantages). Items that often appear on the same order should be located near each other. We will not discuss sophisticated algorithms for institutionalizing item similarities. However, the use of the term "similarities" should suggest to the reader various strategies for "grouping" items. Instead of assigning individual items to storage locations, we can assign groups to storage areas. Mederios and Emamizadeh [1990] propose a mathematical model for automated miniload systems.

<hr>

10.5
ORDER PICKING

We now look at the common situation of picking (small) items from a warehouse to fill orders. The term "small" is implied because we assume that more than one item will be retrieved on each trip from the system I/O location. An overview of important considerations, models, and available results for order picking systems are given in Sharp et al. [1990] and Bozer and White [1990].

The design problem involves choosing a storage medium. The basic decision is whether (1) parts should come to pickers as on a carousel or miniload AS/RS; or, (2) pickers should go to parts as in a man-on-board AS/RS or manual bin or rack storage system. As usual, the optimal answer depends on the application. Important considerations are the desired storage height, throughput volume, desired level of computerization, product weight, and degree of product protection required. A standard economic and qualitative factor analysis should be performed to select the system.

After installation (and debugging) the issue becomes one of operation. Customer orders arrive. Each order lists one or more items to be picked from the warehouse. The basic order-picking problem involves combining customer orders into **pick lists** followed by the ordering of storage location visits on each list. The picker takes the list, and one by one, visits each storage location on the list. At each location, the picker retrieves the indicated number of units. After all locations are visited, the picker returns to the starting point, thus completing a tour. We assume that entire orders must be assigned to pick lists, that is, a pick tour must include all the items from each order serviced on that tour. (Otherwise, treat the orders as multiple single-item orders.)

10.5.1 Forming Pick Lists

The first decision step involves batching customer orders into pick lists. Hwang and Lee [1988] found a clustering algorithm to give solutions within about 5 percent of optimum. Let C be the carrying capacity of the S/R machine. Order k, $k = 1, \ldots, K$ requires I_k items with a total quantity (capacity utilization) of Q_k. The procedure requires computing a similarity coefficient between each pair of orders. Similarity is based on closeness of locations for the two orders.

We begin then by describing a method for determining similarity between orders. S/R machines generally can move in both vertical and horizontal directions simultaneously. We again assume Chebyshev travel along the storage row with travel time measure

$$t_{ij} = \max\left(\frac{|x_i - x_j|}{v_x}, \frac{|z_i - z_j|}{v_z}\right)$$

between locations i and j. Traveling at full speed, the machine travels along a line with slope $s = v_z/v_x$. Such a line L is shown in Figure 10.10. Suppose our destination was point P_1. Since P_1 lies above line L, vertical travel takes longer than horizontal travel. Any path in the parallelogram $0AP_1B$ could be followed to arrive at P_1 in minimum time provided that we maintain velocity v_z in the vertical direction. Likewise, if our destination was P_2, any path in $0CP_2D$ could be traversed in equal time. The lines BP_1 and DP_2 are parallel to L. Suppose we had picked an item at P_1 and had to travel to location P_2 for the next item. Horizontal travel is the limiting factor, and any path in P_1EP_2F could be taken in equal time. P_2E and P_1F have slope opposite in sign to L but with the same magnitude.

Now suppose order 1 required a visit to points P_1 and P_2 and another order required visiting only locations within $0BEP_2D$. It would be efficient to add these orders to the same picking list, since the same travel could be used for both orders. This forms the rationale behind our similarity measure.

In general, the parallelogram (minus the negative x and z regions) that defines the travel area for an order will be determined by three exterior storage locations, as

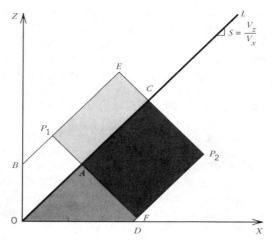

Figure 10.10   Chebyshev travel regions along a row.

shown in Figure 10.11a. Order i must visit locations $P_{i1}, \ldots, P_{i5}$, but its travel region is defined by the boundary points P_{i1}, P_{i2}, and P_{i3}. Let A_i be the area (travel region) of the warehouse that can be visited while serving just order i, without increasing trip time. Figure 10.11b adds the corresponding area for the four locations of order j and the resulting intersection A_{ij}. Using individual areas A_i, A_j, and intersection A_{ij}, our similarity coefficient for orders i and j is

$$s_{ij} = \frac{A_{ij}}{A_i + A_j - A_{ij}}$$

We can find the necessary areas by using the geometry of the regions. Choose an order i. If all of the locations are below (above) line L, then the point $P_{i1}(P_{i3})$ in Figure 10.11a is simply $(0,0)$. The points P_{i1}, P_{i2}, P_{i3} lie on lines with slope s, $-s$, and s respectively. The slopes along with the point coordinates provide all of the data necessary. Let (x_k, z_k) be the coordinates of point (storage location) k, $k = 1, 2, 3$ corresponding to P_{i1}, P_{i2}, and P_{i3}, respectively. The three lines intersect at the points (a_1, b_1) and (a_2, b_2) where

$$a_1 = \frac{z_2 - z_1}{2s} + \frac{x_2 + x_1}{2} \tag{10.12a}$$

$$b_1 = \frac{z_1 + z_2}{2} + \frac{s(x_2 - x_1)}{2} \tag{10.12b}$$

$$a_2 = \frac{z_2 - z_3}{2s} + \frac{x_2 + x_3}{2} \tag{10.12c}$$

and

$$b_2 = \frac{z_2 + z_3}{2} + \frac{s(x_2 - x_3)}{2} \tag{10.12d}$$

For a region of the shape given in Figure 10.11a, the area is found from[2]

$$A_i = \left[a_2^2 s - \frac{b_2^2}{s}\right]/2 + a_1(b_1 - b_2) + a_2(b_2 - a_1 s) \tag{10.13}$$

---

[2]The area can be obtained by projecting the line containing P_{i2} to the vertical and horizontal axes and then using the fact that the area of a triangle is $\frac{1}{2} \cdot$ base $\cdot$ height. From the area of the major triangle containing A_1, subtract the area of the two smaller triangles to be excluded.

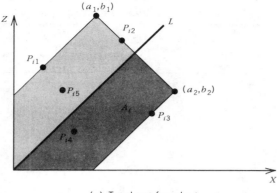

(a) Travel area for order i

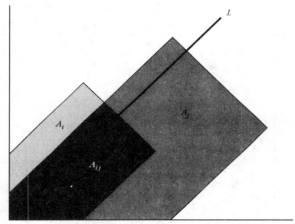

(b) Individual travel areas and intersection of orders

Figure 10.11   Areas for pick orders.

EXAMPLE 10.8

Consider the orders described in Table 10.7. Quantities are measured in cubic feet. The S/R machine will not pick more than 15 cubic feet per trip. Locations are denoted (x, y, z) where x denotes storage bay, z gives storage level, and y indicates the right or left side of the aisle. The S/R vehicle can access either side of the aisle (hence, the y coordinate will not be needed). Vertical velocity is 2 locations per minute. Horizontal velocity is 4 locations per minute. The aisle has 12 bays and 6 levels of locations for a total of $12 \times 6 \times 2 = 144$ storage locations. Find the area of the travel region for each order.

Solution

Consider the first order, illustrated in Figure 10.12a. Slope s is $s = v_z/v_x = 2/4 = 0.5$. With points $P_{i1} = (2, 3), P_{i2} = (6, 2)$, and $P_{i3} = (5, 1)$, we have from equations 10.12

$$a_1 = \frac{2 - 3}{1} + \frac{6 + 2}{2} = 3$$

Table 10.7   Orders to Be Picked for Example 10.8

Order	Item	Location x	y	z	Quantity
1	a	2	R	3	2.0
	b	6	L	2	0.5
	c	5	R	1	0.5
2	a	10	L	4	1.0
	b	9	L	2	2.5
3	a	12	L	1	3.0
	b	9	R	3	6.5
	c	4	L	2	1.0
4	a	2	R	5	3.0
	b	4	L	3	4.5

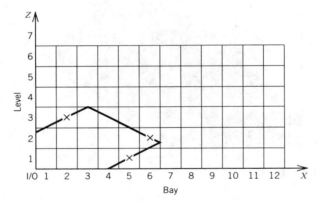

(*a*) First order for example 10.8

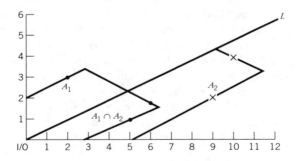

(*b*) Intersection of orders 1 and 2

Figure 10.12   Rack travel regions for Example 10.8.

$$b_1 = \frac{3+2}{2} + \frac{0.5(6-2)}{2} = \frac{7}{2}$$

$$a_2 = \frac{2-1}{1} + \frac{6+5}{2} = \frac{13}{2}$$

$$b_2 = \frac{2+1}{2} + \frac{0.5(6-5)}{2} = \frac{7}{4}$$

Now, using equation 10.13 we find that

$$A_1 = [a_2^2 s - b_2^2/s]/2 + a_1(b_1 - b_2) + a_2(b_2 - a_1 s)$$

$$= \left(\frac{169}{8} - \frac{49}{8}\right)\frac{1}{2} + 3\left(\frac{7}{4}\right) + \frac{13}{2}\left(\frac{7}{4} - \frac{6}{4}\right) = \frac{115}{8}$$

The other areas are found in a similar manner. $A_2 = 19.375$, $A_3 = 25.0$, and $A_4 = 22.75$. Intermediate results are summarized in Table 10.8.

Table 10.8 Results for Example 10.8

Order	P_1	P_2	P_3	a_1, b_1	a_2, b_2	Area
1	(2,3)	(6,2)	(5,1)	(3,3.5)	(6.5,1.75)	14.375
2	(0,0)	(10,4)	(9,2)	(9.0,4.5)	(11.5,3.25)	19.375
3	(4,2)	(9,3)	(12,1)	(7.5,3.75)	(12.5,1.25)	25.000
4	(2,5)	(2,5)	(0,0)	(2,5)	(6,3)	22.750

The next step consists of finding the area of the intersection of the travel regions. Let δ_{i1}, δ_{i2}, and δ_{i3} be the Z-axis intercepts of the lines that form area i. Then the intersection of areas i and j is formed by the lines with intercept closest to the origin. The top boundary line is provided by the order with $\min\{\delta_{i1}, \delta_{j1}\}$. The "northeastern" boundary with slope $-s$ is determined by $\min\{\delta_{i2}, \delta_{j2}\}$, and the lower slope s boundary is determined by the line for $\max\{\delta_{i3}, \delta_{j3}\}$. Geometrically, this is illustrated in Figure 10.11b.

EXAMPLE 10.9

Find the intersection of the travel regions of the orders in Table 10.7. Next, find the similarity coefficients for pairs of orders.

Solution

We first find the intercepts for each of the defining points used in Table 10.8. Defining points are those that are used as P_{i1}, P_{i2}, or P_{i3} for at least one order. Note that for P_{i1} and P_{i3} points, the relevant slope is s, whereas it is $-s$ for P_{i2} points. Using the relation for a line: $z = mx + b$ with slope m, the intercept b is easily found. Table 10.9 contains the intercepts.

For each pair of orders the next step is to find the defining lines or points. For example, take the intersection of orders 1 and 2.

$$\min\{\delta_{11}, \delta_{21}\} = 0.0, \text{ and the associated point is } P_{21} = (0,0)$$

$$\min\{\delta_{12}, \delta_{22}\} = 5.0, \text{ and the associated point is } P_{12} = (6,2)$$

Table 10.9   Relevent Intercepts for Example 10.8

Order i	Point	Position k	Intercept δ_{ik}
1	(2,3)	1	2.0
	(6,2)	2	5.0
	(5,1)	3	−1.5
2	(0,0)	1	0.0
	(10,4)	2	9.0
	(9,2)	3	−2.5
3	(4,2)	1	0.0
	(9,3)	2	7.5
	(12,1)	3	−5.0
4	(2,5)	1	4.0
	(2,5)	2	6.0
	(0,0)	3	0.0

The bottom boundary is defined by

$$\max\{\delta_{13}, \delta_{23}\} = -1.5, \text{ and the associated point is } P_{13} = (5,1)$$

The intersection is that portion of order 1 that lies below the baseline L as shown in Figure 10.12b. Using expressions 10.12 with points (0,0), (6,2), and (5,1) produces

$$a_1 = \frac{2-0}{2(0.5)} + \frac{0+6}{2} = 5.0$$

$$b_1 = \frac{0+2}{2} + \frac{0.5(6-0)}{2} = 2.5$$

$$a_2 = \frac{2-1}{2(0.5)} + \frac{6+5}{2} = 6.5$$

$$b_2 = \frac{2+1}{2} + \frac{0.5(6-5)}{2} = 1.75$$

Then, using equation 10.13

$$A_{12} = \frac{6.5^2(0.5) - 1.75^2/0.5}{2} + 5.0(2.5 - 1.75) + 6.5[1.75 - 5(0.5)] = 6.375$$

Results for the six order combinations are shown in Table 10.10.

Table 10.10   Summary of Intersection Areas for Example 10.9

Orders	$P_{ij,1}$	$P_{ij,2}$	$P_{ij,3}$	a_1, b_1	a_2, b_2	A_{ij}	Similarity
1,2	(0,0)	(6,2)	(5,1)	(5.0,2.5)	(6.5,1.75)	6.375	0.23
1,3	(4,2)	(6,2)	(5,1)	(5.0,2.5)	(6.5,1.75)	6.375	0.19
1,4	(2,3)	(6,2)	(0,0)	(3.0,3.5)	(5.0,2.5)	8.0	0.28
2,3	(0,0)	(9,3)	(9,2)	(7.5,3.75)	(10.0,2.5)	15.625	0.54
2,4	(0,0)	(2,5)	(0,0)	(6.0,3.0)	(6.0,3.0)	0.0	0.00
3,4	(4,2)	(2,5)	(0,0)	(6.0,3.0)	(6.0,3.0)	0.0	0.00

The last step is to compute similarities. For instance,

$$s_{12} = \frac{A_{12}}{A_1 + A_2 - A_{12}} = \frac{6.375}{14.375 + 19.375 - 6.375} = 0.23$$

Other similarities are included in the last column of Table 10.10.   ■

Once similarities are known, orders are batched to form lists. The procedure begins by assigning the order with the most items to the first pick list. The number of items in order k is denoted I_k. To this list we next add the order with highest similarity that does not violate capacity C. This process continues until no more orders can be added. At this point, a new list is started. More formally, let S be the set of unassigned orders, s_{ij} the similarity between orders i and j, and L_t the set of orders assigned to list t. Q_t will be the volume (capacity) currently assigned to list t.

List Formation
STEP 0. INITIALIZE. Set $S = \{1, \ldots, K\}$. $t = 0$.
STEP 1. START NEW LIST. If $S = \phi$, stop. Otherwise, $t = t + 1$, find $k' = \text{argmax}\{I_k : k \in S\}$ and let $L_t = k'$, $Q' = Q_{k'}$, $S = S - k'$ and go to 2.
STEP 2. ADD TO LIST. Let $k^* = \text{argmax}\{s_{k'k} : k \in S, Q' + Q_k \le C\}$. If no such k^*, go to 1.
  $S = S - k^*$, $L_t = L_t + k^*$, $Q' = Q' + Q_{k^*}$. Go to 2.

EXAMPLE 10.10

Combine orders to form pick lists.

Solution

Retrieval capacity was given as 15 in Example 10.8. Order quantitites are shown in Table 10.7 to be 3.0, 3.5, 10.5, and 7.5 units, respectively.

STEP 0. INITIALIZE. Set $S = \{1, 2, 3, 4\}$ and $t = 0$.

STEP 1. START FIRST LIST. Set $t = 1$. Order 1 is tied for the most items, and so we start with order 1. Set $L_1 = \{1\}$, $Q_1 = 3.0$, $S = \{2, 3, 4\}$.

STEP 2. ADD TO LIST 1. Order 4 has the largest similiarity with order 1 and its addition does not exceed S/R machine capacity. Thus, add order 4. $L_1 = \{1, 4\}$, $Q_1 = 10.5$, and $S = \{2, 3\}$.

STEP 2. ADD TO LIST 1. Of orders still in S, order 2 has the largest similiarity with a member of L_1 and its addition does not violate capacity. Thus, add order 2 to list 1. $L_1 = \{1, 2, 4\}$, $Q_1 = 14.0$, and $S = \{3\}$.

STEP 2. ADD TO LIST 1. No members of S can be added to list 1.

STEP 1. START LIST 2. Set $t = 2$, $L_2 = \{3\}$, $Q_2 = 10.5$, and $S = \phi$. Since S is now empty, stop.

The heuristic decided to combine orders 1, 2, and 4 into one trip. Order 3 is left as the second trip. An alternative decision rule could have been used whereby orders are gradually built from the largest similarity coefficients. In this case, we would have first combined orders 2 and 3 and then have clustered 1 and 4. Both methods are heuristic; we leave it to the users to choose their favorite.   ■

10.5.2 Pick Sequencing

The second decision step involves sequencing the visits to retrieval locations on each pick list. Once the picking list is determined, sequencing the visits to the picking locations is another instance of the Traveling Salesman Problem discussed in Chapter 8. If the pick list is short, say 15 or fewer items, optimal routes can be determined by computer. Otherwise, heuristics like the closest insertion procedure can be used. Ratliff and Rosenthal [1983] describe an efficient procedure for constructing the optimal picking tour in a warehouse with multiple aisles and crossover between aisles only at either end of the aisles. The procedure makes use of the fact that if an aisle has one or more picking locations, we must enter the aisle. Assuming aisle sizes and aisle crossover distances are all the same, the optimal tour will either (1) enter and traverse the entire aisle; (2) enter one end of the aisle, go to the farthest location, and then retrace back to the start of the aisle; or (3) enter the aisle from the top (bot-

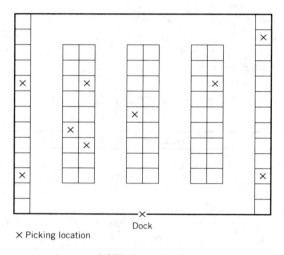

X Picking location

(*a*) Warehouse layout

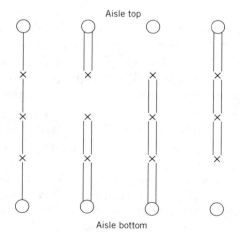

(*b*) Possible aisle trips

Figure 10.13   Possible aisle trips for order picking.

tom), visit the closest locations, and then return to the top (bottom), and then later enter from the bottom (top), visit the remaining locations, and exit at the bottom (top). In the third case, the break between the top and bottom trip locations should be between the two locations on the aisle that are farthest apart. The possible aisle trips are demonstrated in Figure 10.13.

10.6
SUMMARY

Although warehouses are expensive and add no intrinsic value, they can be important or even necessary components of the manufacturing system. Distribution warehouses allow storage of product closer to the marketplace for quicker delivery. If supplies of quality incoming material cannot be assured, a warehouse may be required for receiving. Critical spare parts may also need to be stockpiled.

Warehouses are made up of a building shell, an internal layout or storage pattern, possibly a storage medium such as racks, an S/R mechanism, and control policies. Warehouses with dedicated space for each product tend to require additional space but permit simple inventory control.

The primary rule for assigning loads within a warehouse revolves around the principle that the sooner the load will be retrieved, the closer it should be located to the I/O port. If storage space is dedicated to each product, a generalized assignment model can determine the optimal allocation. If all products use all ports in the same proportion, products are ordered by their throughput frequency and then assigned space.

Interleaving can significantly improve S/R machine throughput. For nondedicated storage, individual loads are placed in a storage class based on their expected time until retrieval.

For order-picking applications, items that are frequently ordered together should be stored together. Pick lists should be compiled from orders that require visits to similar areas of the warehouse.

REFERENCES

Bassan, Y., Y. Roll, and M. J. Rosenblatt [1980], "Internal Layout Design of a Warehouse," *IIE Transactions*, 12(4), 317–322.

Bozer, Y. A., and J. A. White [1990], "Design and Performance Models for End of Aisle Order Picking Systems," *Management Science*, 36(7), 852–866.

Bozer, Y. A., and J. A. White [1984], "Travel-Time Models for AS/RS," *IIE Transactions*, 16(4), 329–338.

De Mars, N. A., J. O. Matson, and J. A. White [1981], "Optimal Block Stacking Patterns," *Proceedings of the 4th Conference on Automation in Warehousing*, Tokyo, Japan.

Elsayed, E. A., and R. G. Stern [1983], "Computerized Algorithms for Order Processing in Automated Warehousing Systems," *Int. J. Prod. Res.*, 21, 579–586.

Elsayed, E. A., and O. I. Unal [1989], "Order Batching Algorithms and Travel-Time Estimation for Automated Storage/Retrieval Systems," *International Journal of Production Research*, 27(7), 1097–1114.

Foley, R. D., and E. H. Frazelle [1991], "Analytical Results for Miniload Throughput and the Distribution of Dual Command Travel Time," *IIE Transactions*, 23.

Francis, R., and J. A. White [1974], *Facility Layout and Location*, Prentice-Hall, Englewood Cliffs, NJ, 1974.

Frazelle, E. A., and G. Sharp [1989], "Correlated Assignment Strategy Can Improve Any Order-Picking Operation," *Industrial Engineering*, 33–37.

Graves, S. C., W. H. Hausman, and L. B. Schwarz [1977], "Storage-Retrieval Interleaving in Automatic Warehousing Systems," *Management Science*, 23(9), 935–945.

Han, Min-Hong, Leon F. McGinnis, Jin Shen Shieh, and John A. White [1987], "On Sequencing Retrievals in an Automated Storage/Retrieval System," *IIE Transactions*, 19(1), 56–66.

Heskett, J. L. [1963], "Cube-per-Order Index: A Key to Warehouse Stock Location," *Transportation and Distribution Management*, 3, 27–31.

Hwang, H., and M. K. Lee [1988], "Order Batching Algorithms for a Man-on-Board Automated Storage and Retrieval System," *Engineering Costs and Production Economics*, 13, 285–294.

Mederios, D. J. and B. Emamizadeh [1990], "Optimal Container Location in Miniload AS/R Systems," *Progress in Material Handling and Logistics*, J. A. White and I. Pence, eds., Springer-Verlag, Heidelberg, Germany, 193–202.

Ratliff, H. D. and A. S. Rosenthal [1983], "Order-Picking in a Rectangular Warehouse: A Solvable Case of the Traveling Salesman Problem," *Operations Research*, 31(3), 507–521.

Rosenblatt, M. J., and Y. Roll [1984], "Warehouse Design with Storage Policy Considerations," *International Journal of Production Research*, 22(5), 809–821.

Schwarz, L. B., S. C. Graves, and W. H. Hausman [1978], "Scheduling Policies for Automatic Warehousing Systems: Simulation Results," *IIE Transactions*, 10(3), 260–270.

Sharp, Gunter P., Kyung Il-Choe, and Chang S. Yoon [1990], "Small Parts Order Picking: Analysis Framework and Selected Results," *Progress in Material Handling and Logistics*, Springer-Verlag, Heidelberg, Germany, 203–219.

Tompkins, J., and J. A. White [1984], *Facilities Planning*, John Wiley & Sons, New York.

White, J. A., and H. D. Kinney [1982], "Storage and Warehousing," in *Handbook of Industrial Engineering*, G. Salvendy, ed., John Wiley & Sons, New York.

PROBLEMS

10.1. Under what conditions is it necessary to have a warehouse?

10.2. Suppose that the warehouse in Example 10.1 was oriented with aisles along the length of the warehouse as shown in Figure 10.5. What are the appropriate α and β? Solve for the optimal dimensions of this warehouse. Find the average travel distance per one-way trip.

10.3. A warehouse is to be built with 3500 storage locations. Racks will be stacked six loads high. All loads will enter and leave from the doors placed in the middle of the west wall. Storage locations will be 50 × 50 inches, including clearance and rack structure. Aisles will be 8 feet wide. Using the layout scheme of Figure 10.3, determine the appropriate size of the warehouse and average travel distance if all storage spaces are visited at the same frequency.

10.4. Resolve Problem 10.3 assuming that the loads enter and leave from the southwest corner instead of the center of the west wall.

10.5. A warehouse is being designed to hold up to 4000 unit loads at a time. Storage locations will be 4 × 4 feet. A layout similiar to Figure 10.3 is being contemplated. Find warehouse dimensions and average trip distance for single-command round trips if

a. Loads are placed three locations high on racks.

b. Loads are placed five locations high on racks.

10.6. A warehouse is being constructed to hold 1800, 4 × 4-foot storage locations. All loads must be directly accessible from 8-foot wide aisles. For security purposes, the warehouse will have a single I/O location. At a height of 20 feet, loads can be stacked four high. Depreciated over 7 years, after-tax building cost is approximately $10.00 per linear foot (based on perimeter) per month. Move cost by forklift is $0.003 per foot and the average storage space will turn over five loads per month.

a. Choose the proper warehouse layout style.

b. Lay out the warehouse (give dimensions and number of rows and bays).

10.7. Repeat problem 10.6 assuming that a 40-foot high-rise warehouse will be built. Perimeter cost is doubled. Loads will be stacked eight high. Narrow-aisle vehicles will be used, requiring only 4-foot wide aisles.

10.8. Consider a triangular warehouse as shown in Figure 10.4*a*. Compute average distance from the I/O point to storage locations.

10.9. A product will be stored in batches of 125 unit loads. Maximum storage height is four loads. Find the number of rows needed if block stacking is used with a storage depth of one load.

10.10. Repeat Problem 10.9 for a depth of five loads. Also find the average amount of wasted space due to honeycomb loss.

10.11. List the advantages and disadvantages of block stacking. Do you see any problem with using block stacking for perishable items?

10.12. Describe the principle of throughput-based storage.

10.13. Given the solution of Example 10.3, find the total distance traveled for storage and retrieval activities. Assume that all commands are single.

10.14. What are the advantages of open storage warehouses as contrasted with dedicated storage space?

10.15. An engineer is trying to decide between an automated high-rise warehouse with a narrow-aisle vehicle for each row and a conventional warehouse with one or more free-roaming fork trucks. What costs and other factors should play an important role in this decision?

10.16. Eight families of products are to be stored in a warehouse. Batch sizes (measured in storage loads) and safety stock for each family are given in Table 10.11. Determine the number of storage locations needed for dedicated and open storage systems. Plan for 85 percent utilization.

10.17. A warehouse has two rows, each with ten bays and five levels. Each row has its own S/R machine. Since rows (aisles) are two-sided, we have $2 \cdot 10 \cdot 5 \cdot 2 = 200$ locations. Travel time is approximated by $10(i + j)$ seconds to bay i level j within a row. Travel time is the same to the location on either side of the aisle. For the products in Table 10.12, assign products to locations.

Table 10.11   Product Loads for Problem 10.16

Family	Batch Size	Safety Stock
1	1200	250
2	305	80
3	550	80
4	625	100
5	125	40
6	235	65
7	1050	225
8	185	45

10.18. Repeat Problem 10.17 if travel time is given instead by $\max(10i, 10j)$, since the S/R machine can move in both directions at once.

Table 10.12   Product Data for Problem 10.17

Product	No. of Locations Required	Average Load Storage Time
1	50	12
2	40	9
3	100	5
4	10	24

10.19. A 100-foot long, 80-foot wide, 10-foot high warehouse is being constructed. Storage locations are being planned for small, medium, and large loads. Loads will be manually retrieved, and travel time is dependent only on distance from the I/O point along rectilinear aisles. The I/O point is in the middle of the east wall. The number of S/R trips per day per cubic foot of storage space is estimated to be 4, 1, and 3, respectively, for the three load sizes. Small loads will occupy 25 percent of the warehouse, medium loads will occupy 40 percent, and large loads 35 percent. Sketch the warehouse and indicate where each load type should be located.

10.20. Six product types will be stored in a 100 × 100 foot warehouse. Each part will have dedicated space. Table 10.13 contains space and flow volume considerations for the warehouse. Each grid occupies 100 square feet of warehouse footprint. Assign products to grids.

Table 10.13   Product Data for Problem 10.20

Product	Grids Required	Total Loads Moved per Day
1	15	60
2	20	100
3	10	75
4	40	25
5	5	25
6	10	400

10.21. Parts are received from two input locations at the north and south corners along the west wall of a warehouse. Loads are shipped through a single loading dock at the middle of the east wall. Parts are received and stored in large batches. Each single load received is broken into five loads for shipping. Using the data in Tables 10.14a and 10.14b, allocate storage grids to each product.

Table 10.14*a*   Product Data for Example 10.21

Product	No. of Grids Required	Loads Received/Day	Receiving Point
1	6	12	Northwest corner
2	2	1	Northwest corner
3	4	6	Southwest corner

Table 10.14*b*   Grid Data for Problem 10.21

					Distance from I/O Point to Grid							
I/O Point	1	2	3	4	5	6	7	8	9	10	11	12
NW corner	1	2	3	4	2	3	4	5	3	4	5	6
SW corner	3	4	5	6	2	3	4	5	1	2	3	4
East wall	5	4	3	2	4	3	2	1	5	4	3	2

10.22. A storage aisle has 12 bays and loads are stacked 10 high. The S/R machine travels 15 locations per minute horizontally and 5 per minute vertically. The machine can move in both directions simultaneously. Assuming that all locations are visited equally often, find the average travel time for single and dual commands. Fixed time for pickup or dropoff including acceleration and deceleration is 30 seconds.

10.23. An S/R machine accesses 20 bays and 8 levels. Vertical and horizonal motion can be simultaneous at the rate of 10 locations per minute in one or both directions. If all locations are used equally, find the average single-command travel time excluding fixed pickup and dropoff times.

10.24. Find average dual-command travel time for Problem 10.23.

10.25. Suppose that a storage row is divided into two classes. The S/R machine moves at the rate v_x horizontally and v_z vertically. Movement can be performed simultaneously in both directions. The row has length X and height Z measured in storage locations. If class I items are assigned to the area within X_1, Z_1 of the I/O point [which is located at coordinates (0,0)], and class II items can be stored anywhere else, find the average single-command time to store a class I item and a class II item. (*Note:* Class I item time can be found using equation 10.10.)

10.26. An S/R machine can retrieve up to five bins of parts per trip. The machine serves both sides of one row. The set of waiting orders is given in

Table 10.15   Order Data for Problem 10.26

Order	Item	Location x	y	z
1	a	4	Right	2
	b	5	Right	1
	c	3	Left	5
2	a	6	Right	2
	b	7	Left	1
3	a	5	Right	1
	b	7	Right	1
4	a	1	Left	7
	b	2	Right	6

Table 10.15. Determine the similarities between order locations. Assume $v_x = v_z$.

10.27. Using the similarity coefficients of Problem 10.26, form pick lists.

10.28. Suppose that after picking an order, individual parts are picked from the retrieved bins and then the bins are returned to their original location. However, trips are dual command; after returning the used bins, new bins are collected before the S/R machine returns to the I/O point. Suggest a strategy for determining the order in which pick lists will be serviced.

Table 10.16   Order Similarities for Problem 10.29

| | Order | | | | | | | |
Order	1	2	3	4	5	6	7	8
1	—	0.5	0.0	0.7	0.0	0.2	0.3	0.5
2	0.5	—	0.9	0.0	0.0	0.4	0.2	0.6
3	0.0	0.9	—	0.5	0.1	0.8	0.0	0.0
4	0.7	0.0	0.5	—	0.6	0.0	0.2	0.3
5	0.0	0.0	0.1	0.6	—	0.8	0.1	0.4
6	0.2	0.4	0.8	0.0	0.8	—	0.3	0.7
7	0.3	0.2	0.0	0.2	0.1	0.3	—	0.5
8	0.5	0.6	0.0	0.3	0.4	0.7	0.5	—

10.29. For the set of orders similarities shown in Table 10.16, determine a set of pick lists. The S/R machine can hold 12 part bins. Orders 1 through 8 require $(2, 3, 6, 1, 5, 3, 4, 2)$ bins to be retrieved, respectively.

10.30. Suppose a pick list has been generated for a storage facility. Describe how the Traveling Salesman Problem could be used to determine the route of the picker.

10.31. Making the same assumptions as in Section 10.4.2—simultaneous travel, random storage and retreival, corner I/O location—derive the average interleave time for an AS/RS system. Show that this can be combined with the T_{sc} result of equation 10.10 to obtain the result of equation 10.11.

PART IV

GENERIC MODELING APPROACHES

Up to this point we have discussed specific systems, and models meant for those systems. We now consider general modeling techniques that can be used in many situations. These techniques are designed to model dynamic systems characterized by the occurence of events that change the state of the system under study.

Chapter 11 covers basic analytical queueing approaches to modeling descrete events. Although various distributions can be used, the models are most convenient when we can assume activities take either a deterministic or exponentially distributed length of time. Simple results for single workstations can be readily extended to "networks" comprising many workstations. Analytical queueing network models have been the subject of much research in recent years, and we can now model a wide variety of systems. However, such models can normally provide only estimates of average, steady-state performance.

Assessing a proposed design of a manufacturing system requires a detailed examination of time-variant system dynamics, including transient effects. Models that meet this requirement must include detailed behavior of a wide variety of system components, unique aspects of system behavior, and system-specific measures of performance. An overview of how simulation is used to address these issues is provided in Chapter 12.

In this textbook we have tried to convey the message that there is no unique, ideal, or correct model of a manufacturing system. Models must be custom fit to the problem, and the process of model building is often iterative. A model is appropriate if it provides valid input necessary to make the proper decision. In the final chapter we present two case studies that illustrate the process and practice of modeling in manufacturing system design and operation. The need for multiple models to address a variety of system design issues is stressed.

GENERAL MANUFACTURING SYSTEMS: ANALYTICAL QUEUEING MODELS

*The number of rational hypotheses that can
explain any given phenomenon is infinite.*
—**Persig's Postulate from *Murphy's Law* by Authur Bloch**

All laws are simulations of reality.
—**Lilly's Metalaw from *Murphy's Law* by Authur Bloch**

11.1
INTRODUCTION

Previous chapters have dealt with specific subsystems of the overall manufacturing system. The design and analysis problems peculiar to those environments were discussed. Although the analytical techniques discussed clearly have broad application, emphasis was on a specific situation. In this chapter we look at queueing network models for quickly evaluating average steady-state performance. Our focus is on finding "rough-cut" or approximate results for the long-term average (steady-state) behavior of static systems. "Static" indicates that process parameters such as mean service time do not change over time. We also assume system stability in the sense that capacity, measured by maximum production rate, exceeds average demand. Demand is measured by the rate at which jobs arrive to the manufacturing system under study. System stability assures finite expected inventory levels and waiting times. We will model manufacturing systems as networks of queues. Each workstation and material handling device can be considered a server in the traditional queueing terminology, and each job (batch of parts) is a customer. The manufacturing system is viewed as a system of service-providing workstations. Customers enter the system of servers, visit their required servers in turn, and then leave the system.

Queueing network models have become very popular in recent years among both academicians and practitioners, particularly for the planning of FMSs and the analysis of computer systems. The queueing network models discussed in this chapter have grown rapidly in sophistication and application in the last 10 years. Indeed, perhaps the most widely researched area of manufacturing modeling in the 1980s was extending queueing networks to more practical situations by relaxing or modifying assumptions of the models. Snowdon and Ammons [1988] survey existing queueing network analysis packages for manufacturing. Descriptions of specific packages are available in Solberg [1980], Suri and Diehl [1985], and Whitt [1983]. Although the models presented in this chapter can provide much important information to the modeler, we nevertheless caution that experience with the models has generally shown that production rate and utilization estimates tend to be more accurate than queue lengths and waiting times.

We begin the chapter with a brief analysis of a single workstation. We then build up to general models of interrelated networks of workstations.

11.2
A SINGLE WORKSTATION

Material arrives in batches to a workstation for processing. The workstation is composed of one or more identical processors or servers. A processor could be a worker, an automated machine, or a team of human and physical resources. A single processor can manufacture the batch. On arrival, each batch enters the workstation's input queue. When a processor becomes idle, a batch is loaded onto the processor and processing begins. When the parts in the batch are finished being processed, the batch is removed from the processor and placed in an output queue to await an available material handler. In this section we are concerned only with the activities from the time the batch arrives at the workstation until it completes processing. The batch is considered to be one job. Processing time includes batch setup time plus time to produce each part in the batch. We will first look at a basic queueing model, which assumes exponential interarrival and service time for batches with a FCFS service policy, and then expand the model to other environments.

11.2.1 Poisson Arrivals, Exponential, FCFS Service

Consider a single workstation that receives jobs from a Poisson arrival process and services jobs according to an exponentially distributed processing time. What might we like to know about this operation? Typical questions include production rate in terms of finished jobs per time, average number of jobs in process, capacity, and throughput time from job arrival to completion. A review of fairly well-known results of Poisson queueing systems will show us how to answer these questions. For an $M/M/c$ system (Poisson distributed interarrival times, exponentially distributed service times, c servers in parallel), we let λ be the average arrival rate, μ the average service rate, and $\rho = \lambda/c\mu$ the utilization factor. L and W indicate expected number of customers at the workstation and the expected throughput time for an arbitrary customer. A subscripted q is used to refer to the queue at the workstation; the workstation system consists of the queue plus jobs being served. The state of the system is given by the number of jobs at the workstation, n. If $n \leq c$, no job is waiting and n servers are busy. We use $p_t(n)$ to denote the probability of n jobs at the workstation at time t. Steady state is defined by the condition $p_t(n) = p_{t+\delta t}(n)$; hence, the time index will be dropped when referring to steady-state results. The Poisson arrival and exponential service time assumptions imply that at most one arrival or service completion can occur in time δt as $\delta t \to 0$. If $c = 1$, our system is described by the relations

$$p_{t+\delta t}(0) = p_t(0)(1 - \lambda\,\delta t) + p_t(1)\mu\,\delta t \qquad (11.1a)$$

and

$$p_{t+\delta t}(n) = p_t(n)(1 - \lambda\,\delta t - \mu\,\delta t)$$
$$+ p_t(n+1)\mu\,\delta t + p_t(n-1)\lambda\,\delta t \qquad n \geq 1 \qquad (11.1b)$$

Expression 11.1a states that to be in state 0 at time $t + \delta t$ we must have either started here at time t and had 0 arrivals during δt, or have started in state 1 and have had a service completion during time δt. Expression 11.1b indicates that to be in state n at time $t + \delta t$ we could have (1) started in this state at time t and had no arrivals or service completions during δt; (2) have had a service completion during δt and lowered the state by one job; or (3) have had an arrival during δt that increased the state by one. By rearranging terms and taking the limit, we convert expression 11.1 to the differential-difference equations

$$\lim_{\delta t \to 0} \frac{p_{t+\delta t}(0) - p_t(0)}{\delta t} = -\lambda p_t(0) + \mu p_t(1) \tag{11.2a}$$

and

$$\lim_{\delta t \to 0} \frac{p_{t+\delta t}(n) - p_t(n)}{\delta t} = -(\lambda + \mu)p_t(n) + \mu p_t(n + 1)$$
$$+ \lambda p_t(n - 1) \tag{11.2b}$$

Steady state is obtained when expressions 11.2 are equal to 0. Noting also that time is irrelevant when discussing steady state, expressions 11.2 can be written

$$\lambda p(0) = \mu p(1)$$

and

$$(\lambda + \mu)p(n) = \mu p(n + 1) + \lambda p(n - 1) \qquad n \geq 1$$

These equations illustrate the intuitive definition that in steady state, the *rate out of any state n is equal to the rate into state n,* that is, state transitions are balanced. We can leave state n by an arrival or by service if the system is nonempty. We can enter state n by completing a service when in state $n + 1$ or, unless $n = 0$, being in state $n - 1$ and having an arrival.

Solving for the state probabilities, we obtain

$$p(1) = \frac{\lambda}{\mu} p(0) \tag{11.3a}$$

$$p(n + 1) = \frac{\lambda + \mu}{\mu} p(n) - \frac{\lambda}{\mu} p(n - 1) \qquad n \geq 1 \tag{11.3b}$$

The reader should be able to show that the relation $p(n) = (\lambda/\mu)p(n - 1), n \geq 1$ satisfies expression 11.3. This is equivalent to

$$p(n) = \left[\frac{\lambda}{\mu}\right]^n p(0) \qquad n \geq 1 \tag{11.4}$$

Now, we also have another constraint, namely,

$$\sum_{n=0}^{\infty} p(n) = 1 \tag{11.5}$$

Substituting equation 11.4 values into equation 11.5 we find that $p(0) = 1 - \rho$. Likewise, $p(n) = (1 - \rho)\rho^n; n \geq 1$. Given the state probabilities, many other results can be found. For instance, the expected number of jobs at the workstation is

$$L = \sum_{n=0}^{\infty} n p(n) = p(1) + 2p(2) + \cdots = \frac{\rho}{1 - \rho}$$

Table 11.1   *M/M/c* Queueing Results

	$M/M/1$	$M/M/c$
$p(0)$	$1 - \rho$	$\left[\dfrac{(c\rho)^c}{c!(1-\rho)} + \displaystyle\sum_{n=0}^{c-1} \dfrac{(c\rho)^n}{n!} \right]^{-1}$
L_q	$\dfrac{\rho^2}{1-\rho}$	$\dfrac{\rho(c\rho)^c\, p(0)}{c!(1-\rho)^2}$
L	$\dfrac{\rho}{1-\rho}$	$L_q + \dfrac{\lambda}{\mu}$
W_q	$\dfrac{\rho}{\mu(1-\rho)}$	$\dfrac{(c\rho)^c\, p(0)}{c!c\mu(1-\rho)^2}$
W	$\dfrac{1}{\mu(1-\rho)}$	$W_q + \mu^{-1}$

The probability that the server is busy is just the probability that the system is not empty, or $1 - p(0) = \rho$. Since the number of jobs in process is 1 when the server is busy and 0 when the server is idle, ρ is also the expected number of jobs in process. Thus, since all jobs are either in process or in queue, the expected number of jobs waiting in the queue is $L_q = L - \rho$. Other measures can be found similarly. Table 11.1 summarizes the analytical results for stable ($\rho < 1$) systems. Remember that $\rho = \lambda/c\mu$.

The reader should note that $L = \lambda W$ in Table 11.1. Little's Law, which was first introduced in Chapter 1, will play a major role in this chapter. This law holds true for all systems in steady state.

EXAMPLE 11.1

Suppose a manufacturing facility operates as a flow shop. Interarrival times are exponentially distributed with an average of 10 orders per week. Orders are processed first come first served (FCFS) with a constant production schedule from week to week. Orders vary in complexity, but the system is capable of processing about 12 orders per week. Find the average time from order arrival to completion.

Solution

We treat the plant as a single server workstation. We have $\rho = 10/12$. According to Table 11.1, the plant is idle $p(0) = 1 - \rho = \frac{1}{6}$ of the time. This time might be used for routine maintenance. Throughput time is

$$W = \frac{1}{\mu(1-\rho)} = \frac{1}{12\left(\frac{1}{6}\right)} = 0.5 \text{ weeks}$$

Processing time averages $\frac{1}{12}$ week and queueing time is $W_q = 0.5 - \frac{1}{12} = \frac{5}{12}$ weeks. ∎

11.2.2 Poisson Arrivals, General, FCFS Service

Although most analytical queueing results rely on the assumptions of exponential interarrival and service times, $M/G/1$ queues have also been widely analyzed. The $M/G/1$

model assumes exponential interarrival times (which can usually be approximately justified if interarrival times are random with a standard deviation nearly equal to the mean), but any general service time distribution is permissible. We will give some basic results for this problem; a more extensive list of results can be found in Lavenberg [1983]. The reader may wish to show that the $M/M/1$ results above are a special case of these more general expressions.

Let S be the random variable for service time, let T be the random variable for throughput time, and let N be the random variable for the number of jobs at the workstation. [In relation to existing notation, $E(S) = \mu^{-1}, E(T) = W$, and $E(N) = L$]. For FCFS service, the following relations are known

$$E(T) = E(S) + \frac{\lambda E(S^2)}{2(1 - \rho)} \tag{11.6}$$

$$V(T) = V(S) + \frac{\lambda E(S^3)}{3(1 - \rho)} + \frac{\lambda^2[E(S^2)]^2}{4(1 - \rho)^2} \tag{11.7}$$

$$E(N) = \rho + \frac{\lambda^2 E(S^2)}{2(1 - \rho)} \tag{11.8}$$

and

$$V(N) = E(N) + \lambda^2 V(S) + \frac{\lambda^3 E(S^3)}{3(1 - \rho)} + \frac{\lambda^4[E(S^2)]^2}{4(1 - \rho)^2} \tag{11.9}$$

It is interesting to note in equations 11.6 and 11.8 that mean throughput time and number of jobs depend only on the first two moments of the service time distribution, not on the·distributional form. Thus, a normally distributed service time with equal mean and standard deviation would yield the same mean results as exponential service time. Variances of throughput time and queue length require knowledge of only the first three moments.

EXAMPLE 11.2

Reconsider Example 11.1 with $\lambda = 10/\text{week}$. Compare performance measures for service distributions of (a) exponential [$E(S) = \frac{1}{12}$]; (b) uniform from $[0, \frac{1}{6}]$; (c) gamma with parameters $r = 10, \Lambda = \frac{1}{120}$, that is, $E(S) = \frac{1}{12}$; and (d) deterministic with ($S = \frac{1}{12}$). All four distributions have the same mean service rate. The gamma distribution would result if the service task was composed of 10 subtasks, each exponentially distributed with mean time one tenth of the mean total service time.

Solution

Using the moments of the various distributions and the expressions 11.6 through 11.9 we compute the results in Table 11.2.

Table 11.2   Performance Comparisons for $M/G/1$ Example

Distribution	$E(S)$	$V(S)$	$E(T)$	$V(T)$	$E(N)$	$V(N)$
Exponential	0.0833	0.00694	0.500	0.250	5.00	30.0
Uniform	0.0833	0.0023148	0.361	0.103	3.61	13.9
Gamma	0.0833	0.000694	0.313	0.0685	3.13	9.97
Deterministic	0.0833	0.0	0.292	0.0550	2.92	8.41

To illustrate, let us look at the gamma case. We still have $\rho = \lambda/\mu = 5/6$. Also, the central moments (about 0) for a gamma distribution with integer r are given by

$$E(S^\alpha) = \frac{(\alpha + r - 1)!}{(r - 1)!} \Lambda^{-\alpha}$$

This yields

$$E(S) = \frac{10!}{9!} \Lambda^{-1} = \frac{10}{120} = 0.08333$$

$$E(S^2) = \frac{11!}{9!} \Lambda^{-2} = \frac{110}{120^2} = 0.0076389$$

and

$$E(S^3) = \frac{12!}{9!} \Lambda^{-3} = \frac{1320}{120^3} = 0.00076389$$

With these three central moments of the service distribution we can compute the performance measures using expressions 11.6 through 11.9. From equation 11.6

$$E(T) = E(S) + \frac{\lambda E(S^2)}{2(1 - \rho)} = 0.08333 + \frac{10(0.0076389)}{2(0.16667)} = 0.3125$$

From equation 11.7,

$$V(T) = V(S) + \frac{\lambda E(S^3)}{3(1 - \rho)} + \frac{\lambda^2 [E(S^2)]^2}{4(1 - \rho)^2}$$

$$= 0.0006944 + \frac{10(0.00076389)}{3(0.16667)} + \frac{100(0.0076389)^2}{4(0.16667)^2} = 0.0685$$

For the number of jobs at the workstation, equation 11.8 yields

$$E(N) = \rho + \frac{\lambda^2 E(S^2)}{2(1 - \rho)} = 0.83333 + \frac{100(0.0076389)}{2(0.16667)} = 3.125$$

[Equivalently, $E(N) = \lambda E(T)$]. Last, from equation 11.9

$$V(N) = 3.125 + 100(0.0006944) + \frac{1000(0.00076389)}{3(0.16667)}$$

$$+ \frac{10^4(0.0076389)^2}{4(0.016667)^2} = 9.974$$

It is clear from Table 11.2 that performance is affected by more than just the mean of the service time distribution. Thus, although we will generally assume exponential service in the remainder of this chapter, this is an assumption that should be validated in practice. If the actual distribution is significantly different, it may be preferable to model that distribution directly, even at the expense of additional computational requirements. We can also note from the table that as the variability of the service time is reduced, so are the mean and variance of throughput times and queue lengths. Expressions 11.6 to 11.9 quantify this relationship. ∎

11.2.3 General, Part-Based Priority Service

To this point we have dealt exclusively with FCFS service. Many workstations work on several types of products. These products may have different priorities. Suppose there are P part types. Type p has arrival rate and service time of λ_p and S_p, respectively. We assume that parts are ordered such that the lower the number the higher the priority. Average service time is the weighted average

$$E(S) = \lambda^{-1} \sum_{p=1}^{P} \lambda_p E(S_p)$$

where $\lambda = \sum_{p=1}^{P} \lambda_p$. Utilization due to the first p types is

$$\rho(p) = \sum_{l=1}^{p} \lambda_l E(S_l)$$

In this instance it can be shown that

$$E(T_p) = E(S_p) + \frac{\lambda E(S^2)}{2[1 - \rho(p-1)][1 - \rho(p)]} \tag{11.10}$$

where it is understood that $\rho(0) = 0$. When priority service is instituted, the lowest numbered (highest priority) part type will have a smaller average throughput time than with FCFS. Average throughput time will increase for the lowest priority item.

EXAMPLE 11.3

Suppose in Example 11.1 that two part types are made at the workstation. Four part type 1's and six part type 2's arrive per week. Service rate is 12 per week for both types. Interarrival and service times are exponential. Determine the difference in throughput time and queue lengths for the two part types if part type 1 is given higher priority.

Solution

First, we have

$$\rho(1) = \frac{4}{12} = 0.33333 \qquad \rho(2) = \frac{4}{12} + \frac{6}{12} = 0.83333$$

We also need $E(S^2)$. $E(S^2) = V(S) + [E(S)]^2$. From Table 11.2, our exponential service has $E(S_p^2) = 0.00694 + 0.0833^2 = 0.01389$ for both types. Then taking a weighted average over the two part types,

$$E(S^2) = \frac{4}{10}(0.01389) + \frac{6}{10}(0.01389) = 0.01389$$

Following equation 11.10,

$$E(T_1) = \frac{1}{12} + \frac{10(0.01389)}{2[1 - 0][1 - 0.33333]} = 0.1875$$

and

$$E(T_2) = \frac{1}{12} + \frac{10(0.01389)}{2[1 - 0.33333][1 - 0.8333]} = 0.7084$$

For an arbitrary job, expected waiting time is

$$E(T) = \frac{4}{10}(0.1875) + \frac{6}{10}(0.7084) = 0.5000$$

This is the same as the 0.5 found earlier for FCFS (no priority). Queue (including jobs in process) lengths follow from Little's Law:

$$E(N_1) = \lambda_1 E(T_1) = 4(0.1875) = 0.750$$

and

$$E(N_2) = \lambda_2 E(T_2) = 6(0.7084) = 4.250$$

Once again we have a total queue average of five jobs; however, the type of jobs at the workstation is much different. Without priority we would have had two type 1 jobs and three of type 2 on average.   ∎

11.2.4 Finite Population Models

Consider a vehicle recharge station in an automated guided vehicle system. If there are only K vehicles in the system, and K is relatively small, then the arrival rate of customers is affected by the number of customers currently in or awaiting service. In fact, if the arrival rate of each operating vehicle is Poisson with rate λ, and there are n vehicles already at the station, then the recharge station sees a Poisson arrival rate with

$$\lambda_n = (K - n)\lambda$$

Peck and Hazelwood [1958] and White et al. [1975] analyze such $(M/M/c) : (GD/K/K)$ systems. The balance equations with state-dependent arrival and service rates can be solved to obtain

$$p(0) = \left[\sum_{n=0}^{c-1} \frac{K!(c\rho)^n}{n!(K-n)!} + \sum_{n=c}^{K} \frac{K!c^c\rho^n}{(K-n)!c!} \right]^{-1} \tag{11.11}$$

and

$$p(n) = \begin{cases} \dfrac{K!(c\rho)^n p(0)}{n!(K-n)!} & n = 0, 1, \ldots, c-1 \\[2ex] \dfrac{K!c^c\rho^n p(0)}{(K-n)!c!} & n = c, \ldots, K \end{cases} \tag{11.12}$$

Equations 11.11 and 11.12 use $\rho = \lambda/c\mu$ as before; however, ρ no longer represents the load on the system, since it fails to account for K or the state-dependent arrival rates.

To illustrate, suppose that there are two recharge stations for six vehicles. Recharging takes one hour (let's assume this is exponentially distributed) and mean time between rechargings for operating vehicles is eight hours. Thus, $c = 2, K = 6, \mu = 1$, and $\lambda = 0.125$. First, $\rho = 0.125/2 = 0.0625$. Then, using equation 11.11,

$$p(0) = [1 + 0.75 + 0.2344 + 0.0293 + 0.0055 + 0.0007 + 0.0000]^{-1} = 0.4951$$

Applying equation 11.12,

$$p(1) = \frac{K!(c\rho)^1 p(0)}{n!(K-n)!} = \frac{6!(0.125)(0.4951)}{5!} = 0.3713$$

Continuing to apply equation 11.12, we obtain the vector of state probabilities $p(\mathbf{n}) = (0.495, 0.371, 0.116, 0.145, 0.003, 0.000, 0.000)$. Summing $p(0) + p(1) + p(2) = 0.982$, we see that there will be more than two vehicles at the recharge station less than 2 percent of the time.

This finite population model is appropriate for many service facilities. It is commonly used to model the load on repair facilities. Another application would be a cell where one or more workers/robots load and unload a number of machines. It can also be thought of as a special case of the more general Machine Interference Problem (see Bunday and Khorram [1988]). Moreover, the state-dependent nature of the arrival process provides a good introduction to the closed queueing network models to be discussed in Section 11.4.

11.3
OPEN NETWORKS

We would also like to be able to evaluate networks of workstations. A network has M workstations with jobs moving between workstations in accord with their route sheets. On completion of the assigned tasks at a workstation, the job is transferred to its next required workstation by the material handling system. In an **open network** an external arrival process generates jobs that arrive at one or more workstations and enter the network.

Before proceeding, we will review several results that will be important in analyzing networks.

FACT 1: POISSON REPRODUCTIVE PROPERTY. The sum of independent Poisson random variables is Poisson.

JUSTIFICATION: This result can be shown using moment-generating functions and is demonstrated in most introductory probability and statistics texts.

FACT 2: If the number of arrivals per time is Poisson distributed, then time between arrivals has an exponential distribution.

JUSTIFICATION: Let X, the arrivals per time, be a Poisson random variable. Then the density function for the number of arrivals in time t is the Poisson density $p(x; \lambda) = [(\lambda t)^x e^{-\lambda t}]/x!$. Let T be the random variable for time between successive arrivals. $F(T)$ is the cumulative distribution, that is, $Pr(T > t) = 1 - F(t)$ where Pr denotes probability. Now, the event of interarrival time exceeding t is equivalent to no arrivals from a Poisson process in time t or

$$Pr(T > t) = p(x = 0; \lambda) = e^{-\lambda t}$$

Thus, $F(t) = 1 - e^{-\lambda t}$ and differentiating, $f(t) = \lambda e^{-\lambda t}, t \geq 0$, precisely the exponential density function.

FACT 3: The interdeparture time from an $(M/M/c)$ system with infinite queue capacity is exponential.

JUSTIFICATION: See Burke [1968].

DISCUSSION: These results lead to a general result on the merging of Poisson processes. If statistically independent Poisson processes are merged together, the resulting process is Poisson. Thus, if a workstation receives
input from several infinite buffer $M/M/c$ workstations, then its arrival
process is also Poisson. This will be very important for analyzing systems of related workstations. It can also be shown that if a Poisson
process of rate λ is split into multiple processes such that each arrival
has probability p_i of being routed to stream i and $\sum_i p_i = 1$, then
the individual streams are Poisson with arrival rates $p_i \lambda$.

EXAMPLE 11.4

As shown in Figure 11.1, six machines stamp part batches and send them to
one inspector. The inspector then sorts batches sending 97 percent to packing
and scrapping 3 percent. Parts are stamped in batches. Each machine has its
own queue and stamps about 10 batches per day. Find the arrival processes at
inspection, packing, and rework.

Solution

The six machines merge their output into one stream for inspection. Assuming
that each machine has exponential interarrival and service times, their output
streams are Poisson by Fact 3. The Reproductive Property then indicates that
input to inspection is a Poisson arrival process with rate 60/day. If inspection
time is exponential, then the output stream is Poisson. Since this output is
randomly divided into two streams, the input to Packing is Poisson with rate

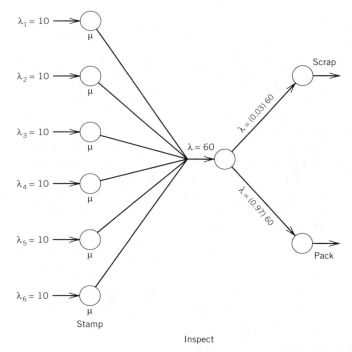

Figure 11.1 An example of the merging and splitting of Poisson processes.

$\lambda_P = (0.97)60 = 58.2/\text{day}$. Input to Scrap is Poisson with rate $\lambda_S = (0.03)60 = 1.8/\text{day}$. Note that this "network" of workstations conserves; at each workstation and for the network as a whole, all jobs that enter eventually leave. ■

11.3.1 Poisson Arrivals and Exponential, FCFS Service

Let us consider another simple system. Parts are withdrawn from a warehouse and kitted into packets for filling production orders. The kit may have enough parts for one finished item or a batch of these items. The kit is sent first to a workstation for assembly. There may be several assemblers working in parallel with the batch going to the first available assembler. The assembler combines parts from the kit to produce a batch of finished parts. Part batches are then transported to an inspection and packing station. The basic progression is shown in Figure 11.2. Let us further assume that there are always orders waiting to be kitted. The dispatching system is designed to keep one hour's worth of work in the input queue for kitters at all times. Kitting turns out an average of 10 kits per hour. Capacities (service rates) at assembly (A) and inspection/pack (I) are 12 and 15 per hour, respectively. Kitting, assembly, and inspection/packing times are exponentially distributed. This system is a serial production system with random processing times.

If we examine our production system closely, we realize that since interdeparture times from kitters are exponential, kitting sends a Poisson stream of jobs to assembly. (Fact 1 says that this is true even if there are multiple kitters!) Thus, assembly sees jobs arriving with an exponential interarrival time. Since A is an ($M/M/c$) workstation, its output is a Poisson stream from Fact 3 above. Thus, I receives jobs with exponential interarrival time. The state of the system can be described by the number of jobs at A and I. Let $p(n_1, n_2) = Pr(n_1 \text{ jobs at } A \cap n_2 \text{ jobs at } I)$. Since the number of jobs at kitting is constant, it need not be included in the state definition. The states and transitions are shown in Figure 11.3.

From the state diagram we can write the steady-state balance equations showing the rate into and out of each state. These expressions assume a single server at each station. The arrival rate λ is the rate at which kits are kitted. (To indicate the generality of the two-stage serial system being modeled, we use the subscript "1" for A and "2" for I.)

$$
\begin{array}{lc}
State\ n_1, n_2 & Rate\ Out = Rate\ In \\
0, 0 & \lambda p(0, 0) = \mu_2 p(0, 1) \\
n_1 > 0, 0 & (\lambda + \mu_1)p(n_1, 0) = \mu_2 p(n_1, 1) + \lambda p(n_1 - 1, 0) \\
0, n_2 > 0 & (\lambda + \mu_2)p(0, n_2) = \mu_2 p(0, n_2 + 1) + \mu_1 p(1, n_2 - 1) \\
n_1 > 0, n_2 > 0 & (\lambda + \mu_1 + \mu_2)p(n_1, n_2) = \lambda p(n_1 - 1, n_2) \\
& \quad + \mu_1 p(n_1 + 1, n_2 - 1) \\
& \quad + \mu_2 p(n_1, n_2 + 1) \quad\quad (11.13)
\end{array}
$$

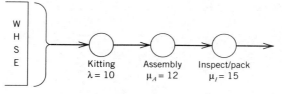

Figure 11.2   The kitting, assembly, inspection/packing system.

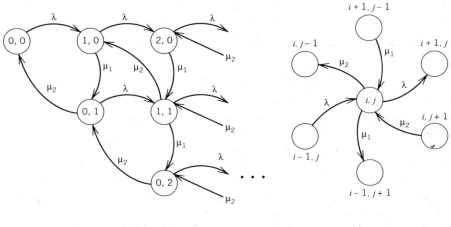

(*a*) "Near-empty" boundary states   (*b*) General state i_j

Figure 11.3   State space for serial process.

Applying the conservation constraint $\sum_i \sum_j p(n_1, n_2) = 1$, we can find the solution to expressions 11.13 as

$$p(n_1, n_2) = (1 - \rho_1)\rho_1^{n_1}(1 - \rho_2)\rho_2^{n_2} \qquad n_1, n_2 \geq 0 \qquad (11.14)$$

where $\rho_j = \lambda/\mu_j$. Result 11.14 is an example of the important **product form solution** for many queueing models. Apart from the constant $(1 - \rho_1)(1 - \rho_2)$, the solution is a product of utilization factors for each workstation replicated by the number of jobs at that workstation. Recall that for the single workstation system in the previous section, we had the result $p(n) = (1-\rho)\rho^n$. Interestingly, the probability of n_1 jobs at the first workstation and n_2 jobs at workstation 2 is simply the product of the probabilities that workstation 1 has n_1 jobs and workstation 2 has n_2 jobs when they are treated as independent, single workstation systems. Probabilistically speaking, the queue lengths are independent, and we can analyze each workstation separately and then combine their results. This is an important result; it will allow us to analyze an M-station network with the same amount of computation as analyzing M single-station systems.

EXAMPLE 11.5

Find the expected number of jobs in the kit, assemble, and pack system along with the expected throughput time. Each station has a single server.

Solution

To find these measures we first analyze each station. Arrival rate is $\lambda = 10/\text{hr}$ for both A and I (jobs are conserved). Service rates are $\mu_A = 12/\text{hr}$ and $\mu_I = 15/\text{hr}$. Accordingly, $\rho_A = 10/12$ and $\rho_I = 10/15$. From Table 11.1,

$$W_A = \frac{1}{\mu_A(1 - \rho_A)} = \frac{1}{12\left(1 - \dfrac{10}{12}\right)} = \frac{1}{2}\ \text{hr}$$

$$L_A = \lambda_A W_A = 10\left(\frac{1}{2}\right) = 5\ \text{jobs}$$

$$W_I = \frac{1}{\mu_I(1 - \rho_I)} = \frac{1}{15\left(1 - \dfrac{10}{15}\right)} = \frac{1}{5} \text{ hr}$$

$$L_I = \lambda_I W_I = 2 \text{ jobs}$$

The next step is to combine results across workstations. System measures include kitting (10 jobs in queue plus one in process on average corresponding to the one-hour workload kept there by the dispatching system), assembly, and inspection values.

$$L = 10 + 1 + L_A + L_I = 18 \text{ jobs}$$

Throughput time can be found either by $L = \lambda W$ or by noting that

$$W = W_{qK} + \lambda^{-1} + W_A + W_I = 1 + 0.1 + 0.5 + 0.2 = 1.8 \text{ hr}$$

where the K subscript represents kitting.

Suppose current plans call for a buffer with a capacity of 10 jobs (including the job in process) in front of the assembly station. Is this sufficient? To answer this question we need to find the probability that an $M/M/1$ station with $\rho = 10/12$ will have more than 10 jobs. From before we have $Pr_A(n) = (1 - \rho_A)\rho_A^n$. Thus

$$Pr_A(n > 10) = (1 - \rho_A) \sum_{i=11}^{\infty} \rho^i$$

Recalling that

$$\sum_{i=0}^{n} r^i = \frac{1 - r^{n+1}}{1 - r} \quad \text{for} \quad r < 1$$

we obtain

$$Pr_A(n > 10) = 1 - Pr_A(n \leq 10) = 1 - (1 - \tfrac{5}{6})\left[\frac{1 - \tfrac{5}{6}^{11}}{1 - \tfrac{5}{6}}\right] = 0.135$$

The buffer will be full 13.5 percent of the time necessitating off-line storage of kits or blocking of the first workstation. A design allowing more buffer space would most likely be preferable.   ■

Ultimately, our interest is with general networks of M workstations, not just two-stage serial systems. Jackson [1957] provided a very powerful generalization of the product form result to this environment. We make the following assumptions:

1. Workstation $j = 1, \ldots, M$ has c_j servers.
2. External arrivals to workstation j are Poisson with mean rate λ_j. External arrivals are new jobs from outside the system of M workstations. The existence of these external arrivals is what makes this an "open" network.
3. Workstations are FCFS.
4. Service rates are exponential with mean service times μ_j^{-1}.
5. A job at station j transfers to station k with probability p_{jk} on completion at j. Jobs leave the system with probability $p_{j0} = 1 - \sum_{k=1}^{M} p_{jk}$.
6. Queue sizes are unlimited (no blocking).

A job arriving to a workstation may have come from outside the system with this being the first operation of the job, or the job may have completed service at another workstation and have been routed here for its next operation. With this model we need to compute the effective arrival rate λ'_j at workstation j. The effective arrival rate is composed of external and internal arrivals. Assumption 2 gave us an external arrival rate of λ_j. Assumption 5 describes internal arrivals. The effective arrival rate can be found by solving the system of linear equations:

$$\lambda'_j = \lambda_j + \sum_{k=1}^{M} \lambda'_k p_{kj} \qquad 1 \le j \le M \qquad (11.15)$$

Internal arrivals to j in equation 11.15 may come from any station k and occur with probability p_{kj} for each job completed at k. Utilization factors should be measured in terms of effective arrival rates for the remainder of this section, that is, $\rho_j = \lambda'_j / c_j \mu_j$.

The system state is described by $(n_1, n_2, \ldots, n_M)$, a vector containing the number of jobs at each workstation. The probability of being in any state $\mathbf{n} = (n_1, \ldots, n_M)$ is denoted $p(\mathbf{n})$. In defining the state of the system, exponential service with its forgetfulness property allows us to ignore when jobs arrived at each workstation. Remaining service time has the same distribution as that of a newly started job. For other service time distributions more information on the state of service at each workstation would be required. The major result is

Jackson's Theorem

Let

$$P_n^j = \begin{bmatrix} P_0^j \left(\dfrac{\lambda'_j}{\mu_j} \right)^n / n! & n = 0, \ldots, c_j \\[3em] P_0^j \left(\dfrac{\lambda'_j}{\mu_j} \right)^n / (c_j! c_j^{n-c_j}) & n = c_j + 1, \cdots \end{bmatrix}$$

where P_0^j is such that $\sum_{n=0}^{\infty} P_n^j = 1$ for all j. Then if $\lambda'_j \le \mu_j c_j$ for all j,

$$p(n_1, \ldots, n_M) = \prod_{j=1}^{M} P_{n_j}^j$$

Proof: We will not present a formal proof (see the Jackson reference) but simply note that the proof can be established by showing that the foregoing product form state definition satisfies all steady-state balance equations for the system. The balance equations state that the rate out of any state due to external arrivals or internal service completions is equal to the rate into that state from external arrivals, internal completions that exit the system, and internal completions that transfer between a pair of stations. In writing the balance equations, note that actual service rates are state dependent and given by $\mu_j[\min(c_j, n_j)]$ at workstation j.

Note that the $P_{n_j}^j$ are precisely the same terms that would result from solving station j as an independent, single-station system with arrival rate λ'_j, service rate μ_j, and c_j servers. In summary then, we have a three-step procedure for analyzing open queueing networks.

1. Determine the effective arrival rates using equation 11.15.
2. Analyze each workstation independently using Table 11.1.
3. Aggregate results across workstations to obtain the desired performance measures.

We will illustrate the procedure for analyzing open Poisson arrival, exponential service, FCFS networks with two examples. The first example is a simple extension of the previous example. The second example is slightly more extensive.

EXAMPLE 11.6

Suppose in Example 11.5 that each time jobs are inspected they fail with probability p_{21} and are returned to assembly. A feedback loop now exists from I to A. Rework averages the same 5 minutes as inital assembly. How is throughput time affected if $p_{21} = 0.1$?

Solution

1. First, we must find the effective arrival rates. Noting that assembly receives both the output from kitting as before plus the rejects at inspection, expression 11.15 becomes

$$\lambda'_A = \lambda + p_{21}\lambda'_I$$

   Every job leaving assembly goes directly to inspection/packing and no jobs arrive at inspection/packing from outside the system; thus

$$\lambda'_I = 1\lambda'_A$$

   This system is easily solved with result $\lambda'_A = \lambda'_I = \lambda/(1 - p_{21})$. Suppose that $p_{21} = 0.1$. With $\lambda = 10$ as before, we obtain

$$\lambda'_I = \lambda'_A = \frac{10}{0.9} = 11.1111$$

2. Second, each workstation is solved independently. $\rho_A = \lambda'_A/\mu_A = 10/0.9(12) = 0.9259$ and $\rho_I = 10/0.9(15) = 0.7407$. From Table 11.1,

$$W_A = \frac{1}{\mu_A(1 - \rho_A)} = 1.1246 \qquad L_A = \lambda'_A W_A = 12.50$$

   and

$$W_I = \frac{1}{\mu_I(1 - \rho_I)} = 0.2571 \qquad L_I = \lambda'_I W_I = 2.857$$

3. The third step involves aggregating over workstations. To find expected throughput time, we must know the expected number of visits to each workstation. For kitting, this is clearly one. In general, if we define v_j as the expected number of visits to workstation j per part produced, we obtain the relations.

$$v_j = \frac{\lambda_j}{\sum_{k=1}^{M} \lambda_k} + \sum_{k=1}^{M} p_{kj} v_k \qquad (11.16)$$

   Expressions 11.16 are similar in form to equation 11.15. Indeed, it can be seen that if λ'_j solve equation 11.15 then $v_j = \lambda'_j/\sum_{k=1}^{M} \lambda_k$ solve equation 11.16. Thus, for our example,

$$v_A = \frac{\lambda'_A}{\lambda} = 1.1111$$

and

$$v_I = \frac{\lambda_I'}{\lambda} = 1.1111$$

We can then find expected throughput time by

$$W = \sum_{j=1}^{M} v_j W_j$$

Each of the v_j visits to workstation j takes W_j time. Thus, for the example

$$W = W_K + v_A W_A + v_I W_I = 1.1 + 1.1111(1.1246) + 1.1111(0.2571)$$
$$= 2.64 \text{ hr} \quad \blacksquare$$

EXAMPLE 11.7

Suppose that we wish to produce the $K = 4$ part families displayed in Table 11.3 with the manufacturing system shown in Figure 11.4. Figure 11.4a shows the system as viewed by the engineer. Four centers, Milling, Drilling, Turning, and Grinding produce four part families. External arrival rates are shown for each family along with internal routing by part type. The (0.5) factors indicate split, probabilistic routing for parts as certain type 2 and 3 parts skip Grinding. Assuming that workstations are available 40 hours per week, evaluate the system as a queueing network.

Solution

Figure 11.4b translates the system into an open queueing network. The same workstations exist but arrival rates are now determined by workcenter. Arc values are routing probabilities, and service rates are weighted averages. We will see how these were computed shortly. Material handling is quick and relatively inexpensive in our system and is excluded from the analysis (otherwise we could include a material handling workstation and reroute all part routing arcs through that workstation).

Table 11.3   Part Data for Example 11.7

Part Family	Demand per Week	Route (Machine, Hr per Part)		
		1	2	3
1	2	M,2	T,1	G,2
2	10	M,4	D,5	G,1(50%)
3	6	M,10	T,1	G,2(50%)
4	3	D,10	T,2	G,2

STEP 1.   COMPUTE EFFECTIVE ARRIVAL RATES. A general technique for determining effective arrival rates is to solve expression 11.15 separately for each part type and then sum arrival rates over part types for each workstation. Consider part type 3. The appropriate expressions are

$$\lambda_{M3}' = 6 + 0$$
$$\lambda_{D3}' = 0 + 0$$
$$\lambda_{T3}' = 0 + 1\lambda_{M3}'$$

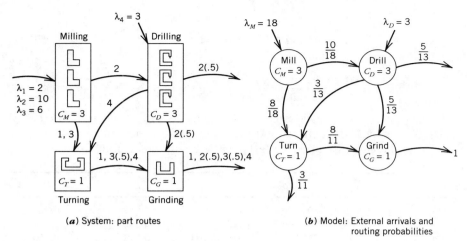

(*a*) System: part routes

(*b*) Model: External arrivals and routing probabilities

Figure 11.4   The job shop for Example 11.7

and

$$\lambda'_{G3} = 0 + 0.5\lambda'_{T3}$$

Part type 3 starts at Milling then goes to Turning. One half of these parts then proceed to Grinding, and the rest exit the system. The added notation "3" in the subscript is just to remind us that we are discussing product 3 only. The solution to this set of equations and those for the other part types is summarized in the following table. The Sum column gives total effective arrival rate for each station. Note that since we were given the part routings, we could have bypassed equations 11.15 and simply have summed the visits shown in Table 11.3. For instance, $\lambda'_M = 18$ results from the two part 1's plus ten part 2's plus six part 3's.

	Part Family				
	1	2	3	4	Sum
λ'_M	2	10	6	0	18
λ'_D	0	10	0	3	13
λ'_T	2	0	6	3	11
λ'_G	2	5	3	3	13

Average processing times are found by $\mu_j^{-1} = \sum_{k=1}^{K} r_{kj}\mu_{kj}^{-1}$, where r_{kj} is the proportion of jobs at station j that are part type k and μ_{kj} is the service rate for part type k at station j. The μ_{kj} are 40(hr/week) divided by the processing times in Table 11.3. Processing times in weeks are then

$$\mu_M^{-1} = \frac{2}{18}\cdot\frac{1}{20} + \frac{10}{18}\cdot\frac{1}{10} + \frac{6}{18}\cdot\frac{1}{4} = 0.1444$$

$$\mu_D^{-1} = \frac{10}{13}\cdot\frac{1}{8} + \frac{3}{13}\cdot\frac{1}{4} = 0.1538$$

$$\mu_T^{-1} = \frac{2}{11} \cdot \frac{1}{40} + \frac{6}{11} \cdot \frac{1}{40} + \frac{3}{11} \cdot \frac{1}{20} = 0.03182$$

and

$$\mu_G^{-1} = \frac{2}{13} \cdot \frac{1}{20} + \frac{5}{13} \cdot \frac{1}{40} + \frac{3}{13} \cdot \frac{1}{20} + \frac{3}{13} \cdot \frac{1}{20} = 0.04038$$

STEP 2.   ANALYZE WORKSTATIONS INDEPENDENTLY. Each workstation is an $M/M/c$ queue. For each we find the basic performance measures using the expressions in Table 11.1. The results are as follows:

Station	λ	μ	c	ρ	W	W_q	L	L_q	P_0
M	18	6.923	3	0.867	0.4185	0.2741	7.534	4.934	0.035
D	13	6.500	3	0.667	0.2222	0.0684	2.889	0.889	0.111
T	11	31.43	1	0.350	0.0489	0.0171	0.538	0.188	0.650
G	13	24.76	1	0.525	0.0850	0.0446	1.105	0.580	0.475

STEP 3.   AGGREGATE FOR PERFORMANCE. The above table provides important information. We know the utilization and average queue lengths at each workstation. We can also determine system performance measures. For instance, the average number of jobs in process is

$$L = \sum_{j=M}^{G} L_j = 7.534 + 2.889 + 0.538 + 1.105 = 12.07$$

Given the number of visits of a part to each workstation, we can also determine its throughput time. This can be accomplished in either of two ways. First, we could multiply the visit counts v_{kj} by W_j and sum over workstations. Alternatively, we could multiply v_{kj} times W_{qj} and add mean (operation and part specific) service times μ_{kj}^{-1} for a slightly more accurate estimate. This latter approach takes advantage of the known mean processing time of specific part types. For part type 1,

$$W_1 = \frac{2}{40} + W_{qM} + \frac{1}{40} + W_{qT} + \frac{2}{40} + W_{qG} = 0.46 \text{ weeks} \qquad \blacksquare$$

11.3.2 Relaxing Assumptions*

The basic model of Section 11.3.1 can be extended in a number of ways. Several extensions are discussed in this section. We first discuss the relatively simple extension of an infinite number of servers. We then look at how the decomposition strategy can be extended to general service time distributions and end with a look at alternatives to FCFS, such as shortest processing time.

Ample Servers

Suppose we always had **ample servers** available. Ample servers implies no waiting and may be appropriate for certain material handling systems with high capacity. A conveyor, for instance, can normally accommodate all parts without queueing. Ample servers can be modeled by setting $W_j = \mu_j^{-1}$ for ample server station j. Waiting time in the queue is zero. As always, $L_j = \lambda_j' W_j$. For a conveyor, if jobs enter in an exponential fashion, and then stay on the conveyor for a constant routing time to the next workstation, they must also depart in an exponential fashion. Each part is simply delayed by the trip length. If travel time is exponential owing either to varying trip lengths or to making a geometrically distributed number of circuits on a closed loop conveyor, then the technique of the previous section may still be used.

General Service Time

The exponential service open network model of Section 11.3.1 provides a valuable tool for analyzing shop performance. Nevertheless, we have learned that it can be important to account for deviations from the exponential service time assumptions. Several approaches have been proposed for evaluating queueing systems when the basic assumptions do not apply. One such approach is to generalize the Decomposition/Recomposition approach of Section 11.3.1. Shanthikumar and Buzacott [1981] demonstrate the application of this approach to our more general problem. We will allow the service time distribution to be arbitrary and to be described by the mean and variance of service time at each workstation. The Decomposition/Recomposition approach involves three basic steps: relate, separate, aggregate. *Relate* refers to the act of describing how external arrivals and flows from within the system combine to produce arrival streams at each workstation. With this knowledge, we may *separate* the workstations and may solve each as an independent queue. Single-station queues were studied in Section 11.2, and accurate approximations to their performance are available under many different interarrival and service distributions. Knowing performance measures at each workstation, we can then use knowledge of part routings and *aggregate* results across workstations to obtain workstation queue lengths and throughput times by part type. These are the same basic steps used for the exponential model; however, as we will observe, the computations in step 1 are now more involved. We now describe the general procedure.

> STEP 1. RELATE. Step 1 involves relating the workstations through their departure and arrival streams. Each workstation is treated as a *GI/G/*1 queue. The method employed assumes that arrivals to workstations constitute a renewal process. Each arrival returns us to an identical situation with respect to the time until the next arrival. Total arrival rates are computed as in Section 11.2. At each workstation, the mean time between arrivals must still be the reciprocal of the effective arrival rate. However, we can no longer assume that arrivals appear as a Poisson process; we must find the variance of interarrival (and accordingly, interdeparture) times. We will do this by modeling the squared coefficient of variance (scv). The scv is defined by $C^2 = \text{Var}(T)/[E(T)]^2$, where T is the random interarrival time. Expressions for the *GI/G/*1 queue usually rely on C_a^2 and C_s^2, the scv for the arrival and service processes at the workstation. C_s^2 is assumed known for each workstation (it depends only on the service time distribution). We must find $C_{a_j}^2$ at workstation j by relating together the arriving job streams. Let $C_{a_{jk}}^2$ be the scv of the arrival process at workstation j for jobs arriving from workstation k. We will assume that external arrivals are exponential and let the scv for this input source accordingly be $C_{a_{j0}}^2 = 1$. $C_{d_j}^2$ is the scv of the time between departures at j. Then,
>
> FACT 1. $C_{a_{jk}}^2 = p_{kj} C_{d_k}^2 + (1 - p_{kj})$.

JUSTIFICATION: We can establish this fact as follows. T_{jk} is the time between arrivals at j from k. Then $T_{jk} = X_1 + X_2 + \cdots + X_n$, where X_u are interdeparture times at k and n is a geometric random variable with mean p_{kj}^{-1}. n represents the number of departures from k until the first routing to j. It can be shown that for any sum of a random number of identically and independently distributed random variables with finite moments

$$E(T_{jk}) = E(n)E(X_u) = \frac{1}{p_{kj}\lambda_k'}$$

$$\begin{aligned} \operatorname{var}(T_{jk}) &= \operatorname{var}(X_u)E(n) + \operatorname{var}(n)E^2(X_u) \\ &= \frac{\operatorname{var}(X_u)}{p_{kj}} + \left(\frac{1-p_{kj}}{p_{kj}^2}\right)E^2(X_u) \end{aligned}$$

Combining these two results

$$\begin{aligned} C_{a_{jk}}^2 = \frac{\operatorname{var}(T_{jk})}{E^2(T_{jk})} &= \frac{\left[\frac{\operatorname{var}(X_u)}{p_{kj}} + \left(\frac{1-p_{kj}}{p_{kj}^2}\right)E^2(X_u)\right]}{\frac{E^2(X_u)}{p_{kj}^2}} \\ &= p_{kj}C_{d_k}^2 + (1-p_{kj}) \end{aligned} \qquad (11.17)$$

and the fact is established.

Of course, equation 11.17 implies that we must know the interdeparture distribution from each workstation. For a stable system, average interdeparture time must equal average interarrival time. An approximation for the scv of interdeparture times is needed. We will not attempt to derive this approximation formally (see the foregoing reference), but will motivate it with the following explanation. When the queue is nonempty, the server is busy and interdeparture times are service times. Thus, in this state interdeparture time is just service time, or $C_{d_j}^2 = C_{s_j}^2$. When the queue is empty, time between departures is equal to some remaining time to wait for arrival plus service time. Thus, components of $C_{a_j}^2$ and $C_{s_j}^2$ are relevant. The chosen approximation is

$$C_{d_j}^2 = (1 - \rho_j^2)C_{a_j}^2 + \rho_j^2 C_{s_j}^2 \qquad (11.18)$$

Now we are getting close to our objective of finding $C_{a_j}^2$. It still remains to combine the scv for external and internal arrivals at j. Inputs are drawn randomly from the streams $k = 0, 1, \ldots, M$. A weighted average of these values gives an estimate for workstation j as

$$C_{a_j}^2 = \sum_{k=1}^{M} \left(\frac{\lambda_k' p_{kj}}{\lambda_j'}\right) C_{a_{jk}}^2 + \frac{\lambda_j}{\lambda_j'} \qquad (11.19)$$

Using equation 11.17 produces

$$C_{a_j}^2 = \sum_{k=1}^{M} \left(\frac{\lambda_k' p_{kj}}{\lambda_j'}\right)[p_{kj}C_{d_k}^2 + 1 - p_{kj}] + \frac{\lambda_j}{\lambda_j'}$$

Using equation 11.18 produces

$$C_{a_j}^2 = \sum_{k=1}^{M} \left(\frac{\lambda_k' p_{kj}}{\lambda_j'}\right)\{p_{kj}[(1 - \rho_k^2)C_{a_k}^2 + \rho_k^2 C_{s_k}^2] + 1 - p_{kj}\} + \frac{\lambda_j}{\lambda_j'} \qquad (11.20)$$

We can multiply through by λ'_j and can rearrange terms to obtain the system of linear equations

$$-\sum_{k=1}^{M} \lambda'_k p_{kj}^2 (1 - \rho_k^2) C_{a_k}^2 + \lambda'_j C_{a_j}^2$$

$$= \sum_{k=1}^{M} \lambda'_k p_{kj} (p_{kj} \rho_k^2 C_{s_k}^2 + 1 - p_{kj}) + \lambda_j \qquad 1 \le j \le M \qquad (11.21)$$

Solving this set of M equations yields the M unknown $C_{a_j}^2$.

STEP 2. SEPARATE. With the total arrival rates and equations 11.21, we have the interactions between workstations. In step 2, we use these relationships and solve each workstation as if it were independent of the others. Each workstation can be treated as a $GI/G/1$ (general independent interarrivals, general service, single server) queue. Such queues have been studied at length, at least for FCFS service. Different approximations are often used depending on the level of server utilization. (As this step is applied in turn to each workstation, we omit for now the index j). One approximation for waiting time in the queue that is useful at moderate utilization levels is

$$E(W_q) \approx \frac{\rho^2 (1 + C_s^2)}{1 + \rho^2 C_s^2} \left\{ \frac{C_a^2 + \rho^2 C_s^2}{2\lambda(1 - \rho)} \right\} \qquad (11.22)$$

Throughput time for the station, per visit, is then

$$E(W) = E(W_q) + \mu^{-1} \qquad (11.23)$$

If multiple part types ($p = 1, \ldots, P$) exist with variable service times at station j, we need only replace μ^{-1} in equation 11.23 by μ_{jp}^{-1}.

STEP 3. AGGREGATE. We now have waiting time per visit for each workstation. Total throughput time for any part type p can be found from

$$E(T_p) = \sum_{j=1}^{M} v_{jp} [E(W_{qj}) + \mu_{jp}^{-1}]$$

where v_{jp} is the expected number of visits for type p parts to workstation j. Little's Law can be used to obtain average queue lengths at each workstation, both total and by part types.

EXAMPLE 11.8

Reevaluate kitting Example 11.6 for the case that inspection and packing are automated and require a constant service time, that is, $C_{s_I}^2 = 0$.

Solution

STEP 1. First, arrival rates, visit counts, and routing probabilities are unchanged. To find the $C_{a_j}^2$ we apply equation (11.21):

$$-\lambda'_I p_{IA}^2 (1 - \rho_I^2) C_{a_I}^2 + \lambda'_A C_{a_A}^2 = \lambda'_I p_{IA} (p_{IA} \rho_I^2 C_{s_I}^2 + 1 - p_{IA}) + \lambda_A$$

and

$$-\lambda_A' p_{AI}^2 \left(1 - \rho_A^2\right) C_{a_A}^2 + \lambda_I' C_{a_I}^2 = \lambda_A' p_{AI} \left(p_{AI} \rho_A^2 C_{S_A}^2 + 1 - p_{AI}\right) + \lambda_I$$

Inserting known quantities, we obtain

$$-11.11(0.01)(1 - 0.7407^2) C_{a_I}^2 + 11.11 C_{a_A}^2$$
$$= 11.11(0.1)[0.1(0.7407^2)(0) + 1 - 0.1] + 10$$

and

$$-11.11(1)(1 - 0.9259^2) C_{a_A}^2 + 11.11 C_{a_I}^2$$
$$= 11.11(1)[1(0.9259^2)(1) + 1 - 1] + 0$$

Solving these two linear equations yields $C_{a_A}^2 = 0.9945$ and $C_{a_I}^2 = 0.9992$. Values are close to one, since deviations from the exponential case occur only for those jobs that fail at inspection and are routed back to assembly.

STEP 2. Solve each workstation. Using equation 11.22,

$$E\left(W_{qA}\right) = \frac{\rho_A^2 \left(1 + C_{S_A}^2\right)}{1 + \rho_A^2 C_{S_A}^2} \left\{ \frac{C_{a_A}^2 + \rho_A^2 C_{S_A}^2}{2\lambda_A' \left(1 - \rho_A\right)} \right\}$$

$$= \frac{0.9259^2(2)}{1 + 0.9259^2(1)} \left\{ \frac{0.9945 + 0.9259^2(1)}{2(11.111)(0.0741)} \right\} = 1.038$$

and similarly,

$$E\left(W_{qI}\right) = \frac{0.7407^2(1)}{1 + 0.7407^2(0)} \left\{ \frac{0.9992 + 0}{2(11.111)(0.2593)} \right\} = 0.09514$$

Adding mean service times at each station, we have $E(W_A) = 1.12$ and $E(W_I) = 0.162$ as compared with previous values of $E(W_A) = 1.12$ and $E(W_I) = 0.26$ for the all-exponential case. Summing time at kitting, assembly, and inspection/packing, average flow time has decreased moderately from 2.64 hours to 2.52 hours. ∎

Shortest Processing Time

Suppose that the scheduling system assigns priorities by part type instead of using FCFS. The effect for a single workstation appeared in Section 11.2. A similar modification could be made here. Buzacott and Shantikumar [1980] suggested the following approximate adjustment to expression 11.22 for modeling the popular shortest processing time (SPT) priority rule. Define $K(s)$ as the fraction of load with processing time less than s and $\rho(s) = \lambda' \int_0^s x \, dS(x)$. For Poisson arrivals it can be shown that a customer with service time s will have an expected wait of $K(s)E(W_q)/[1 - \rho(s)]$ under SPT where $K(s) = (1 - \rho)/[1 - \rho(s)]$. For our problem this translates to

$$E[W_q(SPT)] = (1 - \rho)E(W_q) \int_0^\infty \frac{dS(s)}{[1 - \rho(s)]^2}$$

$E(W_q)$ is found from expression 11.22.

Other Extensions

Many other extensions are possible. Bitran and Tirupati [1991], for instance, model the use of overtime at bottleneck workstations. Service rates are scaled such that all workstations appear to operate on the same schedule. Adjustments to the departure are provided to account for batch arrivals at downstream workstations following overtime. We are rapidly approaching the state that analytical models can be easily developed and quickly solved for a wide variety of realistic manufacturing environments to obtain estimates of important performance measures that are within 10 percent of actual values.

11.4
CLOSED NETWORKS

In the preceeding section no limit was placed on the amount of work in process. Jobs arrived and were added to the manufacturing system. The prudent production control manager, however, attempts to maintain a level of work-in-process inventory that is sufficient to meet demand without cluttering up the aisles and causing excessive throughput times. Hopp and Spearman [1991] and Spearman et al. [1990] describe the use of the CONWIP (CONstant Work In Process) protocol, which releases a new job as one is completed. In this section we examine closed queueing networks. In a closed network, the number of jobs in process is kept at the fixed level N. When a job finishes all its operations and leaves the system, a new job is dispatched to replace it. Thus, whereas work-in-process is an output statistic in open networks, it is a control parameter for closed networks.

As N is increased, both production rate and expected throughput time will increase. Figure 11.5 shows the essential trade-off. Full utilization of the workstation with lowest relative service rate $[\min_j(c_j\mu_j/v_j)]$, defines capacity. Increasing N allows us to approach production capacity asymptotically. Adding jobs up to about the level N_0 in the figure substantially increases output. Thereafter, output increases are minor and are accompanied by significant increases in throughput time.

In solving the network of queues, the closed aspect alters the model structure. Since the number of jobs in the system is fixed, workstations can no longer be independent. Knowing, for instance, that station j is idle implies more jobs at the other stations. If station j happens to have N jobs, all other stations must be empty. Fortunately, the product form of the solution still holds true for basic networks, but we must now solve all workstations jointly. The set of possible states contains all states for which $\sum_{j=1}^{M} n_j = N$, where n_j is the number of jobs at station j. Before we look at how the product form can be used to examine all stations simultaneously, we will descibe an alternative solution methodology. **Mean value analysis** (MVA) offers an alternative approach for model solution and may be preferable for large systems because of numerical problems that can occur with the product form approach. With the exception of the dispatching strategy, our assumptions are as in the open networks of Section 11.3.1.

11.4.1 Mean Value Analysis

Although exact algorithms exist for mean value analysis, they can be slow for large problems. We will present a popular approximate procedure (Suri and Hildebrandt [1985]). We will allow for P part types and will control the system to have N_p, $p = 1, \ldots, P$ parts of type p in process at all times. With our earlier statements, this implies the relation $\sum_{p=1}^{P} N_p = N$. μ_{jp}^{-1} will be the the mean service time for a part of type p at workstation j. Note that this is an advantage over the product form approach,

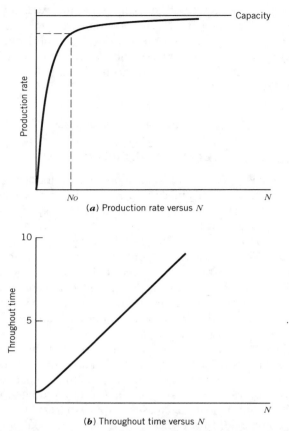

Figure 11.5 Work-in-process versus production rate and throughput trade-offs.

where we were forced to assume that all part types have the same service rate at a workstation.

Throughput Times

MVA relies on three basic equations, two of which are manifestations of Little's Law. First, we can compute the average throughput time per visit of a type p part to station j by

$$W_{jp} = \mu_{jp}^{-1} + \frac{N_p - 1}{N_p} L_{jp} \mu_{jp}^{-1} + \sum_{r \neq p} L_{jr} \mu_{jr}^{-1} \qquad \text{for all } j, p \qquad (11.24)$$

Equation 11.24 reads: Expected throughput time for p at j is composed of part service time (μ_{jp}^{-1}), time in queue waiting for other type p parts $\{[(N_p - 1)/N_p] L_{jp} \mu_{jp}^{-1}\}$, and time spent waiting for other part types ($\sum_{r \neq p} L_{jr} \mu_{jr}^{-1}$). The correction term $(N_p - 1)/N_p$ approximately accounts for the fact that, when a part type p arrives at station j, there are only $N_p - 1$ other parts of its type that are free to be ahead of it in queue. Thus, $[(N_p - 1)/N_p] L_{jp}$ represents the expected number of type p parts in queue when a new part p arrives. With FCFS, the new part must wait an average of μ_{jp}^{-1} for each of these to be served. Other part types could be in queue also. Indeed, by definition we expect L_{jr} parts of type $r \neq p$ to be in the station. We must wait for each of these to finish as well. Since we have assumed exponential service, it does not matter which part is being worked on or for how long when we arrive. Remaining processing time for any job in process has the same distribution as its total processing time when it first arrived.[1]

When we derived results from the state probabilities that solve the steady-state balance equations (see Sections 11.2 and 11.3.1), multiple servers did not present a problem. In fact, any state-dependent service rates could be modeled by including the relevant rate in the product form expression. MVA bypasses the state balance equations, and an alternative model for multiple servers is required. Multiple server stations can be heuristically modeled by dividing the second and third terms on the right-hand side of equation 11.24 by c_j. This states that all servers will always be busy emptying out the queue.[2] Of course, this is not true unless at least c_j customers are at the workstation. This would tend to underestimate the time required to empty the station. This is somewhat offset by the realization that we need not complete all L_j jobs prior to starting the arriving job. The new job can begin service as soon as only $c_j - 1$ jobs are ahead of it, leaving an available server. The overall error in this approximation is small when the average number of jobs at the station, $\sum_{p=1}^{P} L_{jp}$ is large relative to c_j. Seidmann et al. [1987] found that better estimates, particularly for expected waiting times, could be found by modeling a multiserver workstation as a fast single-server workstation followed by an ample server station. Processing time is divided between the two pseudoworkstations. The original service time of μ_j^{-1} is divided such that the fast, single server must work μ_{jp}^{-1}/c_j and the ample server station serves (delays) the customer $[(c_j - 1)\mu_{jp}^{-1}]/c_j$. The fast, single server is an approximation model for the time in wait queue for the multiserver station plus $1/c_j$ of the service time. The subsequent ample server station models the remainder of the service time. With ample servers there is no time spent waiting for service. The replacement expression for equation 11.24 when j has multiple servers is then

$$W_{jp} = \frac{\mu_{jp}^{-1}}{c_j} + \frac{N_p - 1}{N_p} L_{jp} \frac{\mu_{jp}^{-1}}{c_j} + \sum_{r \neq p} L_{jr} \frac{\mu_{jr}^{-1}}{c_j} \tag{11.25a}$$

for the original station and

$$W_{jp} = \frac{c_j - 1}{\mu_{jp} c_j} \tag{11.25b}$$

for the accompanying AS station.[3] At first glance it may seem that expressions 11.25a and 11.25b could be combined to form the original equation 11.24 with the factor c_j^{-1} added to the time in queue terms. Shortly, however, we will define expressions for L_{jp}, the number of type p parts at station j. To combine expressions 11.25a and 11.25b would ignore the fact that the L_{jp} terms are also reduced in expression 11.25a, over equation 11.24, since the L_{jp} expressions that correspond to expression 11.25b will absorb some of the N_p jobs. Thus, waiting time is reduced in expression 11.25a, whereas service time is held constant between expressions 11.25a and 11.25b as compared with equation 11.24.

---

[1]This is the Forgetfulness Property of the exponential, a key factor for analytical tractability in queueing models. Accordingly, exponential distributions are assumed by many researchers.

[2]Another way of looking at this is that the multiple servers are replaced by a single server who works c_j times as fast as the original servers.

[3]Expression 11.25b can be used to model any ample server workstation.

Throughput Rates

The second basic MVA equation gives the overall system production rate. Let X_p be the production rate of part type p. Although production rate is essentially the effective arrival rate, we use X instead of λ to emphasize that we are now dealing with a model output instead of a model input. Then

$$X_p = \frac{N_p}{\sum_{j=1}^{M} v_{jp} W_{jp}} \qquad p = 1, \ldots, P \tag{11.26}$$

As before, v_{jp} are visit counts, the expected number of visits a part type p makes to station j. Expression 11.26 is a form of Little's Law. The arrival (production) rate is the ratio of the number of jobs in the system to the throughput time. The relation must hold true for each part type. The denominator shows total time accumulated by a part p as it progresses through the system.

Queue Lengths

The final set of equations relates production at each station by Little's Law. We have

$$L_{jp} = X_p(v_{jp} W_{jp}) \qquad \text{for all } j, p \tag{11.27}$$

The v_{jp} term in expression 11.27 accounts for all part type ps that are at j for any operation on their process plan.

MVA Algorithm

The relationships 11.24, 11.26, and 11.27 must all be satisfied by the steady-state solution. Unfortunately, this is not a linear system and direct solution is difficult. Our objective is to find a set of L_{jp}, W_{jp}, and X_p that satisfy relationships 11.24, 11.26, and 11.27 given the system parameters N_p, c_j, μ_{jp}, and v_{jp}. A simple iterative scheme will be used. Given an initial guess of station queue lengths L_{jp}, we can solve for W_{jp} estimates in equation 11.24. This permits estimation of production rates in expression 11.26, which in turn makes all the entities on the right-hand side of expression 11.27 known. Thus, we can check expression 11.27 to learn how accurate our initial guesses were. The updated queue lengths from expression 11.27 become our new initial guesses, and the process is continued. We continue the iterative scheme until estimates converge; we take this to mean values change by less than 0.1 percent from the previous iteration.

The algorithm is thus as follows:

STEP 1. INITIALIZE. $\tau = 0$. For all p let Z_p be the number of stations visited by part type p, that is, the sum over j of the number of nonzero v_{jp}. Set $L_{jp}^{(0)} = N_p/Z_p$.

STEP 2. UPDATE W. $\tau = \tau + 1$. For all j, p compute $W_{jp}^{(\tau)}$ from equation 11.24 or 11.25.

STEP 3. UPDATE THROUGHPUT. For all p compute $X_p^{(\tau)}$ by equation 11.26.

STEP 4. UPDATE L. For all j, p compute $L_{jp}^{(\tau)}$ from equation 11.27. If any $[(L_{jp}^{(\tau)} - L_{jp}^{(\tau-1)})/L_{jp}^{(\tau)}] > 0.001$ go to Step 2; otherwise stop.

EXAMPLE 11.9

Consider a small flexible manufacturing system as shown in Figure 11.6. The system manufactures two part types using a turning center (TC) and three hor-

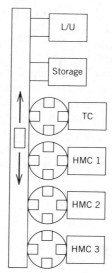

Figure 11.6   An FMS schematic for Example 11.9.

izontal machining centers (HMCs). HMC1 and HMC2 are identically tooled and form a group of interchangeable machines. HMC3 performs different operations. A shuttle cart transports pallets between stations. The shuttle is fast relative to machining operations. Part transfer takes about a minute on the average. The shuttle must reach the origination point, load the pallet, travel to the destination, and off-load the pallet. Part type 1 visits TC, then either HMC1 or HMC2, and lastly HMC3 before returning to L/U. Part type 2 visits TC and either HMC1 or HMC2 before returning to L/U. All machining operations average 10 minutes. To unload a part from the system and to load a new part onto the same pallet and fixture takes about 12 minutes. This time includes a cleaning operation and input of part data into the system controller. Each station has space for three pallets in input queue and one pallet in process. This is expected to be sufficient storage; extra pallets can always be kept at the storage area. Of the nine pallets in the system, six are dedicated to part type 1 and the remaining three are dedicated to part type 2.

Solution

Before beginning the MVA algorithm we must, as was true with the open network, find the relative visit counts to each station. Visit counts follow from the part routings and can be summarized by the following table:

	Visit Counts				
	Station				
Part	L/U	TC	HMC1, HMC2	HMC3	Shuttle
1	1	1	1	1	4
2	1	1	1	0	3

The table reflects our counting loading and subsequent unloading of a part as one visit with mean time of 12 minutes. We chose this modeling approach since the physical reality is one continuous unload and load operation. The alternative would be to use separate load and unload visits of 6 minutes' average processing time. Visits to the shuttle reflect a transport operation after each machining and loading operation. We now begin the MVA algorithm. Since our primary interest is in illustrating the basic technique, we will treat the grouped HMC1 and HMC2 as a single workstation, but we will use the simpler multiserver model. Stations 1 through 5 are L/U, TC, HMC1 and HMC2, HMC3, and the shuttle, respectively.

STEP 1. INITIALIZE. Parts are assumed evenly distributed over the stations visited. Thus

$$L_{11} = L_{21} = L_{31} = L_{41} = L_{51} = \frac{N_1}{5} = \frac{6}{5} = 1.2$$

$$L_{12} = L_{22} = L_{32} = L_{52} = \frac{N_2}{4} = \frac{3}{4} = 0.75, \qquad L_{42} = 0.$$

STEP 2. COMPUTE W_{jp}. We use the basic expression

$$W_{jp}^{(1)} = \mu_{jp}^{-1} + \frac{N_p - 1}{N_p} \frac{L_{jp}}{\mu_{jp}c_j} + \sum_{r \neq p} \frac{L_{jp}}{\mu_{jr}c_j}$$

For the L/U station ($j = 1$) and part type 1, we have

$$W_{11}^{(1)} = 12 + \frac{5}{6}(1.2)12 + 0.75(12) = 33$$

For the turning center and part 1, we have

$$W_{21}^{(1)} = 10 + \frac{5}{6}(1.2)10 + \frac{3}{4}(10) = 27.5$$

A similar calculation is made for each jp combination that occurs. For instance,

$$W_{32}^{(1)} = 10 + \frac{2}{3}\frac{(0.75)10}{2} + \frac{(1.2)10}{2} = 18.5$$

Estimated throughput times for iteration 1 are

	$W_{jp}^{(1)}$ Values				
Part			L/U		
1	33.0	27.5	18.75	20	2.75
2	32.4	27.0	18.50	0	2.70

STEP 3. COMPUTE THROUGHPUT RATES. Using $X_p = N_p / \sum_j v_{jp} W_{jp}$, we have

$$X_1^{(1)} = \frac{6}{33 + 27.5 + 18.75 + 20 + 4(2.75)} = 0.05442 \text{ parts per minute}$$

and

$$X_2^{(1)} = \frac{3}{32.4 + 27 + 18.5 + 0 + 3(2.7)} = 0.03488 \text{ parts per minute}$$

STEP 4. COMPUTE L_{jp}. Using

$$L_{jp} = X_p v_{jp} W_{jp}$$

we have

$$L_{11}^{(1)} = (0.05442)(1)(33) = 1.7959$$

Finding the remaining $L_{jp}^{(1)}$, we have

| | | | $L_{jp}^{(1)}$ Values | | |
| | | | Station | | |
Part	L/U	TC	HMC1, HMC2	HMC3	Shuttle
1	1.7959	1.4966	1.0204	1.0884	0.5986
2	1.1302	0.9419	0.6453	0	0.2826

Note that as fixed by the problem definition $\sum_j L_{j1} = 6$ and $\sum_j L_{j2} = 3$. Comparing the $L_{jp}^{(1)}$ to the initial $L_{jp}^{(0)}$, we find sufficient discrepancy to warrant continuing. Thus we start iteration 2.

STEP 1. COMPUTE $W_{jp}^{(2)}$.

$$W_{11}^{(2)} = 12 + \frac{5}{6}(1.7959)(12) + 1.1302(12) = 43.52$$

All $W_{jp}^{(2)}$ are found similarly resulting in

| | | | $W_{jp}^{(2)}$ Values | | |
| | | | Station | | |
Part	L/U	TC	HMC1, HMC2	HMC3	Shuttle
1	43.52	31.89	17.48	19.07	1.781
2	42.59	25.59	17.25	0	1.787

STEP 2. COMPUTE $X_p^{(2)}$.

$$X_1^{(2)} = \frac{6}{43.52 + 31.89 + 17.48 + 19.07 + 4(1.781)} = 0.05038$$

$$X_2^{(2)} = \frac{3}{42.59 + 25.59 + 17.25 + 3(1.787)} = 0.03304$$

STEP 3. COMPUTE $L_{jp}^{(2)}$. Continuing

$$L_{11}^{(2)} = X_1^{(2)} v_{11} W_{11}^{(2)} = 0.05038(1)(43.52) = 2.1925$$

Similarly, we obtain

	$L_{jp}^{(2)}$ Values				
			Station		
Part	L/U	TC	HMC1, HMC2	HMC3	Shuttle
1	2.1925	1.6066	0.8805	0.9607	0.3590
2	1.4072	0.8456	0.5700	0	0.1771

Comparing $L_{jp}^{(2)}$ with $L_{jp}^{(1)}$, we still have not converged, and iteration 3 is started. We will not continue with the details, but after several more iterations we settle on the solution

$$X_1 = 0.0473 \text{ parts per minute}$$

and

$$X_2 = 0.0278 \text{ parts per minute}$$

Machine utilizations can be found by

$$U_j = \sum_{p=1}^{2} \frac{X_p \, v_{jp}}{\mu_{jp} \, c_j}$$

For L/U this yields

$$U_1 = (0.0473)(1)(12) + (0.0278)(1)(12) = 0.901 \qquad \blacksquare$$

Extensions to MVA: Priority Scheduling

Alternate service discipline strategies can also be modeled by modifications to the time in queue terms on the right-hand side of equation 11.24. Consider a priority scheduling system where part types belong to specified priority classes at each workstation and higher priority parts are serviced first. In equation 11.24 we must eliminate waiting for lower priority part types and add waiting time for the expected arrivals of higher priority while the job is in queue. Typically, jobs are not preempted once processing starts, so higher priority arrivals during processing are not included. The final term of equation 11.24 should be restricted to range over those jobs (part types) with higher priority than p. This accounts for jobs ahead of the new part type p on its arrival. Jobs of higher priority may also arrive before processing of the part type p can begin. These jobs would move ahead of type p in the queue. If we know the expected time in queue $(W_{jp} - \mu_{jp}^{-1})$ for the part and the arrival rate for higher priority parts $(v_{jr}X_r$ for part type r), then we can adjust equation 11.24 by adding the extra term

$$(W_{jp} - \mu_{jp}^{-1}) \sum_{r:pr(j,r)>pr(j,p)} \frac{v_{jr}X_r}{c_j \mu_{jr}} \qquad (11.28)$$

where $pr(r, j)$ is the priority of part type r at j. $(W_{jp} - \mu_{jp}^{-1}) \sum v_{jr}X_r$ indicates the higher priority arrivals while waiting. These jobs are cleared at the rate $c_j \mu_{jr}$. Unfortunately, expression 11.28 places W_{jp} on both sides of equation 11.24, complicating its solution. Reexpression of expression 11.28 for solution along with a discussion of computational issues is given in Shalev-Oren et al. [1985].

11.4.2 Product Form Solution

Baskett et al. [1975] describe a wide class of queueing systems that have a product form solution. In addition to the open networks discussed previously, included are multiclass, multiserver closed systems with a FCFS service discipline provided that the service rate does not depend on product class. More general service time distributions can be used (the distribution need only have a rational Laplace transform) for ample server or processor sharing workstations.[4] These latter station types may also have service rates that depend on the part type. We will briefly describe how the product form nature of the solution can be exploited to find exact state probabilities, $p(\mathbf{n})$, the probability of being in a state $\mathbf{n} = (n_1, \cdots, n_M)$. (For multiple part types we let $\mathbf{n}_j$ be a vector containing the number of jobs of each type at station j.) Closed networks are restricted to the states with $\sum_{j=1}^{M} n_{jp} = N_p$ for all p.

We have stated that multipart systems may also have product form solutions. Sometimes, however, it is more convenient to reduce the size of the model being solved by aggregating part types into a generic part. Now, instead of keeping N_p parts of type p in process, we maintain instead N parts. In addition to computational savings, by stating relative demand rates d_p for part types, we can assure the desired ratio of production. In the previous model we had to be content with a multivariate problem of adjusting the N_p to obtain desired production. Now, when a part finishes processing, we select a new part type p to enter the system with probability $d_p, \sum_{p=1}^{P} d_p = 1$. This strategy fixes long-term relative production rates at the expense of not directly controlling the number of parts of each type in process at any instant of time. Although the former multiclass system made sense for an FMS with limited fixtures for each part type, the present aggregation fits well for many production control environments with demand goals for each part type. We emphasize that the choices of aggregate or multiple class model and MVA or product form algorithm are independent. We aggregate in this section both for convenience and to illustrate the aggregation procedure.

The aggregated generic part is a demand-weighted, linear composite of all part types. Each generic part represents d_p units of each part type p. Station service rates and visit counts must reflect this aggregation. Workstation visit counts for the composite part are

$$v_j = \sum_{p=1}^{P} v_{jp} d_p \qquad (11.29)$$

We may also want to know the total work load relative to the production rate for the composite part. This is measured by

$$\rho_j = \sum_{p=1}^{P} \mu_{jp}^{-1} v_{jp} d_p \qquad (11.30)$$

ρ_j gives the total work required at station j for each composite part produced. Each composite part implies production of d_p part type p's. The service rate at j is found by the ratio of total work per unit produced to the visit count or

$$\mu_j^{-1} = \frac{\rho_j}{v_j} \qquad (11.31)$$

---

[4]Process sharing implies that all n_j units at the workstation at any time receive n_j^{-1} of the server's attention.

Two final parameters are needed for the aggregation. Define $r_j(\mathbf{n})$ as the rate station j is serving customers under state $\mathbf{n}$. Thus, $r_j(\mathbf{n}) = \min(n_j, c_j)\mu_j$. Second, as before, p_{jk} is the probability that a job leaving station j goes next to k. As with the other composite part parameters, p_{jk} can be obtained by averaging over all parts with weights d_p. Of v_j visits into j per composite part produced, we have $p_{jk}v_j$ parts going immediately to k.

Assuming exponential service time for the composite part, we can state the steady-state balance equations for this system. Equating rate out = rate in for each state

$$p(\mathbf{n}) \sum_{j=1}^{M} r_j(\mathbf{n}) = \sum_{k:n_k \geq 1} \sum_{j=1}^{M} p(\mathbf{n}_{jk}) p_{jk} r_j(\mathbf{n}_{jk}) \tag{11.32}$$

where state $\mathbf{n}_{jk} = (n_1, \ldots, n_{j-1}, n_{j+1}, n_{j+1}, \ldots, n_{k-1}, n_{k-1}, n_{k+1}, \ldots, n_M)$, that is, state $\mathbf{n}$ except for one extra job at j and one less job at k. The right hand side of equation 11.32 states that to enter state $\mathbf{n}$ a job must be completed at j and be transferred to k, placing us in state $\mathbf{n}$. There are no external arrivals to our closed system. The model thinks jobs recirculate. The modeler knows jobs finish and leave the system only to be immediately replaced by a new job.

EXAMPLE 11.10

Consider a three-workstation system with a single server at stations 1 and 2 and $c_3 = 2$. Parts are transported between workstations by a closed-loop conveyor that connects all machines. The conveyor has ample capacity and small delay time relative to machine processing time and will be excluded from our analysis. The system is shown in Figure 11.7. Arc values in Figure 11.7b are intermachine transfer probabilities. $N = 4$ jobs are kept in process. List the set of possible states and examine the balance equations.

Solution

The 15 possible states correspond to the ways four jobs can be distributed among three workstations.

Set of Possible States	
(n_1, n_2, n_3)	(n_1, n_2, n_3)
(4,0,0)	(1,1,2)
(3,1,0)	(1,0,3)
(3,0,1)	(0,4,0)
(2,2,0)	(0,3,1)
(2,1,1)	(0,2,2)
(2,0,2)	(0,1,3)
(1,3,0)	(0,0,4)
(1,2,1)	

Next, we will examine the balance equations 11.32. We will not write out the balance equations for all 15 states but consider only states $(4, 0, 0)$ and $(1, 1, 2)$ for illustration. The balance equation for state $(4, 0, 0)$ is

$$p(4, 0, 0)\mu_1 = p(3, 1, 0)(0.5)\mu_2 + p(3, 0, 1)(1.0)\mu_3 \tag{11.33}$$

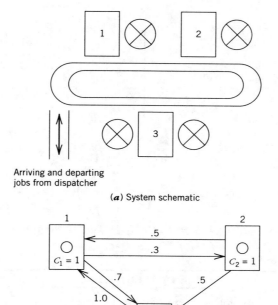

(*a*) System schematic

(*b*) Model schematic with transfer probabilities

Figure 11.7   The system model for Example 11.10.

We can exit state $(4, 0, 0)$ only by having a service completion at the first station. With a single station 1 server, this happens at rate μ_1. We can enter $(4, 0, 0)$ only by having a completion at station 1 or 2 when three jobs were already at station 1. In either case, the completing job must be destined for station 1.

For state $(1, 1, 2)$ the balance equation is

$$p(1, 1, 2)(\mu_1 + \mu_2 + 2\mu_3) = p(2, 1, 1)(0.7)\mu_1 + p(2, 0, 2)(0.3)\mu_1$$
$$+ p(0, 2, 2)(0.5)\mu_2 + p(1, 2, 1)(0.5)\mu_2$$
$$+ p(0, 1, 3)2\mu_3 \qquad (11.34)$$

Equation 11.34 states that we leave state $(1, 1, 2)$ if any of the four servers finishes. We can enter $(1, 1, 2)$ in a number of ways. Station 1 can finish from $(2, 1, 1)$ and can send a job to station 3, or can finish from $(2, 0, 2)$ and can send its job to station 2. Station 2 could finish from $(0, 2, 2)$ and could send its job to station 1 or could finish from $(1, 2, 1)$ and could send its job to station 3. Last, with both station 3 servers busy we could have a completion. Such a job automatically goes to station 1 according to our routing probabilities.   ∎

The major result for product form models is the following fact.

FACT: The balance equations 11.32 are solved by the state probabilities

$$p(\mathbf{n}) = G^{-1}(N) \prod_{j=1}^{M} f_j(n_j) \qquad (11.35)$$

where

$$f_j(n_j) = \begin{cases} \dfrac{\rho_j^{n_j}}{n_j!} & n_j \le c_j \\[2ex] \dfrac{\rho_j^{n_j}}{c_j! c_j^{n_j - c_j}} & n_j > c_j \end{cases}$$

and

$$G^{-1}(N) = \sum_{\mathbf{n}} \prod_{j=1}^{M} f_j(n_j)$$

is a normalizing constant such that $\sum_{\mathbf{n}} p(\mathbf{n}) = 1$

This fact can be established by showing that the solution defined in equation 11.35 solves equation 11.32.

EXAMPLE 11.11

Check whether the solution proposed in equation 11.35 solves the two state equations developed in Example 11.10.

Solution

Substituting the state definitions into equation 11.33 yields

$$G^{-1}(4)\rho_1^4 \mu_1 = G^{-1}(4)\rho_1^3 \rho_2(0.5)\mu_2 + G^{-1}(4)\rho_1^3 \rho_3 \mu_3$$

Dividing by $G^{-1}(4)\rho_1^3$, we must show

$$\rho_1 \mu_1 = \rho_2(0.5)\mu_2 + \rho_3 \mu_3$$

Recall the development of ρ_j in equations 11.29–11.31, namely, $\rho_j = v_j / \mu_j$. Substituting for ρ_j we obtain the requirement

$$v_1 = 0.5v_2 + v_3$$

If we look at Figure 11.7b, we will see that this is precisely the first visit count equation derived from the routing probabilities, and thus this relationship holds. We leave it to the reader to show that the balance equation 11.34 for state $(1, 1, 2)$ and other state balance equations also hold. ∎

The problem of solving for state probabilities now becomes that of finding $G(N)$. Define $G(j, 0) = 1$, $G(1, n) = f_1(n)$, and

$$G(j, n) = \sum_{k=0}^{n} f_j(k)G(j-1, n-k) \qquad 1 \le j \le M, \qquad 1 \le n \le N \quad (11.36)$$

$G(j, n)$ is the cumulative relative probability for the ways to assign n jobs to stations 1 through j. In fact, $G(N)$ is simply $G(M, N)$. To find $G(N)$ we need essentially to fill out an array as in Table 11.4 row by row. Each row represents a station. By the time we reach the Nth column of row j, we know the possible assignments and relative probabilities for assigning the N jobs to the first j stations. The first row of the table is given directly by the definition of

Table 11.4 $G(j, n)$ Values: Recursive Calculation for $G(N)$

Station	Number of Jobs 1	Number of Jobs 2		Number of Jobs N
	1	2	$\ldots$	N
1	$f_1(1)$	$f_1(2)$	$\ldots$	$f_1 N$
2	$f_2(1)f_1(0) - f_2(0)f_1(1)$	$f_2(0)f_1(2) - f_2(1)f_1(1) - f_2(2)f_1(0)$	$\ldots$	$G(2, N)$
M	$G(M, 1)$	$G(M, 2)$	$\ldots$	$G(M, N)$

$G(1, n)$. Thereafter we employ the expression 11.36. For instance, noting $G(j, 0) = f_j(0) = 1$

$$G(2, 1) = f_2(0)G(1, 1) + f_2(1)G(1, 0) = f_1(1) + f_2(1) \qquad (11.37)$$

Equation 11.37 should be interpreted to state that to have $n = 1$ job in $j = 2$ workstations we could have 0 jobs in the second station with 1 job in the first station or 1 job in station 2 with the first station empty. As each new row is encountered, we have an extra station to hold the n jobs. We account in the sum of expression 11.36 for each possible number of jobs at the new station. Given $G(1, n)$ values and $G(2, 1)$, we find that

$$G(2, 2) = f_2(0)G(1, 2) + f_2(1)G(1, 1) + f_2(2)G(1, 0)$$

indicating the possibility of 0, 1, or 2 jobs at station 2. Since $G(1, n)$ values are known, we have

$$G(2, 2) = f_2(0)f_1(2) + f_2(1)f_1(1) + f_2(2)f_1(0)$$

The rest of the table is filled out in the same fashion. The last entry is the desired normalization factor $G(N)$ for each state probability.

We now have all state probabilities. It turns out that relatively simple expressions can be derived for many quantities of interest (Solberg [1981]). For instance, the overall production rate of composite parts is

$$X = \frac{G(M, N - 1)}{G(M, N)} \qquad (11.38)$$

$\lambda_j = X v_j$ would be the arrival rate to station j. The average number of busy servers at a workstation can be shown to be

$$B_j = \rho_j X$$

The probability of i jobs at station j is

$$Pr(n_j = i) = \sum_{\mathbf{n}: n_j = i} p(\mathbf{n})$$

$$= G^{-1}(N) \sum_{\mathbf{n}: n_j = i} \prod_{k=1}^{M} f_k(n_k)$$

$$= G^{-1}(N) f_j(i) \sum_{\mathbf{n}: \sum_{k \neq j} n_k = N - i} \prod_{k \neq j} f_k(n_k) \qquad (11.39)$$

$$= f_j(i) \times \frac{G(M - 1, N - i)}{G(M, N)}$$

$f_j(i)$ is given in equation 11.35. $G(M,N)$ is the final term in the recursive calculation. The $M-1$ index in $G(M-1, N-i)$ of equation 11.39 refers to omitting the row for station j. Thus, by processing station j as the last row, we can find queue length probabilities for any station j.

EXAMPLE 11.12

Solve the system described in Example 11.10. Assume that each job starts at station 1 and that a return to station 1 indicates completion of one real-life job and the insertion of another into the system, that is, let $v_1 = 1$.[5] If $v_1 = 1$, the visit equations

$$v_1 = 0.5v_2 + v_3$$
$$v_2 = 0.3v_1$$
$$v_3 = 0.7v_1 + 0.5v_2$$

imply that $v_2 = 0.3$ and $v_3 = 0.85$.

Next, we fill out the array of $G(j,n)$ values. We also need to know at this point that service rates are 2 jobs per hour at station 1 and 1 job per hour per server at the other stations. As $\rho_j = v_j \mu_j^{-1}$, we compute $\rho_1 = 0.5, \rho_2 = 0.3, \rho_3 = 0.85$. Results for row 1 are $G(1,n) = \rho_1^n$ and are given in Table 11.5. Using equation 11.36 we have

$$G(2,1) = f_2(0)G(1,1) + f_2(1)G(1,0) = 1G(1,1) + \rho_2 G(1,0)$$
$$= 0.5 + 0.3 = 0.8$$

$$G(2,2) = f_2(0)G(1,2) + f_2(1)G(1,1) + f_2(2)G(1,0)$$
$$= 0.25 + 0.3(0.5) + 0.3^2 = 0.49$$

$$G(2,3) = \sum_{k=0}^{3} f_2(k)G(1,3-k)$$
$$= 0.125 + (0.3)(0.25) + (0.3)^2(0.5) + (0.3)^3 = 0.272$$

$$G(2,4) = \sum_{k=0}^{4} f_2(k)G(1,4-k) = 0.1441$$

The last row of Table 11.5 uses the $G(2,n)$ values just found and the

$$f_3(n) = \begin{cases} \dfrac{\rho_3^n}{n!} & n \le 2 \\[2ex] \dfrac{\rho_3^n}{2^n - 1} & n > 2 \end{cases}$$

---

[5]In actuality the v_j can be stated only as relative visit ratios from the data supplied in Figure 11.7. In this event, any of the v_j can be arbitrarily set to some constant. Final production X will be determined proportional to this initial assignment. If the v_j are multiplied by a constant, the ρ_j are also multiplied by this constant and the production rate X is divided by this constant. This follows since each term in the numerator of X contains the constant to the power $N-1$ and the denominator contains the constant to the Nth power. If actual visit counts are known, X is directly meaningful; otherwise, it is a relative output value.

values. For example

$$G(3, 2) = f_3(0)G(2, 2) + f_3(1)G(2, 1) + f_3(2)G(2, 0)$$
$$= 0.49 + 0.85(0.8) + 0.36125 = 1.53125$$

Table 11.5 $G(j, n)$ Values for Example 11.12

Station	Number of Jobs			
	1	2	3	4
1	0.5	0.25	0.125	0.0625
2	0.8	0.49	0.272	0.1441
3	1.65	1.53125	1.13103	0.74039

Production rate 11.38 is then

$$X = \frac{G(3, 3)}{G(3, 4)} = 1.528 \text{ composite parts/hour}$$

Since all $G(3,n)$ values are known, we can easily estimate the change in output if jobs are removed. At $N = 3, X = G(3, 2)/G(3, 3) = 1.354$ composite parts/hour. In terms of server utilization (for $N = 4$), we have

$$B_1 = \rho_1 X = 0.764, \qquad B_2 = \rho_2 X = 0.458, \qquad B_3 = \rho_3 X = 1.300$$

Last, let us look at the distribution of number of jobs at workstation 3. From equation 11.39 we have $Pr(n_3 = k) = f_3(k)[G(2, 4 - k)/G(3, 4)]$. Thus,

$$Pr(n_3 = 0) = f_3(0)\frac{G(2, 4)}{G(3, 4)} = 0.195$$

$$Pr(n_3 = 1) = f_3(1)\frac{G(2, 3)}{G(3, 4)} = 0.312$$

$$Pr(n_3 = 2) = f_3(2)\frac{G(2, 2)}{G(3, 4)} = 0.239$$

$$Pr(n_3 = 3) = f_3(3)\frac{G(2, 1)}{G(3, 4)} = 0.166$$

and

$$Pr(n_3 = 4) = f_3(4)\frac{G(2, 0)}{G(3, 4)} = 0.088$$

The corresponding average number of jobs is

$$L_3 = 0.312 + 2(0.239) + 3(0.166) + 4(0.088) = 1.64 \text{ jobs} \qquad \blacksquare$$

General service time distributions are difficult to model at FCFS stations in closed networks. Yao and Buzacott [1986] describe one approach. The idea is to find a modified set of state-dependent service rates at each workstation that make the workstation behave as if it had a general service time distribution. To illustrate, assume that only one workstation has nonexponential service. The appropriate modified exponential service rates are found by decomposing the original network into the workstation and the remaining network. Essentially, the other workstations are summarized into a single node that interacts with the removed workstation. The station receives input from the network and sends its output to the network. A search is made for exponential service rates for the workstation that produce the same queue length distribution as that of an *M/G/1* workstation receiving the same input. The workstation is then integrated back with the others into an exponential network. The original service rates are replaced by the modified service rates for the general service time station. If several stations have nonexponential service, we can separate out one at a time to find the appropriate modified service rates. The procedure can be iterated until modified rates stabilize. Good approximations result in most cases.

11.5
LIMITED WIP: A HYBRID SYSTEM

Suppose orders arrive to the manufacturing system at the rate λ but are held for dispatching. A simple yet practical dispatching rule might be to enter a job on arrival unless there are already N jobs in the system. This rule satisfies the production controller's objective of trying to keep workstations busy without running the risk of inundating the system to the point of confusion. The critical loading level N may be determined from the production rate versus WIP curves discussed previously. Jobs above the N limit are queued outside the shop floor. When a job finishes, the next job in the queue is dispatched to the shop.

The system just described is a hybrid: it contains aspects of both open and closed systems. Buzacott and Shantikumar [1980] suggested using a hybrid modeling approach for this environment. For all levels of $1 \leq n \leq N$ jobs in the system, we can estimate production rate by a closed network model with n jobs. If we knew the probability of being in each of these load conditions, we could use a weighted average to estimate long-term system production rate. To estimate these state probabilities, we will treat the entire manufacturing shop as an $M/M/1$ queue. Jobs arrive at the rate λ. Jobs are served at the rate $\mu(n)$, where $\mu(n)$ is the production rate determined from the closed network model with n jobs in process. The system is stable provided $\mu(N) > \lambda$. The procedure is then

1. For $1 \leq n \leq N$ solve the closed queueing network model of the production system assuming n jobs in process. Set $\mu(n)$ to the determined aggregate production rate.
2. Solve the $M/M/1$ queue model with arrival rate λ and state-dependent service rate $\mu(n)$. Let $p(i)$ be the probability of i jobs total in the manufacturing system and the dispatching queue. For this model

$$p(i) = p(0) \cdot \prod_{n=1}^{i} \frac{\lambda}{\mu(n)} \tag{11.40}$$

and $\sum_{i=0}^{\infty} p(i) = 1$.

Various performance measures can be computed from the model. For instance, the expected number of jobs in the system plus dispatching queue is

$$L_T = \sum_{i=0}^{\infty} i \, p(i)$$

The expected number of jobs active in the shop is the modified sum

$$L = \sum_{i=0}^{\infty} n_i p(i) \tag{11.41}$$

where n_i is the average number of jobs in process given i jobs in the total system. For the case described above where jobs are immediately dispatched until we reach the limit of N in process

$$n_i = \begin{cases} i & i < N \\ N & i \geq N \end{cases}$$

Expression 11.41 allows for more complex dispatching rules if so desired. Throughput times are known for each state of the system, and long-term average values can be found by

$$W_T = \sum_{i=0}^{\infty} W(i)p(i)$$

where $W(i)$ is the throughput time calculated by the closed network model with i jobs.

EXAMPLE 11.13

Jobs arrive to our small system at the rate of $\lambda = 80$ per period. At most $N = 4$ jobs are allowed in process. Closed network results for the states of $n = 1, \ldots, 4$ jobs in process have been obtained and are shown in Figure 11.8 Evaluate system performance.

Solution

Step 2 involves solving the system as a state-dependent $M/M/1$ queue. From equation 11.40 we have

$$p(1) = \frac{80}{50}p(0) = 1.6p(0)$$

$$p(2) = \left[\frac{80}{50}\right]\left[\frac{80}{80}\right]p(0) = 1.6p(0)$$

$$p(3) = \left[\frac{80}{50}\right]\left[\frac{80}{80}\right]\left[\frac{80}{90}\right]p(0) = 1.4222p(0)$$

$$p(4) = \left[\frac{80}{50}\right]\left[\frac{80}{80}\right]\left[\frac{80}{90}\right]\left[\frac{80}{95}\right]p(0) = 1.1977p(0)$$

Beyond $i = 4$, production rate does not change; thus,

$$p(i) = \frac{80}{95}p(i-1) \qquad i \geq 5$$

Adding the constraint $\sum_{i=0}^{\infty} p(i) = 1$ and substituting, we obtain

$$p(0) \cdot \left[1 + 1.6 + 1.6 + 1.4222 + 1.1977 \sum_{i=0}^{\infty}\left(\frac{80}{95}\right)^i\right] = 1$$

or $p(0) = 0.07571$.

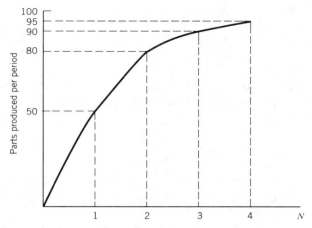

Figure 11.8   The production rate for a closed system.

The average number of jobs in the total system is

$$L_T = \sum_{i=0}^{\infty} i\,p(i)$$

$$= 0(0.07571) + 1(0.1211) + 2(0.1211) + 3(0.1077)$$

$$+ 4(0.09068) + 0.09068 \sum_{i=5}^{\infty} i(0.8421)^{i-4}$$

Rewriting the last term as

$$\frac{0.09068}{0.84211^4} \sum_{i=5}^{\infty} i(0.84211)^i$$

and noting that for $r < 1$

$$\sum_{i=0}^{\infty} i r^i = \frac{r}{(1-r)^2}$$

we obtain

$$L_T = 5.73$$

We can also estimate the average number of jobs in process by

$$L = p(1) + 2p(2) + 3p(3) + 4 \sum_{i=4}^{\infty} p(i)$$

$$= 0.12114 + 2(0.12114) + 3(0.10768) + 4(0.57433) = 2.98$$

On the average $5.73 - 2.98 = 2.75$ jobs are held in storage.   ■

11.6
SUMMARY

Many manufacturing systems can be viewed as networks of queues where customer jobs circulate among workstations. Where assumptions of exponential service times and FCFS service discipline are reasonable, steady-state perfomance estimates can be quickly obtained. The solution to many systems has a product form. Exact average results can be found for such systems. This applies to both open systems, where jobs enter the shop on arrival, and closed systems, where the level of work-in-process is strictly controlled. Open systems have the added advantage of exhibiting independence between workstations. As an alternative to the more complex product form solution algorithm required for closed systems, mean value analysis can be applied. Mean value analysis exploits the forgetfulness property of exponential servers and Little's Law to develop a set of equilibrium equations relating throughput rate, waiting times, and queue lengths. These equations can be solved iteratively. By combining the open and closed model techniques, hybrid systems can also be analyzed.

With increasing modeling and computational complexity, many of the basic model assumptions can be relaxed. Multiple servers can be handled exactly in product form models and heuristically in MVA. Multiple part types can be modeled directly or aggregated into a pseudo-composite item for analysis. Scheduling priorities based on part type are easily incorporated in some models. Nonexponential service times and SPT scheduling can be modeled but require more effort and are approximate.

The models in this chapter provide a quick method for obtaining average results for certain queueing networks. They have the advantage of allowing fast analysis of many system configurations so that a few potentially good configurations can be selected for further study. Trade-off curves graphically illustrate the frontier of production rate versus throughput time, conveying the effect of shop loading policies to decision makers. Transient behavior is not modeled nor are distributions of queue lengths and waiting times in most instances. However, the relative robustness of mean results to underlying service and interarrival distributional forms allows these models to be used for rough-cut estimates in many settings. Throughput rates are often within 5 to 10 percent of actual, with waiting times and queue lengths within 10 to 20 percent. When quick solutions are necessary, these models may be used as the final analytical tool.

REFERENCES

Baskett, Forest, K. Mani Chandy, Richard R. Muntz, and Fernando G. Palacios [1975], "Open, Closed and Mixed Networks of Queues with Different Classes of Customers," *Journal of the Association for Computing Machinery*, 22(2), 248–260.

Bitran, G. R., and D. Tirupati [1991], "Approximations for Networks of Queues with Overtime," *Management Science*, 37(3), 282–300.

Bunday, Brian D., and Esmaile Khorram [1988], "A Closed Form Solution to the G/M/r Machine Interference Problem," *International Journal of Production Research*, 26(11), 1811–1818.

Burke, P. J. [1968], "The Output Process of a Stationary M/M/s Queueing System," *Annals of Mathematical Statistics*, 39, 1144–1152.

Buzacott, J., and J. Shantikumar [1980], "Models for Understanding Flexible Manufacturing Systems," AIIE Transactions, 339–349.

Cox, D. R. [1955], "A Use of Complex Probabilities in the Theory of Stochastic Processes," *Proceedings of the Cambridge Philosophical Society*, 51, 313–319.

Gross, D., and C. M. Harris [1985], *Fundamentals of Queueing Theory*, John Wiley & Sons, Inc., New York.

Hopp, W. J., and M. L. Spearman [1991], "Throughput of a Constant WIP Manufacturing Line Subject to Failures," *International Journal of Production Research*, 29, 635–655.

Jackson, J. R. [1957], "Networks of Waiting Lines," *Operations Research*, 5, 518–521.

Lavenberg, Stephen S., ed. [1983], *Computer Performance Modeling Handbook*, Academic Press, San Diego.

Peck, L. G., and R. N. Hazelwood [1958], *Finite Queueing Tables*, John Wiley & Sons, Inc., New York.

Reiser, M., and S. S. Lavenberg [1980], "Mean-Value Analysis of Closed Multichain Queuing Networks," *Journal of the Association for Computing Machinery*, 27(2), 313–322.

Schweitzer, Paul J., Abraham Seidmann, and Sarit Shalev-Oren [1986], "The Correction Terms in Approximate Mean Value Analysis," *Operations Research Letters*, 4(5), 197–200.

Seidmann, Abraham, Paul J. Schweitzer, and Sarit Shalev-Oren [1987], "Computerized Closed Queueing Network Models of Flexible Manufacturing Systems: A Comparative Evaluation," *Large Scale Systems*, 12, 91–107.

Shalev-Oren, S., A. Seidmann, and P. J. Schweitzer [1985], "Analysis of Flexible Manufacturing Systems with Priority Scheduling: PMVA," *Annals of Operations Research*, 3, 115–139.

Shanthikumar, J. G., and J. A. Buzacott [1981], "Open Queueing Network Models of Dynamic Job Shops," *International Journal of Production Research*, 19(3), 255–266.

Snowdon, Jane L., and Jane C. Ammons [1988], "A Survey of Queueing Network Packages for the Analysis of Manufacturing Systems," *Manufacturing Review*, 1(1), 14–25.

Solberg, James J., [1980], "A User's Guide to CANQ," Tech. Report #9, School of Industrial Engineering, Purdue University, West Lafayette, IN.

Solberg, James J., [1981], "Capacity Planning with a Stochastic Flow Model," *AIIE Transactions*, 13, 116–122.

Spearman, M. L., D. L. Woodruff, and W. J. Hopp [1990], "CONWIP: A Pull Alternative to Kanban," *International Journal of Production Research*, 28, 879–894.

Suri, Rajan, and G. W. Diehl [1985], "Manuplan: A Precursor to Simulation for Complex Manufacturing Systems," *Proceedings of the 1985 Winter Simulation Conference*, D. Gantz, G. Blais, and S. Solomon eds., Institute of Electrical and Electronics Engineers, 411–420.

Suri, Rajan, and R. R. Hildebrandt [1985], "Modeling Flexible Manufacturing Systems Using Mean Value Analysis," *Journal of Manufacturing Systems*, 1(3), 27–38.

Suri, Rajan [1989], "Lead Time Reduction Through Rapid Modeling," *Journal of Manufacturing Systems*, 8, 66–68.

White, J. A., J. W. Schmidt, and G. K. Bennett [1975], *Analysis of Queueing Systems*, Academic Press, New York.

Whitt, Ward [1983], "The Queueing Network Analyzer," *Bell System Technical Journal*, 62(9), 2779–2815.

Yao, David, and John Buzacott [1986], "The Exponentialization Approach to FMS Models with General Processing Times," *European Journal of Operational Research*, 24(3), 410–416.

PROBLEMS

11.1. The queueing models in this chapter provide steady-state results. What is meant by steady-state? Describe what inferences can and cannot be made from steady-state results.

11.2. The models in this chapter normally assume that arrival and service rates are constant through time. Is this a reasonable assumption? Give an example where these assumptions would not apply.

11.3. The shipping department of a large plant receives about 20 shipping orders per day. Orders arrive either because all parts have been completed or the shipment is late and a partial order is being shipped. Most orders are small, requiring little packing and paper preparation. However, some orders are large, and the average number of person-hours to fill an order is 1.5. Five workers are employed; each is available 7.5 hours per day. Find the average number of jobs waiting to be shipped and the average time from receipt of an order until the order is completed.

11.4. Orders arrive at a shop randomly at the rate of 20 per day. The time to process a job is uniformly distributed between $\frac{1}{2}$ and 1 hour. If the shop operates two shifts (16 hours) each day, find the average shop utilization, mean and variance of job throughput time, and mean and variance of the number of jobs in process.

11.5. An assembly workstation receives orders every 10 minutes on average. The time between order arrivals is exponentially distributed. Each order requires five tasks. The time for an individual task is exponentially distributed with a mean of 1.8 minutes. Find the mean and variance of the throughput time and number of jobs at the workstation.

11.6. A shop produces about 40 orders per week. Each order takes an average of one hour of shop resource, but most orders are shorter and some are very long. Workers normally operate 40 hours per week, but overtime is used in busy weeks;

thus, in the long term the shop schedule averages 44 hours per week. The foreman has proposed giving priority to a key customer who has been complaining about long lead times. This customer accounts for 25 percent of the orders. Estimate the effect on throughput time for the key customers and others if the priority system is implemented.

11.7. A flexible assembly system has five workstations. Order arrivals are a Poisson process with mean rate of 18 per hour. Service time is exponential with a rate of 4 per hour per workstation. Workstations place their finished orders on a conveyor, where they are removed by a single material handler.

    a. Determine the distribution of interarrival times at the unload station.

    b. Suppose the unloader sorts the finished orders to two packing stations. If sorting is random, find the interarrival time distribution at each packing station.

    c. Suppose the unloader alternates between packing stations (i.e., every second order goes to the same packer). What is the interarrival distribution for packers?

11.8. Re-solve Example 11.6 with $p_{21} = 0.25$ and an external arrival rate of $\lambda = 8$.

11.9. Three workers "kit" orders by pulling the required number of parts from a warehouse and placing them in a tote. Orders are always ready to be kitted. In fact, the computer scheduling system maintains a one-hour supply of kitting orders in queue at all times in front of the kitters. Time to kit an order is exponential, with mean 40 minutes. Kits then go to an assembly area. Four assemblers are available. Assembly time is exponential, with mean 20 minutes. Assembled kits are then inspected. Two inspectors are available. Inspection time is exponential with mean 15 minutes. Find the average number of orders in process at each station and the average time for an order to go through the system.

11.10. Modify Problem 11.9 such that a kit fails inspection with probability .4. Failed kits are re-

turned to assembly. Rework time is also exponential, with mean 20. Now find the average queue sizes and throughput time.

11.11. A small shop has six machines. Parts are stored in tubs and are transported on pallets. Each pallet holds one batch of parts. All parts are moved between machines by forklift. The shop produces four types of parts. Their routings are shown in Table 11.6. Find the effective arrival rates at each workstation and the forklift.

11.12. A workstation with a single machine and operator receives output from an $M/M/c$ process. Three part types are processed. Arrival rates for the three part types are 10, 5, and 20 per day, respectively. Service time is exponential with mean 0.01 days for part type 1, exponential with mean 0.05 days for part type 2, and deterministic at 0.01 days for part type 3. If part type 1 has highest priority and part type 2 has higher priority than part type 3, find the expected number of each part type and throughput times at the workstation.

11.13. Show that the balance equation 11.34 in Example 11.11 is satisfied by the state probabilities defined in equation 11.35. (*Hint:* Find the v_j visit counts and service rates for workstations. Then replace the ρ_j in the balance equations by v_j / μ_j).

11.14. Reconsider Example 11.6. Suppose jobs that have been returned to assembly because they failed inspection the first time need not be reinspected. After a second visit to assembly, these jobs leave the system. Find the new transfer probabilities and re-solve the example.

11.15. Discuss the different assumptions of open and closed queueing network models. Confronted with a real system, what factors would determine which modeling approach to use?

11.16. Three prospective designs are under consideration for a manufacturing cell. The first uses c parallel machines each with service rate μ, a common queue, and total arrival rate λ. The second option uses the same machines but has a separate queue for each machine. Arriving part batches

Table 11.6   Part Routings and Demand for Problem 11.11

Part Type	Op. 1	Op. 2	Op. 3	Op. 4	Batches/Period
1	M3	M2	M1	M6	100
2	M4	M1	M2	M3	50
3	M5	M6	M1		30
4	M3	M6	M2	M4	50

will be randomly assigned to a machine. The third configuration consists of one fast machine, which serves at the rate 3μ. Ignoring capital and operating costs, which system has the lowest throughput time and WIP levels? You may assume interarrival and service times are exponential and let $\mu = 5$ and $\lambda = 10$.

11.17. A job shop has three types of machines; two mills, one drill press, and one surface grinder. Orders arrive to the shop at a rate of 2 per day. About 60 percent of these go to milling first. The other 40 percent start at the drill. One half of the drilling jobs go next to milling, whereas the other one half leave the system. Thirty percent of jobs being milled are sent for grinding, and the others leave the shop. Jobs always leave the system after grinding. Operation times are exponentially distributed, averaging one day per job for milling, drilling, and grinding. Find the average number of jobs in the system.

11.18. Re-solve problem 11.17 assuming that operation times are uniformly distributed between 0 and 2 days per job.

11.19. Using the closed network of queues model, analyze the following problem. A job shop has three machines. Four products are made according to Table 11.7. One material handler is available, and it takes him about 5 minutes to move a load of material between machines. You may assume that the handler is not needed to deposit new parts to the first operation. Machines 1 and 2 run 40 hours per week, and machine 3 runs 80 hours. Assume that we try to keep one week's demand worth of work in process at all times.
   a. Analyze by explicitly modeling the four separate products.
   b. Aggregate the four products into a generic (weighted-average) product and reanalyze.

11.20. Re-solve Problem 11.19a assuming that ample material handlers exists so that parts never need to wait for material handling.

11.21. Re-solve Problem 11.19b as an open network. Arrival rates are given by weekly demand. In addition to finding average throughput time for an arbitrary job, find the average throughput time for product type 1 specifically.

11.22. Re-solve Problem 11.19a assuming that two machines of type 3 exist and each is scheduled for 40 hours. Use the throughput time equations 11.25 for type 3 machines.

11.23. A four-machine production system (all single servers) is being analyzed. Service rates and routing probabilities are as shown in the following tables. Seven jobs are kept in process at all times. Service times are exponentially distributed.

Machine	1	2	3	4
Mean Service Time (hr)	2.4	5.1	3.2	6.0

Intermachine Routing Probabilities			
From-To	1	2	3	4
1	0.0	0.5	0.2	0.3
2	0.7	0.0	0.1	0.2
3	0.1	0.9	0.0	0.0
4	0.3	0.4	0.3	0.0

   a. Find the relative visit counts and utilizations for each machine.
   b. Solve this network using MVA. Find the production rate.

Table 11.7   Product Routings for Problem 11.19

Product	Operation	Machine	Time in Hr	Weekly Demand
1	1	1	10	1
	2	3	5	
2	1	2	12	2
	2	3	7	
	3	2	2	
3	1	1	5	3
	2	3	10	
4	1	2	5	1
	2	3	7	

11.24. Solve Problem 11.23 using the recursive calculation procedure described in Section 11.4.2 for product form networks. Plot production rate versus the number of jobs in process. Using $L = \lambda W$, also plot throughput time versus the number of jobs in process.

11.25. A three-machine workcell is controlled such that six jobs are kept in process at all times. Four part types are produced. When a job is completed, a new job is dispatched to the system. The new job is selected randomly in proportion to the desired relative production rates of the four part types. The table gives demand and processing data. In each operation in the process plan, the data pair contains (machine, job processing time). Consider an exponential server, single aggregate part, closed queueing network model of the cell.

	Demand	Process Plan		
Part	Rate	Op. 1	Op. 2	Op. 3
1	10	(1,2)	(2,2)	(3,1)
2	5	(2,3)	(1,3)	—
3	20	(3,1)	—	—
4	10	(1,2)	(3,1)	—

a. Specify the possible system states.
b. Find the relative work load factor for each machine. Which machine has the heaviest load?
c. Give an expression for the utilization of machine 1 in terms of the system states defined in part a.

11.26. Solve for the normalization constant, production rate, and average number of jobs in process in Problem 11.25 using the product form model.

11.27. Try varying the number of jobs in process of each part type in Problem 11.19 to find the minimal average work-in-process levels that will achieve the desired demand rate for each product.

11.28. Suppose in Problem 11.19 that all part demands increase proportionally. How much can demand increase before the system reaches capacity?

11.29. Consider the basic $M/M/c$ queueing expressions in Table 11.1. For $c = 1$ and $c = 4$ servers, plot L versus ρ. What does this tell you about loading for single server versus multiple server systems? What does this tell you about the relationship between rapid response time to customers and capacity?

11.30. Show that a FCFS workstation does not lend itself to a product form solution if there are multiple classes of customers and classes have different service rates. (*Hint:* Write the balance equations for a simple system.)

11.31 A three-machine workcell has a limited storage of four jobs. Additional jobs are held outside the cell until space is available. All jobs flow through machine 1, then 2, then 3 before exiting the system. Jobs arrive randomly, six per day. Service times are random also but machines 1, 2, and 3 have average service times of $\frac{1}{8}$, $\frac{1}{10}$, and $\frac{1}{7}$ day, respectively. Find the average number of jobs waiting to enter the system.

11.32. ACE Manufacturing Co. receives an average of 95 custom orders per week, with the actual number of weekly orders being Poisson distributed. The plant employs 23 workers and has more than 40 machines. Jobs are put into production as soon as the order arrives unless there are already 75 jobs in process. In the latter case, jobs are kept in a warehouse until space opens up on the floor. Partial records have been kept on weekly shipments and in-process inventory levels. Using the data in Table 11.8, estimate the average number of jobs in process and in the warehouse, and the average throughput time.

Table 11.8   Historical Shop Data for Problem 11.32

Week	Orders Shipped	Orders in Process plus Warehouse
1	111	88
3	45	10
7	100	50
12	105	60
13	109	78
15	79	25
16	96	40
19	93	38

11.33. Write a computer program to analyze a general open network with exponential arrivals, service, and FCFS discipline. User input should contain external arrival rates to each node, internal routing probabilities, the number of servers, and service rate at each workstation. Program output should include average queue lengths and throughput times for each workstation. The user should be provided with an option of receiving queue length distributions for workstations. Verify your code by re-solving the example problems in Section 11.3.

11.34. Upgrade the program of Problem 11.33 to allow the user to input product demand and routings. Arrival and service rates should be computed internally.

11.35. An approximate result for a $GI/G/1$ queue is

$$E(W_q) = \frac{\rho^2(1 + C_s^2)}{1 + \rho^2 C_s^2} \left\{ \frac{C_a^2 + \rho^2 C_s^2}{2\lambda(1 - \rho)} \right\}$$

where C_s^2 and C_a^2 are the squared coefficients of variation for the service and arrival distributions. A co-worker has suggested that parts should be stored in a central warehouse between operations instead of at the next machine on the process plan. In this way, jobs can be dispatched to machines at fixed intervals. This will make $C_a^2 = 0$ and reduce expected waiting time. What is your response?

11.36. Consider a closed queueing network model of a manufacturing system with two workstations. Both stations operate FCFS. There are two customer classes, and one customer of each type is in the system at all times. Both customer classes require service at station A first, and then at station B. Service times are exponential, but each station has a class-dependent service rate, that is, μ_{ij} for customer class i at station j.

a. Write out the set of possible states for the closed network.

b. Write the steady-state balance equations for all states.

c. If there were a product form solution, what would it be for each state probability?

d. Show that the product form solution is invalid if each customer class has its own service rate.

11.37. Consider the closed network of Figure 11.9. All three workstations are FCFS with exponential service time. Only one class of customers exists. Three jobs will be kept in the system at all times.

a. State the general form for a state probability.

b. For each state, find its probability. Your answer should be the actual value between 0 and 1.

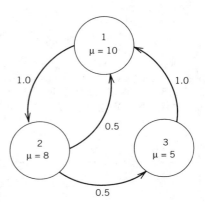

Figure 11.9 The closed network for Problem 11.37.

11.38. A repairman serves five machines. Repair times average 75 minutes. Mean time to failure for machines is 7 hours. Find the percentage of time the repairman is busy and the percent of time one or more machines are waiting for repair.

11.39. A robot is used for loading and unloading three machines. Each machine has a constant supply of parts waiting to be loaded. Part processing times vary but average 6 minutes. It takes the robot an average of 30 seconds to unload a part and load a new part on any machine. Is the robot adequate for handling the load/unload activities?

CHAPTER 12

GENERAL MANUFACTURING SYSTEMS: EMPIRICAL SIMULATION MODELS

Wisdom consists of knowing when to avoid perfection.
—**Horowitz's Rule, from *Murphy's Law* by Arthur Bloch**

12.1
INTRODUCTION

Models that test whether a manufacturing system will achieve the operational goals set by management and will perform as designed by system engineers must be built during the design process. These models help ensure that the components of a manufacturing system, designed with the aid of several other models, like those discussed in the previous chapters, will work together as intended. In addition, the effects of what appears to be random behavior by system components (Law 6 of Chapter 1) must be understood. Meeting these objectives is accomplished both by evaluating average, steady-state behavior using analytic queueing models like those presented in the preceding chapter and by a detailed examination and assessment of time-variant system dynamics.

Simulation (Banks and Carson [1984], Banks and Carson [1988], Balci [1990], Shannon [1975]) is the most common method for constructing models that include random behavior of a large number and a wide variety of components and assess the temporal dynamics of manufacturing systems. Simulations can include detailed information about system components, closely conform to the unique aspects of each particular manufacturing system, evaluate time-variant behavior, and provide system-specific quantities to measure performance. Simulation uses a computer to replicate system behavior over time as expressed in a model and to gather observations of performance metrics of interest, as would be done in any experiment. These observations, and statistical summaries computed from the observations, are used to assess system performance. Thus, simulation is a procedure for experimentation with models and, therefore, requires proper experimental design. Unlike analytic models, results from simulation models are statistical estimates and must be interpreted as such.

Simulation is most preferred when an analytic solution to a model that gives exact values for performance metrics is not feasible. Many manufacturing systems have unique characteristics that vary from the mathematical assumptions that underlie analytic solution procedures. For example, analytic queueing network models have difficulty including such common manufacturing system characteristics as the use of the same robot at multiple workstations or finite buffer space. Furthermore, a manufacturing system may employ a complex, dynamic control strategy based on previous behavior and the current state of system components. In addition, most analytic models evaluate only steady-state conditions and measure behavior by using the expected values of a predefined set of performance metrics. Changes in behavior over time are not taken into account. For example, demands placed on the system may vary over time. The transient effect of extreme conditions or unusual events needs to be assessed. For example, the transient effect of machine failures and their subsequent repair on work-in-process inventory, sequencing and scheduling decisions, and throughput need to be studied. The behavior of individual parts, machines, and workers may be important in evaluating scheduling and control strategies. Therefore, gaining an understanding of and being able to visualize graphically the time-variant dynamic behavior of a system are often goals of a modeling project. Performance metrics may be in the form of time series of values or empirical distributions. In addition, performance metrics unique to individual systems are often needed to meet project requirements or to assess unique behavior. Simulation supports the description of system components using data gathered from the system under study. Thus, including individual data values that characterize behavior over time can be necessary. For example, production requirements may be expressed as a list of orders. Alternatively, such data values may be summarized by a histogram or fit to a distribution function and used to represent component behavior. A wide variety of distributions, including compound distributions, can be employed.

Simulation models include several entity classes, such as **resources, transactions,** and **queues.** A resource is a component of the manufacturing system that performs or assists in processing. **Attributes** of a resource tell its present or past state, such as busy, idle, broken, or under repair. A transaction represents a part, information, or something else processed by the manufacturing system using resources. Attributes of a transaction are values that distinguish it from other transactions, such as time processing began, part code, due date, and so on. A queue or buffer is the place where transactions wait for resources. The number of transactions queued is a typical queue attribute. The model itself is considered an entity and has attributes such as current simulation time.

The state of the model is the current conditions in the model, usually defined by the values of entity attributes. The state is changed by an event that occurs at a particular instant in time. Thus, state changes occur at a discrete set of points in time. This leads to the name **discrete event** simulation. An event is characterized by the time it is to occur, some identifier of the event, and the entities that are processed by the event. Future events are scheduled by preceding events. Times between events may be constants, data values input to the model, or random samples from distribution functions. The set of scheduled events, sorted in order of time of occurrence, form the **event list.** Performance metrics are of two types. Those metrics that have a value at every point in simulation time, such as a queue length or the state of a resource, are called **time-persistent.** Those that can be measured only at event times, such as the time interval a part spends in manufacturing, are called **observed.**

A **simulation language** is a special-purpose computer language that includes constructs for building models and capabilities for experimentally assessing models and reporting the results. Many such simulation languages exist (Banks et al. [1989], Cobbin [1988], Henriksen [1983], Pegden [1982], Pritsker [1982], Pritsker [1986],

Pritsker et al. [1988], Roberts and Flanigan [1990], Russell [1983], Schriber [1974], and Taha et al. [1990]). A **simulation system** (Cox [1988], Grant and Starks [1988], Law and Vincent [1988], Lilegdon and Erlich [1990], Standridge and Pritsker [1987]) integrates a simulation language with a variety of supporting software tools, such as graphical model editors, statistical analysis capabilities, animation developers, and report and graph generators.

This chapter begins with an overview of simulation techniques. Two fundamental viewpoints for simulation model development—discrete event and process—are described. In addition, a manufacturing-specific modeling viewpoint is presented. Evaluation procedures for models are discussed, and an overview of simulation systems is given. Types of simulation results that are particularly useful in manufacturing systems models are discussed and illustrated. Example simulation models and results are described.

12.2

EVENT MODELS

Specifying the events that change the state of a manufacturing system is one way to develop a simulation model. Each state change is modeled by an event as instantaneous. There may be one or more state changes per event. Events may schedule other events, including themselves, to occur in the future or at the current time. Event scheduling may be conditional on the state of the system. Event graphs are a way to identify events and to show their interrelationships (Schruben [1983]). Nodes represent events and show the event ID. The branches represent relationships between events, showing which other events, including itself, each event schedules. The parameters of a branch are the time interval into the future at which the event is scheduled to occur and the condition for scheduling the event, if any.

To illustrate, consider a workstation where arriving parts wait in an infinitely large buffer for service by the single worker. The state of the system is defined by the number of parts waiting in the buffer and the status of the server (busy or idle). The possible state changes and the corresponding events are given in Table 12.1.

Figure 12.1 shows the event graph for the single-worker station system. The three events identified in Table 12.1—part arrives, part starts service, and part completes service—are shown, along with their relationships. The part arrives event schedules itself (the next part arrival) at a future time and schedules the part starts service event at the current time if the worker is idle. The part starts service event schedules the part completes service event in the future. The part completes service event schedules the part starts service event at the current time if the number of parts in the buffer is greater than zero.

Table 12.1   State Changes and Corresponding Events:
Single-Worker Station

State Change	Event
Number of parts in the buffer increases	Part arrives
Number of parts in the buffer decreases	Part starts service
Worker becomes busy	Part starts service
Worker becomes idle	Part completes service

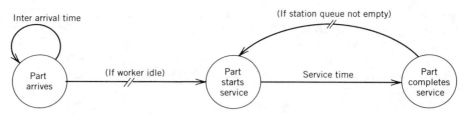

Figure 12.1   An event graph for a single-worker station.

12.2.1 Evaluation of Event Models

Evaluation is the computer-based computational process of experimentally assessing the system behavior represented by the model. Starting at time 0 (or any other specified starting time), a simulation clock is advanced chronologically from event time to next event time. Each event has its own system state change logic. A trace is a chronological record of the events processed in a simulation. Observations of performance metrics, which could include the state of the system, are made at events and are optionally summarized statistically. Traces, performance metric observations, and statistical summaries are reported for review and interpretation.

To illustrate, a partial trace of the simulation of the single-worker station model described by the event graph given in Figure 12.1 is given in Table 12.2. At time 0, the first part arrives. Since the server is idle, the processing of this part is immediately begun by the worker. At time 2, the second part arrives and finds that the worker is processing the first part. Thus, the second part must queue for the worker. At time 3, the worker completes the processing of the first part and thus is able to start processing the second part.

12.3
PROCESS MODELS

Event models are developed by defining the state changes in a system and the points in time, events, where these changes occur. An alternative point of view is the sequence of steps taken by a transaction in its processing by a system. In simulation modeling, this sequence of steps is called a process. A process may contain conditional logic so that the steps taken by any individual transaction may depend on its own attributes or on the current state of the system. A model may consist of multiple processes that may affect each other. Resources may be shared between processes. Transactions in one process may place restrictions on transactions in another process. A process implies an ordered sequence of events. Thus, the event viewpoint

Table 12.2   Example Trace of Single-Worker Station Event Model

Time	Event	Description
0	Part arrives	First part arrives
0	Part starts service	First part begins processing
2	Part arrives	Second part arrives and queues for service
3	Part completes service	First part is finished
3	Part starts service	Second part begins processing

forms the foundation for the process viewpoint. Since the process viewpoint corresponds to the steps in a manufacturing process, this approach is often taken to model manufacturing systems.

To illustrate process modeling, consider again the single-worker station example. The sequence of steps for the processing of a part is a follows:

1. Part arrives.
2. Part enters buffer.
3. Part starts service (and exits buffer).
4. Part completes service (after the service time delay).

The foregoing process consists of generic steps that could be used to represent a wide variety of manufacturing situations. Thus, standard approaches have been developed and have been variously implemented in simulation languages for modeling from the process viewpoint. Transactions are characterized by attribute values, usually numeric. Resources consist of one or more indistinguishable units and are characterized by the number of units in each state. Typical states are busy, idle, blocked, or down. Blocked means that a resource has completed processing of a transaction but for some reason that transaction cannot move to the next processing step and prevents the resource from working on the next transaction. Down means that the resource is receiving maintenance or is broken and under or awaiting repair. Transaction arrival to the system must be modeled. Assignment of values to transaction attributes must be allowed. As a transaction tries to accomplish a step in a process, it is constrained by the availability of required resources. Allocation of resources includes deciding which resource(s) to allocate and how many units of each selected resource to use. On completion of processing, allocated resources become free. Queueing for unavailable resources is necessary. Explicit time delays are included. Logical decision making by transactions for selecting the next step in the process must be supported.

12.3.1 Evaluation of Process Models

Processes are decomposed into a sequence of events for evaluation. Thus, the simulation of a process model is conducted in the same way as that for an event model. Again, the experimental strategy is to compute the logic associated with each event occurrence in time sequence. Traces, performance metric observations, and statistical summaries are reported for review and interpretation.

To illustrate, a part of the trace of the simulation of the single-worker station model described by the process considered in the preceding section is given in Table 12.3.

Table 12.3   Example Trace of Single-Worker Station Process Model

Time	Process Step	Description
0	1. Part arrives	First part arrives
0	2. Part enters buffer	First part enters the buffer
0	3. Part starts service	First part begins processing
2	1. Part arrives	Second part arrives
2	2. Part enters buffer	Second part enters the buffer
3	4. Part completes service	First part is finished
3	3. Part starts service	Second part beings processing

Seven process steps involving two parts occur in the first three time units. The first part arrives and begins service immediately. The second part arrives while service to the first part is ongoing. Thus, it must wait for service on the first part to be completed. When service of the first part is finished, service on the second part is begun.

Note that the traces for the event model and the process model are similar. This indicates that process models are evaluated using a series of events.

12.4
MANUFACTURING SPECIFIC MODELING

Events and processes are generic modeling approaches that are able to describe a wide variety of complex systems. However, their use requires the analyst to perform the cognitive task of translating the objects and information of the system being modeled into events and/or the process modeling constructs of a simulation language. For a complex manufacturing system, this translation may not be direct. Much time and intellectual effort may be spent on model building.

An alternative approach is to specify the objects of a manufacturing system and to give their attributes values. A manufacturing-specific simulation language (Goble [1990], Harrell and Tumay [1990], Murgiano [1990]) provides manufacturing object classes and standard attributes. The system dynamic behavior represented by the model is based on generic manufacturing system behavior embedded in the simulation language. Typical classes would be stations where parts are processed, buffers where parts wait for processing, material handling devices for moving parts from station to station, and process plans for routing parts through the stations. Since such classes correspond more closely to the objects of a manufacturing system, model building should be more straightforward and less time consuming than with an event or process approach. However, since the modeling classes and their attributes are fixed, deviations between the modeling classes and system objects may be difficult to resolve. Many manufacturing-specific simulation languages provide some capabilities for supplementing the modeling classes with events or processes.

Like process models, manufacturing models imply a sequence of events. Thus, as in an event model, the experimental strategy is to compute the logic associated with each event in time sequence. Measures of system performance are computed for reporting, review, and interpretation.

As an example, let us consider a manufacturing-specific model of two single-worker stations in series. Parts are processed by the first station and then by the second. A material handling device moves parts from the first station to the second. A model of this kind would include:

1. Two station objects each of which performs an operation on a part, characterized by a part-processing time and a setup time.
2. Four buffer objects. Each station has two associated buffers, one where inbound parts wait for processing by the station and one where completed parts wait for transportation by the material handling device. All buffer objects have a capacity attribute.
3. A process plan that identifies a part type and the routing of parts of that type through the stations that process them.
4. A material handling device characterized by its speed.

5. An entry port characterized by a time between part arrivals distribution and an exit port.
6. A facility layout indicating the distance between stations.

12.5
SIMULATION SYSTEMS

Simulation languages have been commercially available since the early 1960s. In the late 1970s and early 1980s, capabilities complementary to the simulation languages were developed. Together with a simulation language, these additional capabilities form a simulation system. Figure 12.2 describes the basic strategy of a simulation system. The system is a collection of software tools. A simulation system language helps integrate and gives access to each of the tools. Some of the tools provide for entry and modification of graphical and textual information. Graphical information includes process and manufacturing-specific models and drawings used to display animations. Textual information includes information that controls simulation experiments, scripts that direct animations, and simulation input data. Other tools retrieve data, process it, and store the results. A simulation language retrieves models, experimental control information, and input data and then produces and stores simulation performance metric observations. Statistical analysis tools retrieve data, either simulation inputs or simulation results, perform their analysis, and store resulting statistical estimates. A third group of tools displays information. Thus, a simulation system helps make the statistical analysis and display of simulation results separate steps from the building and simulation of a model. Graph generators display performance metric observations and statistical summaries of the observations. Animation generators map the event occurrences of a simulation into graphical changes shown on a drawing of the system being studied. Database management techniques organize and control the models, data, and other information in a simulation system.

12.6
SIMULATION RESULTS FOR MANUFACTURING SYSTEMS

The flexibility of simulation techniques allows performance metrics to be defined on a model-by-model basis. However, the following performance metrics are common among manufacturing system models:

1. *Throughput:* The number of units completing a production operation in a fixed time interval.
2. *Makespan:* The time interval required to produce a fixed number of units.
3. *Quality:* Percentage of units that are not defective.
4. *Unit time in system or subsystem:* The total time individual units spend in one or more system activities such as processing, transport, or waiting.
5. *Work-in-process inventory:* The number of partially completed units in processing at a given time.
6. *Congestion:* The ratio of waiting time to processing time.
7. *Utilization:* The number of machines, workers, and so on busy over time; the percentage of time spent in each task and idle.
8. *Flexibility:* The number of machines, workers, and so on idle over time.

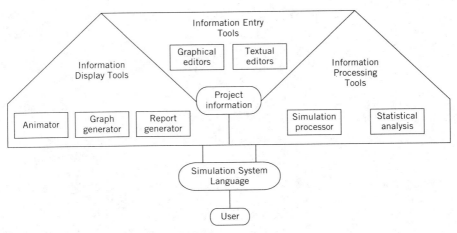

Figure 12.2   Simulation system organization.

Simulations produce several types of estimates of these quantities.

1. Time series of individual values; for instance, the number of busy machines, the work-in-process inventory, or the processing time for each unit.
2. Histograms summarizing the distribution of time series values or showing the distribution of entities across different activities such as the percentage of time a machine spent in setup, in processing, under maintenance, and idle.
3. Statistical summaries of time series values, particularly averages, minimums, and maximums.
4. Single values such as the throughput or the makespan.

A difficulty that arises in the statistical analysis of simulation-generated values is the correlation between these values. For example, consider the time a particular part spends in a buffer waiting for a machine. Suppose the part arrives at the buffer at time t_1 and waits for the preceding part to be removed from the buffer and serviced. Suppose the preceding part arrived at time $t_0 < t_1$. Then the time spent in the buffer by the part w_1, is: $w_0 - (t_1 - t_0) + s_0$, where w_0 is the waiting time and s_0 is the service time of the preceding part. In other words, the waiting time for the part is a portion of the waiting time plus the service time of the preceding part. Thus, w_0 and w_1 are correlated.

In the real world, a particular time period happens only once. In our simulated world, the time period can be **replicated** or repeated as many times as we choose. Replicates differ only in the set of random samples taken from distribution functions used in the model to characterize time intervals between events, decision choices, transaction attribute values, and other quantities. Owing to the stochastic aspects of a simulation model, different behavior can result for each replicate and each replicate can bring new insight into system behavior. Since any particular replicate may represent extreme behavior that occurs with low probability, multiple replicates are usually helpful. Furthermore, values or averages from different replicates are independent, supporting standard statistical analysis of simulation results. In all the stochastic simulations in the examples in this chapter and the case studies in the next, multiple replicates are used to estimate performance measure values of interest.

An important step in studying any experimental results is an examination of the individual data values. For simulation, this examination can be accomplished by using graphs of the time series of performance metric observations (Standridge et al. [1984]). For manufacturing systems, graphs of work-in-process inventory, the number of busy machines or workers, throughput over time, and time spent by units in production are typical. Graphs show patterns and trends over time and help to identify extreme conditions and transient behavior. For example, consider the graph of the work-in-process inventory at a particular station shown in Figure 12.3. Note the buildup of inventory due to machine failure and the time taken to reduce the built-up inventory. Figure 12.4 shows a graph of the number of busy machines at a station. Note that the range of busy machines is large. Thus, the need to level machine utilizations over time is identified. Figure 12.5 shows a graph of the number of units produced over time. The graph is linear except for breaks every 8 hours. This shows that, except for the third shift shutdown included in the model, production occurs at a constant rate, which is good.

Graphs can show statistical summaries and the distribution of time series values or the distribution of time spent in different activities. The histogram in Figure 12.6 shows the distribution of the work-in-process inventory at a particular station. This distribution shows that the inventory was acceptably low most of the time. The bar chart in Figure 12.7 shows on the average how the units spent their time in production. One-half the time was spent in actual processing, which is good for most manufacturing systems. Statistics may be presented graphically as well. The chart in Figure 12.8 shows the average, minimum, and maximum unit time in the system for each of 10 independent replications or repetitions of the same simulation. The variance between replicates shows the danger of making decisions based on one replicate.

Examination of the sequence of event occurrences can yield valuable insights into system dynamics. Animation presents trace information graphically (Earle et al. [1990], Johnson and Poorte [1988], Miles et al. [1988], and Standridge [1986]). A facility diagram, consisting of symbols, provides a static, graphical model of the system. Graphical changes made to the facility diagram show changes in state due to each event occurrence.

The sequence of three animation snapshots shown in Figure 12.9 illustrates the way animation presents information. The structure of the system including the flow of parts is shown by the facility diagram. First parts are processed by one of the four mills at the mill station. Parts wait in the input buffer for an available mill and in the output buffer for a crane. All part movement is performed by overhead cranes, whose routes are shown by the white rectangles. Similarly at the deburr station parts wait for the deburring machine in an input buffer and for a crane in an output buffer. Parts that pass inspection at the inspect station are routed to the assembly area and parts that fail are routed to the rework station.

The animation shows the state of each station and of the overhead cranes. The numbers of busy mills and inspectors are indicated by how much of a rectangle is filled. The mill rectangle is directly below the mill symbol, and the inspect rectangle is directly above the man at the desk symbol. At times 171 and 190 all mills and inspectors are busy. At time 184, three mills and one inspector are busy. Mills are subject to breakdowns. The number of currently broken mills is indicated by the number of mill symbols in the rectangle labeled "down mills." At times 171 and 190, no mills are down. At time 184, one mill is down. The busy/idle status of the single deburr and assembly machines is indicated by the color or crosshatching pattern shown in the deburr and assembly symbols, darker for busy and lighter for idle. Cranes are either transporting a part, moving in response to a request to transport a

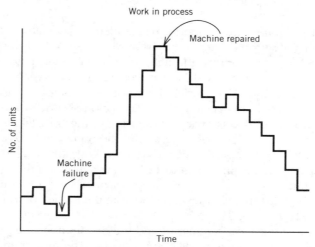

Figure 12.3 An example of work-in-process inventory graph.

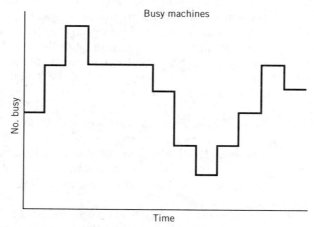

Figure 12.4 An example graph of the number of busy machines.

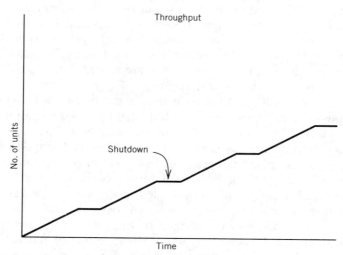

Figure 12.5 A graph of throughput over time.

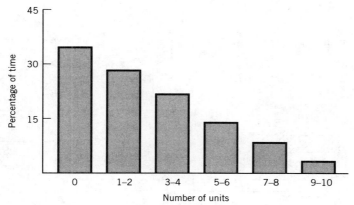

Figure 12.6 A histogram of work-in-process distribution.

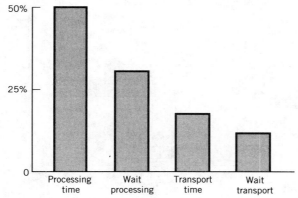

Figure 12.7 A bar chart of unit time in system.

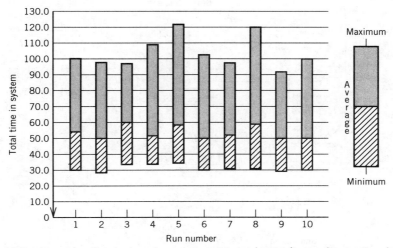

Figure 12.8 A graph of unit time in system statistics by replicate. (From Standridge and Pritsker, *TESS: The Extended Simulation Support System,* copyright © 1987, Halsted Press, New York. Reprinted by permission.)

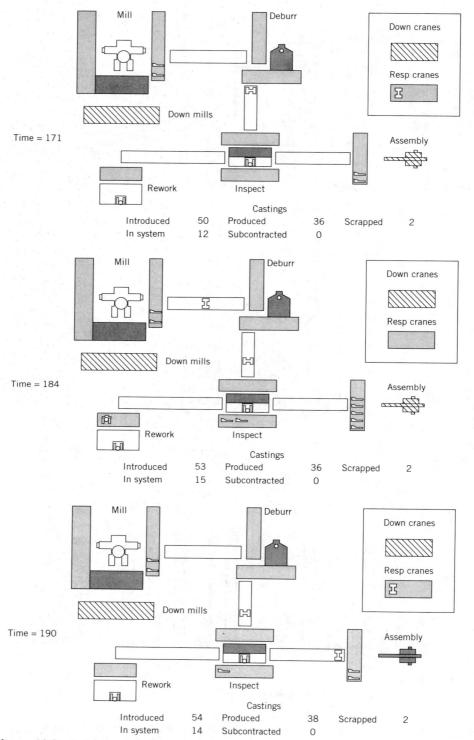

Figure 12.9 Animation snapshots. (From Standridge and Pritsker, *TESS: The Extended Simulation Support System,* Copyright © 1987, Halsted Press, New York. Reprinted by permission.)

part, down, or idle. Cranes transporting a part are indicated by a crane symbol shown on one of the crane routes. Other crane states are indicated by crane symbols located in the rectangles labeled down cranes and resp(onding) cranes. The work-in-process inventory is shown by part symbols in the input and output buffers of each station. The processing of parts is summarized by the numerical data at the bottom of the facility diagram.

12.7
EXAMPLE MANUFACTURING SYSTEM MODELS

The following examples illustrate some of the unique contributions of simulation to manufacturing systems modeling and analysis. Models of more realistic systems are presented in the next chapter to show the role of simulation in a more comprehensive system design process.

12.7.1 Serial Lines

Consider a serial line as discussed in Chapters 2 and 3. The line consists of three stations with two buffers of capacity 10, one between stations 1 and 2 and the other between stations 2 and 3. An infinite capacity for storage exists in front of the first station. The line processes two types of parts. Type A parts arrive every 10 minutes, and the interarrival time for type B parts is exponentially distributed with mean 15 minutes. Processing times at station 1 depend on the part type: uniform with range (3,7) minutes for type A parts and a constant 6 minutes for type B parts. Processing times at station 2 are triangularly distributed, with a minimum of 2 minutes, a maximum of 11 minutes, and a mean of 5.88 minutes, and at station 3 are exponentially distributed with a mean of 5.5 minutes. Station 3 fails once per week on the average with a repair time of 2 hours. Simulation results must show the transient behavior of queue lengths following a failure of station 3, estimate the portion of time a station is blocked, and estimate the distribution of unit time in the system.

A process simulation model of this system is constructed. Transactions represent parts. Transaction attributes are part type, either A or B, and simulation time of arrival to the system. Each station is represented by a resource whose possible states are busy, idle, failed and under repair, or blocked. Each buffer is represented by a resource with the same number of units as the capacity of the buffer. Each of these units is either busy or idle. The process of a part through the system is as follows.

1. Part arrives
    a. Part attributes (part type and simulation arrival time) are assigned values.
2. Part enters buffer of station 1.
3. Part starts service at station 1 (and exits buffer).
    a. The station 1 resource enters the busy state.
4. Part completes service at station 1 (after the service time delay).
5. Part acquires a space in the buffer of station 2.
    a. When a space is available:
        i. Part enters the buffer of station 2.
        ii. The station 1 resource enters the idle state.
    b. If a space is not currently available:
        i. Part remains at station 1.
        ii. The station 1 resource enters the blocked state.

6. Part starts service at station 2 (and exits buffer).
   a. The station 2 resource enters the busy state.
7. Part completes service at station 2 (after the service time delay).
8. Part acquires a space in the buffer of station 3.
   a. When a space is available:
      i. Part enters the buffer of station 3.
      ii. The station 2 resource enters the idle state.
   b. If a space is not currently available:
      i. Part remains at station 2.
      ii. The station 2 resource enters the blocked state.
9. Part starts service at station 3 (and exits buffer).
   a. The station 3 resource enters the busy state.
10. Part completes service at station 3 (after the service time delay).
    a. The station 3 resource enters the idle state.
11. Part exits the system and the part time in the system is observed.

We assume that a failure of station 3 occurs only at a point in time when a part is completed. The failure and repair process for station 3 is as follows:

1. Station 3 fails.
   a. The resource representing station 3 enters the failed state.
2. The repair of station 3 begins.
3. The repair of station 3 is completed after 2 hours.
   a. The resource representing station 3 enters the idle state.

The simulation model straightforwardly incorporates station failure and repair, finite buffer capacity, the blocking of a station preceding a finite capacity buffer, and the various types of interarrival time and service time distributions. The simulation experiment duration corresponds to the system short-term planning period of one week. The analytical queueing models presented in the preceding chapter exclude station failure, finite buffer capacity, and station blocking as well as estimating long-term, steady-state values. As an approximation, however, we use an open network of queues model that assumes exponential interarrival and service time distributions as discussed in Section 11.3.1 to obtain the system performance measure values shown in Table 12.4. The use of the advanced open network of queues model presented in Section 11.3.2 is left as an exercise for the reader.

These same quantities, as well as the percentage of time each station is blocked, are estimated by the simulation model. The model was simulated for five replicates,

Table 12.4   Three–Station Line: Analytic Queueing Model Results

	Station 1	Station 2	Station 3
Arrival rate	1/6	1/6	1/6
Service rate	1/5.4	1/5.88	1/5.5
Utilization	90%	98%	92%
Mean parts at station	9.0	49	11.5
Mean time at station (min)	54.0	294	69.0
Mean parts queued	8.1	48.02	10.6
Mean time in queue (min)	48.6	288.12	63.6

each consisting of one 40-hour week. The simulation results shown in Table 12.5 are averages across the five replicates.

There are differences in station utilizations between the two models. The queueing model indicated that the utilization of station 2 would be 98 percent. In the simulation, this station was busy (94%) or blocked (6%) all of the time. The simulation estimated that the percentage busy time would be 86 percent for station 1 and 85 percent for station 3. In the simulation, station 3 failed once each week and was repaired for a total of two hours out of 40, or 5 percent. Thus, station 3 was busy 85 percent / 0.95 = 89.5 percent of the time it was available. Station 1 and 3 utilization values are less than those computed by the queueing model, most likely because of the effects of blocking and station failure. The effects of finite buffer capacity, blocking, and station failure are seen in mean number of parts and mean time at a station and in the queue values. Values for stations 2 and 3 are lower for the simulation than for the analytic queueing model. Station 1 values are higher. Thus, blocking tended to keep parts at preceding stations, as would be expected.

In addition, simulation gives insight into the transient behavior of each station as a result of a failure at station 3 that occurred 6 hours and 40 minutes into day 2 in the first replicate. The graphs in Figure 12.10 show the number in the queue of each station throughout the week. In about one day, the queue length at station 3 returns to three parts, indicating that the station has recovered from its failure. However, the effect of the failure can be seen at stations 1 and 2 for the rest of the week! For a significant part of that time, the buffer of station 2 is at capacity, blocking station 1. As a result, the number of parts in the buffer of station 1 drops briefly to as few as 10 but then continually grows. Thus, these graphs show the transient behavior of the station that failed as well as the other stations affected by the failure due to blocking. In addition, they indicate that the system is most likely unable to meet weekly production requirements without working overtime.

The distribution of the time a part spends in the system includes all observations from the five replicates and is given in Figure 12.11. Note that the distribution is multimodal due to the failure of station 3 and the resulting blocking. Almost one half of the parts spend from 6 to 8 hours in the system.

As a result of the previous analysis, the effect of doubling the size of each buffer becomes of interest. Figure 12.12 shows the same graphs as Figure 12.10 for this case. Station 1 appears to be unaffected by the failure of station 3, as no increase in the number of parts queued occurs after the failure time. In about two days, the queue of station 3 reaches a value of 0, indicating that the station has recovered from its failure. The number of parts queued at station 2 appears to stay significantly higher after the failure of station 3 than prior to it for about one day. This is expected given

Table 12.5   Three-Station Line: Simulation Model Results

	Station 1	Station 2	Station 3
Arrival rate	1/6	1/6	1/6
Service rate	1/5.4	1/5.88	1/5.5
Utilization	86%	94%	85%
Percentage blocked	13%	6%	N/A
Mean parts at station	46.4	9.9	4.6
Mean time at station (min)	277.9	61.3	27.5
Mean parts queued	45.5	9.0	3.7
Mean time in queue (min)	272.5	55.3	22.0

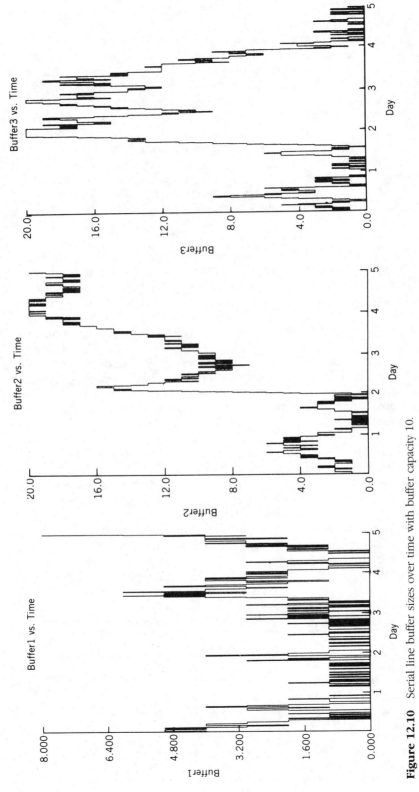

Figure 12.10 Serial line buffer sizes over time with buffer capacity 10.

424

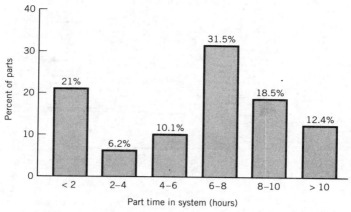

Figure 12.11   Serial line part time in system distribution with buffer capacity 10.

the extra capacity to accept parts from station 1 and the high utilization of station 2. Thus, these graphs indicate that doubling the buffer space helps the system meet weekly production requirements with less or no overtime.

In addition, other quantities show the positive effect of doubling the buffer size. The percentage blocking time for station 1 was reduced from 13 percent to 5 percent and for station 2 from 6 percent to 2 percent. The average throughput for the week increased from 380 to 399. Given that the average processing time at station 2 is 5.88 minutes, the maximum weekly throughput is (40 hours × 60 minutes/hour) / (5.88 minutes/part) = 408 parts. The distribution of part time in the system in Figure 12.13 shows that parts spend less time in the system. No part is in the system for more than 8 hours. More than double the percentage, 73.6 percent, of parts are in the system for 2 to 6 hours.

12.7.2 Flexible Manufacturing

A certain section of a job shop processes a particular type of part that requires three operations. Because of a doubling in demand, an additional, flexible machine capable of performing each of the three operations on a job is to be added to the shop. The other machines will be retained but are fixed and, hence, can perform only one operation, as each does currently. We must assess the utility of the flexible machine by evaluating the policy: perform each operation on a fixed machine if it is available and on the flexible machine if the fixed machine is busy. This policy is implemented by the following rules:

1. When a part is ready for an operation and the corresponding fixed machine is idle, use the fixed machine.
2. When a part is ready for an operation, the corresponding fixed machine is busy, and the flexible machine is idle, use the flexible machine.
3. When a part is ready for an operation, the corresponding fixed machine is busy, and the flexible machine is busy, wait. Use whichever machine (fixed or flexible) becomes idle first.
4. When the flexible machine becomes idle, assign it to operation 3. If no parts are waiting for operation 3, assign it to operation 2. If no parts are waiting for operation 2, assign it to operation 1. This strategy gives priority to operations near the end of the routing instead of operations near the beginning of the routing.

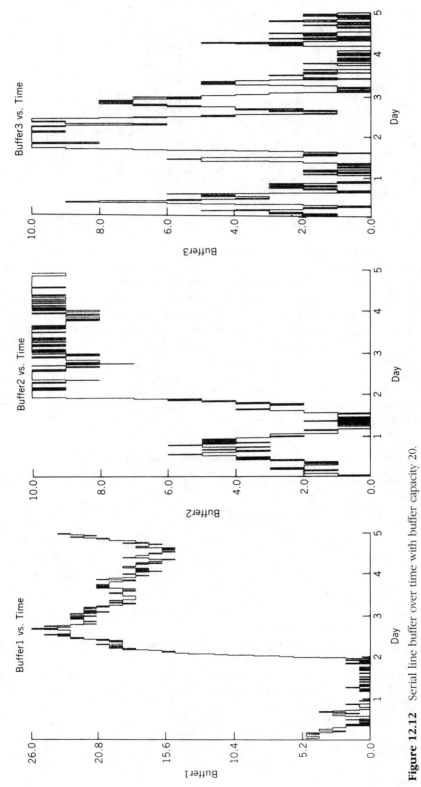

Figure 12.12 Serial line buffer over time with buffer capacity 20.

426

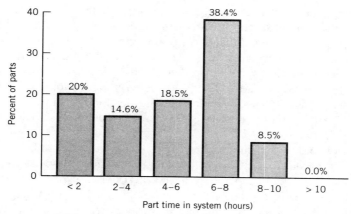

Figure 12.13   Serial line part time in system distribution with buffer capacity 20.

These rules illustrate the need to make complex decisions within models based on the current state of the model (system). Simulation provides the capability to construct and analyze such models.

A process model of this situation is constructed. Transactions represent parts whose attributes are simulation time arrived at the system and the number of operations completed. Resources represent each fixed machine and are busy or idle. A resource representing the flexible machine is busy performing any of three operations or is idle. Part interarrival time is exponentially distributed with a mean of 7.5 minutes. Operations times are a constant 8, 9, and 9.5, respectively. Performance metrics are the percentage of time spent on each of the three operations by the flexible machine and the time to complete a part. The process of a part through the system is as follows.

1. Part arrives.
    a. Part attribute values (simulation time of arrival and number of operations completed) are assigned.
2. Part is assigned to machine for the next required operation using rules 1 to 4 above.
    a. State of machine is set to busy.
3. Part completes operation on machine.
    a. State of machine is set to idle.
4. Part attribute number of operations completed is incremented by 1. If more operations are required return to step 2.
5. Part time in system is observed.

The model was simulated for 100 replicates. Since the only random variable in the model is the time between arrivals, each replicate represented a different part arrival stream. Each replicate consisted of a 40-hour week. Performance measure estimates were obtained by averaging across the replicates. The average time in the system was 55.5 minutes. Replicate averages were in the range [33.6–180]. This indicates that a wide range of streams of parts were considered. A 95 percent confidence interval, [51.4–59.6] minutes, was computed from the 100 independent observations of the average time in the system, one per replicate. The flexible machine spent its time as follows: idle, 7.3 percent; operation 1, 18.4 percent; operation 2, 30.2 percent; and operation 3, 44.2 percent.

Management asked that an alternative to rule 4 be evaluated under which the flexible machine chose its next part to process from the operation buffer with the most parts waiting. Again 100 replicates, corresponding to the same part arrival streams used in simulating the original rule 4, were employed. Simulation results showed that the average time in the system increased to 60.7 minutes with replicate averages in the range [36.7–198] minutes. The 95 percent confidence interval of the average time in the system was [56.3–64.7] minutes. The flexible machine spent its time as follows: idle, 7.1 percent; operation 1, 25.8 percent; operation 2, 32.0 percent; and operation 3, 35.2 percent.

A comparison of the two rule 4 alternatives based on the simulation results was made. The second alternative caused the flexible machine to spend more time on operation 1 and less time on operation 3. Idle and operation 2 times were about the same. In addition, a standard statistical test of the difference of the average time in the system values of each of the two alternatives was conducted. The null hypothesis— average time in the system is the same for both alternatives—could not be rejected at the 0.05 level of significance. However, it was noted that for 97 of the 100 replicates (part arrival streams) the original rule 4 yielded a smaller average time in the system than the new proposal. Thus, the above null hypothesis was retested using a paired-t test on the difference in average time in the system between each pair of replicates. The null hypothesis was rejected at the 0.05 level of significance and the original rule 4 was retained.

12.8
SUMMARY

Simulation is a modeling and experimentation, technique that supports models that closely conform to the structure and dynamics of manufacturing systems. Such models do not have the analytic solutions. Experimental procedures are used to analyze the models instead. Simulation languages provide capabilities for constructing simulation models and analyzing them on a computer. Simulation systems augment these languages by providing information entry, ananysis, and presentation capabilities.

The results of simulation experiments include the time series of values and the statistical summaries of these values. Conclusions about system behavior are drawn by examining these values directly and by further statistical analysis of the values. The structure and dynamics of a modeled system are presented by using animation techniques. Performance measure values or their statistical summaries may be examined using graphs.

Example models illustrate the unique contributions of simulation to manufacturing system analysis. Models straightforwardly incorporate failure and repair, finite buffer capacity, blocking, and a variety of interarrival time and service time distributions. Simulation gives insight into transient system behavior, such as queue length dynamics due to failures and subsequent repairs. Operational policies that depend on the current system state are easily modeled. Performance metrics unique to a particular system can be estimated.

REFERENCES

Balci, Osman [1990], "Guidelines for Successful Simulation Studies," *Proceedings of the 1990 Winter Simulation Conference,* Osman Balci, Randall P. Sadowski, and Richard E. Nance, eds., Institute of Electrical and Electronics Engineers, Piscataway, NJ, 25–32.

Banks, Jerry, and John S. Carson [1984], *Discrete-Event System Simulation,* Prentice-Hall, Englewood Cliffs, NJ.

Banks, Jerry, and John S. Carson [1988], "Applying the Simulation Process," *Proceedings of the 1988 Winter Simulation Conference,* Michael A. Abrams, Peter L. Haigh, and John C. Comfort, eds., Institute of Electrical and Electronics Engineers, Piscataway, NJ, 52–55.

Banks, Jerry, John S. Carson, and John Sy [1989], *Getting Started with GPSS/H,* Wolverine Software Corp, Annandale, VA.

Cobbin, Philip [1988], "The SIMPLE-1 Simulation Environment," *Proceedings of the 1988 Winter Simulation Conference,* Michael A. Abrams, Peter L. Haigh, and John C. Comfort, eds., Institute of Electrical and Electronics Engineers, Piscataway, NJ, 141–145.

Cox, Springer W. [1988], "GPSS/PC Graphics and Animation," *Proceedings of the 1988 Winter Simulation Conference,* Michael A. Abrams, Peter L. Haigh, and John C. Comfort, eds., Institute of Electrical and Electronics Engineers, Piscataway, NJ, 129–135.

Earle, Nancy J., Daniel T. Brunner, and James O. Henriksen [1990], "Proof: The General Purpose Animator," *Proceedings of the 1990 Winter Simulation Conference,* Osman Balci, Randall P. Sadowski, Richard E. Nance, eds., Institute of Electrical and Electronics Engineers, Piscataway, NJ, 106–108.

Goble, John [1990], "Introduction to SIMFACTORY II.5," *Proceedings of the 1990 Winter Simulation Conference,* Osman Balci, Randall P. Sadowski, Richard E. Nance, eds., Institute of Electrical and Electronics Engineers, Piscataway, NJ, 136–139.

Grant, Mary E., and Darrell W. Starks [1988], "A Tutorial on TESS: The Extended Simulation Support System," *Proceedings of the 1988 Winter Simulation Conference,* Michael A. Abrams, Peter L. Haigh, and John C. Comfort, eds., Institute of Electrical and Electronics Engineers, Piscataway, NJ, 136–140.

Harrell, Charles R., and Ken Tumay [1990], "ProModelPC Tutorial," *Proceedings of the 1990 Winter Simulation Conference,* Osman Balci, Randall P. Sadowski, Richard E. Nance, eds., Institute of Electrical and Electronics Engineers, Piscataway, NJ, 128–131.

Henriksen, J. O. [1983], "State-of-the-Art GPSS," *Proceedings of the 1983 Summer Simulation Conference,* The Society for Computer Simulation, San Diego, CA, 918–933.

Johnson, M. Eric, and Jacob P. Poorte [1988], "A Hierarchical Approach to Computer Animation in Simulation Modeling," *Simulation,* 50(1), 30–36.

Kelton, W. David [1989], "Random Initialization Methods in Simulation," *IIE Transactions,* 21(4), December, 355–367.

Kleijnen, J. P. C. [1982], *Statistical Tools for Simulation Practitioners,* Marcel Dekker, Inc., New York.

Kleijnen, Jack P. C. and Charles Standridge [1988], "Experimental Design and Regression Analysis in Simulation: An FMS Case Study," *European Journal of Operations Research,* 33, 257–261.

Law, Averill M. [1988], "Simulation of Manufacturing Systems," *Proceedings of the 1988 Winter Simulation Conference,* Michael A. Abrams, Peter L. Haigh, and John C. Comfort, eds., Institute of Electrical and Electronics Engineers, Piscataway, NJ, 40–51.

Law, Averill M., and W. David Kelton [1991], *Simulation Modeling and Analysis,* McGraw-Hill, New York, NY.

Law, Averill M., and Stephen G. Vincent [1988], "A Tutorial on UNIFIT: an Interactive Computer Package for Fitting Probability Distributions to Observed Data," *Proceedings of the 1988 Winter Simulation Conference,* Michael A. Abrams, Peter L. Haigh, and John C. Comfort, eds., Institute of Electrical and Electronics Engineers, Piscataway, NJ, 188–193.

Lilegdon, William R., and Julie N. Erlich [1990], "Introduction to SLAMSYSTEM," *Proceedings of the 1990 Winter Simulation Conference,* Osman Balci, Randall P. Sadowski, Richard E. Nance, eds., Institute of Electrical and Electronics Engineers, Piscataway, NJ, 77–79.

Miles, Trevor, Randall P. Sadowski, and Barbara M. Werner [1988], "Animation with CINEMA," *Proceedings of the 1988 Winter Simulation Conference,* Michael A. Abrams, Peter L. Haigh, and John C. Comfort, eds., Institute of Electrical and Electronics Engineers, Piscataway, NJ, 180–187.

Murgiano, Charles [1990], "A Tutorial on Witness," *Proceedings of the 1990 Winter Simulation Conference,* Osman Balci, Randall P. Sadowski, Richard E. Nance, eds., Institute of Electrical and Electronics Engineers, Piscataway, NJ, 77–79.

Nadas, A. [1969], "An Extension of a Theorem of Chow and Robbins on Sequential Confidence Intervals for the Mean," *Ann. Math. Stat.,* 40, 667–671.

Norman, Van B. [1990], "AutoMod II," *Proceedings of the 1990 Winter Simulation Conference,* Osman Balci, Randall P. Sadowski, Richard E. Nance, eds., Institute of Electrical and Electronics Engineers, Piscataway, NJ, 94–98.

Pegden, C. Dennis [1982], *Introduction to SIMAN,* Systems Modeling Corporation, Sewickley, PA.

Pritsker, A. Alan B. [1982], "Applications of SLAM," *IIE Transactions,* Institute of Industrial Engineers, Atlanta, GA.

Pritsker, A. Alan B. [1986], *Introduction to Simulation and SLAM II,* 3rd ed. Halsted Press, New York, NY and Systems Publishing Corporation, W. Lafayette, IN.

Pritsker, A. Alan B., C. Elliot Sigal, and R. D. Jack Hammesfahr [1988], *SLAM II Network Models for Decision Support,* Prentice-Hall, Englewood Cliffs, NJ.

Roberts, Stephen D., and Mary Ann Flanigan [1990], "Simulation Modeling and Analysis with IN-SIGHT: A Tutorial," *Proceedings of the 1990 Winter Simulation Conference,* Osman Balci, Randall P. Sadowski, Richard E. Nance, eds., Institute of Electrical and Electronics Engineers, Piscataway, NJ, 80–88.

Russell, Edward C. [1983], *Building Simulation Models with SIMSCRIPT II.5,* CACI International, Los Angeles, CA.

Sargent, Robert G. [1988], "A Tutorial on Validation and Verification of Simulation Models," *Proceedings of the 1988 Winter Simulation Conference,* Michael A. Abrams, Peter L. Haigh, and John C. Comfort, eds., Institute of Electrical and Electronics Engineers, Piscataway, NJ, 33–39.

Schriber, Thomas J. [1974], *Simulation Using GPSS,* John Wiley & Sons, New York.

Schruben, Lee [1983], "Simulation Modeling with Event Graphs," *Communications of the A.C.M.,* 26(11).

Shannon, Robert E. [1975], *System Simulation: the Art and Science,* Prentice-Hall, Englewood Cliffs, NJ.

Standridge, Charles R., John R. Hoffman, and Steven A. Walker [1984], "Presenting Simulation Results with TESS Graphics," *Proceedings of the 1984 Winter Simulation Conference,* Sallie Sheppard, Udo Pooch, and C. Dennis Pegden, eds., Institute of Electrical and Electronics Engineers, Piscataway, NJ, 305–312.

Standridge, Charles R., [1985], "Performing Simulation Projects with The Extended Simulation System (TESS)," *Simulation,* 45(6) 283–291.

Standridge, Charles R., [1986], "Animating Simulations Using TESS," *Computers in Industrial Engineering,* 10(1) 121–134.

Standridge, Charles R., and A. Alan B. Pritsker [1987], *TESS: The Extended Simulation Support System,* Halsted Press, New York.

Starr, N. [1966], "The Performance of a Sequential Procedure of the Fixed-Width Interval Estimation of the Mean," *Ann. Math. Stat.,* 37, 36–50.

Taha, Hamdy A., R. B. Taylor, and Nazar A. Younis [1990], "Simulation and Animation with SIMNET II and ISES," *Proceedings of the 1990 Winter Simulation Conference,* Osman Balci, Randall P. Sadowski, Richard E. Nance, eds., Institute of Electrical and Electronics Engineers, Piscataway, NJ, 99–105.

Wilson, James R., and A. Alan B. Pritsker [1978], "A Procedure for Evaluating Startup Policies in Simulation Experiments," *Simulation,* 31, 79–89.

PROBLEMS

12.1. Give several compelling reasons to use a simulation model.

12.2. List performance metrics that can be estimated using simulation that cannot be estimated using analytic models.

12.3. Tell how the event modeling approach, the process modeling approach, and the manufacturing-specific modeling approach differ from each other.

12.4. List system elements or conditions that simulation models directly incorporate but that require assumptions or approximations in analytic models.

12.5. Develop an event graph model of the three-station line of Section 12.7.1.

12.6. Develop a model of the three-station line of Section 12.7.1 in your favorite simulation language. Estimate the time required to produce 400 parts.

12.7. Embellish and modify Problem 12.6 as follows:

a. There are no station failures but the buffer capacity is 5.

b. The first station is never starved.

c. Both stations 2 and 3 fail on the average once per week and require 2 hours for repair. The time of the failure of each station is uniformly distributed throughout the week.

12.8. Model the three-station line of Section 12.7.1 using an open network of queues model of the type discussed in Section 11.3.2. Estimate the same quantities shown in Table 12.4 and compare these

Item	Value
Unit interarrival distribution	Exponential, mean 3
First inspection time distribution	Exponential, mean 2
Rework time distribution	Exponential, mean 10
Second inspection time distribution	50%: Constant, 1
	50%: Uniform, range (1,3)
Percentage requiring rework	20%
Percentage failing the second inspection	10%

quantities to the simulation results given in Table 12.5.

12.9. Consider again the single-worker station system. Suppose this system moves parts from an entry port to the station and from the station to an exit port using a single material handling device that is not always available, that is, parts may need to queue for the material handling device. Embellish the process model of Section 12.3 to include the material handling device.

12.10. A new inspection station/rework station combination is being installed in an existing manufacturing system. After successfully completing the inspection, a unit is passed to the next step in the production process. If a unit fails the inspection, it is sent to the rework station. After rework, the unit is sent back to the inspection station. If a unit fails the reinspection, it is discarded. Relevant data have been gathered and are summarized in the table at the top of the page. Times are in minutes.

The mixed distribution for the second inspection time reflects the knowledge that one error with a known time for reinspection accounts for 50 percent of the reworked units.

A meeting between decision makers, manufacturing system experts, and modelers identified the objectives of the modeling project as follows:

1. Determine the size of the buffer space at the inspection and rework stations.
2. Assess the adequacy of one worker at the inspection station and one at the rework station.

Based on the project objectives, formulate a base case for analysis. Performance metrics, in addition to those for buffer spaces, are the time between the departure of units from the system and the busy/idle time of the workers at each of the two stations.

Hint: One way to estimate the amount of buffer space needed is as follows. In the first simulation run, let the buffer sizes at the inspection and rework stations be infinite. Thus, simulation results will estimate the distribution of the number of buffer spaces required, as well as the maximum number. From this information, the amount of

buffer space needed by the system can be chosen. Test this maximum buffer size using a second simulation with finite buffer space of size equal to the selected amount. Adjust the maximum buffer size and retest as needed.

a. Write the process model for a unit moving through the inspection rework station.

b. Specify appropriate graphs for displaying simulation results.

c. Design an animation to show the operation of the system.

d. Simulate the model for one 40-hour work week using any available simulation language. Use the simulation results to determine the required buffer sizes. State whether one worker at each station is sufficient.

e. Re-simulate the model using the finite buffer sizes you selected in part d. Are these buffer sizes acceptable? Determine if a smaller buffer size is reasonable.

12.11. Develop an event graph model of the FMS system discussed in Section 12.7.2.

12.12. Suppose a job shop processes an equal number of each of four types of jobs ($A, B, C,$ and D) using three workstations (1, 2, 3). Processing times at workstations do not vary by job type and are 8.0, 9.0, and 9.5 minutes, respectively. Job interarrival time is exponentially distributed with a mean of 7.5. The routes for each of the four job types are as follows:

Job Type	Route (List of Workstations in Order Visited)
A	1, 2, 3
B	3, 2, 1
C	3, 1, 2
D	1, 3, 2

a. Develop a process model for the job shop. *Hints:* Use transaction attributes to keep track of the type of job and the number of stations visited. Use the single-worker station model for each workstation. Develop logic for determining the next station to which to route a job.

Assume that the route information given above is accessible from within the model.

b. What performance metrics would you use for this model and how would you display them?

c. Embellish the process model to include a 2-minute setup time at each station if the current job has a different type than the previous job.

12.13. A simulation model of a warehouse involves keeping track of how much of each product is stored in the warehouse and the time taken to store and retrieve products using a storage/retrieval machine. Suppose a simple, one-aisle warehouse is responsible for storing and retrieving two products using one storage and retrieval machine. The design of the warehouse is such that the time to retrieve a unit load of product, including pickup and dropoff, is uniformly distributed between 9 and 18 minutes. The average demand is 4 unit loads per hour for the first product and 2 unit loads per hour for the second product, Poisson distributed. The warehouse processes demands 5 days per week, 8 hours per day. Once each week, outside of the time the warehouse is processing demands, the weekly expected number of unit loads of each product is delivered. Develop a process model of this situation.

12.14. A just-in-time or pull production strategy seeks to minimize in-process inventory by performing operations on parts and jobs only as required to meet customer demand. The final operation on a product is performed only at the time of a customer demand. The station preceding the final station performs its operation only when the final station needs another part to be prepared to respond to the next customer. In general, every station waits for its following station to need another part before producing that part. The mechanism by which a following station tells its preceding station to produce another part or by which the last station is informed of a customer demand is called a kanban or card.

Develop a model of the three-station system of Section 12.7.1 converted to pull strategy. Assume that there is sufficient raw material available at the first station so that it is never starved.

CHAPTER 13

CASE STUDIES

> *Between the idea*
> *And the reality*
> *Between the motion*
> *And the act*
> *Falls the Shadow.*
> —T. S. Eliot, *The Hollow Men*

13.1
PROBLEM DEFINITION: WXYZ COMPANY PROCESSING LINE

The process of designing a new manufacturing system may require multiple models to design various system components as well as models to assess how the components perform together as a system. This case study uses the models discussed in Chapters 2 and 3 for system component design as well as simulation, as it was presented in Chapter 12 to assess overall system performance. Also one should keep in mind that each model may go through several iterations of formulation, synthesis, analysis, and verification before a suitable model is obtained.

The WXYZ Company has just received an important contract to manufacture four types of parts. The contract will be reviewed after 13 weeks to assure compliance with delivery and quality requirements. If performance is deemed satisfactory, a long-term contract will be negotiated.

Each Friday the customer will inform WXYZ of demand for each product for the following week. The customer utilizes a just-in-time production system and expects 10 percent of weekly demand of each product to be delivered twice a day during the five-day workweek. The only assurance offered ahead of time is that the weekly order quantity for each product will be within the limits shown in Table 13.1. Part demands are thought to be independent and to be uniformly distributed over the permissible range.

The four part types are similar, each requiring the same sequence of 10 operations (OP1–OP10). Standard times in seconds are shown in Table 13.2. The IP column indicates technological constraints (immediate predecessor operations). Weighted average operation times are also provided in the table. Weights are based on average demand. Many of the operations are machine paced; therefore, significant reductions in operation time from learning are not expected during the 13-week period. The equipment is small, but several operations require sophisticated apparatus. The customer has agreed to supply WXYZ with one machine of each type that is used to manufacture and to test the four part types.

Table 13.1   Weekly Order Size Limits

Part	Minimum	Maximum
A	360	440
B	540	660
C	720	880
D	190	210
Total	1800	2200

A 30 × 20-foot area of the WXYZ plant has been set aside for manufacture of the part types. This area has access to a main aisle for material movement. The parts weigh about 5 kg and measure roughly 40 cm × 25 cm × 20 cm.

13.2
WXYZ DESIGN APPROACH

Based on delivery requirements and equipment availability, the company has decided to construct a mixed-product assembly line. With only one unit of each machine available and noting the need to train workers quickly, a decision has been made that each workstation will perform the same set of operations on all four part types. The design of the line requires a series of decision steps as follows:

1. Compute the cycle time accounting for workstation availability.
2. Assign each operation to a workstation.
3. Sequence part production.
4. Determine the number of repair workers needed.
5. Determine buffer sizes.
6. Determine a workplace layout.
7. Assess how well the proposed design will meet customer requirements.

Each of these steps is discussed in turn.

Table 13.2   Operation Times (Seconds) by Part Type

Operation	Part Type				Wt. Average	IPs
	A	B	C	D		
OP1	16	36	17	12	22	—
OP2	14	17	13	19	15	OP1
OP3	41	49	42	53	45	OP2
OP4	5	6	6	8	6	—
OP5	13	15	12	11	13	OP3
OP6	21	31	23	23	25	OP4
OP7	13	16	11	12	13	OP5
OP8	23	40	30	34	32	OP6
OP9	15	21	17	19	18	OP5, OP8
OP10	35	43	44	45	42	OP7, OP9
Total	196	274	215	236	231	

13.2.1 Computing Cycle Time

Computing the cycle time requires knowledge of the weekly number of parts to be produced, in this case a random variable between 1800 and 2200. Management wishes the system to have the capacity to meet weekly demand most of the time but believes design for a production target of 2200 would result in too much idle time. Workers can be requested to work overtime in limited amounts if necessary, although this is expensive. Average demand is 2000 units per week. Since part demands are independent, individual variances can be summed. Noting the relation $\text{var}(X) = (b - a)^2/12$ for X a random variable uniformly distributed between a and b, we find total weekly demand Y to have variance

$$\text{var}(Y) = \frac{(440 - 360)^2 + (660 - 540)^2 + (880 - 720)^2 + (210 - 190)^2}{12} = 3900$$

The standard deviation is $\hat{\sigma}_Y = 62$. From the Central Limit Theorem, we know that the sum of four independent random variables tends toward the normal distribution. If we let 1800 to 2200 represent the commonly used 6σ spread of the normal, we estimate $\tilde{\sigma}_Y = 67$. We will use the compromise choice $\sigma_Y = 65$. Realizing that Y actually has heavier tails than the corresponding normal distribution, a design guideline of regular time capacity equaling mean demand plus two standard deviations was deemed acceptable. Thus target capacity is $2000 + 2(65) = 2130$ parts/week.

Company policy is for 37.5 hours of scheduled work per 40-hour week (two 10-minute breaks and a 10-minute department meeting are planned each day; lunch is unpaid). Available seconds are then

$$60 \; \frac{\text{seconds}}{\text{minute}} \times 60 \; \frac{\text{minutes}}{\text{hour}} \times 37.5 \; \frac{\text{hours}}{\text{week}} = 135,000 \; \frac{\text{seconds}}{\text{week}}$$

Cycle time is then

$$c = \frac{135,000}{2130} = 63.4 \text{ seconds}$$

The average part contains 231 seconds of processing time. The smallest multiple of 63.4 exceeding this value is 253.5 seconds corresponding to four workstations. It was thought that the corresponding 22.5 seconds of idle time per part was too large. Noting also that this would not allow for the expected 5 to 10 percent of downtime expected for workstations, it was decided to try a cycle time of 60 seconds. This would satisfy management efficiency objectives if a four-workstation solution could be obtained.

13.2.2 Assigning Operations to Workstations

The next step is to balance the manufacturing process. The time used for each operation is the demand-weighted average shown in Table 13.2. First, attempt to use the Ranked Positional Weight technique of Section 2.3.2. The precedence diagram is shown in Figure 13.1. The construction procedure is summarized in Table 13.3. Positional weight is the sum of operations times for the operation and its successors. After ranking, operations are iteratively assigned to the first available station. The last two columns of the table indicate the station to which the operation is assigned and the idle time in that station remaining after assignment. Five stations are required. The question arises as to the quality of this balance. Above it was noted that the lower bound on number of workstations is four. Implicit enumeration is used to determine

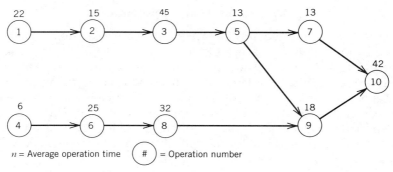

Figure 13.1 A precedence diagram.

if a feasible four-station solution exists. As we start to explore the tree of possible sequences, we fairly readily come on the line balance of Table 13.4.

13.2.3 Part Sequencing

Using average demand, we need 400 part As, 600 part Bs, 800 part Cs, and 200 part Ds per week. Dividing by the greatest common factor 200, we find a production ratio of 2:3:4:1. Thus repetitive lots of 10 units will be planned. As weekly demands are received, these ratios may vary, but it is helpful to evaluate whether a long-run path exists with this ratio. The heuristic of Section 2.5 sequences parts so as to minimize the maximum deviation from desired load on a bottleneck workstation and to spread out all parts over the 10-unit cycle. From Table 13.4, both stations 2 and 4 are fully utilized and thus qualify as bottlenecks. Processing times of each part type on these workstations are shown in Table 13.5.

The sequence is constructed one unit at a time. To encourage a smooth production rate for each part, we will not allow any part type to get more than one unit ahead or behind its hypothetical desired production level. For instance, since 2 of 10 parts should be part type A, if none of the first five parts produced is type A, we will be 1.0 behind ($\frac{2}{10} \cdot 5$) the desired level. Subject to satisfying this constraint, we select the available part that minimizes the largest deviation in cumulative time for either station 2 or 4 from the desired assignment of 60 seconds per cycle. The sequence construction procedure is reviewed in Table 13.6. The first value in the "If Select"

	Table 13.3	Ranked Positional Weight Construction Summary			
Operation	Time	PW	Rank	Station	Idle Time
OP1	22	168	1	1	38
OP2	15	146	2	1	23
OP3	45	131	3	2	15
OP4	6	123	4	1	17
OP5	13	86	7	2	2
OP6	25	117	5	3	35
OP7	13	55	9	4	29
OP8	32	92	6	3	3
OP9	18	60	8	4	42
OP10	42	42	10	5	18

Table 13.4   Improved Line Balance

Workstation	Operations	Assigned Time
1	OP1, OP4, OP6	53
2	OP2, OP3	60
3	OP5, OP7, OP8	58
4	OP9, OP10	60

section under columns A, B, C, and D indicates the time deviation for the bottleneck (station 2 or 4) from the desired 60-second cycle time if the corresponding part is selected. The larger of the deviations for station 2 or 4 is shown. The second value shows the cumulative production level deficit with respect to a constant production rate, if the corresponding part type is not selected. Part C is selected first since no part is a full one unit behind yet and selection of C results in station 2 being 5 seconds underworked and station 4 being 1 second overworked. Parts A, B, and D place more than a 5-second deviation from the desired 60-second workload in station 2 or 4. A would place station 4 ten seconds behind the desired 60 seconds of cumulative work. B would place station 2 six seconds behind, and D places station 2 twelve seconds behind.

13.2.4 Number of Repair Workers

Several operations utilize equipment that can fail. A best guess is that each workstation will cause a major failure about every 1000 cycles. Repair times vary but average 1 to 2 hours. It was believed by management that line workers could take care of the minor problems. If simple solutions failed, a maintenance worker would be called. At 2000 cycles per week, each workstation would require roughly 3 hours of attention per week or less than 15 hours total for all machines. It was not thought that a dedicated repairman was economically justifiable. The existing maintenance department would be expected to handle this load. The maintenance department was supplied with necessary repair documentation. One experienced worker was assigned the task of being the expert for the line. This worker's top priority would be to service the line. A local repair shop could be called for major repairs.

13.2.5 Buffer Allocation

Use of the 93.75 percent = 37.5/40.0 individual workstation availability estimate in the determination of cycle times implied the existence of large buffers that would make

Table 13.5   Bottleneck Loads

	Workstation	
Part	2	4
A	55	50
B	66	64
C	55	61
D	72	64

Table 13.6   Construction of Standard Part Sequence

Step	If Selected A	B	C	D	Part	Selected WS2	WS4
1	10, 0.2	6, 0.3	5, 0.4	12, 0.1	C	55(−5)	61(1)
2	10, 0.4	5, 0.6	10, −0.2	7, 0.2	B	121(1)	125(5)
3	5, 0.6	9, −0.4	6, 0.2	13, 0.3	A	176(−4)	175(−5)
4	15, −0.2	2, 0.2	9, 0.6	8, 0.4	B	242(2)	239(−1)
5	11, 0.0	8, −0.5	3, 1.0	14, 0.5	C	297(−3)	300(0)
6	10, 0.2	4, −0.2	8, 0.4	9, 0.6	B	363(3)	364(4)
7	6, 0.4	—	5, 0.8	15, 0.7	C	418(−2)	425(5)
8	7, 0.6	—	7, 0.2	10, 0.8	A	473(−7)	475(−5)
9	—	—	12, 0.6	5, 0.9	D	545(5)	539(−1)
10	—	—	0, 1.0	—	C	600(0)	600(0)

workstations functionally independent. Buffer sizes of 30 were proposed between workstations to allow production to continue during downtimes. This design will be subjected to further analysis in a later section.

13.2.6 Process Line Layout

A tentative line layout was developed as shown in Figure 13.2. It was believed that this design provided minimal I/O movement of components to workstations, facilitated interstation movement of parts, provided ample work space for each station, and allowed access to workstation for component delivery and maintenance.

13.2.7 Assessing System Performance

As a final step, the ability of the system as designed to meet customer demands was assessed. Specific issues included system performance with respect to various levels

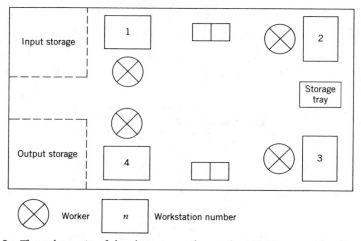

Figure 13.2   The schematic of the department layout for WXYZ processing line.

of demand, the adequacy of buffer sizes, and the ability to cope with machine failures. In practice, alternative operation assignments and sequencing might be proposed as well.

A simulation of the serial line was developed to address the foregoing issues. A process model was constructed. Transactions represented parts and had the attribute part type. Each station was represented by a resource that was either busy, idle, blocked, or failed. Other resources represented the buffer spaces preceding stations 2, 3, and 4. There is one unit of each resource for each space in the buffer. Resource units are either busy or idle. On completion of a part at the first station, processing of the next part is started immediately. Otherwise, the steps in the process are the same as those of the serial line example in Section 12.7.1. Performance measures estimated by the model were makespan for weekly production, number of semidaily deliveries that were on time, maximum buffer levels, and percentage of busy and blocked time for each workstation.

The assessment consisted of two phases. The first phase ignored station failure to determine if the system design was feasible. A full factorial experiment was used. Factors in the experiment were the demand level for each of the four part types. Each factor had two levels: minimum demand and maximum demand. Thus, $2^4 = 16$ simulation runs were made. The simulation runs were deterministic.

Results showed that the system could meet customer demand. Makespan values ranged from 31.7 hours to 38.3 hours, including allowances given the workers for daily meetings and breaks. All semidaily deliveries were on time. Maximum buffer levels at station 3 ranged from 1 to 7 and at station 4 from 1 to 15, both significantly less than the design limit of 30. However, the maximum buffer level at station 2 was always 30 and the average ranged from 23 to 27. This caused the blocking of station 1 from 8.1 percent to 10.6 percent of the time. Station utilization was based on a 40-hour week and included as busy time worker allowances for meetings and breaks. Station 1 utilization ranged from 71 percent to 86 percent and utilization plus blocking ranged from 79 percent to 96 percent. Utilization for station 2 ranged from 82 percent to 97 percent, for station 3 from 78 percent to 94 percent, and for station 4 from 81 percent to 97 percent.

Management decided that preventive maintenance on the machines at least every other evening would significantly extend the average time between major machine failures during the working day. However, an assessment of system response to machine failure was still required. An experiment was constructed as follows. A moderately high demand level was chosen: $A = 360$, $B = 660$, $C = 880$, and $D = 210$. Four experiments were run, each corresponding to the failure of a particular station at the start of the second day. Repair time was a constant 1 hour. Results, including the no-failure case, are shown in Table 13.7. Utilizations are not shown, since they are the same as in the no-failure case.

Failure of station 1 negatively affects system performance. The next three semidaily deliveries are late and makespan increases by 0.6 hours. Since station 1 stops feeding station 2 for an hour, the station 2 buffer is full less of the time, reducing the time station 1 is blocked. Failure of each of the other stations causes the six semidaily deliveries following the failure to be late. Makespan increases by 1.2 hours when station 2 or 3 fails.

The effect of failure of stations 2 and 3 on station 1 is seen in the increase in its block time. Stations 2 and 3 experience some blocking when either station 3 or 4 fails. The reason for this blocking is identified by the graphs in Figure 13.3, which show the number in the buffer of station 3 and station 4. The

**Table 13.7   Simulation Results
for Machine Failure Experiments**

	Failed Station				
	None	1	2	3	4
Blocked percentage					
Station 1	9.1%	7.8%	11.6%	10.3%	9.1%
Station 2	0.0%	0.0%	0.0%	1.2%	0.1%
Station 3	0.0%	0.0%	0.0%	0.9%	3.4%
Semidaily on-time deliveries	10	7	4	4	4
Makespan (hr)	36.9	37.5	38.1	38.1	36.9

buffer of station 3 fills while it is under repair. On resuming processing, station 3 is never again starved for parts. Thus, its output rate exceeds the processing rate of station 4. Eventually, the buffer of station 4 reaches capacity and blocks station 3.

The experiment was repeated with infinite buffer sizes, but no significant changes in performance measure values were noted. Thus, the buffer size of 30 appears to be adequate.

Based on the results of the simulation study, management and industrial engineers are working together to determine policies, such as overtime, for overcoming late semidaily deliveries resulting from station failures.

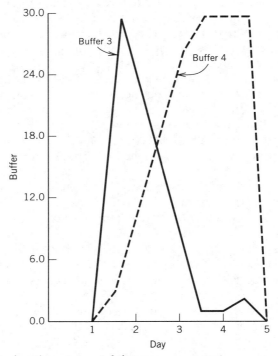

Figure 13.3 Buffer lengths—station 3 fails case.

13.3
PROBLEM DEFINITION: PLANAR COMPANY CELL DESIGN

Analytic queueing network models (Chapter 11) and simulation models (Chapter 12) may both be used in the assessment of alternative system designs. In general, analytical queueing models require less information and may be constructed and analyzed more quickly. Rajan Suri at the University of Wisconsin has coined the term **rapid modeling** to emphasize the advantages and proper use of these techniques in manufacturing system modeling. Rapid modeling is an effective strategy for reducing the number of alternative designs that must be fully evaluated. Final comparisons of a small number of high-potential solutions are then based on more detailed, and therefore more time-consuming, simulation models. The following case study illustrates this approach.

Planar Manufacturing Company is designing a new robotic cell to produce a family of parts. The cell will contain three main pieces of equipment: turret lathe, drill press, and material handling robot, along with supporting hardware and software. Parts will be delivered in trays by conveyor. The family contains six part types. Demands and processing times per unit are given in Table 13.8. Processing times are constant and given in minutes.

All necessary lathe tooling is kept on the turret; thus, tool change time is negligible. However, the robot must change drills for each part type. Setup time is given in the table. Setups reflect a 1-minute setup time per drill bit. Thus, setup time can be subdivided into 1-minute suboperations. Values assume a cell layout where the tools are kept adjacent to the drill press.

Parts may be handled individually or in trays. Trays can hold up to 20 items. The plan is to have an automated cell. Raw materials (part blanks) will be supplied steadily, 16 hours per day, 6 days per week. To avoid the need for storage space, it is desirable to have the cell operate on the same schedule. Arriving part types are random in proportion to demand. The vendor claims that similar cells are running with nearly 100 percent availability provided modest preventive maintenance can be performed daily.

13.4
PLANAR COMPANY DECISION APPROACH

A cell layout that takes the robot's work envelope into account is shown in Figure 13.4. The diagram assumes that parts are processed in trays. Based on the layout and

Table 13.8   Planar Company Part Type Data

Part Type	Weekly Demand	Minutes per Unit		Setup Time Drill Press
		Lathe	Drill Press	
A	800	1.0	0.5	1.0
B	500	2.3	0.9	3.0
C	100	3.5	1.6	2.0
D	240	2.3	1.1	1.0
E	400	2.5	1.6	3.0
F	360	2.3	1.1	1.0

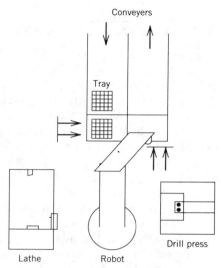

Figure 13.4 A proposed cell layout for planar manufacturing.

machine specifications, it will take the robot 12 seconds to load or unload a part. A part can be unloaded from the lathe and can be transferred directly to the drill press in 20 seconds. Two sequencing strategies are being considered. In the first, individual parts will arrive from the foundry. Parts will be properly oriented for the robot and a vision system or bar code will be used for automatic identification of part type. To reduce setup requirements, 20 parts of the same type will arrive consecutively before a new part type begins arriving. The robots' activity sequence is as follows:

STEP 0. Set up drill press (if necessary).
STEP 1. Load first part onto lathe.
STEP 2. Unload lathe.
STEP 3. Load next part onto lathe.
STEP 4. Load a turned part onto drill press.
STEP 5. Unload drill press.

Repeat steps 2 through 5 until new part type arrives.

The method attempts to keep the lathe busy, since turning times dominate drilling times. A second alternative would be to replace steps 2 and 4 by a transfer of the part directly from the lathe to the drill press.

The last strategy being considered is to use trays of 20 parts each for transport. Parts in one tray would be turned while parts in the second tray, which had already been turned, would be drilled. Parts would arrive in trays, and after the complete tray is turned, the tray would be indexed for access to the drill press. Of course, only one part will be on either machine at a time. The robot would respond to requests from machines with the lathe receiving top priority if both machines needed tending. Once all parts in a tray are completed, the tray leaves by return conveyor.

The basic question concerns the ability of this automated cell to keep up with the feeding process. It will also be of interest to know what queue lengths will develop, since work space is limited. Last, it should be determined if the operating strategies differ in performance.

An initial evaluation of cell feasibility can be performed with an open queueing network approximation. The network is open because trays arrive every 48 minutes (or parts every 2.4 minutes). It is difficult with queueing networks to model the situation that loading, setup, and unloading require simultaneous access to the robot and a machine. Since robot utilization is less than that of the machines, we make the assumption that the robot is always available when needed. This will be checked later by simulation if the design looks promising. Processing times will be inflated to account for load and unload time. Drill setup will be allocated to the individual unit processing times assuming a tray size of 20. Three configurations are tested corresponding to the two sequences for single part handling and the double-tray method. Arrivals and processing times reflect individual parts for the case of individual part handling, and trays for the double-tray policy.

Consider case 1. The lathe is delayed for unloading one part and loading the next. This takes 0.4 minutes, and we add this time to machine processing time. However, on unloading the drill, we must next unload the lathe and then load the lathe before reloading the drill. This is a time span of 0.8 minutes, one half of which the drill is idle. All 0.8 minutes are considered part of drill processing time for the model, since the drill is unavailable for other processing. Knowledge that at least 20 parts of a type will arrive before a new part type appears allows us to inflate each part processing time by only one-twentieth of setup time. Table 13.9 lists the model input for this and the two other cases.

For case 2 (model 1b), both the lathe and drill times are increased by 0.53 minutes for load and unload. For the lathe this includes the 12-second load and the 20-second transfer time. For the drill press this includes the 20-second transfer plus 12-second unload. Activity sequencing will allow the drill to be unloaded while the lathe is processing, owing to the shorter drill times.

Table 13.9   Model Input Parameters for Planar Manufacturing

Model	Part	Arrival Rate	Processing Time (min)	
			Lathe	Drill Press
1a	A	800/5760	1.4	1.35
	B	500/5760	2.7	1.85
	C	100/5760	3.9	2.50
	D	240/5760	2.7	1.95
	E	400/5760	2.9	2.55
	F	360/5760	2.7	1.95
1b	A	800/5760	1.53	1.08
	B	500/5760	2.83	1.58
	C	100/5760	4.03	2.23
	D	240/5760	2.83	1.68
	E	400/5760	3.03	2.28
	F	360/5760	2.83	1.68
2	A	40/5760	28.0	19.0
	B	25/5760	54.0	29.0
	C	5/5760	78.0	42.0
	D	12/5760	54.0	31.0
	E	20/5760	58.0	43.0
	F	18/5760	54.0	31.0

The third model uses trays as processing units. We will assume each of the 20 parts in the tray requires 24 seconds for loading and unloading.

First, analyze model 1a. The part mix will be combined to form a single arrival process. The combination of part types will allow us to approximate service by a random service time distribution. Since the most common part type has the smallest service time and the least common part type has the largest service time, we will settle for the simple exponential service time model. To estimate parameters, the arrival rate is

$$\lambda = \frac{800 + 500 + 100 + 240 + 400 + 360}{5760} = \frac{2400}{5760} = 0.416 \text{ per minute}$$

The weighted average service time for the lathe is

$$\mu_L^{-1} = 1.4 \cdot \frac{800}{2400} + 2.7 \cdot \frac{500}{2400} + 3.9 \cdot \frac{100}{2400}$$
$$+ 2.7 \cdot \frac{240}{2400} + 2.9 \cdot \frac{400}{2400} + 2.7 \cdot \frac{360}{2400}$$
$$= 2.35 \text{ minutes}$$

Likewise, for drilling

$$\mu_D^{-1} = 1.35 \cdot \frac{800}{2400} + 1.85 \cdot \frac{500}{2400} + 2.50 \cdot \frac{100}{2400}$$
$$+ 1.95 \cdot \frac{240}{2400} + 2.55 \cdot \frac{400}{2400} + 1.95 \cdot \frac{360}{2400}$$
$$= 1.85 \text{ minutes}$$

Parameters for models 1b and 2 can be computed in the same fashion from the data in Table 13.8. Using the open network procedure of Section 11.3.1 with the two nodes corresponding to the lathe and drill press, we find the general results shown in Table 13.10. All times are in minutes. Queue lengths are units for models 1a and 1b and trays for model 2. The important measure is feasibility, namely, whether utilizations (ρ values) are less than 1. Drill queue lengths are very suspect, since the model ignores the coordinated robot sequence and overestimates the variability of interarrival times at the drill.

One result is obvious from the table. Method 1b will not work. The load on the lathe is so large that to keep pace the cell must give highest priority to keeping the lathe busy. Thus minimal time is permitted for unloading and reloading the lathe. Model 2 seems to have significantly larger WIP levels, but this may be due to the use of an exponential modeling assumption. It appears that we should acquire more knowledge on how parts will be fed. Knowledge of the actual process indicates that parts, or trays, will arrive with nearly constant interarrival times. We will use this assumption to simulate detailed performance of models 1a and 2.

Table 13.10   Open Network Approximation for Planar Company Alternatives

Model	λ	μ_L	μ_D	ρ_L	L_L	W_L	ρ_D	L_D	W_D
1a	0.417	0.426	0.541	0.978	44.4	107	0.770	3.35	8.04
1b	0.417	0.403	0.632	1.033	∞	∞	0.659	1.933	4.64
2	0.02083	0.02128	0.03443	0.979	46.62	2238	0.605	1.531	73.52

A process model of the cell was constructed. Transactions represent parts and trays and have the attribute part type. Resources represent the robot and each of the two workstations. Each resource is either busy or idle. The robot resource has task priorities as follows: unload the lathe, load the lathe, unload the drill, set up the drill, load the drill. Key steps in the process of transactions through the work cell are as follows:

1. Parts or trays arrive at a constant rate throughout the 96-hour production period. The random sequencing of part types is accomplished by random sampling without replacement from the part type distribution of the 2400 parts or 120 trays.
2. Part processing at the lathe is modeled using the following steps:
   a. Acquire the lathe resource.
   b. Acquire the robot for loading.
   c. Be loaded onto the lathe.
   d. Release the robot.
   e. Be turned on the lathe.
   f. Acquire the robot for unloading.
   g. Be removed from the lathe.
   h. Release the robot.
3. Arrive at the drill. Processing at the drill is modeled as follows:
   a. Acquire the drill resource.
   b. If the part type differs from the preceeding part, set up the drill as follows:
      i. Acquire the robot for setup.
      ii. Set up the drill.
      iii. Release the robot.
   c. The remaining processing on the drill follows the same last seven steps (b–h above) as the lathe.

Consider the lathe, since it is the bottleneck station. Each part must be turned by the lathe in no more than 2.4 minutes to avoid the following part waiting for the lathe. Table 13.8 shows that only part type A meets the requirement. (Note that in the open queueing network model, each part was assigned the average processing time of 2.35 minutes so that all parts met the processing time requirement.) Thus, at least two thirds of the parts will delay the following part at the lathe. The lathe can catch up only when a part of type A is processed. The random sequence of part types could result in many parts waiting for the lathe. In addition, parts may wait at the lathe for loading or unloading while the robot works at the drill press. This delay only adds to the queueing of subsequently arriving parts.

The simulation must assess the number of parts waiting at the lathe and the time parts must wait. Five random sequences of part types were simulated. The results are summarized in Table 13.11.

Note that neither model 1a nor 2 could complete the processing of parts in the required 6 days × 16 hours/day = 96 hours per period. This results from the queueing caused by the random arrival sequence near the end of the 96-hour period. The last column of Table 13.10 gives the average time a part spends waiting for the robot to load it or unload it at the lathe. Add this to the lathe processing time (2.35 + 0.27 or 2.35 + 0.14) to show that contention for the robot makes the processing of all parts in 96 hours infeasible.

Based on the queueing network and simulation results, system design decisions can be made. Simulation results show that model 2 performs better than model

**Table 13.11   Simulation Results
for Planar Company Alternatives**

Model	Time to Complete (hr)	Ave. Max. Lathe Queue Length	Ave. Max. Drill Queue Length	Ave. Robot Queue Time at Lathe
1a	105.1–106.0	216	1	0.27
2	100.3–103.0	6	1	0.14

1a with respect to all lathe-related performance measures. Thus, the use of trays as processing units is preferred. Either one shift will operate on the seventh day to complete the processing of parts, or an extra hour per day must be scheduled. Storage space for waiting trays is set to the average maximum of six at the lathe and one at the drill press.

REFERENCES

Hines, William, and Douglas C. Montgomery [1980], *Probability and Statistics in Engineering and Management Science,* John Wiley & Sons, New York.

Hoel, Paul G., Sidney C. Port, and Charles J. Stone [1972], *Introduction to Stochastic Processes,* Houghton Mifflin Company, Boston.

Lavenberg, Stephen S., ed. [1983], *Computer Performance Modeling Handbook,* Academic Press, New York.

White, J. A., J. W. Schmidt, and G. K. Bennett [1975], *Analysis of Queueing Systems,* Academic Press, New York.

APPENDIX A

REVIEW OF BASIC PROBABILITY

This appendix gives a brief overview of some of the basic probability concepts used in the text. More extensive discussion may be found in the introductory probability and statistics textbooks listed in the references.

A.1
RANDOM VARIABLES AND PROBABILITY DISTRIBUTIONS

A **random experiment** is an experiment for which, though we can describe the set of possible outcomes, the precise outcome on any trial of that experiment cannot be known in advance. Additionally, every possible outcome of the experiment must have a fair chance of occurring. For instance, we pick an arbitrary sheet of metal from a rolling mill and measure its thickness. Even if the measurement system is exact, sheet thickness will vary from the standard, and the thickness of any specific sheet can be known only through measurement after rolling. Each arbitrarily selected sheet constitutes a **random observation** from the **population** of sheets produced at that mill over the study period.

A **random variable** is a function that assigns a value to every outcome of a random experiment. Typical random variables for our purposes include time between consecutive arrivals at a workstation and the number of machines that break down during a work shift. Random variables are either **discrete** or **continuous.** Discrete random variables may take on only a finite or countably infinite number of values. The number of machine failures during a work shift is discrete. Continuous variables may take on any value in one or more intervals. Workstation interarrival time is continuous. Digital measurement devices are often used to give discrete approximations of observations on continuous random variables.

Continuous random variable X is described by its **probability density function** $f(x)$. We can think of $f(x)$ as the relative likelihood that $X = x$. Probability density functions must satisfy

1. $f(x) \geq 0 \quad -\infty < x < \infty$
2. $\int_x f(x) = 1$

The probability that the random variable X takes on a value in the interval $[a, b]$ is

$$\Pr(a \leq x \leq b) = \int_a^b f(x)\, dx$$

447

Similarly, we define the probability mass function $p(x)$ for a discrete random variable such that

1. $p(x) \geq 0$
2. $\sum_x p(x) = 1$

An important concept is **expectation.** The expected value of any random variable or function of random variables is its long-run central tendency (average). If Y is a function of continuous random variable X, then the expected value of Y is

$$E[Y(X)] = \int_{-\infty}^{\infty} Y(x)f(x)\,dx \qquad (A.1)$$

If X is discrete, summation replaces integration in equation A.1. Important quantities include the mean $[E(X)]$ and variance $\{V(X) = E[(X - E(X))^2]\}$ of a random variable. The distribution of X is centered at $E(X)$ and must have at least $\frac{8}{9}$ of its area within $3V(X)^{1/2}$ of $E(X)$. $V(X)^{1/2}$ is called the standard deviation and is useful for visualizing the dispersion of the distribution.

Distributions can also be uniquely described by their moments. The kth moment is defined by $\mu'_k = E(X^k)$. It is sometimes more convenient to remove the location factor and to discuss the central moment $\mu_k = E\{[X - E(X)]^k\}$. Note that $E(X) \equiv \mu'_1$ and $V(X) \equiv \mu_2$. Moments can be found in some instances through knowledge of the **moment-generating function** (MGF) defined as

$$M_X(t) = E\left(e^{tX}\right)$$

Noting the infinite series interpretation of exponential functions, it can be shown that

$$\mu'_k = \frac{d^k}{dt^k} M_X(t)\Big|_{t=0}$$

Two useful relationships for expected values are

$$E(aX) = aE(X)$$

and

$$V(aX) = a^2 V(X)$$

for a constant a.

Two (or more) random variables X_1 and X_2 are **independent** if their joint density is the product of their individual density functions, that is, if $f(X_1, X_2) = f_1(X_1)f_2(X_2)$. For linear combinations of random variables where $Y = a_1 X_1 + \cdots + a_n X_n$,

$$E(Y) = \sum_{i=1}^{n} a_i E(X_i)$$

and, if the X_i are independent,

$$V(Y) = \sum_{i=1}^{n} a_i^2 V(X_i)$$

Several important distributions are summarized in Table A.1.

Table A.1 Summary of Important Distributions

Name	Density	Domain	Mean	Variance	MGF
Poisson	$p(x) = \dfrac{\lambda^x e^{-\lambda}}{x!}$	$x = 0, 1, \ldots$	λ	λ	$e^{\lambda(e^t - 1)}$
Binomial	$p(x) = \dbinom{n}{x} p^x (1-p)^{n-x}$	$x = 0, 1, \ldots$	np	$np(1-p)$	$(pe^t + 1 - p)^n$
Exponential	$f(x) = \lambda e^{-\lambda x}$	$x \geq 0$	λ^{-1}	λ^{-2}	$\left(1 - \dfrac{t}{\lambda}\right)^{-1}$
Gamma	$f(x) = \dfrac{\lambda^r x^{r-1} e^{-\lambda x}}{\Gamma(r)}$	$x > 0$	$\dfrac{r}{\lambda}$	$\dfrac{r}{\lambda^2}$	$\left(1 - \dfrac{t}{\lambda}\right)^{-r}$
Uniform	$f(x) = (b-a)^{-1}$	$a \leq x \leq b$	$\dfrac{a+b}{2}$	$\dfrac{(b-a)^2}{12}$	$\dfrac{e^{tb} - e^{ta}}{t(b-a)}$
Normal	$f(x) = \dfrac{1}{\sigma\sqrt{2\Pi}} e^{-\frac{1}{2}\left(\frac{x-\mu}{\sigma}\right)^2}$	$-\infty < x < \infty$	μ	σ^2	$e^{t\mu + \left[\frac{(\sigma t)^2}{2}\right]}$

A.2
STOCHASTIC PROCESSES

A **stochastic process** is a function over time $t \in T$ whose value is a random variable. The process is continuous or discrete according to whether T is an interval or subset of the integers. An **arrival process** is a continuous stochastic process, say, $X(t)$, $t \geq 0$, where $X(t)$ is the number of arrivals to the system in time $[0, t]$. The time between consecutive arrivals is referred to as the interarrival time. If interarrival times are independent and identically distributed random variables, then the arrival process is called a renewal process. One very important such process occurs when interarrival times are exponentially distributed. This is called a **Poisson arrival process.** The name derives from the close relationship between the exponential and Poisson distributions. If the time between arrivals is exponentially distributed with parameter λ, then the number of arrivals in time t has a Poisson distribution with parameter λt. It can be shown that a Poisson process exists if and only if the following properties hold true:

1. The number of arrivals in nonoverlapping time segments are independent.
2. The number of arrivals in any interval is only a function of the interval length and not its starting time.
3. As $t \rightarrow 0$, the probability of exactly one arrival in t is λt, and the probability of zero arrivals is $1 - \lambda t$.

Two other important results for the Poisson process concern merging and splitting. First, if arrivals come from multiple independent Poisson processes with rates λ_j, then the merged arrivals also form a Poisson process with arrival rate equal to the sum of the individual processes ($\lambda = \sum_j \lambda_j$). If a Poisson process with rate λ is split into multiple streams with probability p_j of any arrival taking stream j, then each stream is also a Poisson process with arrival rate $p_j \lambda$.

Consider a system that at time t can be in any state $x \in S$. $X(t)$ is the random variable representing the state at time t. Some systems satisfy the **Markov Property,** which is defined by

$$\Pr[X(t+1) = x_{t+1} \mid X(0) = x_0, \dots, X(t) = x_t]$$
$$= \Pr[X(t+1) = x_{t+1} \mid X(t) = x_t]$$

for discrete processes. The key idea is that given the current state, past states do not affect future states. Let p_{xy} be the probability of transitioning from state x to state y in one time period. The long-term proportion of time the system is in state x is given by the **stationary distribution,** $\pi(x)$. We may obtain $\pi(x)$ by solving the system of linear equations

$$\pi(x) = \sum_y \pi(y) p_{yx} \qquad \text{for all states } x$$

with the added restriction $\sum_x \pi(x) = 1$.

For continuous processes, the Markov property is

$$\Pr(X(t) = y \mid X(s_1) = x_1, \dots, X(s_n) = x_n, X(s) = x) = \Pr_{xy}(t - s)$$

where $0 \leq s_1 \leq \cdots \leq s_n \leq s \leq t$ and $\Pr_{xy}(t - s)$ is the probability of transitioning from starting state x to y in time $t - s$. This Markov property is satisfied if and only if the time between transitions is exponentially distributed. An important tool in analyzing Markov processes is the **Chapman-Kolmogorov equation**

$$\text{Pr}_{xy}(t + s) = \sum_{z} \text{Pr}_{xz}(t)\text{Pr}_{zy}(s) \qquad s, t \geq 0$$

A state is said to be recurrent if the probability of returning to that state at some time in the future is 1. Furthermore, if the mean return time is finite, the state is positive recurrent. If all states are positive recurrent, the process is called positive recurrent as well. If for all possible starting states, each state has a positive probability of being visited in finite time, the process is called irreducible. Irreducible, positive recurrent processes have a stationary distribution $\pi(x)$, which gives the probability of being in state x. The stationary distribution can be found by solving the set of conservation equations

$$\pi(x) \cdot q_x = \sum_{y} \pi(y) \cdot q_{yx}$$

where q_x is the rate of exiting from state x and q_{yx} is the rate of transition to state x when in state y.

An important application of stochastic processes is the theory of queues. A queue is a waiting line. We are all affected by queues daily, from ensuring that an available transmission line exists when dialing the phone to waiting to see an instructor during office hours. A standard classification scheme for queues is $(a/b/c) : (d/e/f)$ where a is the arrival distribution, b is the service time distribution, c is the number of parallel servers, d is the service discipline, e is the maximum number of customers allowed in the system, and f is the size of the customer population. Poisson arrivals (exponential interarrival time) and exponential service time are denoted M. A general interarrival time distribution is denoted GI, and G is used for general service times. Deterministic times are represented by D. Summary results for $(M/M/c) : (\text{General}/\infty/\infty)$ queues are provided in Chapter 11.

SUBJECT INDEX

AUTHOR INDEX

Printed in Singapore by Kin Keong Printing Co. Pte. Ltd.